Experiments Manual

for

Grob's Basic Electronics

Experiments Manual

for

Grob's Basic Electronics

Thirteenth Edition

Wes Ponick
Technical Consultant

Frank Pugh
Santa Rosa Junior College

Mitch Schultz
Western Technical College

EXPERIMENTS MANUAL FOR GROB'S BASIC ELECTRONICS, THIRTEENTH EDITION

Published by McGraw-Hill Education, 2 Penn Plaza, New York, NY 10121.

Some ancillaries, including electronic and print components, may not be available to customers outside the United States.

This book is printed on acid-free paper.

4 5 6 7 8 9 LHN 24 23 22

ISBN 978-1-260-44530-5
MHID 1-260-44530-5

Managing Director: *Kathleen McMahon*
Product Developer: *Tina Bower and Megan Platt*
Marketing Manager: *Shannon O'Donnell*
Program Manager: *Jolynn Kilburg*
Content Project Managers: *Jane Mohr, Sandra Schnee, and Samantha Donisi-Hamm*
Buyer: *Laura Fuller*
Design: *Beth Blech*
Content Licensing Specialist: *Abbey Jones*
Compositor: *Aptara®, Inc.*
Typeface: *10/12 SRA Serif 1.1 Book*
Printer: *LSC Communications/Harrisonburg*

mheducation.com/highered

Contents

Preface

The *Experiments Manual for Grob's Basic Electronics* gives you step-by-step guidance with electronic components, circuits, and test equipment. The manual will help you develop the skills you need to build, measure, and troubleshoot electronic circuits of all types. It is designed for the beginning students and assumes they have no previous knowledge or experience with electronics technology.

The experiments build on the chapters in *Grob's Basic Electronics*. There are more than 70 experiments, and each one has a separate introduction describing the theory and the objectives. All the necessary components and equipment are listed. The procedures include all the necessary steps and drawings.

The experiments begin with electronics math basics, lab safety, component recognition, and details on how to use basic lab equipment. Afterward, the labs progress to simple resistive circuits and then on to reactive circuits. Circuits using diodes and transistors are also included, so you learn to work on power supplies, amplifiers, and much more. Of course, you will be using lab equipment such as digital multimeters (DMM), DC power supplies, signal generators, and oscilloscopes.

In addition to the experiments, the MultiSim circuit simulation software (available on the textbook's associated website) can also be used with this lab manual. You can use it to simulate the circuit experiments and view the results or to see what happens with circuit failure. MultiSim runs on your computer so that it does not require using real circuits or equipment. Your instructor can provide more information about MultiSim and how it applies to this set of experiments.

The experiments usually take less than 2.5 hours to complete, but some are longer than others. Every experiment begins with an introduction (theory) before detailing the step-by-step procedures that you will follow. Each experiment has exercises or questions and tables or graphs for recording your data. Finally, a blank report form is included, where you describe how the objectives were met or how the theory was verified.

The Appendix lists all the required components and equipment and has extra graphs, plots, and example reports to get you started.

This *Experiments Manual* is dedicated to every student who needs a working knowledge of basic electronics.

Experiments Manual

for

Grob's Basic Electronics

INTRODUCTION

EXPERIMENT I

INTRODUCTION EXPERIMENT—ELECTRONICS MATH

LEARNING OBJECTIVES

At the completion of this experiment, you will be able to:

- Understand electronics math concepts.
- Use power of 10, scientific, and engineering notation.
- Understand precision and accuracy in electronic measurements.

SUGGESTED READING

Introduction, *Basic Electronics*, Grob/Schultz, twelfth edition.

INTRODUCTION

The topics covered in this introduction may also be covered in separate math courses or in your electronics courses. They are presented here to enhance, review, or introduce you to electronics math, which you will use in the lab to predict or verify measurement results. The material here assumes you have basic math skills to add, subtract, multiply, and divide, including working with fractional values (using decimal points). In addition, the ability to do some basic algebra or geometry is recommended. You can also go online to learn about or review any math concepts that are unfamiliar. However, most math concepts you need will be explained in this lab manual. If you are skilled at mathematics, this introduction will only be a review, except for the introduction of electronic values and units of measure.

The first section begins with *power of 10* notation, which is based on our number system of tens. Next, scientific notation and then engineering notation (other variations of the power of 10 notation) will be introduced. The last section will introduce you to significant numbers, metric prefixes, and Ohm's law, which is used with calculations and measurements.

Calculators: It is expected that you will be using a scientific calculator. You should be familiar with using the add, subtract, multiply, and divide functions on your calculator as a minimum, and be sure you can clear the entries using the C (clear) or CE (clear entry) keys. These keys will be specific to your calculator; for example, some calculators use AC (all clear) or DEL (clear entry), and so on. Other functions such as using exponents (EXP on most calculators), square roots ($\sqrt{}$), and others may also be required. So be sure to read the information in your calculator's manual.

EQUIPMENT

Calculator, paper, and pencil.

PROCEDURE

Read through the four sections and be sure you understand each one before you move on. The sections have examples of calculating values of circuit voltage, current, and resistance, which are common practice in electronics. At the end of each section are exercise problems.

Exercise Problems: Do the exercise problems after each section and record the results in the tables at the end of this introduction. Also, check with your instructor for specifics about which exercises may require calculations turned in on a separate sheet of paper.

Power of 10 Notation

In most sciences, like electronics, very large or very small numbers are used in measurements and calculations. Therefore, it is necessary to have a method for displaying and using numbers in a convenient manner. For example, a small number like 0.00000254 or a large number like 512,000,000 can be more easily written using a notation we call **power of 10.** This means using the number 10 with a multiplier (the exponent) instead of using lots of zeros. This is often referred to as 10 raised to the nth power. So, the small raised or superscript number on the right of the 10 is the exponent. The exponent indicates how many times you multiply the base number (10), as shown here:

Large or Positive Numbers in Powers of 10

1	=	1 (not multiplied by 10)	or 10^0
10	=	10×1	or 10^1
100	=	10×10	or 10^2
1,000	=	$10 \times 10 \times 10$	or 10^3
10,000	=	$10 \times 10 \times 10 \times 10$	or 10^4
100,000	=	$10 \times 10 \times 10 \times 10 \times 10$	or 10^5
1,000,000	=	$10 \times 10 \times 10 \times 10 \times 10 \times 10$	or 10^6

Therefore, a number such as 16,000, which is $16 \times 10 \times 10 \times 10$, can be simply written as 16×10^3, or a larger number such as 512,000,000 can be simply written as 512×10^6. Also, when speaking about a number such as 1,000, we say 10 to the third power.

Decimal Fractions Expressed Using Negative Powers of 10

Numbers less than 1 can be expressed using negative powers of 10. As you will see, negative powers of 10 are actually just reciprocals of positive powers of 10.

$$0.1 = \frac{1}{10^1} = 10^{-1}$$

$$0.01 = \frac{1}{10^2} = 10^{-2}$$

$$0.001 = \frac{1}{10^3} = 10^{-3}$$

$$0.0001 = \frac{1}{10^4} = 10^{-4}$$

$$0.00001 = \frac{1}{10^5} = 10^{-5}$$

$$0.000001 = \frac{1}{10^6} = 10^{-6}$$

Therefore, a number like 0.016 can be written as 16×10^{-3}. A smaller number like 0.000512 can be written as 512×10^{-6}.

Note that the number before the multiple of 10 (such as 16 or 512 here) is known as the *coefficient* or *factor* because it is used in the multiplication. Regardless, you should now be able to express large or small numbers using exponents for power of 10 notation. Next, let's look at using a calculator with these numbers.

Using the Calculator (positive exponent): On your calculator, you should have an exponent button: **Exp, EXP**, or it may use a caret ^ or second function with **EE**. This button will allow you to enter a value and then enter the exponent value (power of 10 value). The calculator should also display the letter **e** or **E** afterward. If your calculator does not have this type of button or function, you should replace it with one that does. Otherwise, you will have more difficulty in this course and your work. Note that the Exp button (function) is not the same as the X^y or the 10^x button, both of which perform slightly different exponential functions. For example, the 10^x button means you enter the number of times that 10 is multiplied by itself (10). Also, you can use the C or CE button to clear mistakes.

If you are not sure about your calculator, read its manual or ask an instructor for help. A good inexpensive scientific calculator or even a mobile phone application of one can be easily obtained to ensure your success. The following steps show how to use power of 10 notation (with exponents) on a generic scientific calculator that has an Exp button. Try this:

1. Enter the number 16,000 in exponential (power of 10) notation: $\mathbf{16 \times 10^3}$. For this type of calculator, press the buttons in this order (see diagrams below): 1, 6, Exp, and then 3. You should now see the entry is 16.e + 3. The plus sign should appear automatically to indicate it is a positive number. See Fig. I-1.

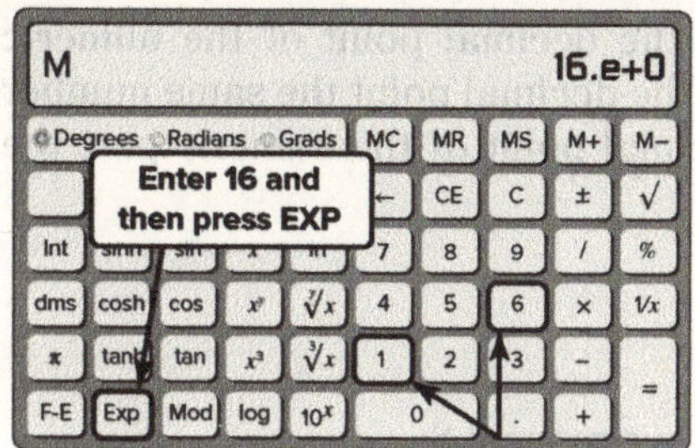

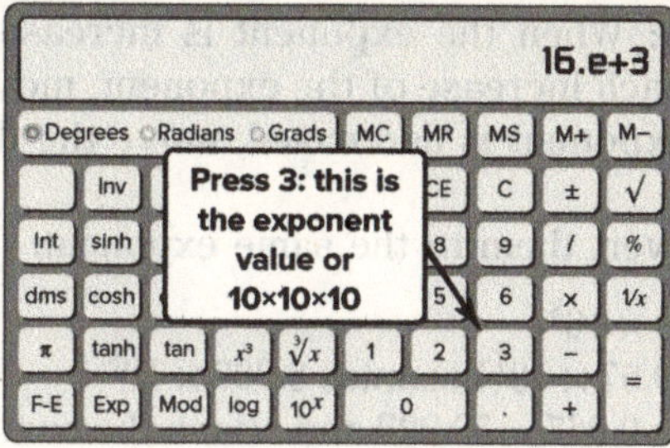

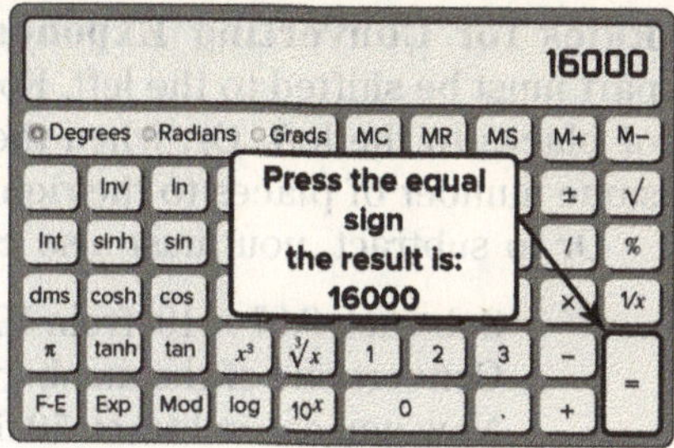

Fig. I-1

2. Now, press the equal button, and you will see the result in regular notation: 16000 (see the diagram above). By using this method, you can do all the necessary power of 10 math on your calculator using the exponent button or function. In fact, this will allow you to add, subtract, multiply, and divide exponential values easily. As you'll see, doing it by hand is much more difficult and time consuming.

Also, notice that the number you enter may automatically appear with a decimal point. This happens because scientific calculations typically use decimal points, and you will be using the decimal point often with electronic values.

3. Locate the decimal point on your calculator (usually near the zero button) and enter 16,050 in exponential form: 16.05×10^3. When you press the Exp or EXP button, the letter **e** or **E** should appear to indicate that the following number is the exponent. Your calculator should read: 16.05e+3. Always know the buttons on your calculator and how to use them.

Using the Calculator (negative exponent): On your calculator, type in the value 16×10^{-3}. This would be pressing the buttons 1, 6, EXP, and then 3 like before. However, you need to press the **Change Sign** button (±) to make the exponent negative. Finally, press the equal button and you will see the result in regular notation: 0.016 (see Fig. I-2 below).

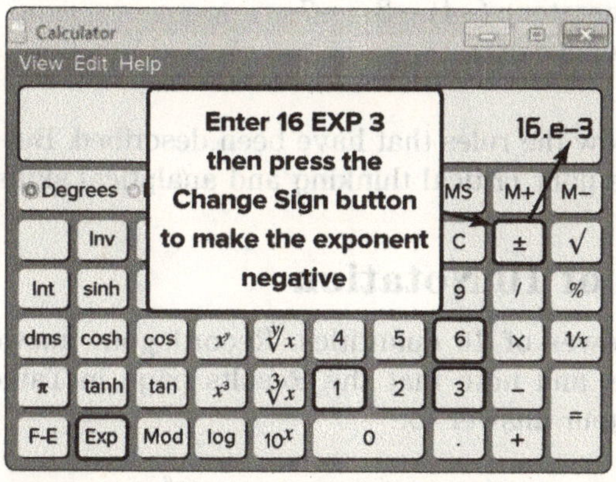

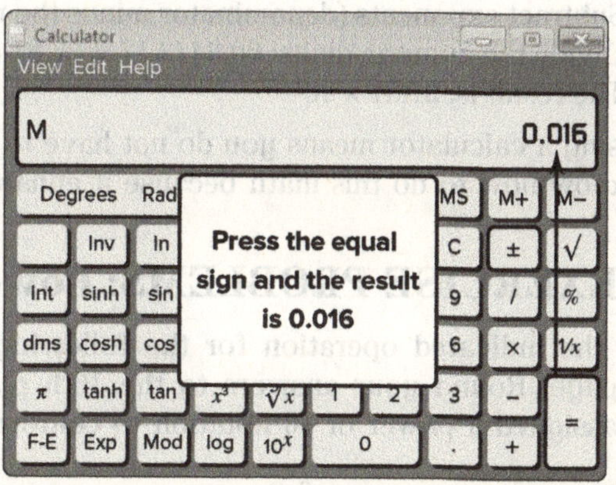

Fig. I-2

Using the F-E Key: The **F-E** button (fixed-exponential) on scientific calculators will toggle between fixed-point notation (regular numbers with a specific number of digits) and the same number in exponential form.

4. If you have an F-E key (or similar) on your calculator, enter the number 16,050 and then press the F-E key. You should see the fixed digit number become an exponential number: 1.605e + 4. Press F-E again it should toggle back to 16,050. Try it with some other numbers and then clear the calculator using C or Clear key. Remember that CE only clears the last entry.

Adding and Subtracting Power of 10 Notation: Without a calculator, adding or subtracting power of 10 values means you must convert them to the same power of 10 (same exponent). It does not matter which number you convert as long as they are the same. For example:

Add: 50.69×10^8 and 3.67×10^7
Convert one of the numbers: 3.67×10^7 becomes 0.367×10^8
Now you can add: $50.69 + 0.367 = 51.057$ and attach the common exponent 8
The result is: 51.057×10^8

Rules for Converting Exponents: When the exponent is increased, the decimal point of the numeric part must be shifted to the left. For each increase of the exponent, move the decimal point the same number of places to the left. Or, when the exponent is decreased, move the decimal point of the numeric part the same number of places to the right.

Or to subtract, you must also convert them to the same exponent:

Subtract: 9.67×10^7 from 50.69×10^8
Convert one of the numbers: 9.67×10^7 becomes 0.967×10^8
Now you can subtract: $50.69 - 0.967 = 50.323$ and attach the common exponent 8
The result is: 49.723×10^8

Of course, using a calculator means you do not need to convert different exponents—the calculator does it for you. But it's good to know how it's done by hand. You should also know the rules for adding or subtracting negative and positive numbers, such as how adding negative numbers results in a greater negative number, but subtracting negative numbers can result in either negative or positive values. These are rules of algebra and are available online or in most math books. We will not cover them here because you will use your calculator most of the time.

Multiplying Power of 10 Notation: Without your calculator, you multiply the numeric digits (coefficients) and algebraically add the exponents. For example:

Multiply: 12.19×10^{-7} by 1.97×10^9
Multiply the numeric values: $12.19 \times 1.97 = 24.0143$
Algebraically add the exponents: -7 and $9 = 2$
The result is: 24.0143×10^2

Dividing Power of 10 Notation: Without your calculator, you subtract the exponents (subtract the denominator from the numerator). Then divide the numeric digits (coefficients). For example:

Divide: $60.34 \times 10^{-4}/2.17 \times 10^3$
Subtract exponents (denominator minus the numerator): $(-4) - 3 = -7$
Divide the numeric digits: $60.34/2.17 = 27.81$ (rounded off)
The result is: 27.81×10^{-7}

Again, using a calculator means you do not have to know the rules that have been described. But it's a good idea to know how to do this math because it enhances your critical thinking and analytical skills.

I-1.1 EXERCISE PROBLEMS: Power of 10 Notation

Perform the indicated operation for the following power of 10 quantities. Record your answers on the Results page. Round your answers to the 10th place, and note that the Results page in Table I-1.1 has already assigned a power of 10 notation to conform your answer to.

1. $4.24 \times 10^2 \times 6.87 \times 10^3$

2. $3.21 \times 10^7 \times 8.66 \times 10^9$

3. $5.35 \times 10^4 \times 4.33 \times 10^3$

4. $72.8 \times 10^1 \times 8.62 \times 10^2$

5. $6.45 \times 10^5 \times 0.12 \times 10^7$

6. $18.4 \times 10^6 \times 7.34 \times 10^6$

7. $34.5 \times 10^9 \times 15.1 \times 10^8$

8. $7.24 \times 10^2 \times 6.37 \times 10^1$

9. $35.8 \times 10^5 \times 23.7 \times 10^5$

10. $3.84 \times 10^7 \times 1.21 \times 10^7$

11. $42.2 \times 10^1 \times 0.02 \times 10^3$

12. $7.65 \times 10^8 \times 0.27 \times 10^9$

13. $1.00 \times 10^2 \times 2.00 \times 10^5$

14. $3.46 \times 10^{-2} \times 7.25 \times 10^4$

15. $18.3 \times 10^4 \times 20.1 \times 10^{-5}$

16. $2.67 \times 10^{-7} \times 16.8 \times 10^{-2}$

17. $5.55 \times 10^1 \times 3.46 \times 10^2$

18. $3.14 \times 10^{-3} \times 1.22 \times 10^{-3}$

19. $\dfrac{4.84 \times 10^2}{2.22 \times 10^4}$

20. $\dfrac{8.00 \times 10^3}{2.00 \times 10^3}$

21. $\dfrac{7.50 \times 10^{-4}}{0.50 \times 10^3}$

22. $\dfrac{6.40 \times 10^{-5}}{2.00 \times 10^{-4}}$

23. $\dfrac{12.6 \times 10^2}{3.00 \times 10^{-3}}$

24. $\dfrac{8.88 \times 10^3}{1.00 \times 10^{-4}}$

Scientific Notation

Another form of *power of 10 notation* used in electronics is scientific notation. Like power of 10 notation, it uses exponents. The difference is that scientific notation usually has only one number followed by a decimal point and then the other numbers and exponent. For example, 17,645 is written in scientific notation as 1.7645×10^4 or 1.7645e4. So, if you understand power of 10 notation, then scientific notation is an even simpler version of it. As you can see here, all these power of 10 numbers (all the same value) can be written as a single value in scientific notation:

$$\left.\begin{array}{l} 17{,}645 \times 10^0 \\ 1764.5 \times 10^1 \\ 176.45 \times 10^2 \\ 17.645 \times 10^3 \\ 1.7645 \times 10^4 \\ 0.17645 \times 10^5 \\ 0.017645 \times 10^6 \end{array}\right\} = 1.7645 \times 10^4$$

Note: Scientific notation does not always require an exponent. For example, 3.14 is correct scientific notation. You do not need to add the 10^0 because it is understood. Also, the single number to the left of the decimal will always be between 1 and 10 or between −10 and −1.

Converting to Scientific Notation: Move the decimal point of the number being converted to the right of the first non-zero digit. Then count the number of places you moved it. This will become the exponent. If you move the decimal point to the left, the exponent will be positive (+). If you move the decimal point to the right, the exponent will be negative (−). Here are some examples:

$2970 = 2.97 \times 10^3$ Decimal point moved left 3 places so the exponent is +3
$0.0016 = 1.60 \times 10^{-3}$ Decimal point moved right 3 places so the exponent is −3
$52.9 = 5.29 \times 10^1$ Decimal point moved left one place so the exponent is +1

Note again that you can also use **e** instead of ×10, for example: 2.97e3. You can check these with your calculator (enter them and use the F-E button). Remember that you need the **change sign button** for entering negative values (as change 3 to −3).

Converting from Scientific Notation: If the number has a positive exponent, move the decimal point to the right the same number of places as the exponent. If the number has a negative exponent, then move the decimal point to the left the same number of places. You may also need to add zeros. Here are some examples:

$4.02 \times 10^3 = 4020$ Decimal point moved right 4 places
$3.11 \times 10^{-3} = 0.00311$ Decimal point moved left 3 places
$0.61 \times 10^6 = 610{,}000$ Decimal point moved right 6 places

You can check these with your calculator (enter them and use the F-E button). Again, scientific notation has one number to the left of the decimal point, and this makes it easy to list a lot of numbers when taking measurements.

Adding and Subtracting Scientific Notation: Without a calculator, adding or subtracting numbers in scientific notation uses the same rules as power of 10 notation: the exponents must be the same number. The difference is that the result will be in scientific notation: one number to the left of the decimal point. Here are some examples:

Add: 4.61×10^5 and 7.31×10^7
Convert one of the numbers: 4.61×10^5 becomes 0.0461×10^7
Now you can add: $0.0461 + 7.31 = 7.3561$ and attach the common exponent 7
The result is: 7.3561×10^7 or $7.3561\ e^7$

Also,

Subtract: 6.42×10^{-3} from 8.70×10^{-2}
Convert one of the numbers: 6.42×10^{-3} becomes 0.642×10^{-2}
Now you can subtract: $8.70 - 6.42 = 8.058$ and attach the common exponent -2
The result is: 8.058×10^{-2} or $8.058\ e^{-2}$

Multiplying Scientific Notation: Without your calculator, you multiply the numeric digits (coefficients) and algebraically add the exponents. This is the same as multiplying power of 10 notation except that the final result must be in scientific notation. For example:

Multiply: 2.19×10^{-7} by 1.97×10^9
Multiply the numeric values: $2.19 \times 1.97 = 4.3143$
Algebraically add the exponents: $(-7) + 9 = 2$
The result is: 4.3143×10^2

Dividing Scientific Notation: Without your calculator, you subtract the exponents (subtract the denominator from the numerator) and then divide the numeric digits (coefficients). This is the same as multiplying power of 10 notation except that the final result must be in scientific notation. For example:

Divide: $6.34 \times 10^{-4}/2.17 \times 10^3$
Subtract exponents (denominator minus numerator): $(-4) - 3 = -7$
Divide the numeric digits: $6.34/2.17 = 2.92$ (rounded off)
The result is: 2.92×10^{-7}

Note on rounding-off numbers: Most of the time, if the value is less than 5, you can ignore it. If the value is greater than 5, increase the preceding number by 1. This will be described later in more detail.

Of course, you will be using your calculator most of the time; therefore, you do not have to remember the rules for performing these calculations. But it is a good idea to know how to do this math with confidence because it enhances your critical thinking and analytical skills. These qualities are valuable in any science, especially electronics.

I-1.2 EXERCISE PROBLEMS: Scientific Notation

Convert the following numbers to scientific notation. Record your answers on the Results page in Table I-1.2.

1. 34
2. 139
3. 1,296
4. 12.65
5. 0.067

Convert the following numbers to regular notation. Record your answers on the Results page.

6. 2.04×10^6
7. 3.45×10^{-8}
8. 7.24×10^2
9. 3.779×10^{-7}
10. 3.66×10^6

Perform the indicated operation for the following power of 10 quantities. All results should be in scientific notation. Record your answers on the Results page.

11. $4.28 \times 10^1 + 1.02 \times 10^1$

12. $7.45 \times 10^8 + 1.17 \times 10^9$

13. $3.00 \times 10^2 + 2.00 \times 10^3$

14. $5.46 \times 10^{-2} + 7.25 \times 10^{-4}$

15. $1.93 \times 10^1 + 2.91 \times 10^{-1}$

16. $3.65 \times 10^7 - 1.68 \times 10^6$

17. $5.33 \times 10^3 - 3.46 \times 10^2$

18. $4.24 \times 10^{-3} - 1.22 \times 10^{-3}$

19. $2.90 \times 10^{-3} - 1.90 \times 10^{-4}$

20. $5.73 \times 10^2 - 3.67 \times 10^{-1}$

21. $(3.91 \times 10^2)\ (4.67 \times 10^5)$

22. $(2.19 \times 10^3)\ (7.63 \times 10^{-4})$

23. $(4.45 \times 10^{-5})\ (3.83 \times 10^{-6})$

24. $(6.34 \times 10^7)\ (2.56 \times 10^{-3})$

25. $(5.55 \times 10^4)\ (2.78 \times 10^4)$

26. $(9.00 \times 10^2)\ /\ (3.00 \times 10^1)$

27. $(2.56 \times 10^{-2})\ /\ (5.44 \times 10^5)$

28. $(7.71 \times 10^4)\ /\ (1.00 \times 10^{-4})$

29. $(9.31 \times 10^{-3})\ /\ (2.99 \times 10^{-3})$

30. $(1.10 \times 10^2)\ /\ (3.35 \times 10^6)$

Engineering Notation and Metric Prefixes

Engineering notation and the use of metric prefixes are what you will usually use to describe values in an electronic circuit. Engineering notation is simply a variation of scientific notation. The difference is that engineering notation only uses **multiples of three** for the exponent: 10^3, 10^6, 10^{-3}, 10^{-6}, and so on. Therefore, it's convenient to use metric prefixes (symbols or text) to represent the exponent in engineering notation which is a multiple of 3 (e.g., 10^3 is kilo, 10^{-6} is micro). In electronics, you will commonly hear technicians and engineers refer to kilo ohms, micro amps, and so on.

Another way to describe engineering notation or metric prefixes is that they represent a power of 10 in a grouping of three (1,000, 100,000, 0.001, 0.000001, etc.). For example, 5 milliamps means: 5×10^{-3} amps, or 1 kilo ohm means 1,000 ohms. When talking about milliamps, it's understood that it means thousandths (10^{-3}) of an amp.

Shown here are metric prefix names, metric symbols, and their power of 10 values in groupings of three. Some of these are rarely used, but those near the middle are used all the time. You may also see these referred to as SI units (International System of Units).

Metric Prefixes, Symbols, Power of 10 Value

Metric Name	Metric Symbol	Power of 10 Value
yotta	Y	10^{24}
zetta	Z	10^{21}
exa	E	10^{18}
peta	P	10^{15}
tera	T	10^{12}
giga	G	10^9
mega	M	10^6
kilo	k	10^3
None (base unit)	None	10^0
milli	m	10^{-3}
micro	μ	10^{-6}

(continued)

Metric Name	Metric Symbol	Power of 10 Value
nano	n	10^{-9}
pico	p	10^{-12}
femto	f	10^{-15}
atto	A	10^{-18}
zepto	z	10^{-21}
yocto	y	10^{-24}

Common Units of Measure in Electronics: A, V, Ω, W: As you become familiar with electronics terminology, you will usually see the following units of measure with engineering notation used to specify their values. For example, 10 kΩ means 10,000 ohms, where ohm (Greek symbol Ω) is the unit of measure for resistance to electrical current. Or, 20 μV where volts (Latin letter V) is the unit of measure for voltage, which is the potential difference between two electrical points. You do not have to know these units at this time; however, they will be used in the examples that follow for engineering notation. Here is a table of some common electronic units of measure and their symbols.

Electronic Units and Symbols

Unit Name	Unit Symbol	Quantity
Ampere (amp)	A	Electric current (I)
Volt	V	Voltage (V, E) Electromotive force (E) Potential differene ($\Delta\varphi$)
Ohm	Ω	Resistance (R)
Watt	W	Electric power (P)

Here are some examples of electronic units of measure, such as amps, volts, ohms, and watts. These examples show engineering notation and their equivalent values using metric prefixes. Notice how the metric prefixes are used after the units of measure:

μ (micro) is 1^{-6}	Therefore, 0.0000165 A	$= 16.5 \times 10^{-6}$ A	= 16.5 μA
k (kilo) is 10^{3}	Therefore, 27,000 V	$= 27 \times 10^{3}$ V	= 27 kV
M (mega) is 10^{6}	Therefore, 1,200,000 Ω	$= 1.2 \times 10^{6}$	= 1.2 MΩ
m (milli) is 10^{-3}	Therefore, 0.039 W	$= 39 \times 10^{-3}$ W	= 39 mW

Converting to Engineering Notation: To convert a non-exponential number to engineering notation, you move the decimal point to a power of 10 notation where the exponent is a multiple of 3 (such as −3, 6, 12, etc.).

Note: Use the same general rules as scientific notation for moving the decimal point right or left and assigning the value of the exponent. However, use a multiple of 3.

Here are some examples of converting a regular notation number to engineering notation and also the metric prefix used instead of the exponent. Notice that you can have more than one correct answer with multiples of three:

0.00005 A:	Move the decimal right 3 places	$= 0.05 \times 10^{-3}$	= 0.05 m (milli)
	Move the decimal right 6 places	$= 50 \times 10^{-6}$	= 50 μ (micro)
3,200,000 Ω:	Move the decimal left 3 places	$= 3{,}200 \times 10^{3}$	= 3,200 k (kilo)
	Move the decimal left 6 places	$= 3.2 \times 10^{6}$	= 3.2 M (mega)

Adding and Subtracting Engineering Notation or Metric Prefixes: Without a calculator, adding or subtracting numbers in engineering notation uses similar rules as scientific notation. That is, you convert the numbers to the same exponential form, except the exponents should be in common multiples of three. If they have the same metric prefix (m, M, k, etc.), it is even easier to add or subtract. Here are some examples, including metric prefixes instead of the exponent:

Add:	$0.003 + 0.0204$	$= 0.0234$	
Or,	$3 \times 10^{-3} + 20.4 \times 10^{-3}$	$= 23.4 \times 10^{-3}$	using engineering notation
Or,	3 m + 20.4 m	= 23.4 m (milli)	where m or milli is the metric prefix
Subtract:	2,600 − 1,000	= 1,600	
Or,	$2.6 \times 10^{3} - 1 \times 10^{3}$	$= 1.6 \times 10^{3}$	using engineering notation
Or,	2.6 k − 1k	= 1.6 k (kilo)	where k or kilo is the metric prefix

Note: If your calculator has an ENG button, pressing it can convert the displayed number to engineering notation, but not all scientific calculators have this button. If your calculator has this button, then you can use it.

I-1.3 EXERCISE PROBLEMS: Engineering Notation

For each of the following problems, perform the indicated operation and record your answers on the Results page in Table I-1.3.

Place the following numbers in engineering notation.

1. 7.94×10^{2}
2. 2.21×10^{7}
3. 5.35×10^{4}
4. 72.8×10^{1}

Replace the power of 10 notation with the appropriate metric prefix.

5. 6.55×10^{6}
6. 98.1×10^{-3}
7. 32.7×10^{3}
8. 9.24×10^{-9}

Add the following numbers, and place the answer in engineering notation.

9. $56.8 \times 10^{-3} + 33.7 \times 10^{-6}$
10. $384 \times 10^{3} + 121 \times 10^{3}$
11. $42.2 \times 10^{-6} + 1.02 \times 10^{-3}$
12. $7.65 \times 10^{-12} + 274 \times 10^{-9}$

Subtract the following numbers, and place the answer in engineering notation.

13. $1.00 \times 10^{3} - 0.10 \times 10^{3}$
14. $3.46 \times 10^{-6} - 2.25 \times 10^{-3}$
15. $18.3 \times 10^{9} - 16.1 \times 10^{12}$
16. $2.67 \times 10^{-3} - 1.68 \times 10^{0}$

Multiply the following numbers, and place the answer in engineering notation.

17. $(7.50 \times 10^{-3}) \times (3.46 \times 10^{3})$
18. $(3.44 \times 10^{-3}) \times (1.22 \times 10^{-3})$
19. $(6.84 \times 10^{9}) \times (2.12 \times 10^{9})$
20. $(8.00 \times 10^{-3}) \times (2.00 \times 10^{-3})$

Divide the following numbers, and place the answer in engineering notation.

21. $(7.50 \times 10^{-3})/(0.50 \times 10^{3})$
22. $(6.00 \times 10^{-3})/(2.00 \times 10^{-6})$
23. $(12.6 \times 10^{3})/(3.00 \times 10^{-3})$
24. $(8.88 \times 10^{3})/(1.00 \times 10^{-6})$

Multiplying Engineering Notation or Metric Prefixes: Use the same rules as those used for scientific notation. Without your calculator, you first multiply the numeric digits (coefficients) and then algebraically add the exponents. Afterward, you can convert the result to engineering notation if the exponent is not a multiple of 3. For example:

Multiply: 14.5×10^{-3} A by 2.07×10^{6}

Multiply the numeric values: $14.5 \times 2.07 = 30.15$

Algebraically add the exponents: −3 and 6 = 3
The result is: 30.15×10^3 using engineering notation
Using the metric prefix: 30.15 m (milli as in mA)

Dividing Engineering Notation or Metric Prefixes: Without your calculator, you subtract the exponents (subtract the denominator from the numerator), then divide the numeric digits (coefficients). This is the same as multiplying power of 10 notation except that the final result must be in scientific notation. For example:

Divide: $6.34 \times 10^2 / 2.18 \times 10^{-4}$.
Subtract exponents (numerator minus denominator): 2 − (−4) = 6.
Divide the numeric digits: 6.34/2.18 = 2.91 (rounded off).
The result is: 2.91×10^6.

Significant Figures, Accuracy, and Precision

When doing electronics math on your calculator, you use exact or finite numbers (not infinite). Finite numbers are like the number of people in a room or the number of pages in a book: exact or finite values. But when measurements are taken with meters and instruments, the values you obtain are not finite and may not match your calculations exactly. Remember that no measuring instrument is perfect, even if it has a digital display. But first, it's important to introduce you to types of calculations you will need to use with your measurements—and that begins with a quick description of Ohm's law for calculating values in an electronics circuit.

Ohm's Law: Ohm's law describes the relationships between current, voltage, and resistance in an electronic circuit. As a technician, you will be measuring and calculating these values using Ohm's law. Ohm's law is based on the work of German physicist George Ohm in 1827 and states: $I = V/R$, where I is the current in amperes, V is the voltage in volts, and R is the resistance in ohms. Therefore, to find the current I in a circuit, you divide the voltage V by the resistance R. The result will be in ohms using the Ω symbol.

Notice that current is written as I in the formula ($I = V/R$) because I refers to the Intensity of current (this is an old scientific term). Resistance (R) is how the circuit resists or opposes the flow of current and is always in ohms (Ω). And, V is easy to recognize as voltage (in volts). Finally, another early scientist, James Watt from Scotland, determined that electronic consumed power can also be determined by Ohm's law using V, I, and R (more about this later).

As you continue, you will see these units of current, voltage, and resistance used in the problems that follow. And Ohm's law will be one of your most used formulas.

Measurement Accuracy and Precision

The accuracy and precision of a measurement are always provided in the manufacturer's specification sheet. Here is a simplified example: a digital multimeter (DMM) specifies that voltage measurements are accurate ±5% when in a range between 1 and 5 volts. This means that a measurement of 3 volts can be +0.15 V or between 2.85 and 3.15 volts. This is also known as the uncertainty of the measurement. But, if you take the measurement 100 times exactly the same way and it always reads the same number, such as 2.95, then it's precise but not necessarily accurate. It's like a scale where when you weigh something three times the same way within a few minutes, and each time you get a slightly different reading.

The terms are usually defined this way:

- **Accuracy** is how close a measurement is to the true value.
- **Precision** is the consistency of the measured value in the same circumstances.
- **Significant figures** are a number of digits used to describe accuracy or precision.

When you use a meter or instrument, such as a DMM, it may be accurate but not precise. This means it may display a measured value slightly different each time. For example, three voltage measurements made exactly the same way might be displayed as 1.43 V, then 1.45 V, and then 1.47 V; the displayed value varied by 0.02 V each time. Therefore, the accuracy appears to be ±0.02 V (not very accurate by some standards). However, it may be precise if it displays exactly the same number of digits to the right of the decimal point each time: 1.43 V, 1.47 V, and 1.45 V all use exactly two digits after the decimal. Therefore, the measurement is precise to two digits each time (never uses 3 or 4, etc.). So the measurement may be precise but not accurate.

Or it may do just the opposite, showing you the same value each time you measure the voltage but with a different number of digits to the right of the decimal point. For example, 1.45 V, then 1.4499 V, and then 1.4500001 V are all similar. But, the precision varies because the number of digits to the right of the decimal is not repeatable—therefore it's not precise. In summary, all these values are close, but instruments do have differences in their accuracy or precision or both. Therefore, we need to know which values are significant and which are not.

Significant Figures or Digits: A standard way to describe accuracy or precision is to use significant figures (digits). Here are the rules for significant figures (digits):

Rule 1: All non-zero digits are significant. For example:

215 has three significant digits.

18.2 has three significant digits.

5.6 × 103 has two significant digits. (Note: disregard the exponential. There is no exponential in the problem.)

Rule 2: All zeros between significant digits are significant. For example:

1,405 has four significant digits.

20,005 has five significant digits.

10.06 has four significant digits.

Rule 3: All zeros to the right of a decimal point are significant. For example:

94.0 has three significant digits.

34.00 contains four significant digits.

10.0×10^{-3} contains three significant digits.

Note: Some instruments may display zeros to the right of a decimal point as placeholders. This may affect the number of significant digits in a measurement.

Rule 4: Zeros to the right of a whole number are NOT significant. For example:

600 has one significant digit.

12,000 has two significant digits.

150 has two significant digits.

Rule 5: Zeros to the left of a number less than one are NOT significant. For example:

0.2 has one significant digit.

0.056 has two significant digits.

0.44×10^{-3} has two significant digits.

Significant Digits and Measured Values: For measurements with a DMM, you can use the least significant digit to describe the precision. The least significant digit is the smallest displayed significant digit. Here are some examples:

Example 1: When a DMM is used as a voltmeter, it displays a measured 12,000 volts. The least significant digit is 2, which appears in the 1000's place. Therefore, the precision is to the nearest 1,000 volts. The DDM may be capable of measuring 12,005 volts accurately, but the displayed value is to the nearest 1,000 volts in this case.

Example 2: When a DMM is in the ohmmeter setting (Ω), it displays a measured 625 ohms of resistance. The least significant digit is 5, which appears in the 1's place. Therefore, the precision is to the nearest ohm or Ω. Or, if 625.1 were displayed, the precision would be to the nearest tenth of an ohm.

Example 3: When a DMM is in the A or uA setting, it displays a measured value of 0.007896 amps. The least significant digit is 6, which is in the one-millionth place. Therefore, the precision is to the nearest microamp (μA) or 10^{-6} amps.

Rounding: In calculations or measurements, rounding is used to simplify values. Also, for a DMM, displayed numbers are often rounded to avoid showing insignificant figures or to make it easier to read—and don't forget that many basic measurements are OK if they are not exact. For example, if you calculate a value of current using Ohm's law ($I = V/R$) and the answer is 16.501 μA (microamps), your DMM might display only 16.5 in the μA setting. That is OK, especially because it's a very small number.

General rules for rounding: If the last digit is less than 5, you can eliminate it and use the remaining digits. For example, 163.92 can be rounded to 163.9, or the number 0.0447 can be rounded to 0.045.

Adding, Subtracting, Multiplying, and Dividing Significant Digits: When doing any of these, your final answer should have no more significant figures than the value with the least number of significant digits. In other words, the answer can never be more precise or accurate than the least accurate or precise number measured. Here are some examples:

Adding voltages: 10.1 V + 123.41 V + 56.01 V = 189.52 V
The least precise measured value is 10.1 V.
Therefore, the rounded answer = 189.5 V.

Multiplying resistance and current: ($I \times R = V$, derived from Ohm's law: $I = V/R$)
1.4 A × 15.68 Ω = 21.952 V
The least precise measure value is 1.4 A (2 significant digits).
Therefore, the answer is 21.9 V or rounded to 22 V (2 significant digits).

Because these are measured values, a final answer may depend upon some established protocol or lab practice. In the last example, 22 V is more precise (using significant digits), but it may be more accurate to use 21.9 V, especially if you are using that value in another calculation with a very small or large number (such as 21.9/20 μA). Check with your instructor for any established protocol on what number of significant digits should be used in the lab experiments.

I-1.4 EXERCISE PROBLEMS: Significant Figures, Accuracy, Precision

For each of the following problems, perform the indicated operation and record your answers on the Results page in Table I-1.4.

How many significant figures are contained in the following measured quanitities?

1. 8.001
2. 721
3. 0.632

How many significant figures are contained in the following measured quanitities?

4. 14.11
5. 738.6
6. 0.0012

What is the accuracy of each of the following measured quantities?

7. 12.63
8. 7,684
9. 0.6117

Add the following measured quantities, and reduce the answer accordingly usually 2 or 3 significant digits.

10. 16.3 + 17.14
11. 10,031 + 6,763
12. 101 + 12

Subtract the following measured quantities, and reduce the answer accordingly.

13. 15.6 + 3.33

14. 1,002 – 863

15. 12.000 + 11,110

Add the following measured quantities, and reduce the answer accordingly.

16. 6.23 × 1.23

17. 6.1 × 21.21

18. 34 + 12.1

Divide the following measured quantities, and reduce the answer accordingly.

19. 73/6.1

20. 108/12.4

21. 17.3/10.03

Round each of the following measured quantities to three significant figures.

22. 763.62

23. 64.762

24. 1.27721

SUMMARY AND REVIEW—BE FAMILIAR WITH THESE TERMS

- Coefficient and exponent (math terms)
- Power of 10 notation (×10)
- Scientific notation (one digit left of the decimal)
- Engineering notation (exponent in multiples of three)
- Metric prefixes (symbols for multiples of three: m, k, etc.)
- Units of measure in a circuit: amps, volts, ohms, watts (A, V, Ω, W)
- Ohm's law: $V = I \times R$ (where I is used for current and R used for resistance)
- Accuracy of a meter or instrument (manufacturer specifications)
- Precision of a measurement
- Significant figures (digits) and the least significant figure (digit)

Know how to use these buttons or functions on your scientific calculator:

- Exponent function (Exp, EE, etc.)
- Change sign (±) function
- FE or fixed-point change to exponential
- Eng (engineering notation) button if available

Important Note: Do not be discouraged by the math or the terminology if it is new to you. The math and the terminology will become easier to use in time if you work at it.

NAME ______________________ DATE ____________

RESULTS TABLES FOR INTRODUCTION TO ELECTRONICS MATH

TABLE I-1.1 Power of 10 Notation

1. ________ $\times 10^{3}$	9. ________ $\times 10^{3}$	17. ________ $\times 10^{3}$
2. ________ $\times 10^{9}$	10. ________ $\times 10^{3}$	18. ________ $\times 10^{-6}$
3. ________ $\times 10^{3}$	11. ________ $\times 10^{0}$	19. ________ $\times 10^{-2}$
4. ________ $\times 10^{3}$	12. ________ $\times 10^{6}$	20. ________ $\times 10^{0}$
5. ________ $\times 10^{6}$	13. ________ $\times 10^{7}$	21. ________ $\times 10^{-7}$
6. ________ $\times 10^{9}$	14. ________ $\times 10^{2}$	22. ________ $\times 10^{-1}$
7. ________ $\times 10^{9}$	15. ________ $\times 10^{-1}$	23. ________ $\times 10^{5}$
8. ________ $\times 10^{0}$	16. ________ $\times 10^{-9}$	24. ________ $\times 10^{7}$

TABLE I-1.2 Scientific Notation

1. ________	11. ________	21. ________
2. ________	12. ________	22. ________
3. ________	13. ________	23. ________
4. ________	14. ________	24. ________
5. ________	15. ________	25. ________
6. ________	16. ________	26. ________
7. ________	17. ________	27. ________
8. ________	18. ________	28. ________
9. ________	19. ________	29. ________
10. ________	20. ________	30. ________

TABLE I-1.3 Engineering Notation

1. ______________
2. ______________
3. ______________
4. ______________
5. ______________
6. ______________
7. ______________
8. ______________
9. ______________
10. ______________
11. ______________
12. ______________
13. ______________
14. ______________
15. ______________
16. ______________
17. ______________
18. ______________
19. ______________
20. ______________
21. ______________
22. ______________
23. ______________
24. ______________

TABLE I-1.4 Significant Figures, Accuracy, Precision

1. ______________
2. ______________
3. ______________
4. ______________
5. ______________
6. ______________
7. ______________
8. ______________
9. ______________
10. ______________
11. ______________
12. ______________
13. ______________
14. ______________
15. ______________
16. ______________
17. ______________
18. ______________
19. ______________
20. ______________
21. ______________
22. ______________
23. ______________
24. ______________

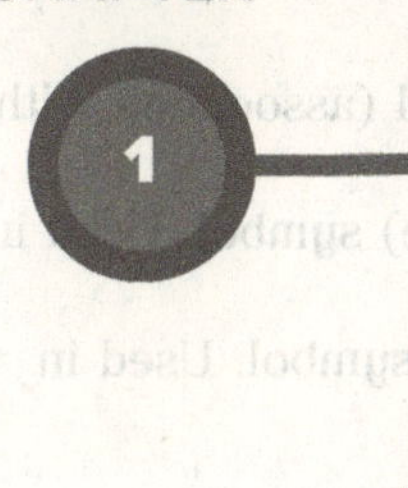

CHAPTER

EXPERIMENT 1-1

LAB SAFETY, EQUIPMENT, AND COMPONENTS (Resistor Color Code)

LEARNING OBJECTIVES

At the completion of this experiment, you will be able to:

- Understand the basic safety concepts used in a lab.
- Be familiar with lab equipment and components.
- Understand the resistor color code.

SUGGESTED READING

Chapters 1 and 2, *Basic Electronics*, Grob/Schultz, twelfth edition

INTRODUCTION

An electronics lab combines a classroom with a technician's workbench. Most classroom labs have similar equipment. You should become familiar with the equipment in your classroom lab, including safety equipment, safety policies, and practices.

SAFETY

Safety always comes first—in a lab class and at work. In fact, most large companies train their technicians in safety measures, including good ergonomic practices. You should know not only how to use the equipment safely but also how to sit at a lab station without stressing your arms, wrists, or eyes when working. Most of the time, this is a matter of common sense—avoid any stressful position for a prolonged time.

But most safety issues in an electronics lab come from the possibility of getting an electric shock. It is your responsibility to avoid this hazard by following these basic rules:

1. Do not plug in or turn on the power to any equipment without knowing how to use the equipment. This means asking for help when you do not know how to use something safely. Do not be afraid to ask for help.

2. Always check power cords. They should not be frayed, loose, or damaged in any way. In general, always inspect your equipment before using it.

3. Never wear loose jewelry or rings when working on equipment. Most jewelry is made of metal and may conduct hazardous current accidentally.

4. Do not keep beverages, open liquids, or food near electronic equipment.

5. Do not touch any circuit or components unless you know that it is safe to do so. For example, touching some components (especially some integrated circuits) can damage them. Other components (large capacitors) can discharge a dangerous current through your body if you touch them—even if they are not connected to a circuit.

6. Never use a soldering iron without proper training.

7. Never disturb another person who is using equipment.

8. Always wear safety glasses or any other required safety equipment.

Your lab class may have a safety test, signs, or some other information that you should read or respond to as required. Although most electronics courses are not known to cause injury, it is still a good idea to be cautious and prepared. Be sure you know where first aid is available and what to do in an emergency.

EQUIPMENT

Every electronics lab, and every school, has different types of equipment. However, there are some basic meters that are common to all electronics labs. In one form or another, your school will have these meters. They may appear to be different only because they are made by different manufacturers. Typical equipment is shown in Grob/Schultz, *Basic Electronics*, twelfth edition.

Simple Meters (Single Purpose)

Ohmmeter: Used to measure resistance (in ohms). Remember that resistance is the opposition to current flow in a circuit. Resistors usually have color stripes on them to identify their values. One way to be sure your resistor is the correct value is to measure it with an ohmmeter.

Ammeter: Used to measure current (in amperes). Remember that current is an electric charge in motion. Ammeters usually measure the current in milliamperes (0.001 A) because most electronics lab courses do not require large amounts of current.

Voltmeter: Used to measure voltage (in volts). Remember that voltage (force) refers to units of potential difference. Most voltmeters use more than one range, or scale, because voltage, unlike current, is often measured in larger values.

Multimeters (Multipurpose)

VOM (volt-ohm-milliammeter): A multipurpose meter used to measure DC and AC voltage, DC current, and resistance (see Fig. 1-1.1*a*).

DMM (digital multimeter): A digital multipurpose meter, also known as a DVM (digital voltmeter) (see Fig. 1-1.1). These can be used as ohmmeters, voltmeters, or ammeters.

Other Equipment

DC Power Supply: A source of potential difference, like a battery. It supplies voltage and current, and it can be adjusted to provide the required voltage for any experiment.

Soldering Iron: A pointed electric appliance that heats electric connections so that solder (tin and lead) will melt on those connections. Used to join components and circuits.

Breadboard/Springboard/Protoboard: Used to assemble basic circuits, by either soldering, inserting into springs, or joining together in sockets. These boards are the tools of designers, students, and hobbyists.

Component Familiarization (Symbols)

The symbol for resistance or a resistor.

The symbol for a battery (cells) or a power supply.

+ Positive potential symbol (associated with the color red).

– Negative potential symbol (associated with the color black).

Capacitor (or capacitance) symbol. Used in the later study.

Inductor (or inductance) symbol. Used in the later study.

Leads are simply insulated wires used to join the meters to the circuits, the power supply to the circuit, and so on. They are conductors and have no polarity of their own. Leads have different types of connectors on the ends, such as alligator clips, banana plugs, and BNC connectors.

Ω = ohms (unit of resistance), Greek letter omega

∞ = infinity (used to indicate infinite resistance)

Common Symbols for Multipliers

Lowercase k = kilo = 1000 or 1×10^3

Uppercase M = mega = 1,000,000 or 1×10^6

Lowercase m = milli = 0.001 or 1×10^{-3}

Greek letter mu (μ) = micro = 0.000001 or 1×10^{-6}

COMPONENTS

Certain basic components are used in electronics all the time. They are the resistor, the capacitor, the inductor (see Fig. 1-1.2), the diode, and the transistor. Along with these are also hundreds of variations and customized devices and chips.

Capacitors have the ability (capacity) to store electrical energy, and inductors have the ability to induce a voltage; these will be studied later. But the practical use of resistors is studied early in most lab classes. Learning about resistors and learning (memorizing) the resistor color code are critical for any technician.

In general, resistors limit or resist the flow of current in a circuit. That is why, they are called *resistors*. Made of a carbon compound, resistors come

Handheld DMM

(*a*)

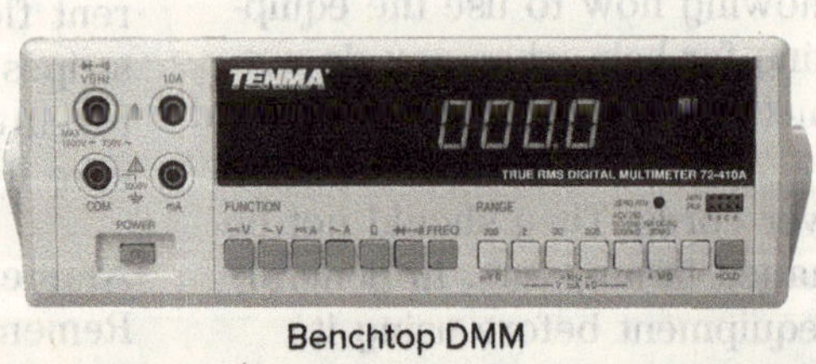

Benchtop DMM

(*b*)

Fig. 1-1.1 Basic meters. (*a*) Handheld DMM. (*b*) Benchtop DMM.

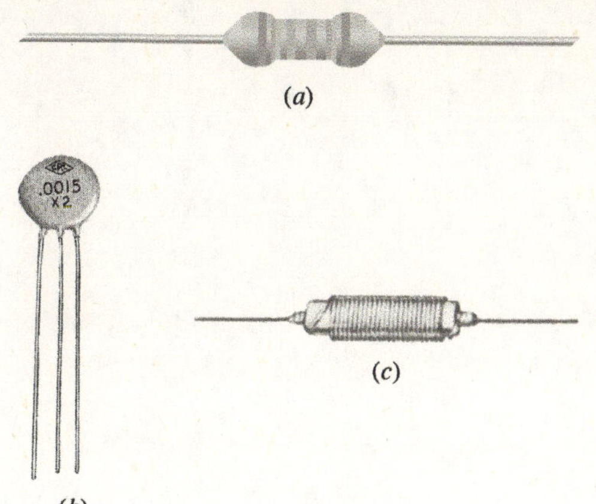

Fig. 1-1.2 Typical resistor (*a*), capacitor (*b*), and inductor (*c*).

in various sizes and power ratings. The most common resistors on circuit boards have the value of resistance color-coded (painted) on them. The value of resistance is the *ohm,* named for Georg Simon Ohm, the scientist who determined a law for resistance in 1827. The resistance value in ohms refers to the resistor's ability to resist the flow of electricity. The lower the value, the less resistance. In practical applications, one ohm (1 Ω) of resistance is very small and has little effect on most circuits. However, 1 million Ω has so much resistance that it resists all but the smallest amount of current flow.

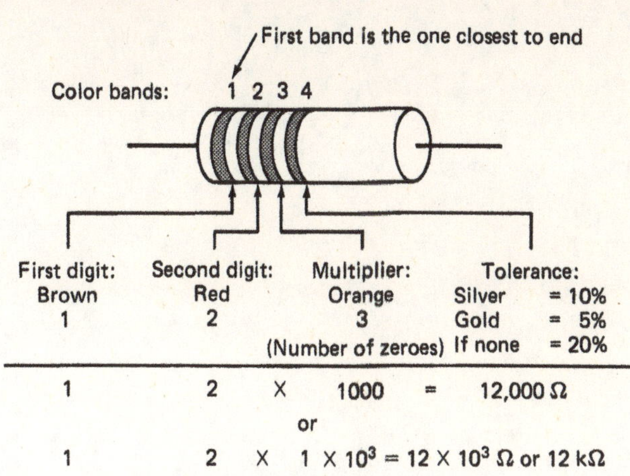

Remember: Multiplier means adding zeros. Here, add three zeros.

Note: If the multiplier (third band) is gold, multiply by 0.1. Here, a gold band multiplier would mean 1.2 Ω. If the multiplier is black, multiply by 1.

COLOR CODE

Color	Value	Color	Value
Black	0	Green	5
Brown	1	Blue	6
Red	2	Violet	7
Orange	3	Grey	8
Yellow	4	White	9

Fig. 1-1.3 How to read color bands (stripes) on carbon resistors. Resistors come in various shapes and sizes. This is only one common form. The larger the physical size, the greater the wattage rating. Refer to Appendix A for other component codes, including the five-band carbon-film resistors.

PROCEDURE

Four-Band Resistor Color Code Investigation

1. Answer questions 1 to 15.

2. After studying the resistor color code in Fig. 1-1.3, fill in Tables 1-1.1 and 1-1.2.

3. List the types of meters that you already know how to use and compare them to any meters in your lab.

After you finish, turn in your answers to procedure steps 1, 2, and 3.

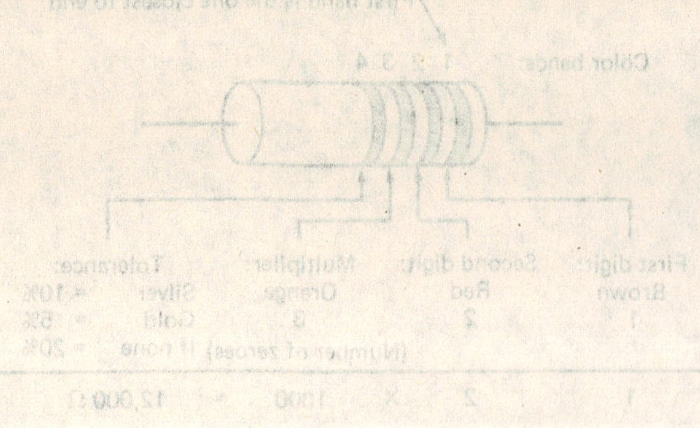

NAME ____ DATE ____________

QUESTIONS FOR EXPERIMENT 1-1

Fill in the blanks (1–15) with the letter of the correct answer.

____ **1.** An instrument used to measure potential difference.

____ **2.** An instrument used to measure current.

____ **3.** An instrument used to measure resistance.

____ **4.** A passive component that opposes the flow of current.

____ **5.** An instrument used to heat solder and join components.

____ **6.** A source of DC voltage other than a battery.

____ **7.** A symbol for DC voltage source.

____ **8.** A symbol for resistance.

____ **9.** A color used to represent negative polarity.

____ **10.** An item used to temporarily build circuits on.

____ **11.** A Greek letter used to represent a unit of resistance.

____ **12.** An English letter used to represent 1000.

____ **13.** An English letter used to represent 0.001.

____ **14.** An instrument used as an ohmmeter, a voltmeter, or an ammeter. capable of measuring both voltage and current.

____ **15.** A symbol for infinity.

a. ammeter

b. —/\/\/—

c. power supply

d. voltmeter

e. −|||+

f. black

g. ohmmeter

h. breadboard

i. soldering iron

j. resistor

k. DMM, DVM, VOM

l. m

m. k

n. Ω

o. ∞

TABLES FOR EXPERIMENT 1-1

TABLE 1-1.1 Resistor Color Codes

First Digit Band 1	Second Digit Band 2	Multiplier Band 3	Tolerance Band 4	Resistor Value
Red	Brown	Brown	Gold	
Brown	Brown	Black	Gold	
Green	Blue	Red	Silver	
Blue	Green	Yellow	Silver	
Red	Red	Orange	Silver	
Orange	White	Brown	Gold	
Blue	Green	Black	Silver	
Brown	Black	Red	Gold	
Yellow	Violet	Green	Gold	
Brown	Black	Orange	Silver	
Orange	Orange	Orange	Silver	
Brown	Black	Gold	Gold	
White	Blue	Red	Silver	
Brown	Black	Yellow	Silver	
Brown	Green	Green	Gold	

TABLE 1-1.2 Resistor Color Codes

Band 1 Color	Band 2 Color	Band 3 Color	Band 4 Color	Resistor Value
				680 kΩ, 5%
				10 kΩ, 10%
				100 kΩ, 5%
				3.3 MΩ, 5%
				1.2 kΩ, 10%
				820 Ω, 10%
				47 kΩ, 5%
				330 Ω, 10%
				470 kΩ, 5%
				560 Ω, 10%
				1.5 MΩ, 10%
				220 Ω, 5%
				56 Ω, 10%
				12 kΩ, 5%
				560 kΩ, 5%

EXPERIMENT RESULTS REPORT FORM

Experiment No.: ______________________ Name: ______________________

Date: ______________________

Experiment Title: ______________________ Class: ______________________

______________________ Instr: ______________________

Explain the purpose of the experiment:

List the first Learning Objective:

OBJECTIVE 1:

After reviewing the results, describe how the objective was validated by this experiment.

List the second Learning Objective:

OBJECTIVE 2:

After reviewing the results, describe how the objective was validated by this experiment.

List the third Learning Objective:

OBJECTIVE 3:

After reviewing the results, describe how the objective was validated by this experiment.

Conclusion:

If required, attach to this form: ☐ Answers to Questions, ☐ Tables, and ☐ Graphs.

CHAPTER

2

EXPERIMENT 2-1

RESISTANCE MEASUREMENTS

LEARNING OBJECTIVES

At the completion of this experiment, you will be able to:

- Operate the ohmmeter of a digital multimeter or volt ohmmeter.
- Measure resistance (resistors) in ohms.
- Understand the difference between meters.

INTRODUCTION

This experiment will familiarize you with resistance measurements and the ohmmeter. Resistance is the opposition to the flow of current and is measured in ohms (Ω). The ohmmeter is usually a function of a multimeter. Before modern digital meters, there were individual dedicated ohmmeters that only measured resistance.

An ohmmeter works by sending a known small value of current into one end of a component and measuring the value that comes through the other end. Because the value of current sent is known, the value that returns is used to calculate the resistance in ohms. In addition, the current is very small so it cannot damage the component or circuit. Also, ohmmeter measurements are done with no other power applied to the component.

For example, if all the current sent into one end of a component is returned through the other end, the ohmmeter reading should be zero ohms (0Ω). This means there was no opposition to current flow. This is also known as short or short circuit (no resistance), like a short piece of copper wire. Or, if no current returns, the ohmmeter reading will be very large, such as millions or billions of ohms. This means that there is infinite resistance to the flow of current. This is also known as an open circuit, such as open air or an opened switch. Anything between zero and infinite ohms is determined by the meter using its internal circuitry and the applied voltage.

In general, dedicated single measurement meters are no longer used. The only example is a **continuity checker,** which is like an ohmmeter except that it only verifies that a circuit is closed (has continuity) or open (no continuity). It does not give the value in ohms. Continuity checkers are often used by technicians, mechanics, and electricians to check for blown fuses or breaks in wiring.

Types of Meters: For measuring resistance, voltage, and current, the **DMM** (digital multimeter) is the instrument of choice. Some are handheld (portable and small) and others are mounted or kept on a bench (benchtop). The DMM is an ohmmeter, voltmeter, ammeter and more, all in one instrument. The first DMMs were known as **DVMs** (digital voltmeters) and are still used, but some do not have all the functions of the modern DMM. The volt ohmmeter (**VOM**) is another type and can be digital or analog. In general, the terms DMM, DVM, and VOM are often used interchangeably because they are multimeters.

The older analog VOM and the vacuum tube voltmeter (VTVM) are disappearing. They used magnetic movements and pointers (needles) with scales on the face for reading the values. VTVMs will not be shown in this lab manual because they are not manufactured anymore and they require vacuum tubes for operation. The analog VOMs use batteries.

Today, most technicians use digital DMMs because they are accurate and easy to read. Regardless of what meter you have, the ability to measure resistance, voltage, and current are absolutely necessary in electronics.

DMM (digital multimeter): The DMM (Fig. 2-1.1) is the modern standard instrument of choice for measuring resistance, voltage, and current. Some DMMs also have special functions to measure other components and values, such as diodes.

Throughout this lab manual, the DMM will be the equipment of choice for measuring all the basic circuits. Every DMM is different, but all require power. A handheld DMM uses a battery, and a bench-mounted (benchtop) DMM uses power from an AC (alternating current) outlet. All DMMs have LED (light-emitting diode) displays or readouts. DMMs have buttons or knobs for adjustments and/or range settings. Each DMM also has various inputs (terminals) for the leads or probes that make contact with the component or circuit being tested.

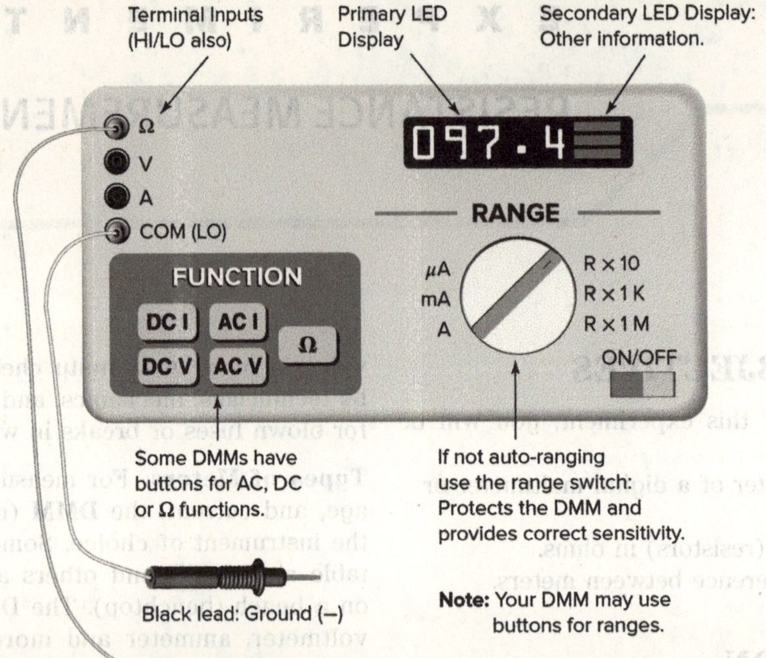

Fig. 2-1.1 Generic benchtop DMM.

Handheld DMMs (Fig. 2-1.2) are often inexpensive and are easier to use than benchtop models because they have fewer functions.

Again, the terms DMM, DVM (digital voltmeter), or VOM (volt ohmmeter—if digital) are often used interchangeably. The difference is in individual functions, ranges, displays, and so on. For the purpose of the experiments in this book, the abbreviation DMM will refer to any digital display meter that can measure resistance, current, and voltage.

Remember, if a resistance measurement (in ohms) is required, the DMM is simply functioning as an ohmmeter. Or if voltage measurements are required, then the DMM is functioning as a voltmeter. As you will see, this experiment has generic instructions for using a DMM and an older analog VOM. You should check with your instructor or the meter manual for more information if you are unsure or are new to electronics.

Analog VOM and Calibration: The analog VOM (volt ohmmeter) is an older meter that can measure current, voltage, and resistance. The analog VOMs use a magnetic meter movement with a needle or pointer. They have scales on the face and sometimes have a mirror to avoid parallax (side viewing). VOMs are usually battery operated and have an adjustment for calibration of resistance between zero and infinity. The zero adjust is also used to set the pointer to the zero position. Because of the movement, the pointer, and the scale, analog VOMs are not as reliable or as accurate as DMMs. Figure 2-1.3 shows an older VOM with the range switch in the Ω

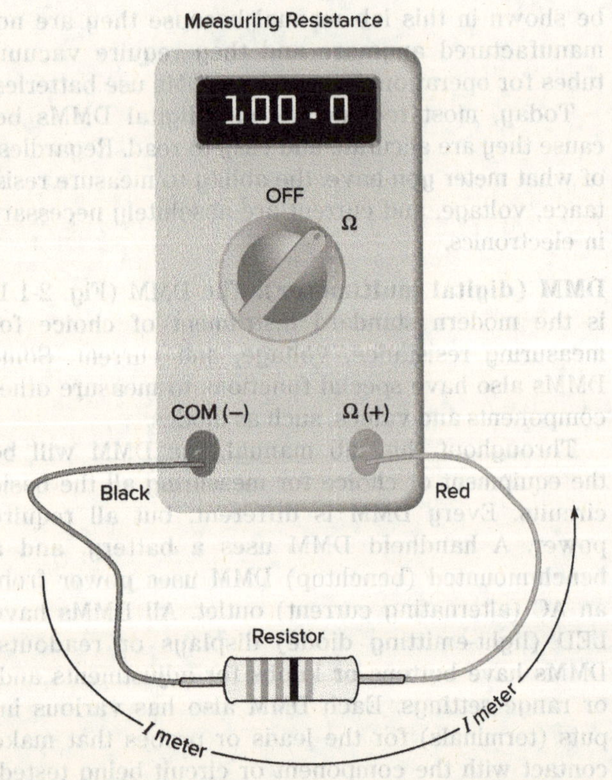

Fig. 2-1.2 Circuit Building Aid: Generic Handheld DMM.

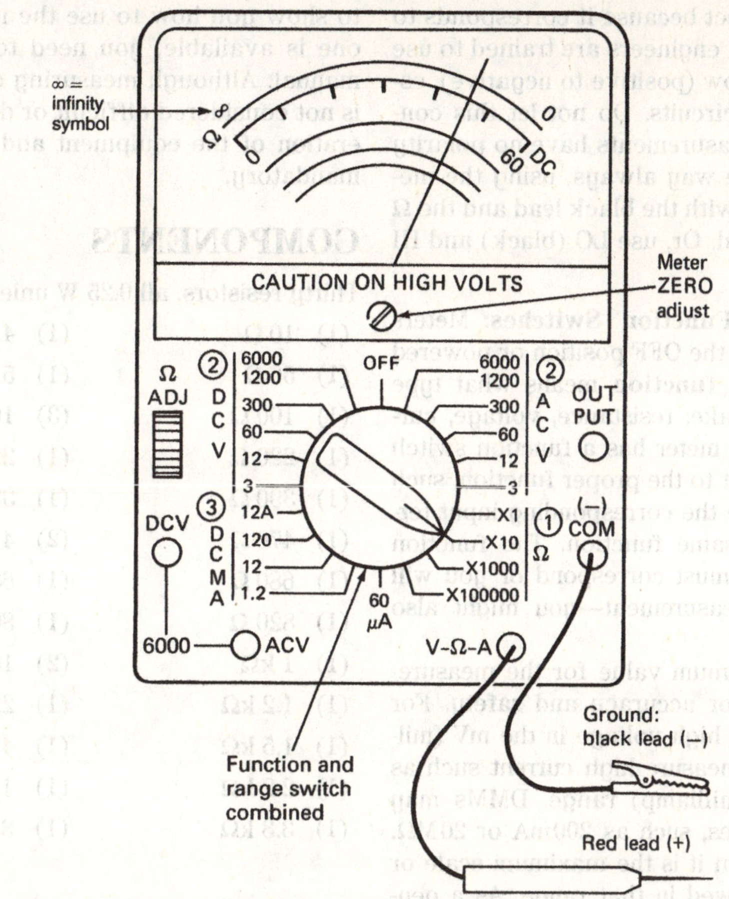

Fig. 2-1.3 Analog VOM. Linear and nonlinear scale (ohms). A generic VOM measures resistance in ohms (Ω), AC and DC voltage, and current in amperes (A). The VOM shown is set for resistance: $R \times 10$.

(resistance) area set to X10. This means the scale value is multiplied by 10 for the measured value.

To calibrate the analog VOM: In the ohms position and proper terminal connection, separate the leads (open), and the needle should move toward the infinity marker ∞, which is the starting point for calibration. Next, with the leads connected together (short), the needle should move toward 0—use the ZERO adjust to align the needle with 0. Repeat the process (∞ and zero) as necessary until they both align correctly when the leads are opened or shorted. The mirror that may be on the meter face can be used to avoid parallax errors due to reading the scale from an angle. If you use older model analog meters, you will have to calibrate them often.

Note: If you only have access to an analog VOM, you can still do most of the experiments in this lab manual. However, DMMs are recommended, and handheld models are inexpensive and available worldwide.

Leads, Probes, and Input Terminals: In general, leads and probes are similar because they both conduct current. Leads often have alligator clips, and probes are usually pointed. So, leads or probes do the same thing: they make contact from the meter to the component.

COM and Ω terminal inputs: The black lead or probe is considered negative polarity (–). The black lead is plugged into the COM (–) terminal of the meter. COM refers to common ground or common return path for current. The red lead or probe is considered positive polarity (+). It is plugged into the ohm Ω terminal. Sometimes Ω (resistance) and V (voltage) share the same terminal. For resistance measurements, most meters will have a COM for the black (–) lead and a Ω (ohm) terminal for the red (+) lead. But some meters use separate HI and LO terminals.

LO/HI terminal inputs: If your DMM has a **LO** input terminal for ohms, then use it like the COM (it is the LO side of the measurement). The **HI** terminal should have the Ω symbol or ohms. Use it for the red (+) lead.

Flow of Current: Current flow for ohmmeters is from the COM (–) or LO side, then through the resistor, and then back into the Ω (ohm) terminal or HI (+) side. This is the way most technicians are taught, and this is correct because it corresponds to electron flow. However, engineers are trained to use conventional current flow (positive to negative), especially for designing circuits. Do not let this confuse you: resistance measurements have no polarity and are made the same way always, using the meter's COM (−) terminal with the black lead and the Ω terminal for the red lead. Or, use LO (black) and HI (red) for those DMMs.

Power, Range, and Function Switches: Meters should always be left in the OFF position or powered OFF if not in use. The **function** means what type of measurement you make: resistance, voltage, current, and so on. If your meter has a function switch or button, then you set it to the proper function, such as VDC, Ω, mA, and use the corresponding input terminals that have the same function. The function and the input terminal must correspond or you will not get the correct measurement—you might also damage the meter.

The **range** is a maximum value for the measurement. It is important for accuracy and safety. For example, don't measure high voltage in the mV (millivolt) range; or don't measure high current such as 10 amps in the mA (milliamp) range. DMMs may also have specific ranges, such as 200 mA or 20 MΩ. If a value is shown, then it is the maximum scale or measureable value allowed in that range. As a general rule, you should always start in the maximum range if you are unsure—this will protect the meter from damage.

If your meter has values such as X1, X10, and so on, then you multiply the reading by the range setting. This is often used for resistance measurements on some meters. For DMMs, you usually press a button or change the lead terminal if it shows a range. You may see specific terminals such as 200 mA or A. In general, you set the measurement and then the range before making contact. If you are using a DMM with **auto-range**, then the range is automatic and no setting is required—but the input terminals must be correct regardless of the meter you use.

Now it's time to start measuring resistors and using your meter.

EQUIPMENT

Ohmmeter: DMM or VOM
Leads or probes
Proto board or spring board if desired

NOTE on Equipment: The general instructions in this experiment do not replace the meter's instruction manual and are only intended as a guide. If you are in a laboratory classroom, please ask the staff to show you how to use the meters provided. If no one is available, you need to read the instruction manual. Although measuring disconnected resistors is not considered difficult or dangerous, the safe operation of the equipment and your safety are both mandatory.

COMPONENTS

Thirty resistors, all 0.25 W unless indicated otherwise:

(1) 10 Ω	(1) 4.7 kΩ
(1) 56 Ω	(1) 5.6 kΩ (0.5 W or less)
(1) 100 Ω	(3) 10 kΩ (0.5 W or less)
(1) 220 Ω	(1) 22 kΩ (0.5 W or less)
(1) 390 Ω	(1) 33 kΩ (0.5 W or less)
(1) 470 Ω	(2) 47 kΩ (0.5 W or less)
(1) 680 Ω	(1) 68 kΩ (0.5 W or less)
(1) 820 Ω	(1) 86 kΩ (0.5 W or less)
(1) 1 kΩ	(2) 100 kΩ (0.5 W or less)
(1) 1.2 kΩ	(1) 220 kΩ (0.5 W or less)
(1) 1.5 kΩ	(1) 470 kΩ (0.5 W or less)
(1) 2.2 kΩ	(1) 1.2 MΩ (0.5 W or less)
(1) 3.3 kΩ	(1) 3.3 MΩ (0.5 W or less)

PROCEDURE

1. Meter check: With power on, set the meter for measuring ohms with the leads in the correct input terminals. If you have range settings or inputs, use a value near 10 K ohms ($R \times 10K$) if possible—but it's not necessary for this step. Just be sure you are ready to measure resistance in ohms and have the leads in the proper input terminals.
2. With the leads shorted (connected together), verify the meter shows 0 Ω or approximately zero ohms. Then open the leads (disconnected) and verify the meter shows infinite Ω or a very large value (many millions of ohms). Try this same check in other ranges, if available.
3. Measure all the resistor values in the components list and record the values in Table 2-1.1. Refer to the circuit-building aid figure. If you do not have an auto-ranging DMM, you must set the range or leads to the range value that is slightly above the resistance you are measuring—this will give you the most accurate measurement. Finally, always remember that no measurements are exact, especially when measuring resistors.
4. Answer the questions for this experiment and write a report if required. Refer to the appendix for examples of writing a report.

NAME ______________________________ DATE ______________

QUESTIONS FOR EXPERIMENT 2-1

Answer true (T) or false (F) to the following:

_____ **1.** The DMM can measure current.

_____ **2.** An ohmmeter will show zero ohms when the leads are not connected together (open circuit).

_____ **3.** Linear scales are used for resistance measurements.

_____ **4.** It is necessary to adjust infinity (∞) and zero ohms whenever changing ranges on an ohmmeter.

_____ **5.** A continuity check gives the value in ohms.

_____ **6.** shorting the leads together on an ohmmeter results in zero ohms.

_____ **7.** Parrallax is an error resulting from reading meter scales with needles or pointers from an angular view.

_____ **8.** An ohmmeter cannot be damaged by measuring voltage.

_____ **9.** The range switch is used only for voltage measurements.

Short Answer:

_____ **10.** Refer to Appendix A, and write an explanation of the differences between 4-band and 5-band resistors.

TABLE FOR EXPERIMENT 2-1

TABLE 2-1.1 Resistance Measurements

Nominal Value, Ω	Measured Value, Ω	Nominal Value, Ω	Measured Value, Ω

EXPERIMENT RESULTS REPORT FORM

Experiment No.: ______________________ Name: ______________________

Date: ______________________

Experiment Title: ______________________ Class: ______________________

______________________ Instr: ______________________

Explain the purpose of the experiment:

List the first Learning Objective:

OBJECTIVE 1:

After reviewing the results, describe how the objective was validated by this experiment.

List the second Learning Objective:

OBJECTIVE 2:

After reviewing the results, describe how the objective was validated by this experiment.

List the third Learning Objective:

OBJECTIVE 3:

After reviewing the results, describe how the objective was validated by this experiment.

Conclusion:

If required, attach to this form: ☐ Answers to Questions, ☐ Tables, and ☐ Graphs.

CHAPTER 2

EXPERIMENT 2-2

RESISTOR *V* & *I* MEASUREMENTS

LEARNING OBJECTIVES

At the completion of this experiment, you will be able to:

- Measure voltage with a voltmeter.
- Measure current with an ammeter.
- Use a DC power supply.

SUGGESTED READING

Chapters 1 and 2, *Basic Electronics*, Grob/Schultz, twelfth edition

INTRODUCTION

Ammeters measure the amount of current (electron flow) in a circuit. Current is measured in units called *amperes* or *amps*. Ammeters are almost always used in basic electronics lab courses and in physics lab courses because they are excellent for teaching and verifying the flow of current. But they are rarely used by technicians for repair work because ammeters usually measure only direct current (DC) and are not useful where alternating current (AC) is found. In this experiment, you will be inserting the ammeter in simple circuits and measuring small and safe amounts of DC current.

Voltmeters are the most useful tool for testing and troubleshooting DC circuits. They measure the amount of electrical force (charge or a potential difference between two points). The measurement unit is called the *volt*. Wherever there is current, there is also voltage. Most voltmeters measure DC voltage, and many also measure AC voltage, which is covered later. For this experiment, you will be measuring only DC voltage.

Common Values of Current and Voltage

The *milliampere* is a common value of current. The milliampere is one-thousandth of an amp, and that is enough current to do a lot of useful work on a circuit board. Throughout this course, you will mostly be measuring milliamperes and sometimes *microamperes*. In general, any value of current greater than a few amperes is considered high and can present a danger or hazard. Current in the area of 10 amperes and above is high enough to be deadly.

The *millivolt* is also a common value of voltage in DC electronics. It is one-thousandth of a volt. Values of hundreds of millivolts are common on circuit boards, as are values of voltage below 10 or 20 volts (V). Voltage is somewhat different from current, because large amounts of voltage can exist without a lot of current and may not be dangerous if it cannot supply current. However, it is always best to be cautious and safe.

Types of Ammeters and Voltmeters

The most commonly used voltmeters are digital handheld meters with liquid crystal displays, often referred to as *DVMs* (*digital voltmeters*) or *DMMs* (*digital multimeters*). However, older meters such as the *VOM* (*volt-ohm-milliammeter*) and the *VTVM* (*vacuum tube voltmeter*) are sometimes used in schools. And some schools may use dedicated single-function ammeters and voltmeters that may have been built especially as teaching tools. For information about the various meters, refer to Experiment 2-1, on resistance measurements, where the meters were shown.

DC Power Supply

A source of DC power is essential for activating any circuit. In fact, a battery is the most common source of DC power, but it is not a variable type of supply, and it will run out of power if not recharged. For lab work, a variable and constant source of DC power is essential. For that reason, your lab is equipped with a DC power supply that may be capable of supplying 10 or more volts and is also variable. See Fig. 2-2.1 for a typical DC power supplies.

The DC power supply is similar to a battery but is adjustable. Most bench supplies can maintain a constant voltage and, sometimes, a constant current. Such power supplies usually have their ranges or maximum values listed on the face, for example, 0 to 10 V or 0 to 30 V. Also, bench supplies can often be adjusted to values less than a volt, including millivolts.

Because every lab has a different type and manufacturer of power supplies, it is necessary to learn how to safely operate the power supply in your situation. One way to do this is to read the manual or simply to look at the controls. The simplest of power supplies may have only positive and negative terminals, a display, and a knob to adjust the voltage.

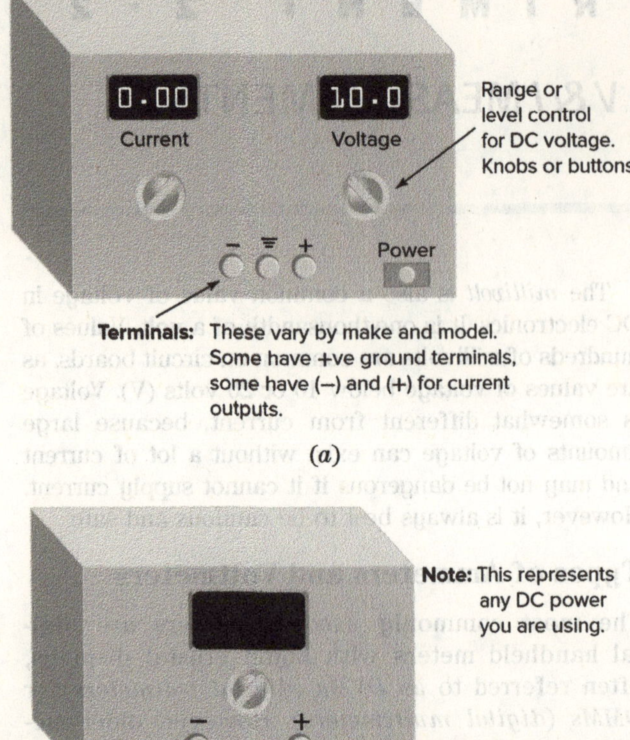

Fig. 2-2.1 (*a*) Generic DC Power Supply: Current and Voltage. (*b*) Simplified DC Power Supply: Voltage only.

However, there are sophisticated supplies that have both positive and negative voltages, constant or variable current adjustments, and even swept outputs. For this experiment you will need a supply that adjusts between 0 and 10 V.

EQUIPMENT

Voltmeter (DMM)
Ammeter (DMM)
DC power supply

COMPONENTS

Resistors (all 0.25 W unless indicated otherwise):

(1) 100 Ω
(1) 1 kΩ
(1) 10 kΩ
(1) 330 Ω
(1) 560 Ω

PROCEDURE

1. **Setting the voltmeter (DMM):** Set the function switch to the appropriate VOLTS position. You will be measuring less than 10 V throughout this experiment. Be sure that the leads are properly

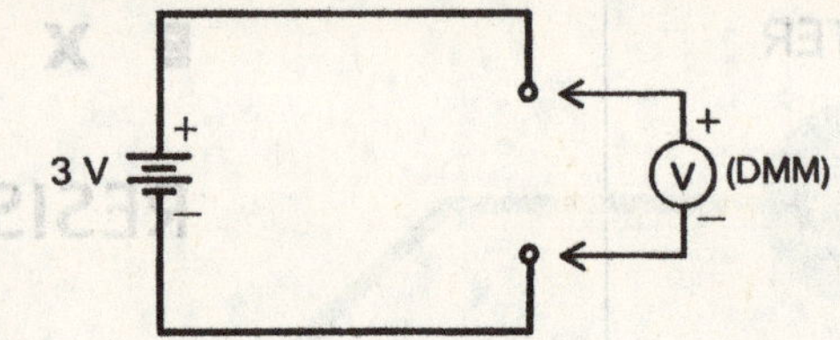

Fig. 2-2.2 Simple schematic of a DC voltage measurement. (A schematic is a type of electronic road map or blueprint.) The 3 V on the left means that the power supply is set at 3 V. The V in a circle on the right indicates a voltmeter (DMM or VOM or DVM).

connected. Digital meters are usually easy to operate and require only a minimal amount of adjustment.

2. **Setting the power supply:** Locate the power supply and locate its terminals (+) and (–). These are similar to the ends of a battery. If you are using a power supply with a variable knob, turn it counterclockwise to start at zero. Also, if the power supply has any setttings, be sure to set it for 10 V or less with minimal current.

Turn on the power supply.

3. **Measuring power supply voltage:** Connect the voltmeter directly to the power supply as shown in Fig. 2-2.2 (schematic) or Fig. 2-2.3 (circuit-building aid). First, connect the negative lead or ground side to the negative output of the power supply. This is standard practice. Then connect the positive lead (probe) to the positive (+) side of the power supply.

Note: When a *voltmeter* is shown connected in a schematic, the circled symbol Ⓥ is used.

Adjust the power supply to read 3 V by slowly increasing the voltage control knob. If you are using a digital supply, adjust it in a similar manner. When your voltmeter reads 3 V, stop. Even if the power supply display is a little more or less than the meter reading, it does not matter. The voltmeter is making the measurement, and that is the data you want.

Try increasing and decreasing the power supply in minimal amounts between 0 and 5 V, watching the voltmeter. If you are using a meter movement (needle), never allow the needle to bump up to full-scale deflection. This may ruin the meter movement or bend the needle pointer. For example, if you are going to measure 10 V, set the meter to a higher scale, such as 20 V.

Obtain any battery less than 9 V and try measuring it with the voltmeter. Because a voltmeter has a very high resistance (millions of ohms), it cannot drain the battery of its current.

4. **Measuring resistor voltage:** Refer to the circuit in Fig. 2-2.4, where a 1-kΩ resistor is used in

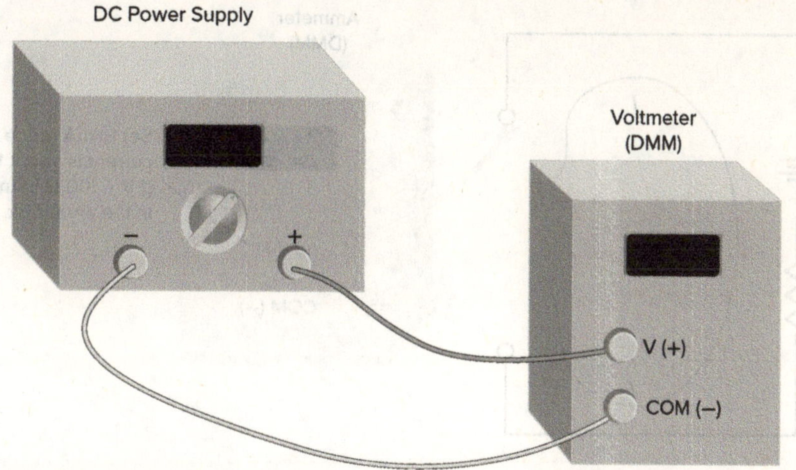

Fig. 2-2.3 Circuit Building Aid: Measuring the output of a DC power supply using a voltmeter (DMM).

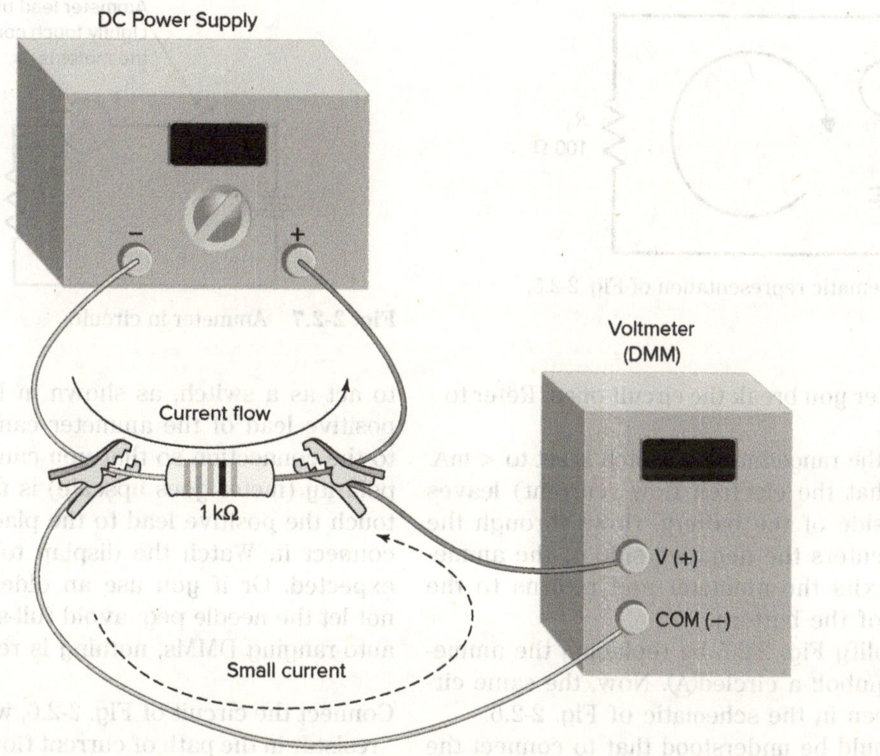

Fig. 2-2.4 Current Building Aid: Resistor Voltage Measurement: Voltmeter is across the resistor.

the path of current flow. Notice that the leads of the voltmeter are connected *across* the resistor like a bridge. This allows a very small (microamps) amount of current to travel into the highly resistive voltmeter to be measured with little or no effect on the rest of the circuit.

Connect the circuit of Fig. 2-2.4 with 3-V output from the power supply. Record the voltage reading in Table 2-2.1.

Replace the 1-kΩ resistor with a 100-Ω resistor without adjusting the power supply, and record the voltage reading on the voltmeter.

Replace the 100-Ω resistor with a 10-kΩ resistor without adjusting the power supply, and record the voltage reading on the voltmeter.

5. Measuring current with an ammeter: You have already learned how to use an ohmmeter to measure resistance and a voltmeter to measure voltage. For both meters, it was necessary to place the two (+ and −) leads *across* the component you measured. However, an ammeter is actually connected *in* the circuit—in the path of current (electron) flow. This is also known as *in series*. Therefore, the leads are used to connect the ammeter into a

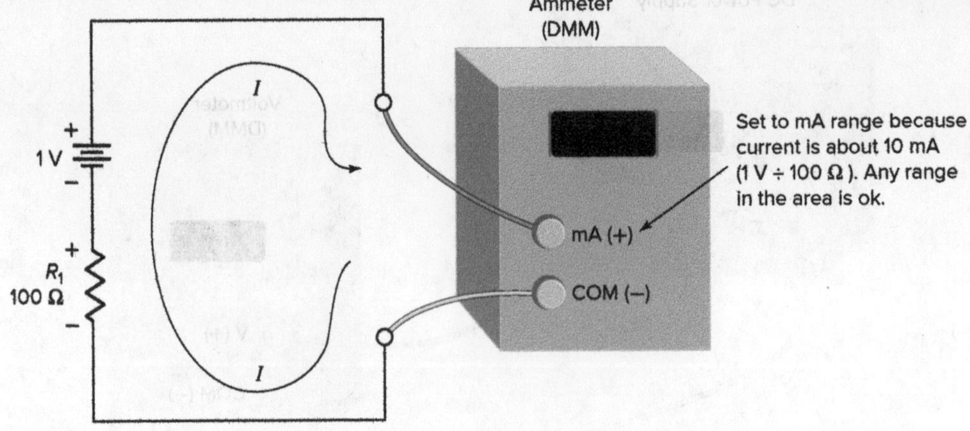

Fig. 2-2.5 Circuit Building Aid: Measuring current with an ammeter (in series with a battery and resistor).

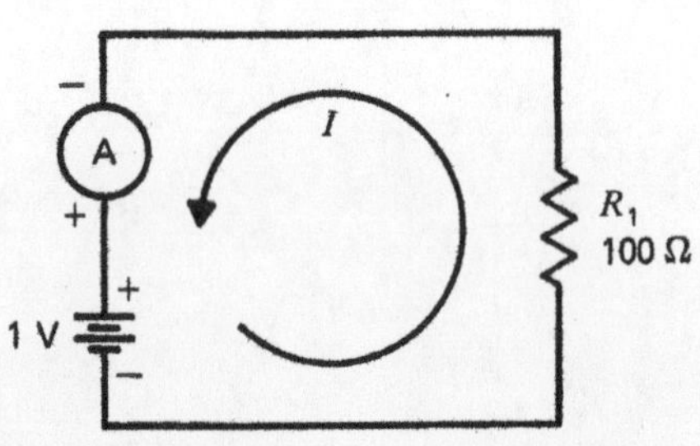

Fig. 2-2.6 Schematic representation of Fig. 2-2.5.

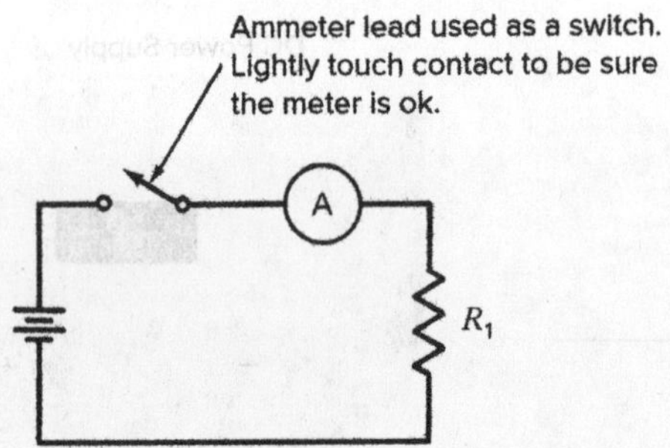

Fig. 2-2.7 Ammeter in circuit.

circuit only after you break the circuit open. Refer to Fig. 2-2.5.

Notice that the range/function switch is set to < mA. Also, notice that the electron flow (current) leaves the negative side of the battery, flows through the resistor, and enters the negative side of the ammeter. Then it exits the ammeter and returns to the positive side of the battery.

Let us simplify Fig. 2-2.5 by replacing the ammeter with its symbol: a circled Ⓐ. Now, the same circuit can be seen in the schematic of Fig. 2-2.6.

Here, it should be understood that to connect the ammeter, the circuit was broken between the positive side of the battery and the resistor. Also, to determine where the polarities are, notice that wherever electron flow enters a component, that side of the component is then considered negative. Where electron flow leaves the component, closer to the positive side of the battery source, it is considered positive. Thus, the resistor and ammeter, in Fig. 2-2.6, are shown where electron flow determines polarity. Of course, electron flow must enter the common, or negative, side of the ammeter and exit from the positive side.

Be sure that power is turned off before making any connections and then turned on again.

Finally, an easy way to protect an ammeter, or any meter, is to allow one lead (positive, for example) to act as a switch, as shown in Fig. 2-2.7. Here, the positive lead of the ammeter can be touched lightly to the connection so that you can be sure the proper polarity (meter goes upscale) is there. Lightly tap or touch the positive lead to the place you are going to connect it. Watch the display to see if it reacts as expected. Or if you use an older analog meter, do not let the needle peg; avoid full-scale deflection. For auto-ranging DMMs, nothing is required.

Connect the circuit of Fig. 2-2.6, which is a 100-Ω resistor in the path of current flow with the ammeter, but do not turn the power supply on yet.

Slowly adjust the power supply voltage for 1 V across the resistor. You can use a voltmeter to measure the voltage across the resistor and adjust to 1 V if necessary. Note the value of current on the ammeter and record it in Table 2-2.2.

Replace the 100-Ω resistor with a 330-Ω resistor. The voltage should still be 1 V across the resistor. Note the value of current on the ammeter and record it in Table 2-2.2.

Replace the 330-Ω resistor with a 560-Ω resistor. The voltage should still be 1 V across the resistor. Note the value of current on the ammeter and record it in Table 2-2.2.

Turn off the power supply and disconnect the circuit.

NAME

DATE

QUESTIONS FOR EXPERIMENT 2-2

In your own words, describe the different ways an ohmmeter, an ammeter, and a voltmeter are connected to make measurements.

TABLES FOR EXPERIMENT 2-2

TABLE 2-2.1 Voltage Measurements

Resistor Value	Voltage
1 kΩ	______
100 Ω	______
10 kΩ	______

TABLE 2-2.2 Current Measurements

Resistor Value	Voltage	Current, *I*
100 Ω	1 V	______
330 Ω	1 V	______
560 Ω	1 V	______

Note for Report: Write your conclusion based on data from these two tables.

EXPERIMENT RESULTS REPORT FORM

Experiment No.: ____________________ Name: ____________________

Date: ____________________

Experiment Title: ____________________ Class: ____________________

____________________ Instr: ____________________

Explain the purpose of the experiment:

List the first Learning Objective:
OBJECTIVE 1:

After reviewing the results, describe how the objective was validated by this experiment.

List the second Learning Objective:
OBJECTIVE 2:

After reviewing the results, describe how the objective was validated by this experiment.

List the third Learning Objective:
OBJECTIVE 3:

After reviewing the results, describe how the objective was validated by this experiment.

Conclusion:

If required, attach to this form: ☐ Answers to Questions, ☐ Tables, and ☐ Graphs.

CHAPTER 3

EXPERIMENT 3-1

OHM'S LAW

LEARNING OBJECTIVES

At the completion of this experiment, you will be able to:

- Validate the Ohm's law expression, where
  - $V = I \times R$
  - $I = \frac{V}{R}$
  - $R = \frac{V}{I}$

SUGGESTED READING

Chapter 3, *Basic Electronics*, Grob/Schultz, twelfth edition

INTRODUCTION

Ohm's law is the most widely used principle in the study of basic electronics. In 1827, Georg Simon Ohm experimentally determined that the amount of current I in a circuit depended upon the amount of resistance R and the amount of voltage V. If any two of the factors V, I, or R are known, the third factor can be determined by calculation: $I = V/R$, $V = IR$, and $R = V/I$. Also, the amount of electric power, measured in watts, can be determined indirectly by using Ohm's law:

$$P = \frac{V^2}{R} \quad \text{or} \quad P = I^2R$$

Ohm's law is usually represented by the equation $I = V/R$. Because it is a mathematical representation of a physical occurrence, it is important to remember that the relationship between the three factors (I, V, and R) can also be expressed as follows:

Current is directly proportional to voltage if the resistance does not vary.

Current is inversely proportional to resistance if the voltage does not vary.

This experiment provides data that will validate the relationships between current, voltage, and resistance.

EQUIPMENT

Ohmmeter (DMM)
Voltmeter (DMM)
Ammeter (DMM)
Power supply
Connecting leads
Circuit board

COMPONENTS

Resistors (all 0.25 W):

(1) 100 Ω (1) 820 Ω
(1) 330 Ω (1) 1 kΩ
(1) 560 Ω

PROCEDURE

1. Connect the circuit of Fig. 3-1.1*a* with the power supply turned off. The polarity of the power supply, the voltmeter, and the ammeter must be correct in order to avoid damage to the equipment.

Note: Use the circuit-building aid of Fig. 3-1.1*b* if you have trouble.

2. Refer to Table 3-1.1. Calculate the current for the first voltage setting (1.5 V, 100 Ω nominal value). Record the value in Table 3-1.1 under the heading Calculated Current.

3. Check your circuit connection by tracing the path of electron flow. Be sure the ammeter reading is correct for the value of calculated current. Also, be sure the voltmeter range is correct.

4. Before turning the power supply on, turn the voltage adjust to zero. Now, turn the power supply on. Use the voltmeter to monitor the power supply, and adjust for 1.5-V applied voltage.

5. Read the value of measured current and record the results in Table 3-1.1.

6. Repeat procedures 2 to 5 for each value of applied voltage listed in Table 3-1.1.

Note: It is not necessary to turn off the power supply unless you are disconnecting the circuit in order to change meter ranges.

7. After recording all the measured and calculated values of current for Table 3-1.1, turn off the power supply. Be sure that the voltage adjusted is set to zero. Disconnect the circuit. The calculated values of power can be done later. Indicate which formula you used.

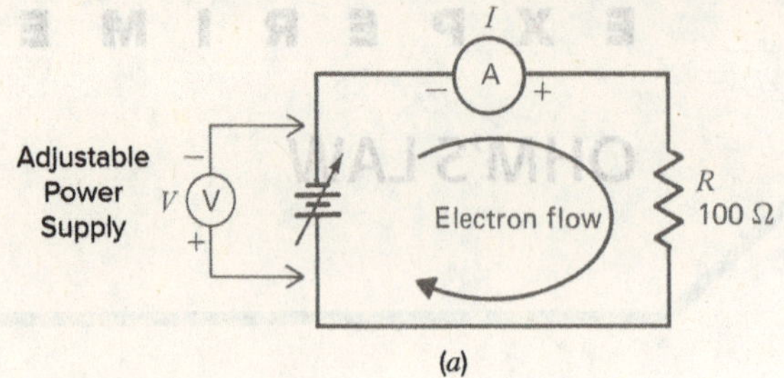

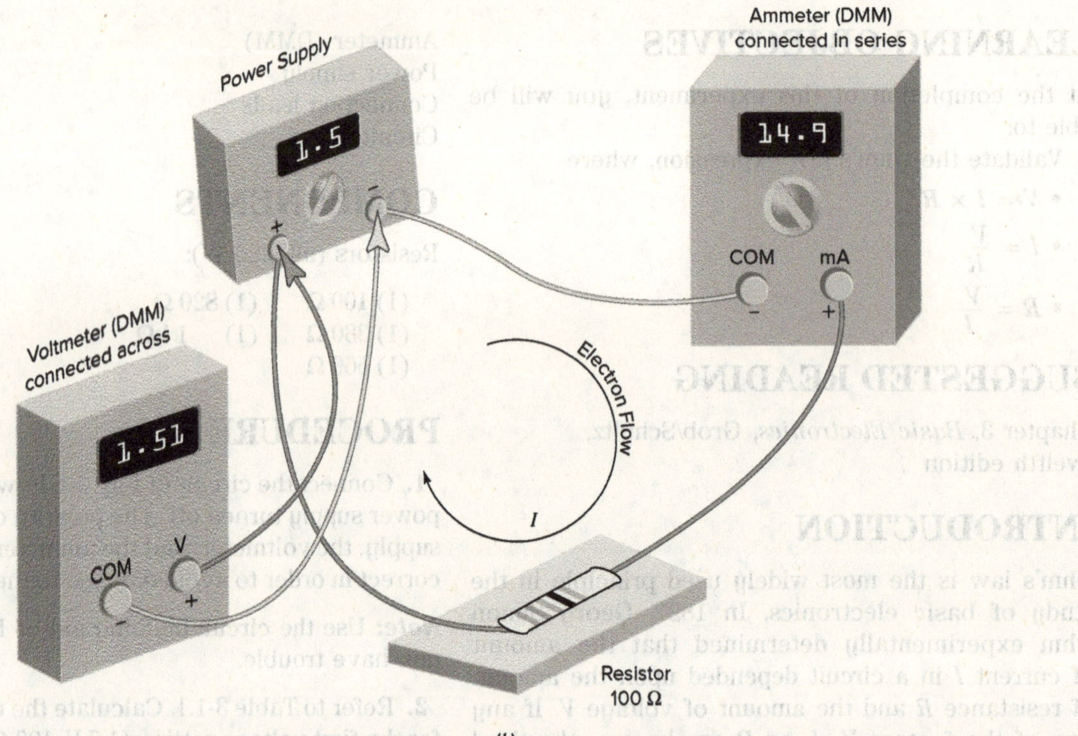

Fig. 3-1.1 (*a*) Ohm's law schematic. (*b*) Circuit-building aid.

8. Refer to Table 3-1.2. Measure and record the values of each resistor listed in Table 3-1.2.

9. Connect the circuit of Fig. 3-1.2. Keep the power supply turned off, and begin with the first value of resistance in Table 3-1.2, $R = 1000\ \Omega$ (1 kΩ).

Note: This is the same basic circuit as Fig. 3-1.1. However, the ammeter is located in a different part of the circuit's current path. Use the same precautions as you did with Fig. 3-1.1.

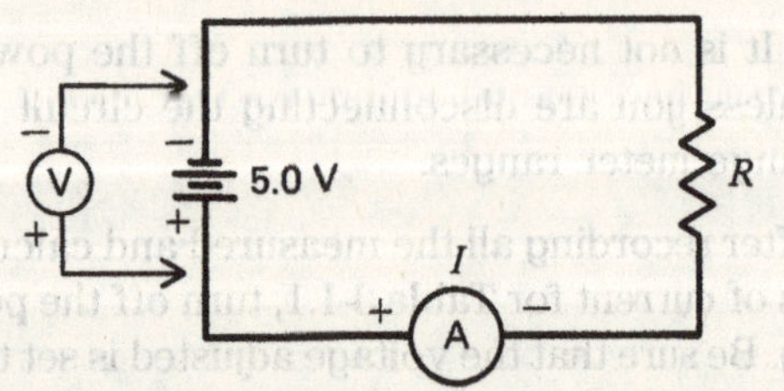

Fig. 3-1.2 Ohm's law circuit.

10. Refer to Table 3-1.2. Calculate the current for the value of resistance listed in Table 3-1.2 (5.0 V, 1 kΩ nominal value). Record the calculated value in Table 3-1.2 under the heading Calculated Current.

11. Check your circuit. Be sure that the ammeter range is correct for the value of calculated current. Also, be sure that the voltmeter range is correct.

12. Turn on the power supply and adjust for 5.0 V applied voltage. Monitor the meters as you did for the circuit of Fig. 3-1.1.

13. Read the value of measured current and record the results in Table 3-1.2.

14. Turn off the power supply. Replace the resistor with the next value of resistance listed in Table 3-1.2.

15. Repeat steps 10 to 14 for each value of resistance listed in Table 3-1.2.

16. Turn off the power supply. Disconnect the circuit.

NAME ______________________________ DATE ________

QUESTIONS FOR EXPERIMENT 3-1

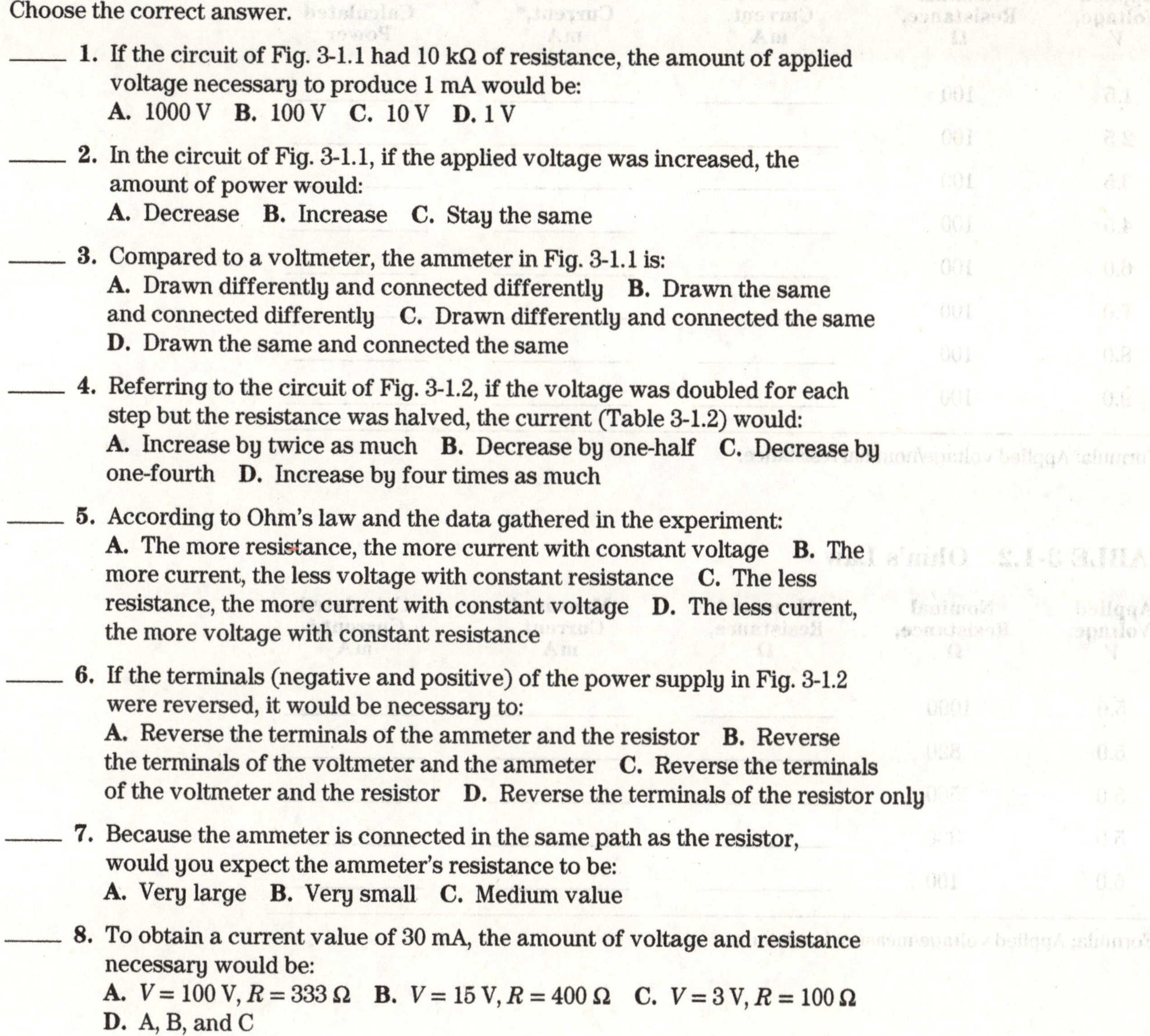

Choose the correct answer.

_____ **1.** If the circuit of Fig. 3-1.1 had 10 kΩ of resistance, the amount of applied voltage necessary to produce 1 mA would be:
A. 1000 V **B.** 100 V **C.** 10 V **D.** 1 V

_____ **2.** In the circuit of Fig. 3-1.1, if the applied voltage was increased, the amount of power would:
A. Decrease **B.** Increase **C.** Stay the same

_____ **3.** Compared to a voltmeter, the ammeter in Fig. 3-1.1 is:
A. Drawn differently and connected differently **B.** Drawn the same and connected differently **C.** Drawn differently and connected the same **D.** Drawn the same and connected the same

_____ **4.** Referring to the circuit of Fig. 3-1.2, if the voltage was doubled for each step but the resistance was halved, the current (Table 3-1.2) would:
A. Increase by twice as much **B.** Decrease by one-half **C.** Decrease by one-fourth **D.** Increase by four times as much

_____ **5.** According to Ohm's law and the data gathered in the experiment:
A. The more resistance, the more current with constant voltage **B.** The more current, the less voltage with constant resistance **C.** The less resistance, the more current with constant voltage **D.** The less current, the more voltage with constant resistance

_____ **6.** If the terminals (negative and positive) of the power supply in Fig. 3-1.2 were reversed, it would be necessary to:
A. Reverse the terminals of the ammeter and the resistor **B.** Reverse the terminals of the voltmeter and the ammeter **C.** Reverse the terminals of the voltmeter and the resistor **D.** Reverse the terminals of the resistor only

_____ **7.** Because the ammeter is connected in the same path as the resistor, would you expect the ammeter's resistance to be:
A. Very large **B.** Very small **C.** Medium value

_____ **8.** To obtain a current value of 30 mA, the amount of voltage and resistance necessary would be:
A. $V = 100$ V, $R = 333$ Ω **B.** $V = 15$ V, $R = 400$ Ω **C.** $V = 3$ V, $R = 100$ Ω
D. A, B, and C

TABLES FOR EXPERIMENT 3-1

TABLE 3-1.1 Ohm's Law

Applied Voltage, *V*	Nominal Resistance, Ω	Measured Current, mA	Calculated Current,* mA	Calculated Power
1.5	100			
2.5	100			
3.5	100			
4.5	100			
6.0	100			
7.0	100			
8.0	100			
9.0	100			

***Formula: Applied voltage/nominal resistance.**

TABLE 3-1.2 Ohm's Law

Applied Voltage, *V*	Nominal Resistance, Ω	Measured Resistance, Ω	Measured Current, mA	Calculated Current,* mA
5.0	1000			
5.0	820			
5.0	560			
5.0	330			
5.0	100			

***Formula: Applied voltage/measured resistance.**

EXPERIMENT RESULTS REPORT FORM

Experiment No.: ______________________ Name: ______________________
Date: ______________________
Experiment Title: ______________________ Class: ______________________
______________________ Instr: ______________________

Explain the purpose of the experiment:

List the first Learning Objective:
OBJECTIVE 1:

After reviewing the results, describe how the objective was validated by this experiment.

List the second Learning Objective:
OBJECTIVE 2:

After reviewing the results, describe how the objective was validated by this experiment.

List the third Learning Objective:
OBJECTIVE 3:

After reviewing the results, describe how the objective was validated by this experiment.

Conclusion:

If required, attach to this form: ☐ Answers to Questions, ☐ Tables, and ☐ Graphs.

CHAPTER 3

EXPERIMENT 3-2

APPLYING OHM'S LAW

LEARNING OBJECTIVES

At the completion of this experiment, you will be able to:

- Calculate V, I, or R from measured values.
- Determine the difference between calculated and measured power.
- Design a simple circuit using Ohm's law.

SUGGESTED READING

Chapter 3, *Basic Electronics*, Grob/Schultz, twelfth edition

INTRODUCTION

As you have already learned in the previous experiment, Ohm's law proves that there is a proportional relationship between voltage, current, and resistance. This law forms the basis for much of the troubleshooting work that technicians do every day. Regardless of the complexity of a circuit, engineers and technicians still use Ohm's law to determine voltage, current, or resistance. This experiment will reinforce your understanding of Ohm's law and show you how to apply it to simple circuits.

Ohm's Law Review

To review, Ohm's law states that

$V = I \times R$	Voltage equals current multiplied by resistance
$I = V/R$	Current equals voltage divided by resistance
$R = V/I$	Resistance equals voltage divided by current

The relationship between V and I is *directly proportional* because as one factor is increased, the other factor is increased at the same rate, but only if R does not change. But the relationship between R and I is *inversely proportional* because as one factor is increased, the other factor is decreased at the same rate, but only if V does not change (fixed value). For example, if the resistance R is increased two times, the current I will decrease by one-half.

In Fig. 3-2.1, two circuits have the same value of current although the values of V and R are different for each circuit.

Ohm's law can also be used to calculate power using these formulas:

$$P = V \times I$$
$$P = I^2R$$
$$P = V^2/R$$

These relationships are used to calculate the amount of power (in watts) dissipated in a circuit. This is important for protecting circuits from overheating and in determining whether a component is operating within its allowed range or wattage specification. In general, whenever current travels through a resistance, power is used, resulting in heat or light being created. For example, when current passes through a resistor, friction causes heat to be released or dissipated. If too much current passes through a resistor, exceeding its power rating, it may burn or melt the resistor. But most circuits are designed with fuses to prevent excessive current from causing too much damage. By applying Ohm's law, it is easy to determine the levels of current in a circuit.

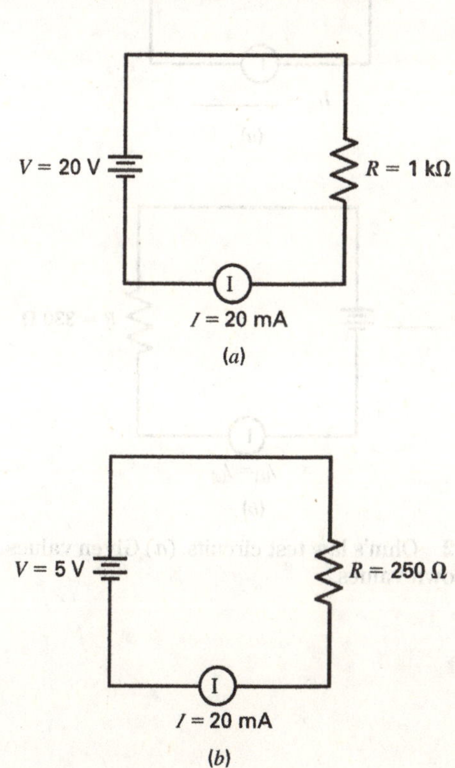

Fig. 3-2.1 Two circuits with the same current but different values of V and R.

EQUIPMENT

DC power supply
Ohmmeter (DMM)
Ammeter (DMM)
Voltmeter (DMM)

COMPONENTS

Resistors: (1) 330 Ω, (1) 470 Ω, and (1) 1 kΩ or less for design (see procedure step 11)

PROCEDURE

1. Measure the 330-Ω and 470-Ω resistors to be sure that they are in tolerance, and record their values in Table 3-2.1. If any resistor is out of tolerance, replace it with a good one.

2. Refer to the circuit schematic of Fig. 3-2.2*a*. Use Ohm's law to calculate the current in the circuit. Record the value in Table 3-2.1.

3. Connect the circuit of Fig. 3-2.2*a*.

4. Measure the voltage across the resistor and the current in the circuit. Record both values, *V* and *I*, in Table 3-2.1.

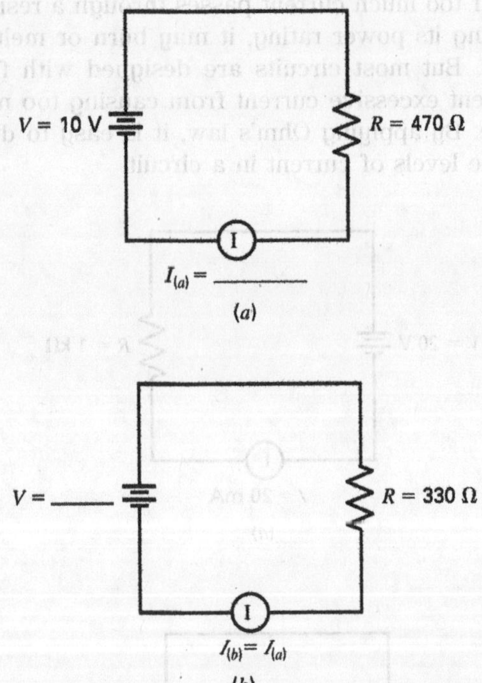

Fig. 3-2.2 Ohm's law test circuits. (*a*) Given values. (*b*) Unknown values.

5. Calculate the power dissipated, using the nominal values of resistance and voltage and the formula $P = V^2/R$. Record the calculated power dissipation in Table 3-2.1.

6. Calculate the power dissipated in the resistor, using the measured values of *I* and *R* only and the formula $P = I^2 \cdot R$.

7. Connect the circuit of Fig. 3-2.2*b*, where *R* = 330 Ω and there is also an ammeter connected in the circuit.

8. Carefully adjust the power supply voltage so that the current in the circuit in Fig. 3-2.2*b* is approximately equal to the current in the circuit in Fig. 3-2.2*a*. Record the value of voltage in Table 3-2.1.

9. Repeat steps 4 through 6 above for the circuit in Fig. 3-2.2*b*.

10. Percentage of error in power measurments: Calculate the percentage of error between the calculated and measured values of power in Table 3-2.1. To do this, find the difference between the two values and divide that difference by the calculated value: % error = difference between measured and calculated value÷calculated value. It does not matter whether the measured or calculated value is greater; simply find the difference between them and divide by the calculated value. Record the percentage of error in Table 3-2.1.

11. Design a circuit that has the approximate value of current (±10 percent) as the circuits of Fig. 3-2.2*a* and *b* but with the following three specification restrictions: (1) voltage must be between 1 and 15 V; (2) resistance must be less than 1 kΩ; and (3) the resistor used must be a standard value available in your lab (listed in Appendix I).

12. On a separate sheet of paper, draw the schematic for the circuit, showing the design values *V*, *I*, and *R*. Be sure to label all components carefully and show the meter connections. Also, show the design calculations (Ohm's law calculations) you used, and be sure to indicate the calculated values of *V*, *I*, and *R*.

13. Measure the resistor you have chosen and record the value in Table 3-2.2.

14. Repeat steps 4 through 6 above for the circuit you designed. Record all the values in Table 3-2.2.

Note: You may want to record the values on a separate sheet of paper before recording them in the table in case the design is invalid

NAME DATE

QUESTIONS FOR EXPERIMENT 3-2

1. In the circuits of Fig. 3-2.2*a* and *b*, what aspects of Ohm's law are validated?

2. Describe any difference between the power calculations using the measured values and the calculated values. Refer to the percentage of error values.

3. Explain why you think an ammeter can be connected in the circuit and not affect the flow of current.

4. Explain why the voltage measured across the resistor is the same as the applied voltage.

5. In the circuit you designed, explain how you could reduce the dissipated power by one-half.

TABLES FOR EXPERIMENT 3-2

TABLE 3-2.1 Approximate Values Are Acceptable

Circuit	Measured Value, R	Measured Value, V	Measured Value, I	Calculated Value, I	Calculated Value, $P = V^2/R$	Measure Value, $P = I^2R$	% Error, P
a (470 Ω)							
b (330 Ω)							

TABLE 3-2.2

Measured Value, R	Measured Value, V	Measured Value, I	Calculated Value, $P = V^2/R$*	Measured Value, $P = I^2R$	% Error P

*Refers to calculated values on a separate sheet.

EXPERIMENT RESULTS REPORT FORM

Experiment No.: ______________________ Name: ______________________

Date: ______________________

Experiment Title: ______________________ Class: ______________________

______________________ Instr: ______________________

Explain the purpose of the experiment:

List the first Learning Objective:

OBJECTIVE 1:

After reviewing the results, describe how the objective was validated by this experiment.

List the second Learning Objective:

OBJECTIVE 2:

After reviewing the results, describe how the objective was validated by this experiment.

List the third Learning Objective:

OBJECTIVE 3:

After reviewing the results, describe how the objective was validated by this experiment.

Conclusion:

If required, attach to this form: ☐ Answers to Questions, ☐ Tables, and ☐ Graphs.

CHAPTER **4**

EXPERIMENT 4-1

SERIES CIRCUITS

LEARNING OBJECTIVES

At the completion of this experiment, you will be able to:

- Recognize the basic characteristics of series circuits.
- Compare the mathematical relationships existing in a series circuit.
- Compare the mathematical calculations to the measured values of a series circuit.

SUGGESTED READING

Chapter 4, *Basic Electronics*, Grob/Schultz, twelfth edition

INTRODUCTION

An electric circuit is a complete path through which electrons can flow from the negative terminal of the voltage source, through the connecting wires or conductors, through the load or loads, and back to the positive terminal of the voltage source.

If the circuit is arranged so that the electrons have only one possible path, the circuit is called a *series circuit.* Therefore, a series circuit is defined as a circuit that contains only one path for current flow. Figure 4-1.1 shows a series circuit with several resistors.

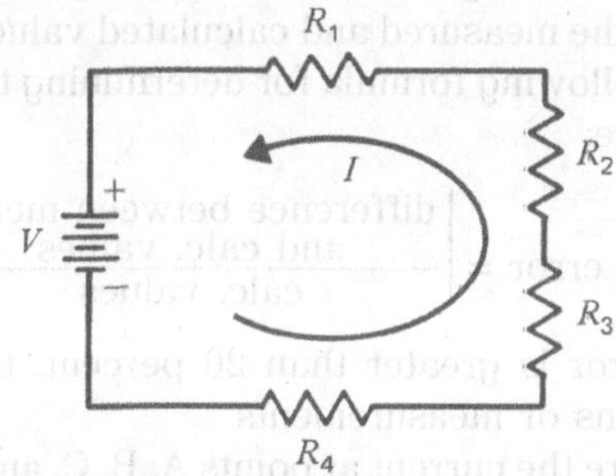

Fig. 4-1.1 Series circuit.

One of the most important aspects of a series circuit is its relationship to current. Current in a series circuit is determined by Ohm's law, which defines a proportional relation between voltage and the total circuit resistance. This relationship between total voltage and total circuit resistance results in the current being the same value throughout the entire circuit. In other words, the measured current will be the same value at any point in a series circuit.

Figure 4-1.2 shows a series circuit consisting of a battery and two resistors. The battery is labeled V_t and provides the total voltage across the circuit. The resistors are labeled R_1 and R_2. The resistance values are $R_1 = 100\ \Omega$ and $R_2 = 2.7\ k\Omega$. The power supply has been adjusted to a level of 20 V.

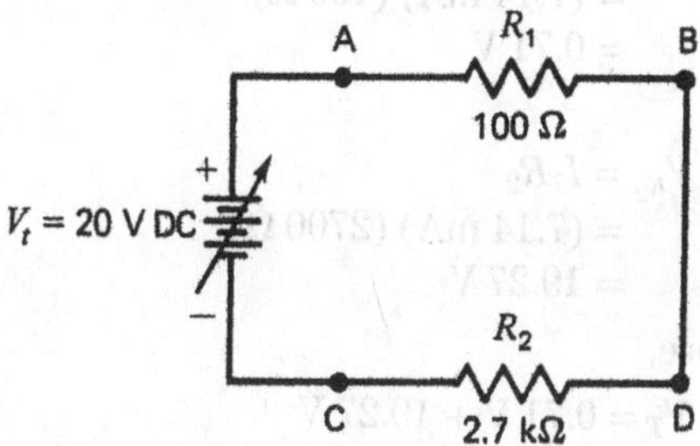

Fig. 4-1.2 Series circuit for analysis.

Figure 4-1.2 shows several points labeled A, B, C, and D. These are the points at which current could be measured. To measure the current at these points, it is necessary to break into the circuit and insert an ammeter in series. Remember, in a series circuit the amount of current measured will be the same at each of these points.

The total current, labeled I_T, flowing throughout the series circuit depends upon two factors: the total resistance R_T and the applied total voltage V_T. The applied total voltage is given in Fig. 4-1.2 such that

$$V_T = 20\ \text{V}$$

To determine the current, the total resistance R_T must be calculated by

$$R_T = R_1 + R_2$$

The total resistance for the circuit shown in Fig. 4-1.2 is

$$R_T = 100\ \Omega + 2700\ \Omega = 2800\ \Omega$$

For circuits that have three (3) or more series resistors, the above formula may be modified as shown to include the additional resistors:

$$R_T = R_1 + R_2 + R_3 + \cdots$$

When solving for the current I, Ohm's law states that

$$I = \frac{V}{R}$$

When solving for current I_T,

$$I_T = \frac{V_T}{R_T}$$
$$= \frac{20\text{ V}}{2800\ \Omega}$$
$$= 0.00714\text{ A}$$
$$= 7.14\text{ mA}$$

Applied voltage will be dropped proportionally across individual resistances, depending upon their value of resistance. As shown in Fig. 4-1.3, the IR voltage drops can be solved as follows. The sum of the voltage drops will be the total applied voltage. This is stated by

$$V_T = IR_1 + IR_2$$

or $$V_T = V_{R_1} + V_{R_2}$$

where

$$V_{R_1} = I_T R_1$$
$$= (7.14\text{ mA})\ (100\ \Omega)$$
$$= 0.71\text{ V}$$

and

$$V_{R_2} = I_T R_2$$
$$= (7.14\text{ mA})\ (2700\ \Omega)$$
$$= 19.27\text{ V}$$

Therefore,

$$V_T = 0.71\text{ V} + 19.27\text{ V}$$
$$= 19.98\text{ V}$$
$$= 20\text{ V (approximately)}$$

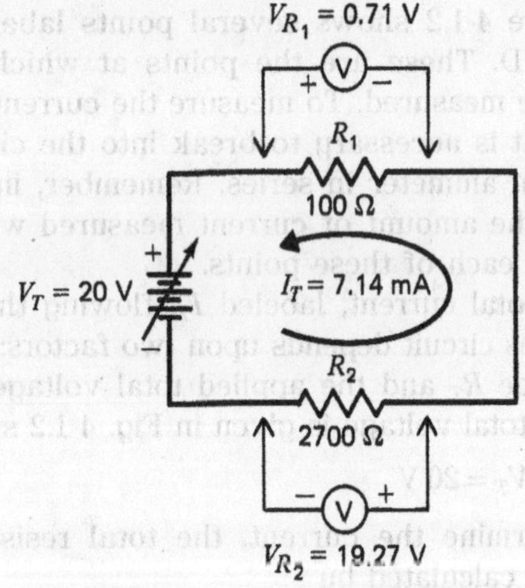

Fig. 4-1.3 Determining the IR (voltage) drops.

EQUIPMENT

Breadboard
DC power supply
Voltmeter (DMM)
Ammeter (DMM)
Ohmmeter (DMM)

COMPONENTS

Resistors (all 0.25 W unless indicated otherwise):

(1) 10 Ω (1) 15 Ω (1) 150 Ω, 1 W

PROCEDURE

1. With an ohmmeter, measure each resistor value for the resistors required for this experiment. Connect the resistors in a series and measure the total resistance R_T. Record the results in Table 4-1.1.

2. Using the information in the introduction and Fig. 4-1.4, calculate the total resistance R_T, the series current I_T, and the IR voltage drops across R_1, R_2, and R_3. Record the results in Table 4-1.2.

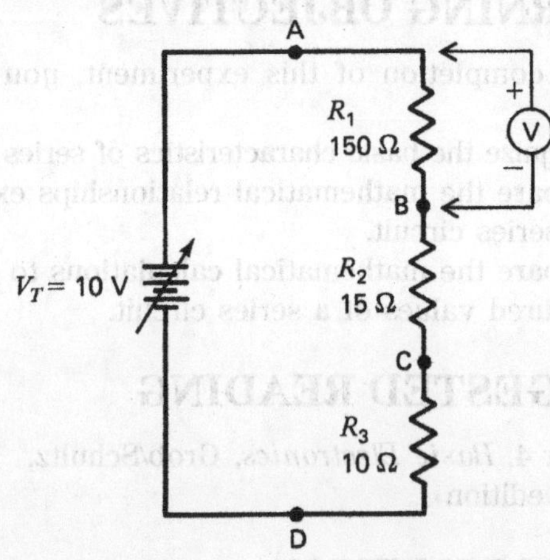

Fig. 4-1.4 Series circuit.

3. Connect the circuit in Fig. 4-1.4 and turn on the DC power supply. Using a voltmeter, adjust the power supply level to 10 V DC.

4. Measure the IR voltage drops across each resistance as indicated in Fig. 4-1.4. For example, the voltmeter is shown connected across R_1 to measure V_{R_1}. Record the results in Table 4-1.2.

5. After recording the measured voltage drops, compare the measured and calculated values. Use the following formula for determining this percentage:

$$\%\text{ error} = \left|\frac{\text{difference between meas. and calc. values}}{\text{calc. values}}\right| \times 100$$

If any error is greater than 20 percent, then repeat calculations or measurements.

6. Measure the current at points A, B, C, and D in Fig. 4-1.4. Record this information in Table 4-1.3.

Note 1: To measure current, you must actually break the circuit at these points and insert your ammeter.

Note 2: Check with your instructor in case you need to leave this circuit connected. When all the results have been recorded, call your instructor over to your lab station to verify your results if required.

NAME ______________________________ DATE ______________

QUESTIONS FOR EXPERIMENT 4-1

Choose the correct answer.

_____ **1.** In a series circuit, total current I_T is equal to:
A. $R_T \times I_T$ **B.** $V_T \times R_T$ **C.** V_T/R_T **D.** R_T/V_T

_____ **2.** In a series circuit, the current is:
A. Different through every resistor in series **B.** Always the same through every resistor in series **C.** Calculated by using Ohm's law as $I = V \times R$ **D.** Found only by using the voltmeter

_____ **3.** The total voltage in a series circuit is:
A. Equal to total resistance **B.** Found by adding the current through each resistor **C.** Equal to the sum of the series *IR* voltage drops **D.** Found by using an ohmmeter

_____ **4.** In a series circuit with 10 V applied:
A. The greater the total resistance, the greater the total current
B. The greater the total current, the greater the total resistance
C. The *IR* voltage drops will each equal 10 V **D.** The sum of the *IR* voltage drops will equal 10 V

_____ **5.** When an *IR* voltage drop exists in a series circuit:
A. The polarity of the resistor is equal to positive **B.** The polarity of the resistor is equal to negative **C.** The polarity of the resistor is less than the total current on both sides **D.** The polarity of the resistor is positive on one end and negative on the other because of current flowing through it

Write a short answer to the following questions.

6. Does $I_T = I_{R_1} + I_{R_2} + I_{R_3}$ in a series circuit? Explain your answer.

7. Is the measurement current the same at all parts of the series circuit? Explain your answer.

8. Does $R_T = R_1 + R_2 + R_3$ in a series circuit? Explain your answer.

TABLES FOR EXPERIMENT 4-1

TABLE 4-1.1

Nominal Resistance, Ω	Measured, Ω
$R_1 = 150$	
$R_2 = 15$	
$R_3 = 10$	
$R_t = 175$	

TABLE 4-1.2

	Calculated	Measured	% Error
R_T			
I_T			
V_{R_1}			
V_{R_2}			
V_{R_3}			

TABLE 4-1.3

Point	Current, mA
A	
B	
C	
D	

EXPERIMENT RESULTS REPORT FORM

Experiment No.: ______________________ Name: ______________________

Date: ______________________

Experiment Title: ______________________ Class: ______________________

______________________ Instr: ______________________

Explain the purpose of the experiment:

List the first Learning Objective:

OBJECTIVE 1:

After reviewing the results, describe how the objective was validated by this experiment.

List the second Learning Objective:

OBJECTIVE 2:

After reviewing the results, describe how the objective was validated by this experiment.

List the third Learning Objective:

OBJECTIVE 3:

After reviewing the results, describe how the objective was validated by this experiment.

Conclusion:

If required, attach to this form: ☐ Answers to Questions, ☐ Tables, and ☐ Graphs.

CHAPTER 4

EXPERIMENT 4-2

SERIES CIRCUITS—RESISTANCE

LEARNING OBJECTIVES

At the completion of this experiment, you will be able to:

- Recognize the basic characteristics of series circuits; for example, the total resistance R_T of a series string is equal to the sum of the individual resistances.
- Understand that a combination of series resistances is often called a *string* and that the string resistance equals the sum of the individual resistances.
- Compare the mathematical calculations to the measured values of a series circuit and recognize that the amount of current between two points in a circuit equals the potential difference divided by the resistance between these points.

SUGGESTED READING

Chapter 4, *Basic Electronics*, Grob/Schultz, twelfth edition

INTRODUCTION

When a series circuit is connected across a voltage source, such as a power supply, as shown in Fig. 4-2.1, the electrons forming the current must pass through all the series resistances. This path is the only way the electrons can return to the power supply. With two or more resistances in the same current path, the total resistance across the voltage source is the total of all individual resistances found in what is referred to as a *series string.* This "total" resistance is referred to as R_T.

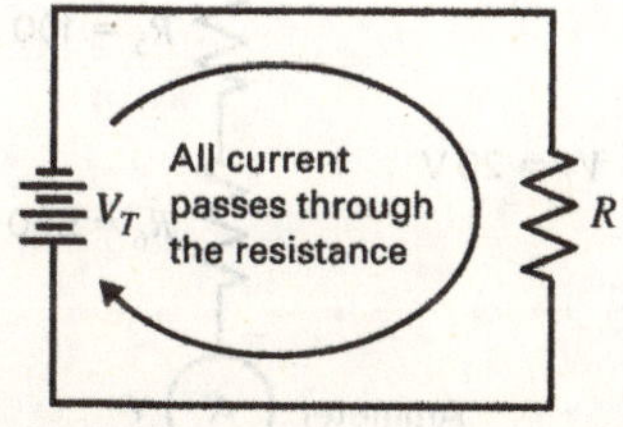

Fig. 4-2.1 Series resistance.

For example, if the circuit in Fig. 4-2.2 is analyzed, the total resistance R_T is the sum of the two individual resistances R_1 and R_2. In this case the total resistance is equal to 1150 Ω because the string resistance of R_1 (470 Ω) is added to R_2 (680 Ω). The total resistive opposition to current flow is the same as if a 1150 Ω resistor were substituted for the two resistors, as shown in Fig. 4-2.3.

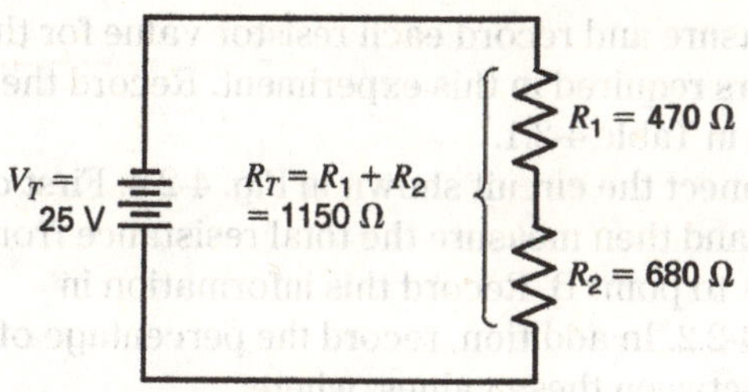

Fig. 4-2.2 R_T equals the sum of R_1 and R_2.

According to Ohm's law, the amount of current between two points in a circuit equals the potential difference divided by the resistance between these points. Since the entire string is connected across the voltage source, the current equals the voltage applied across the entire string divided by the total series resistance of the string (between points A and B in Fig. 4-2.3). In this case a power supply applies 25 V across 1150 Ω to produce 21.7 mA. This is the same amount of current flow through R_1 and R_2, as shown in Fig. 4-2.2.

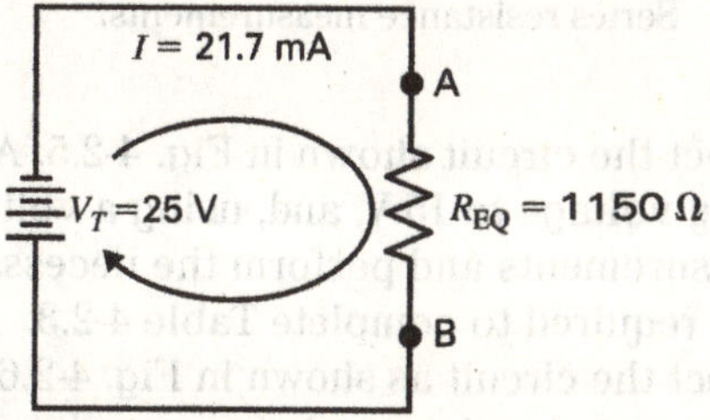

Fig. 4-2.3 Equivalent series resistance.

EQUIPMENT

Breadboard
DC power supply
Voltmeter (DMM)
Ammeter (DMM)
Ohmmeter (DMM)

COMPONENTS

All resistors are 0.25 W unless indicated otherwise:

- (1) 100-Ω resistor
- (1) 150-Ω resistor
- (1) 220-Ω resistor
- (1) 390-Ω resistor
- (1) 560-Ω resistor

PROCEDURE

1. Measure and record each resistor value for the resistors required in this experiment. Record the results in Table 4-2.1.

2. Connect the circuit shown in Fig. 4-2.4. First calculate and then measure the total resistance from point A to point B. Record this information in Table 4-2.2. In addition, record the percentage of error between these values, where

$$\% \text{ error} = \left| \frac{\text{difference between meas. and calc. values}}{\text{calc. values}} \right| \times 100$$

If the error is greater than 10 percent, repeat the calculations and measurements.

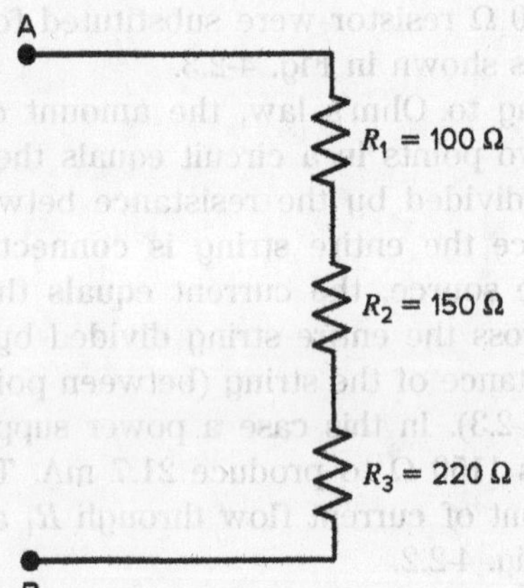

Fig. 4-2.4 Series resistance measurements.

3. Connect the circuit shown in Fig. 4-2.5. Adjust the supply voltage to 15 V, and, using a voltmeter, take measurements and perform the necessary calculations required to complete Table 4-2.3.

4. Connect the circuit as shown in Fig. 4-2.6. Calculate the expected resistance between points X and Y and record the results in Table 4-2.4. With the *power turned off and totally disconnected* from the circuit, measure and record, in Table 4-2.4, the resistance between X and Y.

5. With the ohmmeter disconnected from the circuit, connect the DC power supply to the circuit in Fig. 4-2.6 and adjust to 20 V. With an ammeter, measure the current at point Y and record the result in Table 4-2.4. Break the circuit at this point, and insert the ammeter into the circuit, as shown in Fig. 4-2.7.

6. Measure the voltage across R_5 and R_6 and record it in Table 4-2.4.

7. Calculate the current for R_5 and R_6 and record the results.

8. Determine and record the percentage of error for current using current at point Y. Record the results.

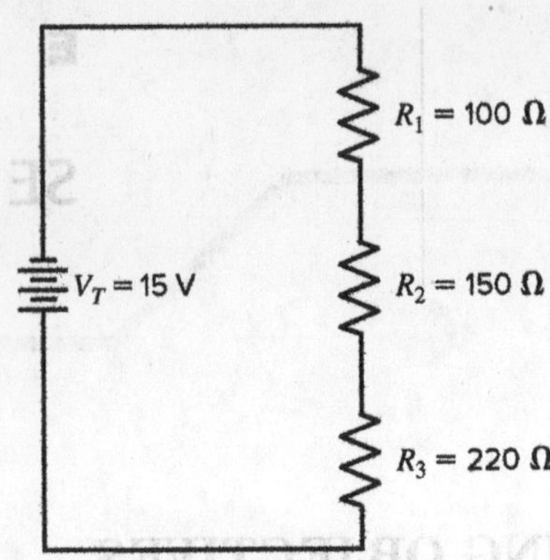

Fig. 4-2.5 Series string resistance.

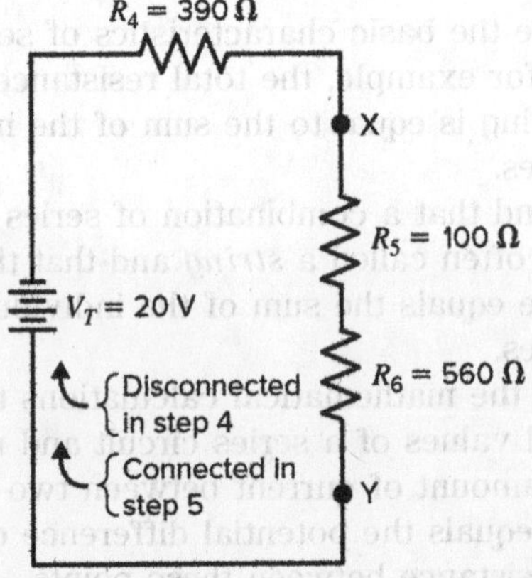

Fig. 4-2.6 Series resistance with an applied voltage source.

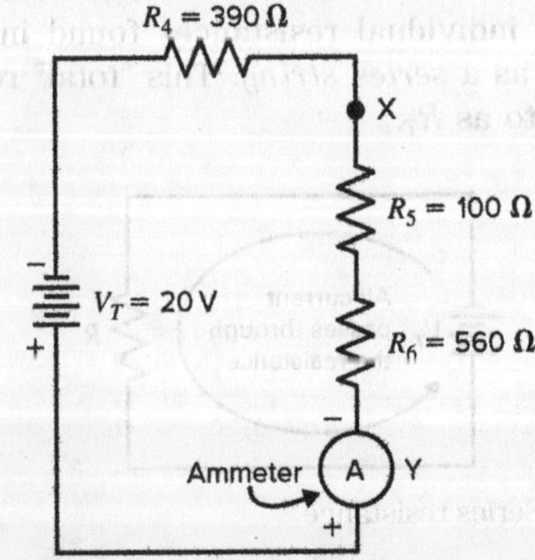

Fig. 4-2.7 Inserting the ammeter into the circuit.

NAME DATE

QUESTIONS FOR EXPERIMENT 4-2

1. Write the formula that determines the total resistance R_T that is found in a series resistance string.

2. Write the formula that determines the amount of current flow for the circuit shown in Fig. 4-2.5 if the potential difference is 85 V.

3. In your estimation would $V_T = IR_1 + IR_2 + IR_3 + \cdots$ be true for all series resistive strings? Explain.

4. Do you believe that the measured current is the same at all points of a series resistive string when a potential difference is present across the circuit? Explain.

5. Could you always expect that $R_T = R_1 + R_2 + R_3 + \cdots$ in a series resistive circuit? Explain.

CRITICAL THINKING QUESTIONS

Note: The following questions are designed to help you analyze the previous laboratory experiment in a complete and in-depth fashion. To answer these questions, you should review the related material in Grob/Schultz, *Basic Electronics,* twelfth edition.

1. Explain where "sources" of error exist in this experiment. How could these sources of error be reduced?

2. The purpose of a series circuit is to connect different components that need the same current. After reviewing your results for this experiment, explain how this purpose supports, or agrees with, your findings.

3. Explain why the current is the same in all parts of a series circuit.

4. Is it the case that the mathematical calculations of the measured values of a series circuit are such that the amount of current between two points in a circuit equals the potential difference divided by the resistance between these points? Explain.

5. Explain what is meant by the term *series string*.

TABLES FOR EXPERIMENT 4-2

TABLE 4-2.1 Individual Resistor Values

Resistors	Measured	% Tolerance
100		
150		
220		
390		
560		

TABLE 4-2.2 Total Resistance R_T and Percentage of Error

	Calculated	Measured	% Error
Total Resistance			

TABLE 4-2.3 Series Circuit Measurements

Resistors	Value	Measured*	Measured Voltage	Calculated Current
R_1	100			
R_2	150			
R_3	220			

*Values from previous table.

TABLE 4-2.4 Series Circuit Measurements and Percentage of Error

Resistance	Expected	Measured	Measured Voltage	Calculated Current	% Error
X to Y					
Current Y					
R_5					
R_6					

EXPERIMENT RESULTS REPORT FORM

Experiment No.: ______________________ Name: ______________________

Date: ______________________

Experiment Title: ______________________ Class: ______________________

______________________ Instr: ______________________

Explain the purpose of the experiment:

List the first Learning Objective:

OBJECTIVE 1:

After reviewing the results, describe how the objective was validated by this experiment.

List the second Learning Objective:

OBJECTIVE 2:

After reviewing the results, describe how the objective was validated by this experiment.

List the third Learning Objective:

OBJECTIVE 3:

After reviewing the results, describe how the objective was validated by this experiment.

Conclusion:

If required, attach to this form: ☐ Answers to Questions, ☐ Tables, and ☐ Graphs.

CHAPTER 4

EXPERIMENT 4-3

SERIES CIRCUITS—ANALYSIS

LEARNING OBJECTIVES

At the completion of this experiment, you will be able to:

- Recognize that when you know the current I for one component connected into a series circuit, this value is used for the current I in all components, since the current is the same in all parts of a series circuit.
- Determine that in order to calculate I, the total V_T can be divided by the total R_T, or an individual IR drop can be divided by its resistance.
- Compare the mathematical calculations to the measured values of a series circuit to the extent that when the individual voltage drops around the series circuit are known, they can be added to equal the applied V_T. Further, you will demonstrate that a known voltage drop can be subtracted from the total V_T to find the remaining voltage drop.

SUGGESTED READING

Chapter 4, *Basic Electronics*, Grob/Schultz, twelfth edition

INTRODUCTION

It is useful to remember general methods for analyzing series circuits. For example, when analyzing series resistive circuits, if the circuit current for one component is known, you can use this value for currents in all the remaining series components, since the current is the same in all parts of a series circuit.

In order to calculate the circuit current I, the total applied voltage V_T can be divided by the total circuit resistance R_T, or the voltage drops (IR drops) of an individual component can be divided by its resistance R.

When the individual voltage drops around a circuit are known, they can be added to equal the applied voltage V_T. This also means that a known voltage drop can be subtracted from the total applied voltage in order to find the remaining voltage drop.

A common application of series circuits is to use a resistance to drop the voltage from the voltage source to a lower value, as shown in Fig. 4-3.1. The load resistance represented here is a transistor amplifier which would normally operate at 10 V DC with a constant DC load current of 15 mA. The load requirements are 10 V DC at 15 mA. The available voltage source can be lowered only to 15 V DC minimum. A suitable series voltage-dropping resistor R_s must be designed so that the voltage-dropping resistor is inserted in series to provide a voltage drop V_s that will make V_L equal to 10 V DC. The required voltage drop across V_s is the difference between V_L and the higher V_T. For example:

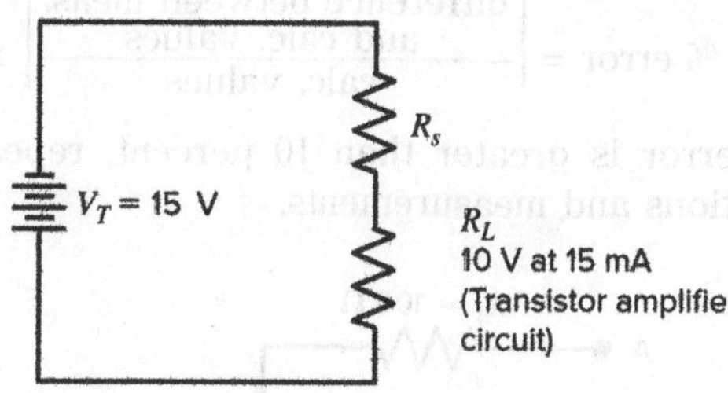

Fig. 4-3.1 Series-dropping resistor.

$$V_s = V_T - V_L = 15 - 10 = 5 \text{ V}$$

This voltage drop of 5 V must be provided with a current of 15 mA, because the current is the same through R_s and R_L. To calculate R_s

$$R_s = 5 \text{ V}/15 \text{ mA} = 333.33\ \Omega$$

EQUIPMENT

Breadboard
DC power supply
DMM:
- Voltmeter (DMM)
- Ammeter (DMM)
- Ohmmeter (DMM)

COMPONENTS

All resistors can be 0.25 W or 0.5 W unless indicated otherwise:

(1) 100-Ω resistor
(1) 150-Ω resistor
(1) 220-Ω resistor
(1) 390-Ω resistor
(1) 560-Ω resistor

PROCEDURE

1. Measure and record each resistor value for the resistors required in this experiment. Record the results in Table 4-3.1.

2. Connect the circuit shown in Fig. 4-3.2. First, calculate and then measure the total resistance from point A to point B. Record this information in Table 4-3.2. Then, record the percentage of error between these values, where

$$\% \text{ error} = \left| \frac{\text{difference between meas. and calc. values}}{\text{calc. values}} \right| \times 100$$

If the error is greater than 10 percent, repeat the calculations and measurements.

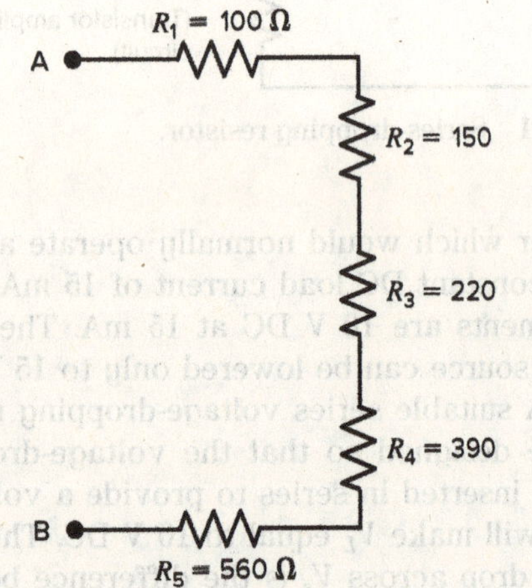

Fig. 4-3.2 Series resistive circuit.

3. Connect the circuit as shown in Fig. 4-3.3 and apply 20 V DC. With the supply voltage on, open the circuit at point A and, using an ammeter, measure and record the current in Table 4-3.3. Disconnect the ammeter and reconnect the circuit.

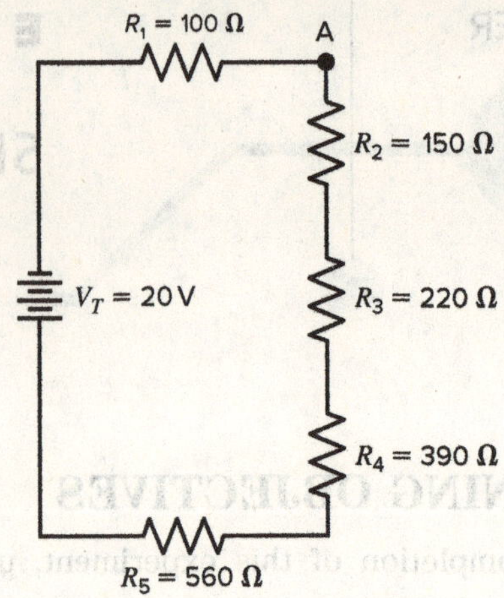

Fig. 4-3.3 Voltage applied to the series resistive circuit.

4. Using a voltmeter, measure the *IR* voltage drop across the resistors R_1, R_2, R_3, and R_4, and record this information in Table 4-3.3.

5. By mathematical calculation, record the current in Table 4-3.3. If the series current *I* is known for one component, this value is the current *I* found in all the components in series.

6. Subtract the *IR* voltage drops of R_1, R_2, R_3, and R_4 from the total applied voltage and record the results in Table 4-3.4 as IR_5.

7. Complete Table 4-3.4. Calculated *IR* drops are calculated current times resistance. Use the formula shown in step 2 in determining the percentage of error. If the error is greater than 10 percent, repeat the calculations and measurements.

8. Using a voltmeter, in Table 4-3.4, measure and record the *IR* voltage drop across R_5. Does this step demonstrate that a known voltage drop can be subtracted from the total V_T to find the remaining voltage drop? Explain in Table 4-3.5.

NAME DATE

QUESTIONS FOR EXPERIMENT 4-3

1. **Explain the significance of the following statement: In a series resistive circuit, in order to calculate *I*, the total V_T can be divided by the total R_T, or an individual *IR* drop can be divided by its resistance.**

2. **Describe what a zero voltage drop could be.**

3. **Describe the nature of the *IR* drop and current value of a component that displays the characteristics of a short.**

4. **In a series circuit, if the *IR* voltage drop is measured to be the same across two resistors, what can be said about these two resistors?**

5. **In a series circuit, if the current is measured to be 0 mA and is the same in all resistances, what would you suspect is taking place in this circuit?**

CRITICAL THINKING QUESTIONS

Note: **The following questions are designed to help you analyze the previous laboratory experiment in a complete and in-depth fashion. To answer these questions, you should review the related material in Grob/Schultz, *Basic Electronics,* twelfth edition.**

1. **As described in the introduction of this experiment, a common application of series circuits is to use a resistance to drop the voltage from the voltage source to a lower value. Design a series-dropping resistor R_s so that the load requires 5 V at 50 mA, while the supply voltage is limited to 14 V DC.**

2. Determine the resistor's wattage rating for the design problem as described in critical thinking question 1. Using the manufacturer's specification sheets and/or supplier catalogs, select a resistor that closely matches your component design. Using this value, and assuming that your design is not a perfect match, how will the voltage drops and circuit current values be affected? Explain.

3. The purpose of a series circuit is to connect different components that need the same current. After reviewing your results for this experiment, explain how this purpose supports, or agrees with, your findings.

4. Explain why the current is the same in all parts of a series circuit.

5. Explain where the sources of errors exist within this experiment. How could these sources of error be reduced?

TABLES FOR EXPERIMENT 4-3

TABLE 4-3.1 Individual Resistor Values

Resistors	Measured	% Tolerance
100		
150		
220		
390		
560		

TABLE 4-3.2 Total Resistance R_T and Percentage of Error

	Calculated	Measured	% Error
Total Resistance			

TABLE 4-3.3 Current and Voltage Measurements

Current = ______ at point A.

Point A Resistors	Value	Measured*	Measured Voltage	Calculated Current
R_1	100	______	______	______
R_2	150	______	______	______
R_3	220	______	______	______
R_4	390	______	______	______

*Values from the previous table.

TABLE 4-3.4 The Total Voltage *IR* Drops and Percentage of Error

Voltage Applied = ______

	Measured *IR* Drops*	Calculated *IR* Drops	% Error
IR_1	______	______	______
IR_2	______	______	______
IR_3	______	______	______
IR_4	______	______	______
Sum of *IR* drops	______		
V_T − sum = IR_5	______		

*Values from the previous table.

TABLE 4-3.5 Analyzing *IR* Drops

Calculated IR_5*		Explain:
Measured IR_5		

*Use the value of IR_5 from Table 4-3.4.

EXPERIMENT RESULTS REPORT FORM

Experiment No.: ____________________ Name: ____________________

Date: ____________________

Experiment Title: ____________________ Class: ____________________

____________________ Instr: ____________________

Explain the purpose of the experiment:

List the first Learning Objective:

OBJECTIVE 1:

After reviewing the results, describe how the objective was validated by this experiment.

List the second Learning Objective:

OBJECTIVE 2:

After reviewing the results, describe how the objective was validated by this experiment.

List the third Learning Objective:

OBJECTIVE 3:

After reviewing the results, describe how the objective was validated by this experiment.

Conclusion:

If required, attach to this form: ☐ Answers to Questions, ☐ Tables, and ☐ Graphs.

CHAPTER 4

EXPERIMENT 4-4

SERIES CIRCUITS—WITH OPENS

LEARNING OBJECTIVES

At the completion of this experiment, you will be able to:

- Recognize that an open circuit is a break in the current path. Since the current is the same in all parts of a series circuit, an open in any part results in no current flow for the entire circuit.
- Understand that the resistance of an open circuit path is very high because an insulator such as air takes the place of the conducting path of the circuit.
- Verify that at the point of open in a series circuit, the value of current is practically zero, although the supply voltage is considered maximum.

SUGGESTED READING

Chapter 4, *Basic Electronics*, Grob/Schultz, twelfth edition

INTRODUCTION

An open circuit is a break in the current path. The resistance of the open path is very high because an insulator, in this case air, takes the place of a conducting part of the circuit. Since the current is the same in all parts of a series circuit, as shown in Fig. 4-4.1*a*, an open in any part results in no current throughout the entire circuit. This is illustrated in Fig. 4-4.1*b*. The open that exists between points A and B provides an infinite resistance to the series circuit. Since the open circuit resistance at this point is extremely large when compared to the resistances of R_1 and R_2, and since the supply voltage remains unchanged, the resulting current flow is considered to be zero. The current flow in Fig. 4-4.1*b* is zero and therefore results in there being no *IR* voltage drop across any of the series resistances.

It is extremely important to note, however, that even though there is no current flow in an open series circuit and therefore no *IR* drops, the voltage source still maintains its output voltage and can be a potential shock hazard for the technician working on such open circuits.

EQUIPMENT

Breadboard
DC power supply
- Voltmeter (DMM)
- Ammeter (DMM)
- Ohmmeter (DMM)

COMPONENTS

All resistors can be 0.25 W or 0.5 W unless indicated otherwise:

(1) 100-Ω resistor
(1) 150-Ω resistor
(1) 220-Ω resistor
(1) 390-Ω resistor

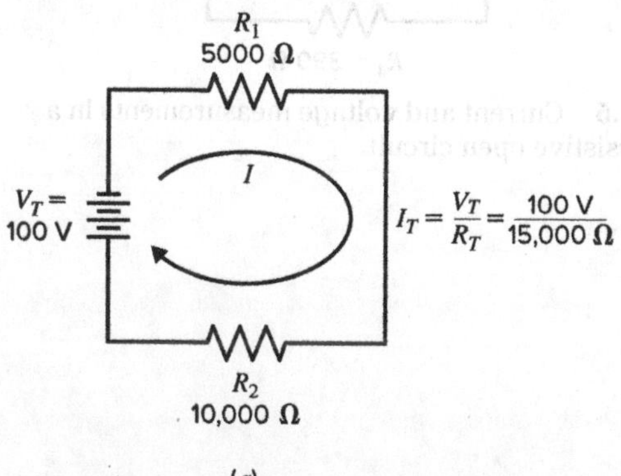

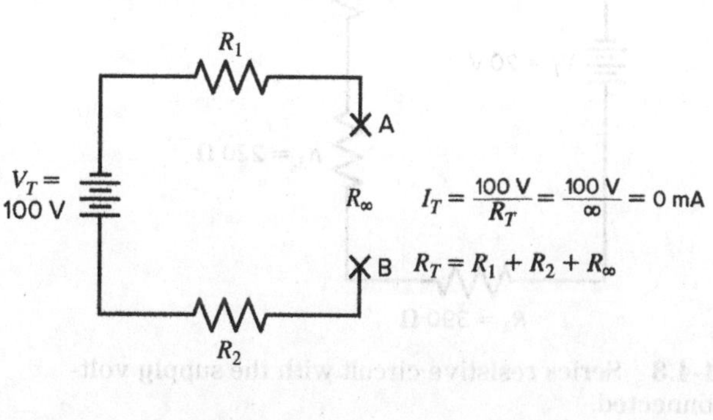

Fig. 4-4.1 Series resistive circuits. (*a*) Series circuit with a complete path for current flow. (*b*) Series circuit with an open that establishes an infinite series resistance.

PROCEDURE

1. Measure and record each resistor value for the resistors required in this experiment. Record the results in Table 4-4.1.

2. Connect the circuit shown in Fig. 4-4.2. First, calculate and then measure the total resistance from point A to point B. Record this information in Table 4-4.2. Then, record the percentage of error between these values, where

$$\% \text{ error} = \left| \frac{\text{difference between meas. and calc. values}}{\text{calc. values}} \right| \times 100$$

If the error is greater than 10 percent, repeat the calculations and measurements.

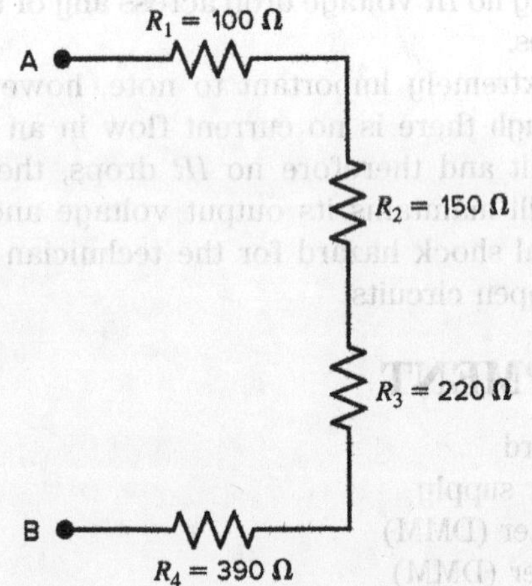

Fig. 4-4.2 Series resistive circuit without a supply voltage.

3. Connect the supply voltage to the circuit as shown in Fig. 4-4.3. Adjust this voltage to 20 V DC.

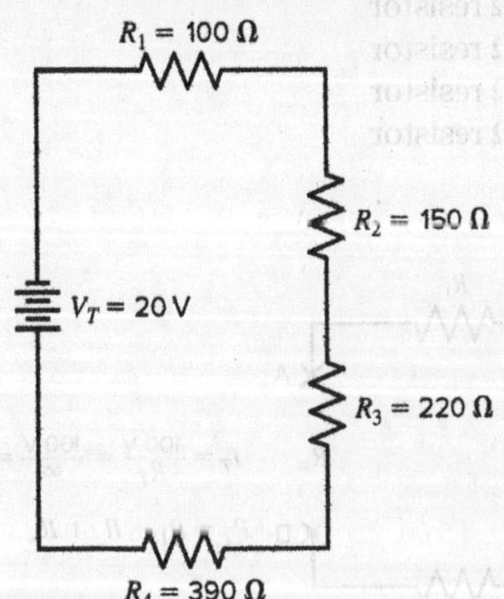

Fig. 4-4.3 Series resistive circuit with the supply voltage connected.

4. Using a voltmeter, measure and record, in Table 4-4.3, the voltage drops across each of the series resistances.

5. Using an ammeter and with the supply voltage on, open the circuit at each of the points, as shown in Fig. 4-4.4, and measure and record the currents in Table 4-4.3.

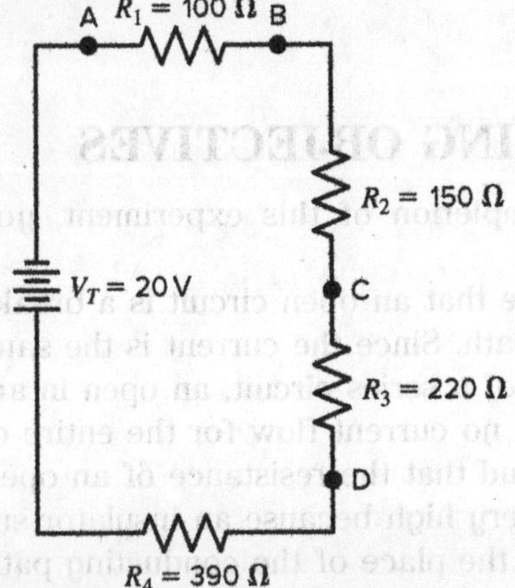

Fig. 4-4.4 Current and voltage measurements in a series resistive circuit.

6. With the supply voltage on, open the circuit as shown in Fig. 4-4.5, and repeat steps 4 and 5. Again, record your measurements in Table 4-4.3.

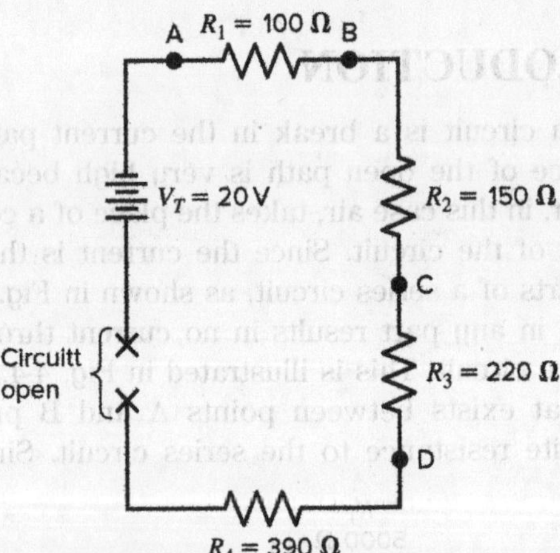

Fig. 4-4.5 Current and voltage measurements in a series resistive open circuit.

NAME DATE

QUESTIONS FOR EXPERIMENT 4-4

1. In your own words, define the unique characteristics of a series resistive circuit.

2. Describe what happens to the current in a series resistive circuit if an open in the circuit path occurs.

3. Describe what happens to the voltage drops in a series resistive circuit if an open in the circuit path occurs.

4. In a series resistive circuit that provides a given amount of current flow, when the resistors are compared, why does the resistor with a higher ohmic value also have a larger voltage drop across it? Is this also true for a circuit with an open?

5. Give an application of a series circuit.

CRITICAL THINKING QUESTIONS

***Note:* The following questions are designed to help you analyze the previous laboratory experiment in a complete and in-depth fashion. To answer these questions, you should review the related material in Grob/Schultz, *Basic Electronics*, twelfth edition.**

1. Explain where the sources of error exist in this experiment. How could these sources of error be reduced?

2. How could the idea of open circuits apply to the 120-V AC voltage from the power line in your home?

3. Explain why the current is the same in all parts of a series circuit. Further, describe why the *IR* drops are zero in an open series resistive circuit.

4. Demonstrate mathematically why the current of an open circuit is practically zero.

5. Why is the source voltage in an open series resistive circuit present with or without current flow in the external circuit?

TABLES FOR EXPERIMENT 4-4

TABLE 4-4.1 Individual Resistor Values

Resistors	Measured	% Tolerance
100	______	______
150	______	______
220	______	______
390	______	______

TABLE 4-4.2 Total Resistance R_T and Percentage of Error

	Calculated	Measured	% Error
Total Resistance			

TABLE 4-4.3 Current and Voltage Measurements

	Circuit without Open			Circuit with Open		
Resistor Values	Measured Voltage	Circuit Points	Measured Current	Measured Voltage	Circuit Points	Measured Current
R_1	______	A	______	______	A	______
R_2	______	B	______	______	B	______
R_3	______	C	______	______	C	______
R_4	______	D	______	______	D	______

EXPERIMENT RESULTS REPORT FORM

Experiment No.: ______________________ Name: ______________________

Date: ______________________

Experiment Title: ______________________ Class: ______________________

______________________ Instr: ______________________

Explain the purpose of the experiment:

List the first Learning Objective:

OBJECTIVE 1:

After reviewing the results, describe how the objective was validated by this experiment.

List the second Learning Objective:

OBJECTIVE 2:

After reviewing the results, describe how the objective was validated by this experiment.

List the third Learning Objective:

OBJECTIVE 3:

After reviewing the results, describe how the objective was validated by this experiment.

Conclusion:

If required, attach to this form: ☐ Answers to Questions, ☐ Tables, and ☐ Graphs.

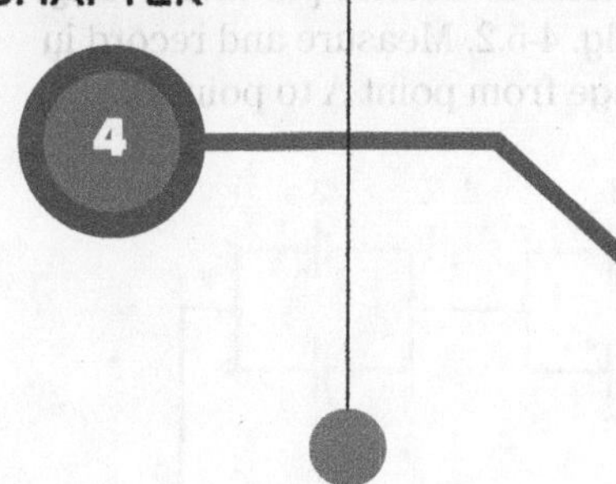

EXPERIMENT 4-5

SERIES-AIDING AND SERIES-OPPOSING VOLTAGES

LEARNING OBJECTIVES

At the completion of this experiment, you will be able to:

- Define the terms *series-aiding* and *series-opposing*.
- Construct a series-aiding circuit and a series-opposing circuit.
- Measure current and voltage; anticipate correct polarity connections.

SUGGESTED READING

Chapter 4, *Basic Electronics*, Grob/Schultz, twelfth edition

INTRODUCTION

In many practical applications, a circuit may contain more than one voltage source. Voltage sources that cause current to flow in the same direction are considered to be *series-aiding,* and their voltages add. Voltage sources that tend to force current in opposite directions are said to be *series-opposing,* and the effective voltage source is the difference between the opposing voltages. When two opposing sources are inserted into a circuit, current flow would be in a direction determined by the larger source. Examples of series-aiding and series-opposing sources are shown in Fig. 4-5.1*a* and *b*.

Series-aiding voltages are connected such that the currents of the sources add, as shown in Fig. 4-5.1*a*. The 10 V of V_1 produces 1 A of current flow through the 10-Ω resistance of R_1. Also, the voltage source V_2, of 5 V, creates 0.5 A of current flowing through the 10-Ω resistance of R_1: The total current would then be additive and be 1.5 A.

When voltage sources are connected in a series-aiding fashion, where the negative terminal of one source is connected to the positive terminal of the next, voltages V_1 and V_2 are added to find the total voltage.

$$V_T = 5 \text{ V} + 10 \text{ V} = 15 \text{ V}$$

The total current can then be determined by

$$I_T = 15 \text{ V}/10\Omega = 1.5 \text{ A}$$

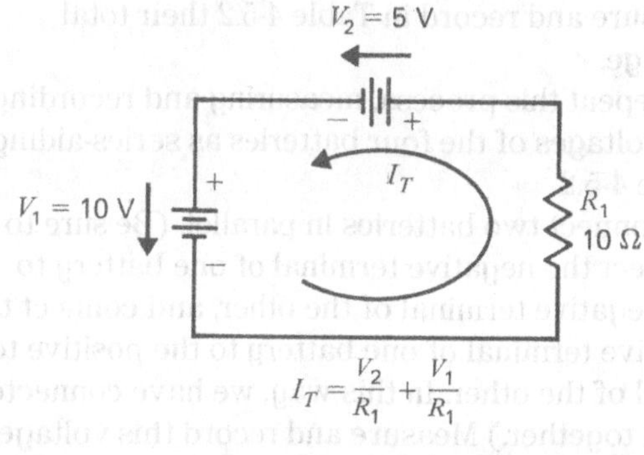

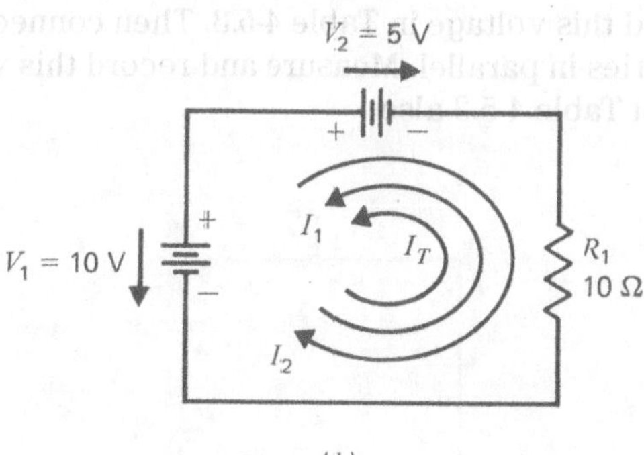

Fig. 4-5.1 (*a*) Series-aiding voltages. (*b*) Series-opposing voltages.

Series-opposing voltages can be subtracted, as shown in Fig. 4-5.1*b*. The currents that are generated are opposing each other. These voltages can still be algebraically added, keeping in mind their algebraic sign. Here, V_2 is smaller than V_1, and the difference between them is +5 V, resulting in an I_T of 0.5 A.

EQUIPMENT

DMM
Test leads

COMPONENTS

(Four) D cells (1.5-V batteries)

PROCEDURE

1. Connect the negative lead of the voltmeter to the negative terminal of the battery, and connect the positive lead of the meter to the positive terminal of the battery. Identify and number each battery 1 through 4. Measure and record in Table 4-5.1 the voltage of each of the dry cells supplied to you.
2. Connect batteries 1 and 2 as series-aiding. Measure and record in Table 4-5.2 their total voltage.
3. Connect batteries 1, 2, and 3 as series-aiding. Measure and record in Table 4-5.2 their total voltage.
4. Repeat this process, measuring and recording the voltages of the four batteries as series-aiding, in Table 4-5.2.
5. Connect two batteries in parallel. (Be sure to connect the negative terminal of one battery to the negative terminal of the other, and connect the positive terminal of one battery to the positive terminal of the other. In this way, we have connected them together.) Measure and record this voltage in Table 4-5.3.
6. Connect three batteries in parallel. Measure and record this voltage in Table 4-5.3. Then connect four batteries in parallel. Measure and record this voltage in Table 4-5.3 also.

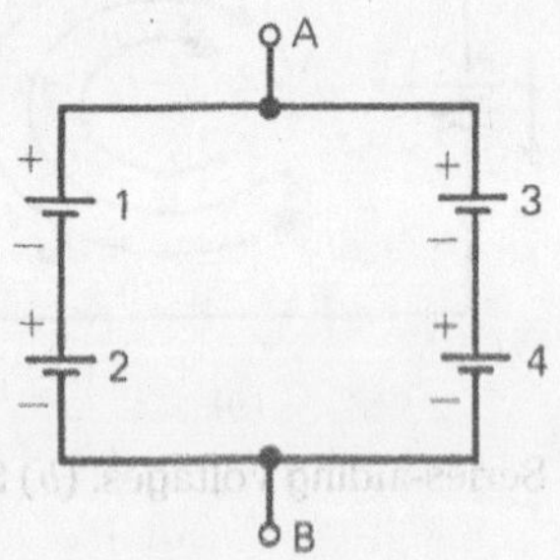

Fig. 4-5.2 Series-parallel batteries.

7. Connect the batteries in a series-parallel arrangement, as shown in Fig. 4-5.2. Measure and record in Table 4-5.4 the voltage from point A to point B.

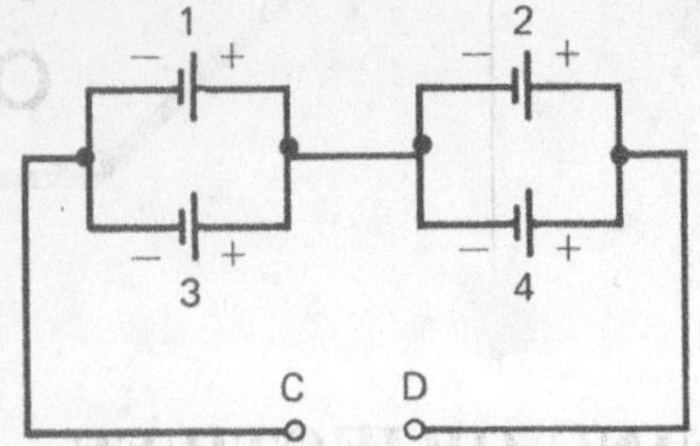

Fig. 4-5.3 Series-aiding circuit.

8. Connect the circuit as shown in Fig. 4-5.3. Measure and record in Table 4-5.4 the voltage from point C to point D.
9. Connect the series-aiding–series-opposing arrangements shown in Fig. 4-5.4*a* to *c*. Measure and record in Table 4-5.5 the voltage across each circuit. Before connecting the voltmeter, determine which are the probable positive and negative terminals.

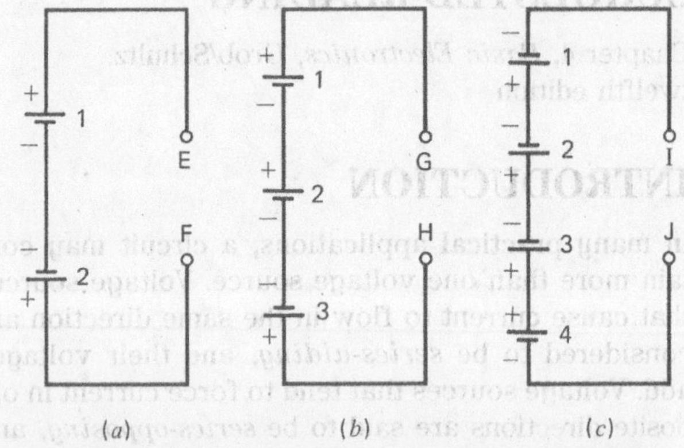

Fig. 4-5.4 (*a*) Series-opposing circuit. (*b*) Series-aiding and series-opposing circuit. (*c*) Series-aiding and series-opposing circuit.

NAME ______ DATE ______

QUESTIONS FOR EXPERIMENT 4-5

1. Name four precautions that must be observed in measuring voltages.

2. What would happen to a dry cell or battery if the positive and negative terminals were short-circuited?

3. What arrangement of six dry cells gives the maximum voltage?

4. Draw a practical arrangement of ten 1.5-V dry cells to give a battery of 7.5 V.

5. Explain the difference in connection between two dry cells connected in series-aiding and in series-opposing.

TABLES FOR EXPERIMENT 4-5

TABLE 4-5.1

Battery	Measured Voltages
1	______
2	______
3	______
4	______

TABLE 4-5.2

Series-Aiding Battery	Measured Voltages
1 + 2	________
1 + 2 + 3	________
1 + 2 + 3 + 4	________

TABLE 4-5.3

Parallel-Arrangement Battery	Measured Voltages
1 + 2	________
1 + 2 + 3	________
1 + 2 + 3 + 4	________

TABLE 4-5.4

Circuit	Voltages	Terminals
Fig. 4-5.2	________	A to B
Fig. 4-5.3	________	C to D

TABLE 4-5.5

Circuit	Voltages	Terminals
Fig. 4-5.4*a*	________	E to F
Fig. 4-5.4*b*	________	G to H
Fig. 4-5.4*c*	________	I to J

EXPERIMENT RESULTS REPORT FORM

Experiment No.: ______________________ Name: ______________________

Date: ______________________

Experiment Title: ______________________ Class: ______________________

______________________ Instr: ______________________

Explain the purpose of the experiment:

List the first Learning Objective:

OBJECTIVE 1:

After reviewing the results, describe how the objective was validated by this experiment.

List the second Learning Objective:

OBJECTIVE 2:

After reviewing the results, describe how the objective was validated by this experiment.

List the third Learning Objective:

OBJECTIVE 3:

After reviewing the results, describe how the objective was validated by this experiment.

Conclusion:

If required, attach to this form: ☐ Answers to Questions, ☐ Tables, and ☐ Graphs.

CHAPTER 4

EXPERIMENT 4-6

POSITIVE AND NEGATIVE VOLTAGES TO GROUND

LEARNING OBJECTIVES

At the completion of this experiment, you will be able to:

- Calculate circuit current and voltage drops found in a voltage divider circuit.
- Determine the polarity of voltages found in a circuit with a common ground.

SUGGESTED READING

Chapters 4, *Basic Electronics*, Grob/Schultz, twelfth edition

INTRODUCTION

In the wiring of practical circuits, one side of the voltage source is usually grounded. In electronic equipment the ground often indicates a metal chassis, which is used as a common return for connections to the voltage source. Where printed-circuit boards are used, usually a common ground path is run around the outside perimeter of the circuit board. In other cases the entire back side of a two-sided board may be used as a common-return path. Note that the chassis ground may or may not be connected to earth ground.

When a circuit has a chassis as a common return, measure the voltages with respect to chassis ground. Consider the voltage divider in Fig. 4-6.1. In Fig. 4-6.1, the circuit shows no ground system. It is

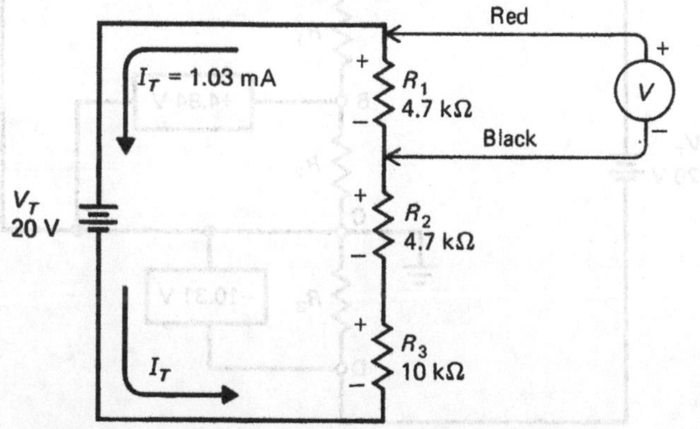

Fig. 4-6.1 Voltage divider circuit without a common ground.

instead a closed series circuit. This circuit has an applied power supply voltage V_T of 20 V. To determine the total circuit current, the total circuit resistance R_T must be determined. And R_T can be calculated as

$$\begin{aligned} R_T &= R_1 + R_2 + R_3 \\ &= 4.7\ \text{k}\Omega + 4.7\ \text{k}\Omega + 10\ \text{k}\Omega \\ &= 19.4\ \text{k}\Omega \end{aligned}$$

The circuit current can now be calculated from the Ohm's law relationship:

$$\begin{aligned} I_T &= \frac{V_T}{R_T} \\ &= \frac{20\ \text{V}}{19.4\ \text{k}\Omega} \\ &= 1.03\ \text{mA} \end{aligned}$$

After the current has been calculated, the individual voltage drops V_1, V_2, and V_3 of the voltage divider can be found from the Ohm's law relationship of

$$V = I \times R$$

For V_1

$$\begin{aligned} V_1 &= I_T \times R_1 \\ &= 1.03\ \text{mA} \times 4.7\ \text{k}\Omega \\ &= 4.84\ \text{V} \quad (\text{or } +4.84\ \text{V}) \end{aligned}$$

For V_2

$$\begin{aligned} V_2 &= I_T \times R_2 \\ &= 1.03\ \text{mA} \times 4.7\ \text{k}\Omega \\ &= 4.84\ \text{V} \quad (\text{or } +4.84\ \text{V}) \end{aligned}$$

For V_3

$$\begin{aligned} V_3 &= I_T \times R_3 \\ &= 1.03\ \text{mA} \times 10\ \text{k}\Omega \\ &= 10.31\ \text{V} \quad (\text{or } +10.31\ \text{V}) \end{aligned}$$

Also, the sum of the voltage drops will equal the applied voltage:

$$\begin{aligned} V_T &= V_1 + V_2 + V_3 \\ &= 4.84\ \text{V} + 4.84\ \text{V} + 10.31\ \text{V} \\ &= 19.99\ \text{V}\ (\text{round to } 20\ \text{V}) \end{aligned}$$

Refer to Fig. 4-6.1 and note that the polarities are included on this schematic. The polarity is determined by how the circuit current and the individual voltmeters are connected. The polarity of the resistors,

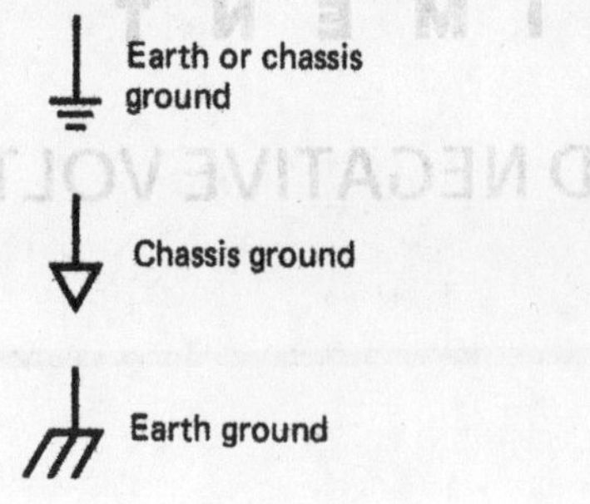

Fig. 4-6.2 Schematic symbols for ground.

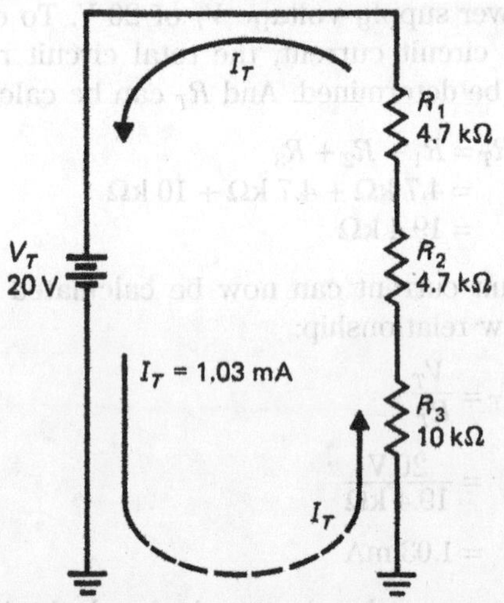

Fig. 4-6.3 Voltage divider circuit using a ground return.

which indicates the direction of current flow, and the color of the test leads are also indicated in Fig. 4-6.1.

Figure 4-6.2 shows the schematic symbols for ground. The ground symbol is used in Fig. 4-6.3. Here the same circuit with the same component values and applied voltage is shown. The only addition is the "ground return." Also note that the voltage drops of this circuit are equivalent to those shown in Fig. 4-6.1. The 1.03 mA generated by the battery is pushed out into ground point A and returns to ground point B. This, in effect, means that the ground symbols are connected, perhaps through a cable, foil pattern, or metal chassis.

The circuit shown in Fig. 4-6.4 details the same circuit of Figs. 4.6.1 and 4.6.3. The difference here is that all voltages are taken "with respect to ground." In this case, the voltage from point C to D is +10.31 V. The voltage from point B to D is +15.15 V (where $V_{BD} = VR_2 + VR_3$). The voltage from point A to D is +19.99 V (or 20 V), where $V_{AD} = VR_1 + VR_2 + VR_3$).

The circuit in Fig. 4-6.5 is similar to the one in Fig. 4-6.4. The main difference is that the ground has been moved to point C. This then indicates that all measurements should be taken "with reference" to this point. In this case V_{AC} = +9.68 V (where $V_{AC} = VR_1 + VR_2$). The voltage V_{BC} = +4.84 V (where $V_{BC} = VR_2$). The voltage V_{DC} = −10.31 V (where $V_{DC} = VR_3$).

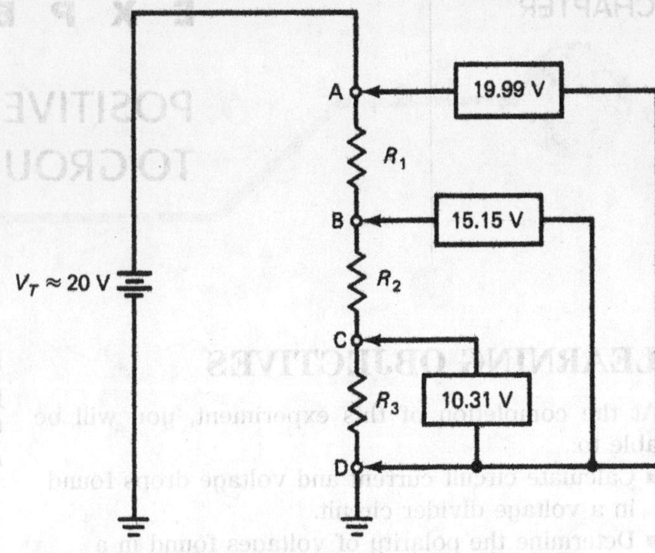

Fig. 4-6.4 Voltage divider circuit where voltages are measured to common ground.

The voltage measured from point D to C results in a negative voltage. This circuit is known as a *positive and negative voltage divider.*

In summary, these DC circuits operate in the same way with or without the ground symbol shown in the schematic. The only factor that changes is the reference point for measuring the voltage.

While this experiment focuses on the use of ground as a reference point, keep in mind that there are several different symbols for ground, as shown in Fig. 4-6.2. The earth ground symbol usually indicates that one side of the power supply is connected to the earth, usually by a metal pipe in the ground (this is the third prong on the AC wall plug). The other two ground symbols are chassis grounds and may or may not be connected to earth (an automobile and an airplane are good examples).

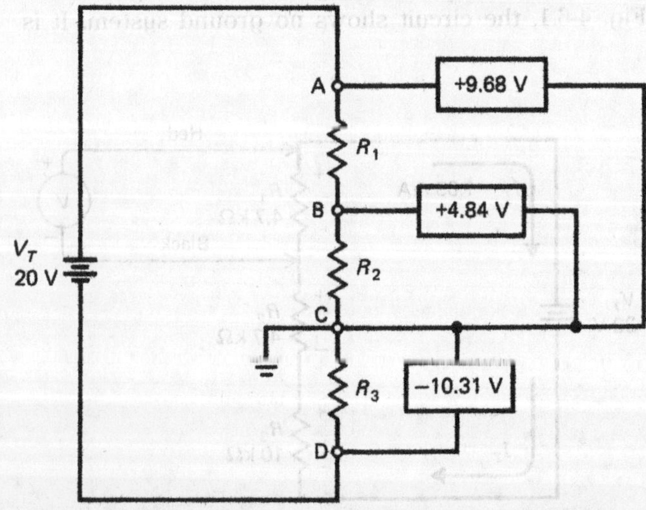

Fig. 4-6.5 Voltage divider displaying both positive and negative voltages to ground.

EQUIPMENT

DC power supply, 0 to 20 V
- Leads
- Breadboard
- DMM

COMPONENTS

Resistors (all 0.25 W):

- (2) 4.7 kΩ
- (1) 10 kΩ

PROCEDURE

1. Connect the circuit of Fig. 4-6.6 to the DC power supply as shown.

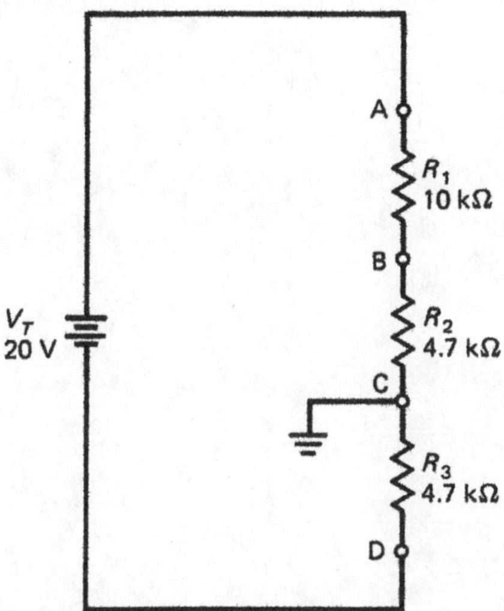

Fig. 4-6.6 Positive and negative voltage divider circuit with ground.

2. With the power supply turned off and disconnected, measure and record in Table 4-6.1 the resistance values of R_1, R_2, and R_3.
3. With the power supply turned off and disconnected, measure and record in Table 4-6.1 the resistive value of R_T.
4. Reconnect the power supply to the circuit shown in Fig. 4-6.6.
5. Calculate and record in Table 4-6.1 the values of R_T, I_T, V_{AC}, V_{BC}, and V_{DC} referenced to ground.
6. Turn on the power supply and adjust its voltage value to 20 V DC.
7. Measure and record in Table 4-6.1: I_T, V_T, V_{AC}, V_{BC}, and V_{DC}. Also note the polarities of V_{AC}, V_{BC}, and V_{DC} in Table 4-6.1.
8. Turn off the power supply, and reverse its polarity.
9. Reconnect the circuit with $V_A = -20$ V, repeat steps 2–7, and record the results in Table 4-6.2.

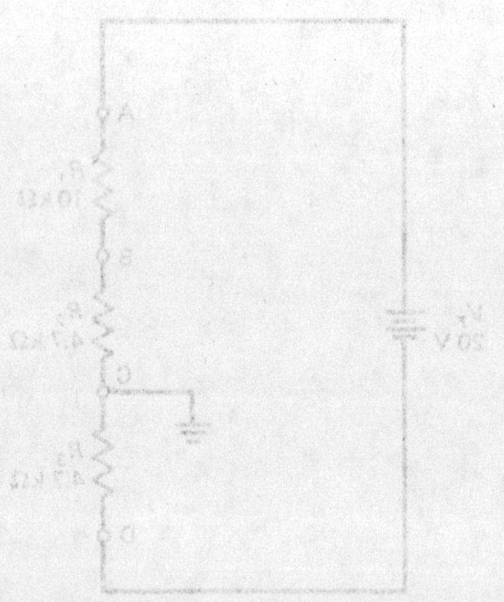

NAME ______________________ DATE ______________

QUESTIONS FOR EXPERIMENT 4-6

1. Explain the circuit function of a voltage divider.

2. What is the purpose of a circuit ground?

3. Draw an example of a circuit where a voltage is negative with respect to ground.

TABLES FOR EXPERIMENT 4-6

TABLE 4-6.1 (Steps 2–7)

	Measured	Calculated	Polarity*
R_1			
R_2			
R_3			
R_T			
I_T			
$V_{A\text{-}C}$			
$V_{B\text{-}C}$			
$V_{D\text{-}C}$			
V_T			

TABLE 4-6.2 (Steps 8 and 9)

	Measured	Calculated	Polarity*
R_1			
R_2			
R_3			
R_T			
I_T			
$V_{A\text{-}C}$			
$V_{B\text{-}C}$			
$V_{D\text{-}C}$			
V_T			

***Note: Polarity with respect to common ground.**

EXPERIMENT RESULTS REPORT FORM

Experiment No.: ______________________ Name: ______________________

Date: ______________________

Experiment Title: ______________________ Class: ______________________

______________________ Instr: ______________________

Explain the purpose of the experiment:

List the first Learning Objective:

OBJECTIVE 1:

After reviewing the results, describe how the objective was validated by this experiment.

List the second Learning Objective:

OBJECTIVE 2:

After reviewing the results, describe how the objective was validated by this experiment.

List the third Learning Objective:

OBJECTIVE 3:

After reviewing the results, describe how the objective was validated by this experiment.

Conclusion:

If required, attach to this form: ☐ Answers to Questions, ☐ Tables, and ☐ Graphs.

EXPERIMENT 5-1

PARALLEL CIRCUITS

LEARNING OBJECTIVES

At the completion of this experiment, you will be able to:

- Identify a parallel circuit.
- Accurately measure current in a parallel circuit.
- Use Ohm's law to verify measurements taken in a parallel circuit.

SUGGESTED READING

Chapter 5, *Basic Electronics, Grob*/Schultz, twelfth edition

INTRODUCTION

A *parallel circuit* is defined as one having more than one current path connected to a common voltage source. Parallel circuits, therefore, must contain two or more load resistances which are not connected in series. Study the parallel circuit shown in Fig. 5-1.1.

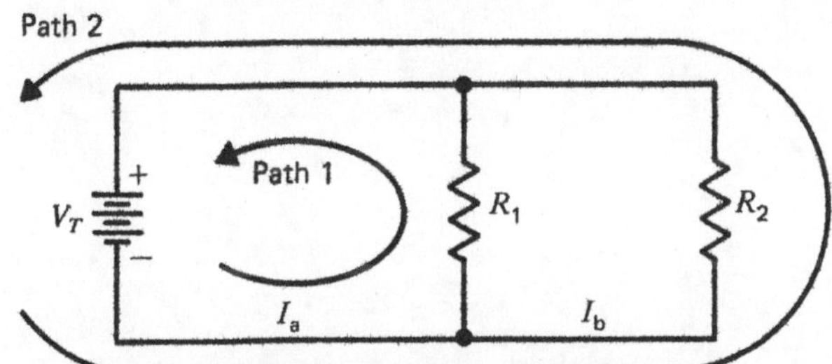

Fig. 5-1.1 Parallel circuit with two branches.

Beginning at the voltage source V_T and tracing counterclockwise around the circuit, two complete and separate paths can be identified in which current can flow. One path is traced from the source through resistance R_1 and back to the source; the other is traced from the source through resistance R_2 and back to the source.

The source voltage in a series circuit divides proportionately across each resistor in the circuit. In a parallel circuit, the same voltage is present across all the resistors of a parallel bank. In other words,

$$V_T = V_{R_1} = V_{R_2} = \cdots$$

The current in a circuit is inversely proportional to the circuit resistance. This fact, obtained from Ohm's law, establishes the relationship upon which the following discussion is developed. A single current flows in a series circuit.

In summary, when two or more electronic components are connected across a single voltage source, they are said to be *in parallel.* The voltage across each component is the same. The current through each component, or branch, is determined by the resistance of that branch and voltage across the bank of branches. Adding a parallel resistance of any value increases the total current. The total resistance of a circuit (R_T) can be found by dividing the total voltage applied (V_T) by the total current (I_T).

When measuring currents and resistances in the following procedure, some important precautions should be observed. When measuring currents, be sure to install the ammeter in a series configuration, with the ammeter connected in series with the individual branch resistances through which the current is flowing. Also, before measuring branch resistances, be sure to turn off all voltage sources.

EQUIPMENT

DC power supply, 0–10 V
DMM
Voltmeter
Protoboard or springboard
Test leads

COMPONENTS

Resistors (all 0.25 W):

(2) 1 kΩ (1) 2.2 kΩ

PROCEDURE

1. Measure the resistance values of R_1, R_2, and R_3, and record in the required locations of Tables 5-1.1 to 5-1.3, where the nominal values are

$$R_1 = 1.0\ \text{k}\Omega$$
$$R_2 = 1.0\ \text{k}\Omega$$
$$R_3 = 2.2\ \text{k}\Omega$$

2. Connect the circuit as shown with R_1 only in Fig. 5-1.2. Adjust the power supply voltage to 10 V.

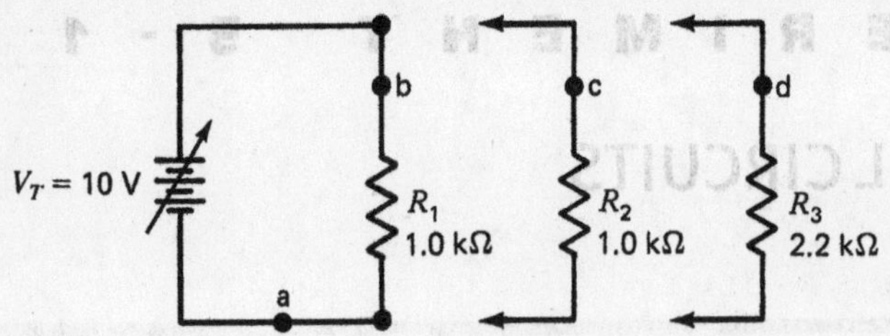

Fig. 5-1.2 Three-branch parallel circuit.

3. Measure and record in Table 5-1.1 the current through points a and b.

4. Measure and record in Table 5-1.1 the voltage across R_1.

5. To this circuit, add R_2 across (meaning that R_2 is connected in parallel) R_1. Measure and record in Table 5-1.2 the current through points a, b, and c.

6. Measure and record in Table 5-1.2 the voltages across R_1 and R_2.

7. Finally, add R_3 across R_2. Measure and record in Table 5-1.3 the current through points a, b, c, and d.

8. Measure and record in Table 5-1.3 the voltages across R_1, R_2, and R_3.

9. Calculate the total resistances and the current for each of the three previous circuits, and record this information in Tables 5-1.1 to 5-1.3.

NAME DATE

QUESTIONS FOR EXPERIMENT 5-1

1. What is a parallel circuit? What circuit characteristics indicate that a parallel circuit condition exists?

2. In the circuit of Fig. 5-1.2, determine the power being dissipated by each resistor using the values of current determined in procedure step 7. Use the formula I^2R = power (watts).

3. Are the voltages the same across each resistor in a parallel circuit?

4. Are the currents the same through each resistor in a parallel circuit?

5. Suppose in procedure step 7 that R_3 developed a short-circuited condition. How would the current flowing through each resistor change? Would the voltage drops across each resistor change? How?

TABLES FOR EXPERIMENT 5-1

TABLE 5-1.1

	Measured	Calculated
R_1	________	
R_T		________
V_{R_1}	________	
I_a	________	________
I_b	________	

TABLE 5-1.2

	Measured	Calculated
R_1	________	
R_2	________	
R_T		________
V_{R_1}	________	
V_{R_2}	________	
I_a	________	________
I_b	________	
I_c	________	

TABLE 5-1.3

	Measured	Calculated
R_1	________	
R_2	________	
R_3	________	
R_T		________
V_{R_1}	________	
V_{R_2}	________	
V_{R_3}	________	
I_a	________	________
I_b	________	________
I_c	________	________
I_d	________	________

EXPERIMENT RESULTS REPORT FORM

Experiment No.: ______________________ Name: ______________________

Date: ______________________

Experiment Title: ______________________ Class: ______________________

______________________ Instr: ______________________

Explain the purpose of the experiment:

List the first Learning Objective:

OBJECTIVE 1:

After reviewing the results, describe how the objective was validated by this experiment.

List the second Learning Objective:

OBJECTIVE 2:

After reviewing the results, describe how the objective was validated by this experiment.

List the third Learning Objective:

OBJECTIVE 3:

After reviewing the results, describe how the objective was validated by this experiment.

Conclusion:

If required, attach to this form: ☐ Answers to Questions, ☐ Tables, and ☐ Graphs.

CHAPTER 5

EXPERIMENT 5-2

PARALLEL CIRCUITS—RESISTANCE BRANCHES

LEARNING OBJECTIVES

At the completion of this experiment, you will be able to:

- Recognize that for a parallel resistor circuit each branch current equals the applied voltage divided by the branch resistance.
- Recognize that with a parallel circuit arrangement, any branch that has less resistance allows more current to flow.
- Determine that current can have different values in different parts of parallel circuits.

SUGGESTED READING

Chapter 5, *Basic Electronics,* Grob/Schultz, twelfth edition

INTRODUCTION

A parallel circuit is formed when two or more components, or resistances, are connected across a voltage source, as shown in Fig. 5-2.1. In this figure, R_1, R_2, R_3, and R_4 are in parallel with each other and the 10-V supply source. This figure can be redrawn, as shown in Fig. 5-2.2, to readily show that each component connection is really equivalent to a direct connection at the terminals of the battery, because the connecting wires exhibit almost no resistance. Since R_1, R_2, R_3, and R_4 are connected directly across the two terminals of the battery, all resistances must have the same potential differences as the battery. It therefore follows that the voltage is the same across each of the components that are connected in a parallel fashion. Thus, the parallel circuit arrangement is used to connect components that require the same voltage from a common supply voltage.

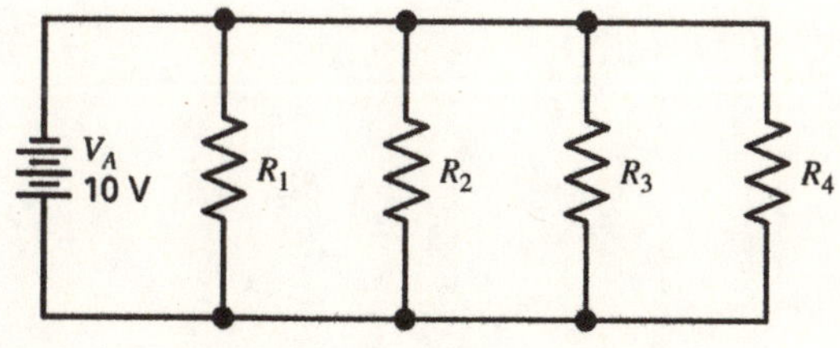

Fig. 5-2.1 A parallel circuit with four resistors.

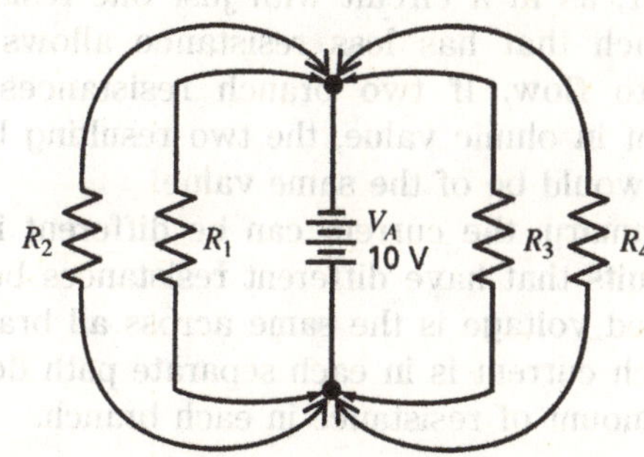

Fig. 5-2.2 Reconfiguration of a parallel circuit with four resistors.

Again (with reference to Fig. 5-2.1), R_1, R_2, R_3, and R_4 can be referred to as *branch* resistances. Although the voltage is the same across each branch resistance, the current divides into branches according to Ohm's law.

In each branch circuit the amount of branch current is equivalent to the applied voltage divided by the individual branch resistor. In other words, when Ohm's law is applied to the parallel circuit, the current equals the voltage applied across the circuit divided by the resistance between the two points at which that voltage is applied. As shown in Fig. 5-2.3, a 15-V source is applied across the 3900-Ω resistor (R_3). This results in a calculated current flow of 0.0038 A (3.8 mA)

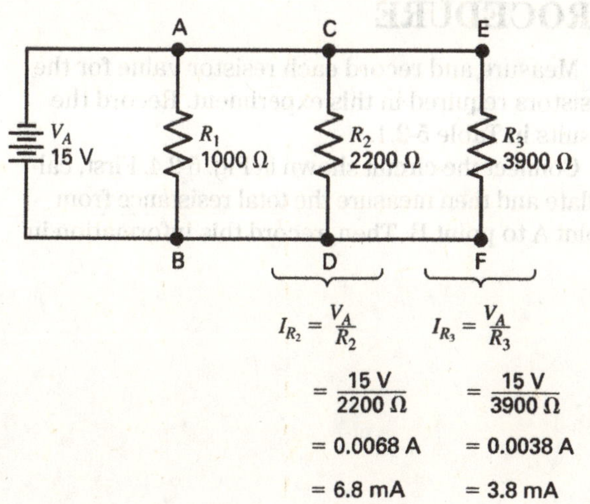

Fig. 5-2.3 Ohm's law and the parallel circuit.

between points E and F. The battery voltage is *also* applied across the parallel resistance of R_2, which applies 15 V across a 2200-Ω resistor. This results in a calculated current of 0.0068 A (6.8 mA) between points C and D. This current provides a different value through R_1 with the same applied voltage, because the resistance is different. Remember that the current in each parallel branch equals the applied voltage divided by each individual branch resistance.

Further, as in a circuit with just one resistance, any branch that has less resistance allows more current to flow. If two branch resistances were equivalent in ohmic value, the two resulting branch currents would be of the same value.

In summary, the current can be different in parallel circuits that have different resistances because the applied voltage is the same across all branches. How much current is in each separate path depends on the amount of resistance in each branch.

EQUIPMENT

Breadboard
DC power supply
DMM

COMPONENTS

All resistors are 0.25 W unless indicated otherwise:

(1) 390 Ω
(1) 1 kΩ
(1) 1.2 kΩ
(1) 2.2 kΩ
(1) 3.9 kΩ

PROCEDURE

1. Measure and record each resistor value for the resistors required in this experiment. Record the results in Table 5-2.1.
2. Connect the circuit shown in Fig. 5-2.4. First, calculate and then measure the total resistance from point A to point B. Then, record this information in

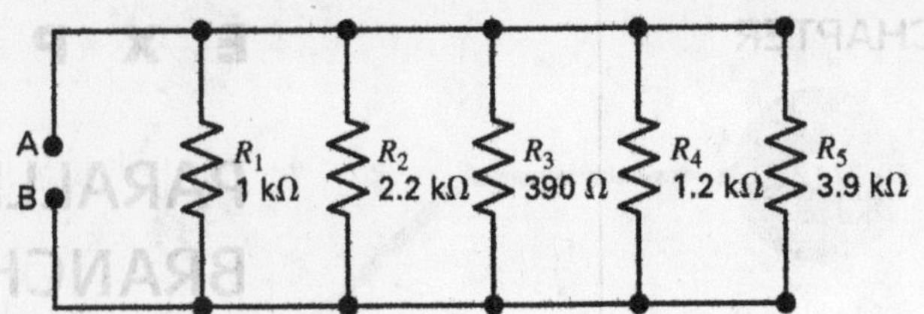

Fig. 5-2.4 Calculate and measure the total circuit resistance.

Table 5-2.2. In addition, record the percentage of error between these values, where

$$\% \text{ error} = \left| \frac{\text{difference between meas. and calc. values}}{\text{calc. values}} \right| \times 100$$

If the error is greater than 10 percent, repeat the calculations and measurements.

3. Connect the circuit shown in Fig. 5-2.5. Adjust the supply voltage to 15 V, and, using a voltmeter, take measurements required to complete Table 5-2.3 (Part A).

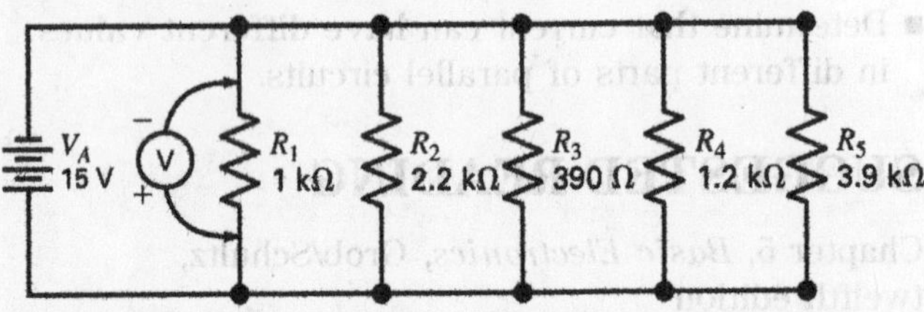

Fig. 5-2.5 Measuring the voltage existing in a parallel branch circuit.

4. Referring to Fig. 5-2.6, adjust the supply voltage to 15 V, and, using an ammeter, take the current measurements at points A through F and perform the necessary calculations required to complete Table 5-2.3 (Part B).
5. Using the formula as shown in step 3, record in Table 5-2.3 (Part C) the percentage of error between the calculated and measured currents.

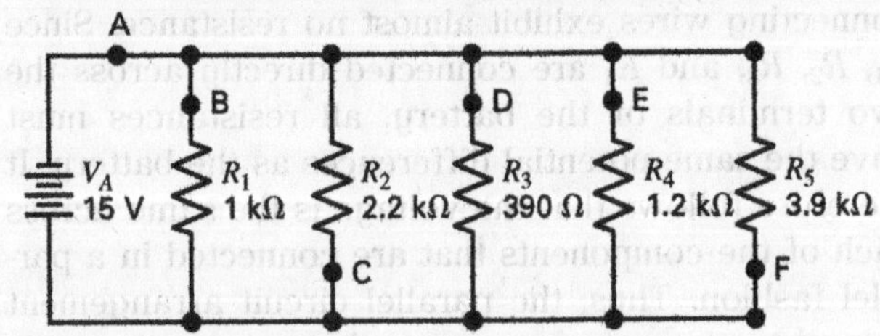

Fig. 5-2.6 Measuring the current existing in a parallel branch circuit.

NAME DATE

QUESTIONS FOR EXPERIMENT 5-2

1. Write the formula that determines the total resistance R_T of the circuit depicted in Fig. 5-2.3.

2. Mathematically determine the branch current passing through R_1, as shown in Fig. 5-2.3. Show all calculations.

3. In your estimation would $V_T = IR_1 = IR_2 = IR_3 = \cdots$ be true for all parallel resistive branches? Explain.

4. In a parallel circuit, the amount of current in each separate branch depends upon which factor?

5. Could you always expect that $R_T = R_1 + R_2 + R_3 + \cdots$ in a series resistive circuit? Explain.

CRITICAL THINKING QUESTIONS

Note: **The following questions are designed to help you analyze the previous laboratory experiment in a complete and in-depth fashion. To answer these questions, you should review the related material in Grob/Schultz, *Basic Electronics*, twelfth edition.**

1. Explain where the sources of errors exist in this experiment. How could these sources of error be reduced?

2. The purpose of a parallel circuit is to connect different components that need the same voltage. After reviewing your results for this experiment, explain how this purpose supports, or agrees with, your findings. Explain.

3. Explain why the voltage is the same across all branches in a parallel circuit.

4. Explain why, for a parallel resistive circuit, each branch current equals the applied voltage divided by the branch resistance.

5. How is it possible for current to be different in various parts of parallel circuits when the applied voltage is found to be the same across all the branches?

TABLES FOR EXPERIMENT 5-2

TABLE 5-2.1 Individual Resistor Values

	Nominal Value	Measured Value
$R_1 =$	1 kΩ	
$R_2 =$	2.2 kΩ	
$R_3 =$	390 Ω	
$R_4 =$	1.2 kΩ	
$R_5 =$	3.9 kΩ	

TABLE 5-2.2 Total Resistance R_T and Percentage of Error

	Nominal Calculated Total Resistance $R_1 \| R_2 \| R_3 \| R_4 \| R_5$	Total Resistance Measured	% Error
$R_T =$			

TABLE 5-2.3 Recording Voltage, Current, and Error Measurements

Part A				Part B				Part C
	Measured Voltage		Tables 5-2.1 and 5-2.2, Measured Resistance		Calculated, $I = V/R$		Measured, I	% Error
V_A		R_T		I_T		Point A		
V_{R_1}		R_1		I_{R_1}		Point B		
V_{R_2}		R_2		I_{R_2}		Point C		
V_{R_3}		R_3		I_{R_3}		Point D		
V_{R_4}		R_4		I_{R_4}		Point E		
V_{R_5}		R_5		I_{R_5}		Point F		

EXPERIMENT RESULTS REPORT FORM

Experiment No.: ____________________ Name: ____________________

Date: ____________________

Experiment Title: ____________________ Class: ____________________

____________________ Instr: ____________________

Explain the purpose of the experiment:

List the first Learning Objective:

OBJECTIVE 1:

After reviewing the results, describe how the objective was validated by this experiment.

List the second Learning Objective:

OBJECTIVE 2:

After reviewing the results, describe how the objective was validated by this experiment.

List the third Learning Objective:

OBJECTIVE 3:

After reviewing the results, describe how the objective was validated by this experiment.

Conclusion:

If required, attach to this form: ☐ Answers to Questions, ☐ Tables, and ☐ Graphs.

CHAPTER 5

EXPERIMENT 5-3

PARALLEL CIRCUITS—ANALYSIS

LEARNING OBJECTIVES

At the completion of this experiment, you will be able to:

- Recognize that, for a parallel circuit, the main-line total current equals the sum of the individual branch currents.
- Use the formula $I_T = I_1 + I_2 + I_3 + I_4 + \cdots$, applied to any number of parallel branches, regardless of whether the individual branch resistances are equal or unequal.
- Recognize that a circuit's equivalent resistance across a common voltage source is found by dividing the common voltage by the total current of all the branches.

SUGGESTED READING

Chapter 5, *Basic Electronics*, Grob/Schultz, twelfth edition

INTRODUCTION

Components connected in parallel are usually wired directly across each other so that the entire parallel combination is connected across the voltage source, as shown in Fig. 5-3.1. A combination of parallel branches is often called a *bank*, and a bank may contain two or more parallel resistors. The connecting wires provide no resistance and have almost no effect on circuit operation. One advantage of having circuits wired in parallel is that since only one pair of connecting leads is attached to the voltage source, less wire is used. It is common to find a pair of leads connecting all the branches to the terminals of the voltage source.

The main-line current is also known as the *total circuit current* and is equivalent to the sum of the individual branch currents. In Fig. 5-3.2, the supply voltage is 35 V, and each of the three resistors establishes an independent current flow according to Ohm's law. The sum of the three currents is equivalent to the total (main-line) current, as demonstrated by the formula $I_T = I_1 + I_2 + I_3$. This formula applies for any number of parallel branches, whether the resistances are equal or unequal.

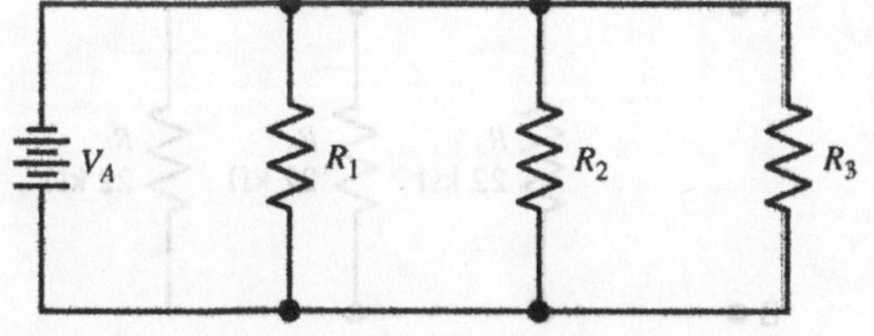

Fig. 5-3.1 A parallel circuit with three resistors.

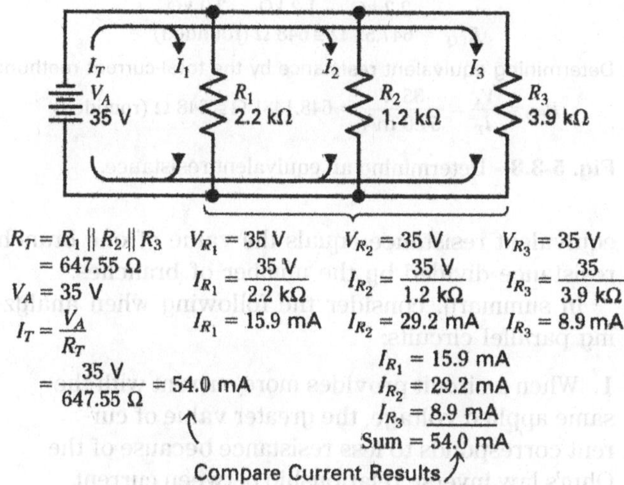

Fig. 5-3.2 A parallel circuit with three resistors.

The main-line current is equal to the total of the branch currents. All current in the circuit must originate from one side of the voltage source and return to its opposite side in order to form a complete path. Therefore, the main-line current I_T equals the sum of the branch currents.

Analysis of the parallel resistive circuit reveals that the combined equivalent resistance across the supply voltage can be found by Ohm's law, so that the common voltage across the parallel resistances is divided by the total current of all the branches. Again referring to Fig. 5-3.2, note that the parallel equivalent resistance of R_1, R_2, and R_3 is 648 Ω and results in an opposition to the total current flow in the main line.

Two techniques are used to determine the equivalent resistance of a parallel circuit. These are referred to as the *reciprocal resistance formula* and the *total-current method* (see Fig. 5-3.3).

There are unique occasions when the resistance is equal in all branches. At those times, the combined

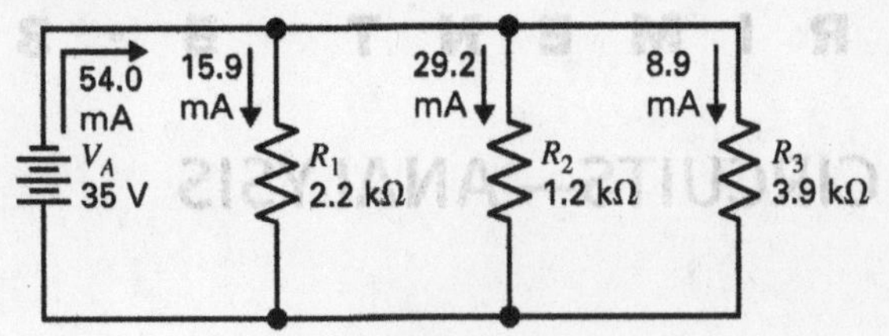

Determining equivalent resistance by the reciprocal resistance formula:

$$\frac{1}{R_{EQ}} = \frac{1}{R_1} + \frac{1}{R_2} + \frac{1}{R_3}$$

$$= \frac{1}{2.2\text{ k}\Omega} + \frac{1}{1.2\text{ k}\Omega} + \frac{1}{3.9\text{ k}\Omega}$$

$$R_{EQ} = 647.55\ \Omega \cong 648\ \Omega \text{ (rounded)}$$

Determining equivalent resistance by the total-current method:

$$R_{EQ} = \frac{V_A}{I_T} = \frac{35\text{ V}}{54.0\text{ mA}} = 648.1481\ \Omega \cong 648\ \Omega \text{ (rounded)}$$

Fig. 5-3.3 Determining an equivalent resistance.

equivalent resistance equals the value of one branch resistance divided by the number of branches.

In summary, consider the following when analyzing parallel circuits:

1. When a circuit provides more current with the same applied voltage, the greater value of current corresponds to less resistance because of the Ohm's law inverse relationship between current and resistance.
2. The combination of parallel resistances for a parallel bank is always less than the smallest individual branch resistance.

EQUIPMENT

Breadboard
DC power supply
DMM

COMPONENTS

All resistors are 0.25 W unless indicated otherwise:

(1) 680 Ω (1) 1.2 kΩ
(1) 1.0 kΩ (1) 22 kΩ

PROCEDURE

1. Measure and record each resistor value for the 680-Ω, 1.0-kΩ, and 1.2-kΩ resistors required in the first part of this experiment. Record the results in Table 5-3.1.
2. Connect the circuit shown in Fig. 5-3.4. First, calculate and then measure the total resistance from point A to point B. Then record this information in Table 5-3.1. In addition, record the percentage of error between these values, where

$$\%\text{ error} = \left|\frac{\text{difference between meas. and calc. values}}{\text{calc. values}}\right| \times 100$$

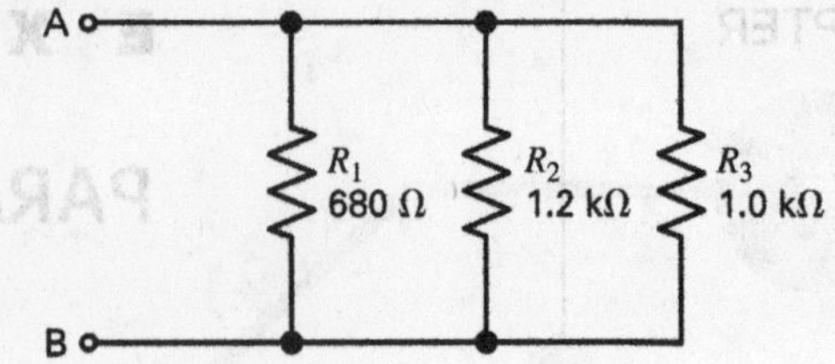

Fig. 5-3.4 Finding the total resistance and percentage of error.

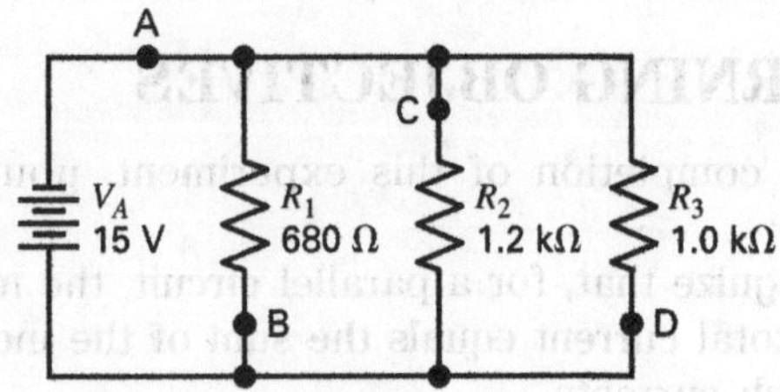

Fig. 5-3.5 Main-line and branch current.

If the error is greater than 10 percent, repeat the calculations and measurements.

3. Connect the circuit shown in Fig. 5-3.5. Adjust the supply voltage to 15 V, and, using a voltmeter, make voltage measurements across the supply voltage and the individual resistors. Record this information in Table 5-3.2.
4. Again referring to Fig. 5-3.5, while the voltage is connected and turned on, break the circuit at point A and, with an ammeter, measure the current passing through this point. Record this information in Table 5-3.2.
5. Repeat step 4 and make current measurements at points B, C, and D. Record this information in Table 5-3.2.
6. Complete all calculations required in Table 5-3.3. Be certain that you use the measured resistance values from Table 5-3.1 in all calculations. Determine the percentage of error between the calculated and measured currents and record the results in Table 5-3.3.
7. Connect the circuit in Fig. 5-3.6. Measure and record, in Table 5-3.4, each individual resistance value as well as the total resistance from points A and B.
8. Complete and record the calculations required for Table 5-3.4. Explain in the table how the measured and calculated values compare.

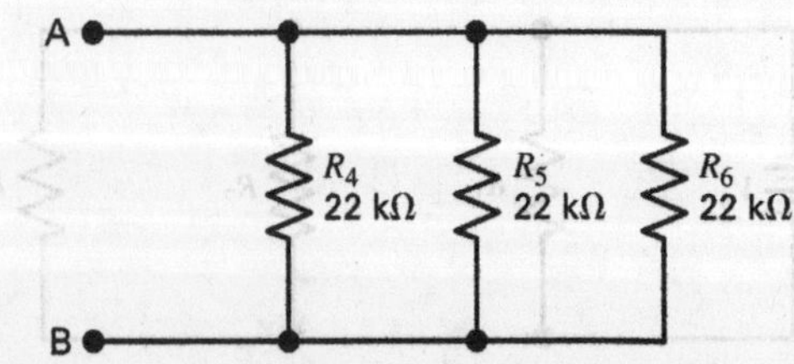

Fig. 5-3.6 Three equal resistances in parallel.

NAME DATE

QUESTIONS FOR EXPERIMENT 5-3

1. Why is it true that the combined resistance of a parallel branch circuit equals the value of one branch resistance divided by the number of branches?

2. Define the term *bank* with regard to the parallel resistive circuit.

3. In your estimation would $V_T = IR_1 = IR_2 = IR_3 = \cdots$ be true for all parallel resistive branches? Explain.

4. Does the formula $I_T = I_1 + I_2 + I_3 + I_4 + \cdots$ apply for any number of parallel branches? What if the individual branch resistances were equal or unequal? Explain.

5. What is meant by the *total-current method?*

CRITICAL THINKING QUESTIONS

Note: The following questions are designed to help you analyze the previous laboratory experiment in a complete and in-depth fashion. To answer these questions, you should review the related material in Grob/Schultz, *Basic Electronics,* twelfth edition.

1. Explain where the sources of errors exist in this experiment. How could these sources of error be reduced?

2. One purpose of this experiment is to validate that the formula for the total current I_T in the main line of a parallel resistive circuit is $I_T = I_1 + I_2 + I_3 + I_4 + \cdots$. After reviewing your results for this experiment, explain how this purpose supports, or agrees with, your findings.

3. Explain why the voltage is the same across all branches in a parallel circuit.

4. Describe the two methods of analyzing main-line current flow as demonstrated in Fig. 5-3.2.

5. Explain what is meant by the statement, "When a circuit provides more current with the same applied voltage, the greater value of current corresponds to less resistance because of the Ohm's law inverse relationship between current and resistance." Show an example of this statement.

TABLES FOR EXPERIMENT 5-3

TABLE 5-3.1 Individual Resistor Values

	Nominal Value	Measured Value	% Error
$R_1 =$	680 Ω	________	________
$R_2 =$	1.2 kΩ	________	________
$R_3 =$	1.0 kΩ	________	________
		Step 1	Step 2

		Sum Measured	% Error
Sum =	302.7 Ω	________	________
		Step 2	

TABLE 5-3.2 Voltage Measurements

V_A	________	I_A	________
V_{R_1}	________	I_B	________
V_{R_2}	________	I_C	________
V_{R_3}	________	I_D	________
	Step 3		Step 4

TABLE 5-3.3 Comparison of Calculated and Measured Currents

Voltage Measurements from Table 5-3.2		Resistance Measurements from Table 5-3.1		Calculated Currents Columns 1 and 2		Measured Currents from Table 5-3.2		Current % Error
V_{R_1}	______	R_1	______	I_1	______	I_1	______	______
V_{R_2}	______	R_2	______	I_2	______	I_2	______	______
V_{R_3}	______	R_3	______	I_3	______	I_3	______	______

TABLE 5-3.4 Tabulations for Three Equal Resistances in Parallel

	Nominal Values	Measured Values	Calculated by Reciprocal Method	Calculated by Other Method Described in Introduction
$R_4 =$	22 kΩ	______		
$R_5 =$	22 kΩ	______		
$R_6 =$	22 kΩ	______		
		$R_T =$ ______	$R_T =$ ______	$R_T =$ ______

EXPERIMENT RESULTS REPORT FORM

Experiment No.: ____________________ Name: ____________________

Date: ____________________

Experiment Title: ____________________ Class: ____________________

____________________ Instr: ____________________

Explain the purpose of the experiment:

List the first Learning Objective:

OBJECTIVE 1:

After reviewing the results, describe how the objective was validated by this experiment.

List the second Learning Objective:

OBJECTIVE 2:

After reviewing the results, describe how the objective was validated by this experiment.

List the third Learning Objective:

OBJECTIVE 3:

After reviewing the results, describe how the objective was validated by this experiment.

Conclusion:

If required, attach to this form: ☐ Answers to Questions, ☐ Tables, and ☐ Graphs.

CHAPTER 5

EXPERIMENT 5-4

PARALLEL CIRCUITS—OPENS AND SHORTS

LEARNING OBJECTIVES

At the completion of this experiment, you will be able to:

- Validate that an open in a parallel circuit results in an infinite resistance that allows no current to flow, whereas the effect of a short circuit is excessive current flow.
- Recognize that short-circuited components are typically not damaged, since they do not have greater than normal current passing through them.
- Recognize that the amount of current resulting from a short circuit is limited by the small resistance of the wire conductors.

SUGGESTED READING

Chapter 5, *Basic Electronics*, Grob/Schultz, twelfth edition

INTRODUCTION

Parallel Open Circuits: An open in any circuit creates an infinite resistance that permits no current to flow. However, in parallel circuits there is a difference between an open circuit in the main line and an open circuit in a parallel branch, as shown in Fig. 5-4.1. An open circuit in the main line prevents any electrons from flowing in the main line or in any branch. An open in a branch results in no current flow in that particular branch. The current in all the other parallel branches remains the same because each is connected directly to the voltage source. However, the total main-line current will change, since the total resistance of the circuit changes because one of the branches is open.

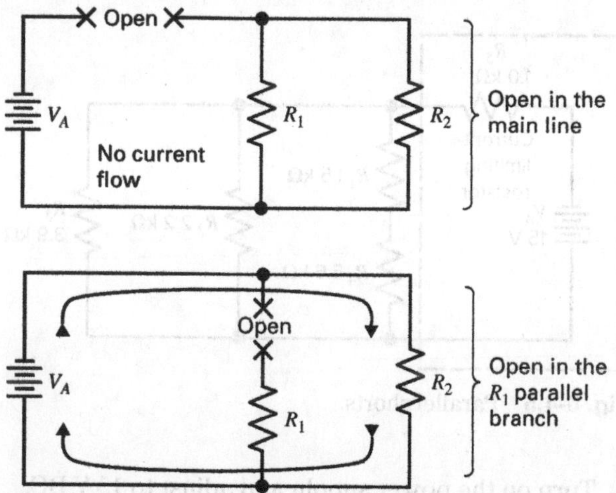

Fig. 5-4.1 An open in the main line versus an open in a branch.

One advantage of wiring circuits in parallel is that an open in one component will open only that branch, allowing the other parallel branches to operate with normal voltages and currents.

Parallel Short Circuits: A short circuit has almost a zero resistance level. Its effect, therefore, is to allow excessive current to flow in the shorted circuit, as shown in Fig. 5-4.2. If a wire conductor were connected across a component, the effective resistance would be essentially zero and the current would be extremely high. This high amount of current is limited by the current-carrying capability of the wire.

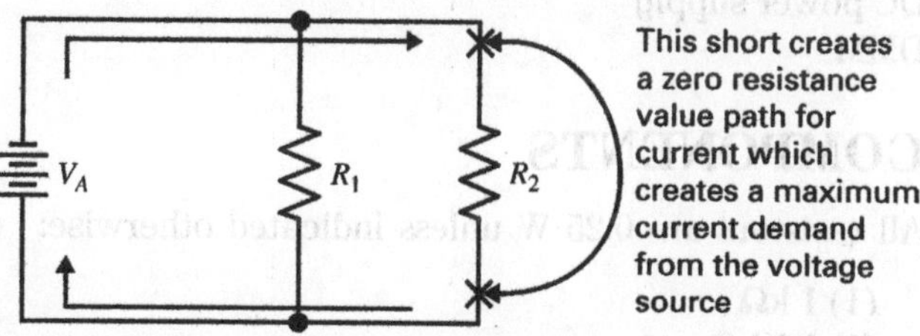

Fig. 5-4.2 A short circuit and excessive current flow.

Typically, in circuits that develop a shorted component, the voltage source cannot usually maintain its voltage level while providing the excessive current demand to the shorted component. The amount of current flowing can be dangerously high to the extent that wire can become hot enough to burn insulation and other components. To prevent the damaging effects of excessively high current flow, fuses are installed. A *fuse* is a component that intentionally burns open when there is too high a level of current demand from the supply voltage.

Short-circuited components, such as resistors, have no current passing through them. This is because the short circuit is a parallel path with almost no resistance. When this is the case, all current flows

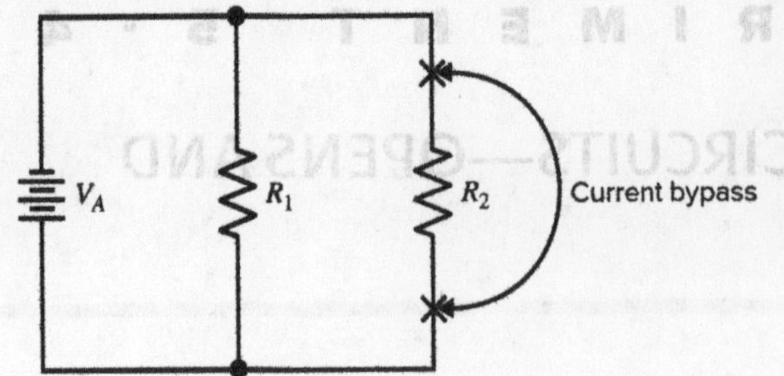

Fig. 5-4.3 A shorted component's current is bypassed.

in this path, as shown in Fig. 5-4.3. In this figure the short bypasses the parallel resistances, and these resistances are referred to as being *shorted out.*

When a parallel branch circuit is analyzed, if any individual resistor or component is shorted, all resistive branches are also shorted. This idea can be extended to a short existing across the voltage source; in this case the entire circuit is shorted as well.

Short-circuited components are not usually damaged, since the current passing through the actual component is less than it is under normal operating conditions. If the short circuit does not damage the voltage source and the associated wiring for the circuit, the components will function again when the short is removed and the circuit is restored to normal.

EQUIPMENT

Breadboard
DC power supply
DMM

COMPONENTS

All resistors are 0.25 W unless indicated otherwise:

(1) 1 kΩ
(1) 1.5 kΩ
(1) 2.2 kΩ
(1) 3.9 kΩ
(1) 5.6 kΩ

For Critical Thinking Question 1:

(1) Clear glass functioning fuse (any current rating)
(1) Clear glass nonfunctioning fuse (any current rating)

PROCEDURE

1. Measure and record each resistor value for the resistors required in this experiment. Record the results in Table 5-4.1.
2. Connect the circuit shown in Fig. 5-4.4. First calculate and then measure the total resistance from point A to point B. Record the results in Table 5-4.2.

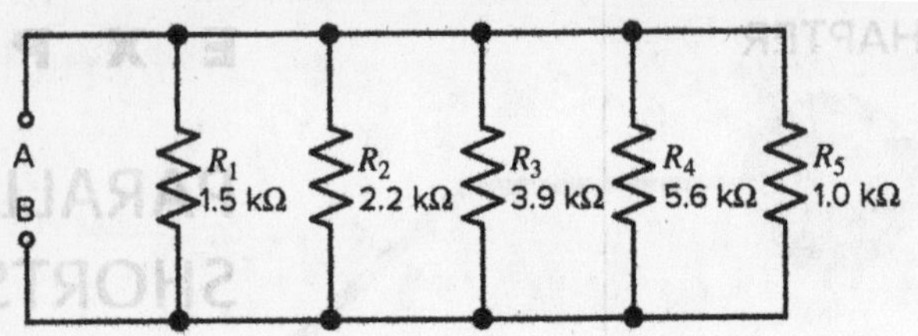

Fig. 5-4.4 Determining the total resistance.

In addition, record the percentage of error between these values, where

$$\% \text{ error} = \left| \frac{\text{difference between meas. and calc. values}}{\text{calc. values}} \right| \times 100$$

If the error is greater than 10 percent, repeat the calculations and measurements.

Short Circuit

3. Connect the circuit shown in Fig. 5-4.5, but *do not* turn on the power supply. When constructing this circuit, notice that this configuration uses a resistor in series with the supply voltage. This is done so that there will always be a reference level of current being maintained from the power supply. For example, if in this circuit a short exists across the entire parallel branch arrangement, this series resistance will maintain a minimal circuit resistance and a maximum circuit current. This is necessary because if the entire parallel branch arrangement were shorted and there were no series resistance, a theoretically infinite amount of current could be demanded from the power supply. In fact, the power supply that you will be using cannot provide an infinite amount of current, and if a short were provided across the supply, the internal circuitry would be damaged or a fuse opened or a circuit breaker tripped (opened).

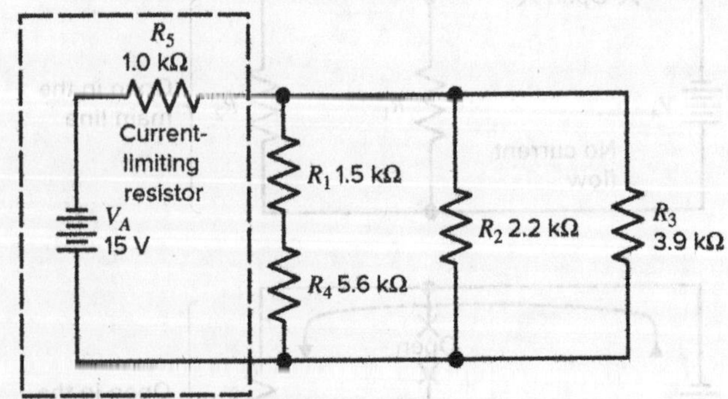

Fig. 5-4.5 Parallel shorts.

4. Turn on the power supply and adjust to 15 V DC. Measure and record the voltage drops and currents as needed to complete Table 5-4.3.

5. Using a clip lead of a small piece of wire, place a short across resistance (R_1, R_2, R_3, and R_4) one at a time, and retake voltage and current measurements necessary to complete Table 5-4.4. Be sure to answer the questions in Table 5-4.4.

Open Circuit

6. Connect the circuit shown in Fig. 5-4.6, but *do not* turn on the power supply. Refer to step 3, and notice that when the *previous* circuit was constructed, this configuration used a resistor in series with the supply voltage. This was done so that there would always be a reference level of current maintained from the power supply. In the analysis of open circuits, it is *not* necessary to include a series resistor in the test circuit. This is because when an open occurs in a parallel resistive circuit, the parallel branch resistance becomes practically infinite. If the open occurs in the main line of the power supply, the total resistance becomes infinite as well. The addition of infinite resistances does not harm the power supply. Because of Ohm's law, an infinitely large resistance creates an infinitely small current in relation to the constant voltage provided by the power supply.

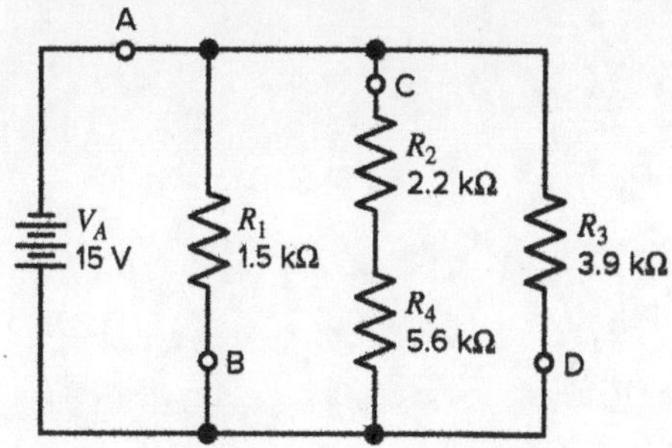

Fig. 5-4.6 Parallel opens.

7. Turn on the power supply and adjust it to 15 V DC. Measure and record the voltage drops and currents as needed to complete Table 5-4.5.

8. Open the circuit at the identified points (A, B, C, and D). Retake voltage and current measurements and complete Table 5-4.6. Be sure to answer the question that appears in Table 5-4.6.

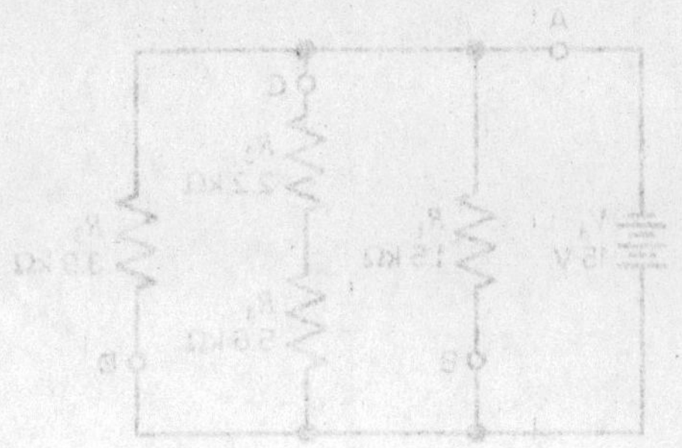

NAME DATE

QUESTIONS FOR EXPERIMENT 5-4

1. **What is the result of an open circuit? Explain what happens to the levels of voltage and current in a basic parallel branch circuit.**

2. **How will an open in a parallel branch affect the main-line current?**

3. **What is the main advantage of wiring a circuit in a parallel fashion?**

4. **Why does a short display practically no resistance? Provide a schematic drawing explaining your answer.**

5. **What limits the current resulting from a short?**

CRITICAL THINKING QUESTIONS

Note: **The following questions are designed to help you analyze the previous laboratory experiment in a complete and in-depth fashion. To answer these questions, you should review the related material in Grob/Schultz, *Basic Electronics*, twelfth edition.**

1. **Ask your laboratory instructor for two fuses, one that functions and one that does not. After studying these components, describe how they are similar and how they are different. Explain in your own words how you think these components operate.**

2. **Explain in your own words why short-circuited components are rarely damaged.**

3. If any individual resistor is shorted in a parallel circuit, why are all other resistive branches shorted as well?

4. Explain what is meant by the term *bypass* with reference to a shorted circuit.

5. When a parallel circuit develops a shorted component, the voltage source cannot usually maintain its voltage level. Explain.

TABLES FOR EXPERIMENT 5-4

TABLE 5-4.1 Individual Resistor Values

		Nominal Resistance	Measured Resistance
Step 1	R_1	1500 Ω	
	R_2	2200 Ω	
	R_3	3900 Ω	
	R_4	5600 Ω	
	R_5	1000 Ω	

TABLE 5-4.2 Total Resistance R_T and Percentage of Error

		Calculated	Measured	% Error
Step 2	R_T			

TABLE 5-4.3 Voltage Drops and Current Measurements

	Measured R Values from Table 5-4.1	Measured Voltage Drop	Current Measured through Resistor
R_1			
R_2			
R_3			
R_4			

TABLE 5-4.4 Voltage Drops and Current Measurements with Shorted Components

With Only R_1 Shorted		With Only R_2 Shorted		With Only R_3 Shorted		With Only R_4 Shorted	
V_{R1} ______	I_1 ______	V_{R1} ______	I_1 ______	V_{R1} ______	I_1 ______	V_{R1} ______	I_1 ______
V_{R2} ______	I_2 ______	V_{R2} ______	I_2 ______	V_{R2} ______	I_2 ______	V_{R2} ______	I_2 ______
V_{R3} ______	I_3 ______	V_{R3} ______	I_3 ______	V_{R3} ______	I_3 ______	V_{R3} ______	I_3 ______
V_{R4} ______	I_4 ______	V_{R4} ______	I_4 ______	V_{R4} ______	I_4 ______	V_{R4} ______	I_4 ______

How do the above measurements reinforce the characteristics of a short?

Verify by calculations the expected voltages and currents for this circuit when R_2 is shorted.

TABLE 5-4.5 Voltage Drops and Current Measurements

	Measured R Values from Table 5-4.1	Measured Voltage Drop	Current Measured through Resistor
R_1	______		
R_2	______	______	______
R_3	______	______	______
R_4	______	______	______

TABLE 5-4.6 Voltage Drops and Current Measurements with Opens

Point A Open		Point B Open		Point C Open		Point D Open	
V_{R1} ______	I_1 ______	V_{R1} ______	I_1 ______	V_{R1} ______	I_1 ______	V_{R1} ______	I_1 ______
V_{R2} ______	I_2 ______	V_{R2} ______	I_2 ______	V_{R2} ______	I_2 ______	V_{R2} ______	I_2 ______
V_{R3} ______	I_3 ______	V_{R3} ______	I_3 ______	V_{R3} ______	I_3 ______	V_{R3} ______	I_3 ______
V_{R4} ______	I_4 ______	V_{R4} ______	I_4 ______	V_{R4} ______	I_4 ______	V_{R4} ______	I_4 ______

How do the above measurements validate the characteristics of an open circuit?

EXPERIMENT RESULTS REPORT FORM

Experiment No.: ______________ Name: ______________

Date: ______________

Experiment Title: ______________ Class: ______________

______________ Instr: ______________

Explain the purpose of the experiment:

List the first Learning Objective:

OBJECTIVE 1:

After reviewing the results, describe how the objective was validated by this experiment.

List the second Learning Objective:

OBJECTIVE 2:

After reviewing the results, describe how the objective was validated by this experiment.

List the third Learning Objective:

OBJECTIVE 3:

After reviewing the results, describe how the objective was validated by this experiment.

Conclusion:

If required, attach to this form: ☐ Answers to Questions, ☐ Tables, and ☐ Graphs.

CHAPTER 6

EXPERIMENT 6-1

SERIES-PARALLEL CIRCUITS—RESISTANCE

LEARNING OBJECTIVES

At the completion of this experiment, you will be able to:

- Recognize the basic characteristics of series-parallel resistive circuits.
- Confirm that, within a series-parallel circuit, main-line currents are the same through series resistances.
- Confirm that voltage drops across parallel resistive components in a series-parallel circuit are equivalent.

SUGGESTED READING

Chapter 6, *Basic Electronics*, Grob/Schultz, twelfth edition

INTRODUCTION

All components in a parallel configuration with the voltage supply have the same voltage drop shown in Fig. 6-1.1. However, if there is a need to have a voltage less than what is created across one of the parallel resistance branches, an additional resistor can be placed in series within one of the branches, as shown in Fig. 6-1.2.

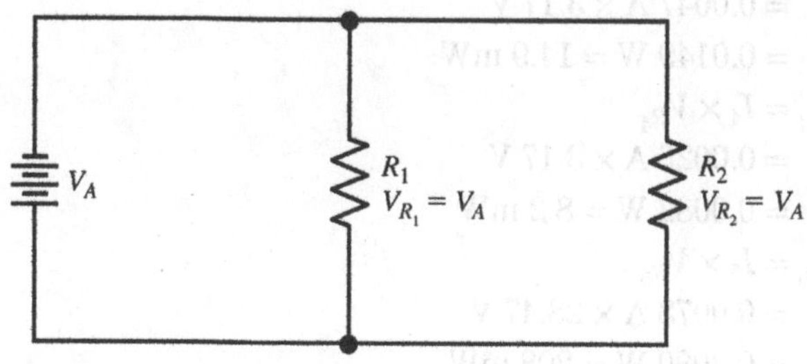

Fig. 6-1.1 Parallel circuit.

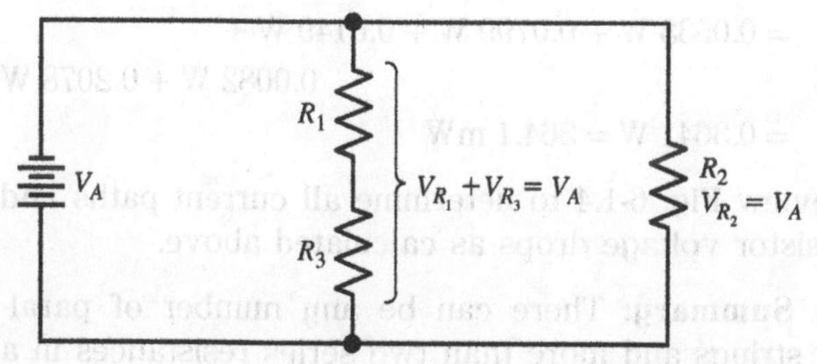

Fig. 6-1.2 Adding a series resistor to a parallel circuit.

Components in series have the same level of current passing through them. If there is a need to have a component with the same voltage level, another component can be placed across (or in parallel with) one of the series resistors.

Series-parallel circuits are used when it is necessary to provide different amounts of voltage and current from components that use only one source of supply voltage.

Finding R_T: Figure 6-1.3 depicts a series-parallel circuit. In order to determine the total resistance (R_T) of this combination, it is necessary to follow the current from the negative side of the supply voltage, through the resistor network and back to the power supply.

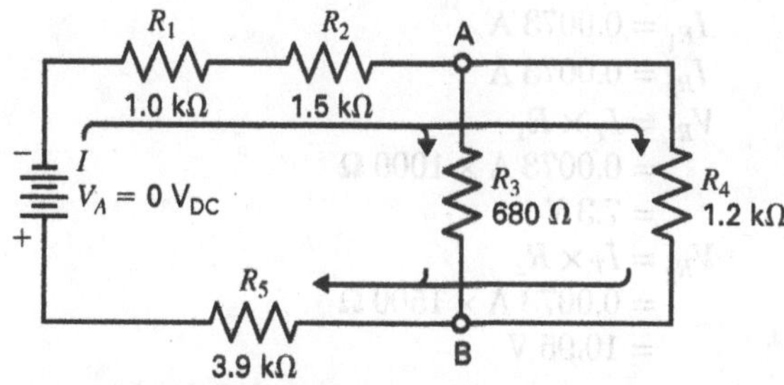

Fig. 6-1.3 Analysis of a series-parallel circuit.

When this series-parallel circuit is analyzed in this fashion, it is seen that all the current is passing through both R_1 and R_2, and because of this, R_1 and R_2 are seen in series with each other. From this series arrangement, the current then divides at junction point A. At this point, some of the current will pass through R_3, and the remainder through R_4. The amount of current that will pass will depend upon Ohm's law, and because at this junction point there is current division, resistors R_3 and R_4 are in parallel with each other. The divided currents will reunite at junction joint B and pass through R_5. In this case, all the current will pass through resistor R_5, and because of this fact, R_5 is seen to be in series with the parallel combination of R_3 and R_4, and then in series with R_1 and R_2. Expressed as a formula, total resistance R_T is

$$R_T = R_1 + R_2 + R_3 \parallel R_4 + R_5$$
$$= 1\text{ k}\Omega + 1.5\text{ k}\Omega + \frac{1}{1/680\ \Omega + 1/1.2\text{ k}\Omega} + 3.9\text{ k}\Omega$$
$$= 1\text{ k}\Omega + 1.5\text{ k}\Omega + 434\ \Omega + 3.9\text{ k}\Omega$$
$$= 6834\ \Omega$$

Total Current: The total current can be calculated if the level of the supply voltage is known. If the power supply in Fig. 6-1.3 were adjusted to 50 V DC, the current could be calculated by Ohm's law to be approximately 7.3 mA. There is 7.3 mA of main-line current present, as determined by

$$R_T = 6834\ \Omega$$
$$V_T = 50\text{ V DC}$$
$$I_T = \frac{V_T}{R_T}$$
$$= \frac{50\text{ V}}{6834\ \Omega}$$
$$\approx 0.0073\text{ A} = 7.3\text{ mA}$$

Current and Voltage Calculations: Now that the main-line current is known, it is seen that it passes through R_1 and R_2. The current passing through these resistors creates a voltage drop of 7.3 and 10.95 V, respectively. This is determined as follows:

$$I_T = 7.3\text{ mA} = 0.0073\text{ A}$$
$$I_{R_1} = 0.0073\text{ A}$$
$$I_{R_2} = 0.0073\text{ A}$$
$$V_{R_1} = I_T \times R_1$$
$$= 0.0073\text{ A} \times 1000\ \Omega$$
$$= 7.3\text{ V}$$
$$V_{R_2} = I_T \times R_2$$
$$= 0.0073\text{ A} \times 1500\ \Omega$$
$$= 10.95\text{ V}$$

When the 7.3 mA enters junction A, the current divides through R_4 and R_5. Since R_4 and R_5 are in parallel, the voltage drop is the same across each of these parallel resistors. From the above, the equivalent resistance R_{eq} of R_3 and R_4 was determined to be 434 Ω. The total main-line current of 7.3 mA passes through this equivalent resistance and creates a voltage drop of 3.17 V. Again, since R_3 and R_4 are in parallel, the voltage drop across R_3 is 3.17 V and the voltage drop across R_4 is the same (3.17 V). These calculations are as follows:

$$R_{eq(R_3 \| R_4)} = 434\ \Omega$$
$$I_T = 7.3\text{ mA} = 0.0073\text{ A}$$
$$V_{eq(R_3 \| R_4)} = I_T \times R_{eq(R_3 \| R_4)}$$
$$= 0.0073\text{ A} \times 434\ \Omega$$
$$= 3.1628\text{ V} \cong 3.17\text{ V}$$
$$V_{R_3} \cong 3.17\text{ V}$$
$$V_{R_4} \cong 3.17\text{ V}$$

$$I_3 = \frac{V_{R_3}}{R_3}$$
$$= \frac{3.17\text{ V}}{680\ \Omega}$$
$$= 0.0047\text{ A} \cong 4.7\text{ mA}$$
$$I_4 = \frac{V_{R_4}}{R_4}$$
$$= \frac{3.17\text{ V}}{1.2\text{ k}\Omega}$$
$$= 0.0026\text{ A} \cong 2.6\text{ mA}$$

At junction B the two currents from R_3 and R_4 are reunited, and this 7.3 mA passes through R_5. The voltage drop on R_5 is 28.6 V. This is determined as follows:

$$I_3 + I_4 = 0.0047\text{ A} + 0.0026\text{ A}$$
$$I_5 = 0.0073\text{ A} = 7.3\text{ mA}$$
$$V_{R_5} = I_5 \times R_5$$
$$= 0.0073\text{ A} \times 3900\ \Omega$$
$$= 28.47\text{ V}$$

Note that

$$V_T \cong V_{R_1} + V_{R_2} + V_{eq(R_3 \| R_4)} + V_{R_5}$$
$$50\text{ V} \cong 50\text{ V}$$

Power demands of each component can now also be determined, since the resistance values, voltage drops, and currents are all known. The wattage dissipation needs of the components are determined as follows:

$$P_{R_1} = I_1 \times V_{R_1}$$
$$= 0.0073\text{ A} \times 7.3\text{ V}$$
$$= 0.0533\text{ W} = 53.3\text{ mW}$$
$$P_{R_2} = I_2 \times V_{R_2}$$
$$= 0.0073\text{ A} \times 10.95\text{ V}$$
$$= 0.0799\text{ W} = 79.9\text{ mW}$$
$$P_{R_3} = I_3 \times V_{R_3}$$
$$= 0.0047\text{ A} \times 3.17\text{ V}$$
$$= 0.0149\text{ W} = 14.9\text{ mW}$$
$$P_{R_4} = I_4 \times V_{R_4}$$
$$= 0.0026\text{ A} \times 3.17\text{ V}$$
$$= 0.0082\text{ W} = 8.2\text{ mW}$$
$$P_{R_5} = I_5 \times V_{R_5}$$
$$= 0.0073\text{ A} \times 28.47\text{ V}$$
$$= 0.2080\text{ W} = 208\text{ mW}$$
$$P_T = P_1 + P_2 + P_3 + P_4 + P_5$$
$$= 0.0533\text{ W} + 0.0799\text{ W} + 0.0149\text{ W} + 0.0082\text{ W} + 0.2078\text{ W}$$
$$= 0.3641\text{ W} = 364.1\text{ mW}$$

Review Fig. 6-1.4 to determine all current paths and resistor voltage drops as calculated above.

In Summary: There can be any number of parallel strings and more than two series resistances in a string. Still, Ohm's law can be used in the same way for

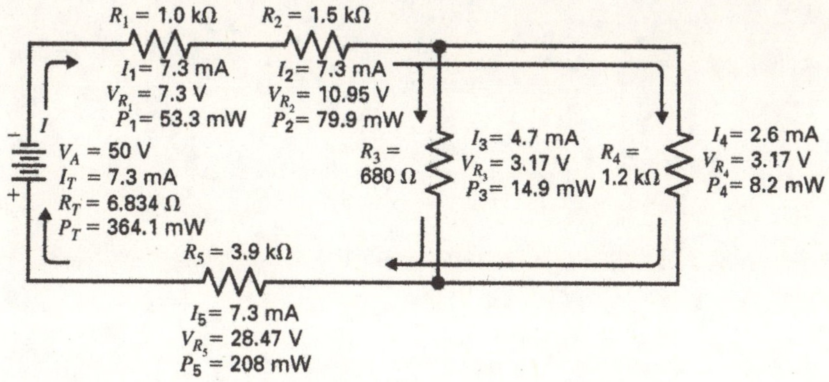

Fig. 6-1.4 Voltage drops, currents, and power dissipations of a series-parallel circuit.

the series and parallel parts of the circuit. The series parts have the same current; the parallel parts have the same voltage. Remember that to find the total current V/R, the total resistance must include all the resistances present across the two terminals of applied voltage.

EQUIPMENT

Breadboard
DC power supply
DMM

COMPONENTS

All resistors are 0.25 W unless indicated otherwise:

(1) 100 Ω
(1) 150 Ω
(1) 220 Ω
(1) 390 Ω
(1) 560 Ω

PROCEDURE

1. Measure and record each resistor value for the resistors required in this experiment. Record the results in Table 6-1.1.

2. Connect the circuit shown in Fig. 6-1.5. First calculate and then measure the total resistance from point A to point B. Record this information in Table 6-1.2. In addition, record the percentage of error between these values, where

$$\% \text{ error} = \left| \frac{\text{difference between meas. and calc. values}}{\text{calc. values}} \right| \times 100$$

If the error is greater than 10 percent, repeat the calculations and measurements.

3. Connect the circuit shown in Fig. 6-1.6. Perform the necessary calculations as required in Table 6-1.2.

4. For the circuit shown in Fig. 6-1.6, adjust the supply voltage to 15 V, and, using a voltmeter, take voltage measurements as required in Table 6-1.2.

5. Again, for the circuit shown in Fig. 6-1.6, and with the power supply adjusted to 15 V, use an ammeter and, by breaking the current path at the identified points, measure the current and record your results in Table 6-1.2.

6. Using the formula described in step 2, calculate the percentages of error between your calculations and voltage and current measurements, as required in Table 6-1.2.

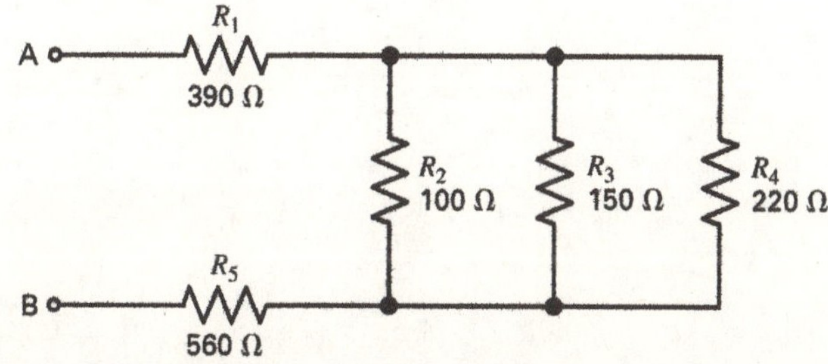

Fig. 6-1.5 Finding the total resistance.

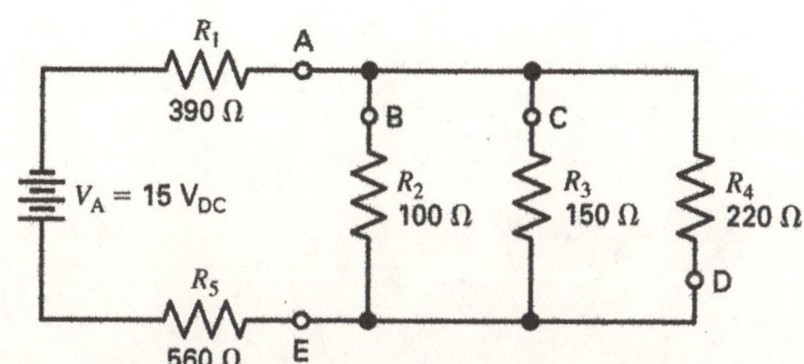

Fig. 6-1.6 Series-parallel circuit under analysis.

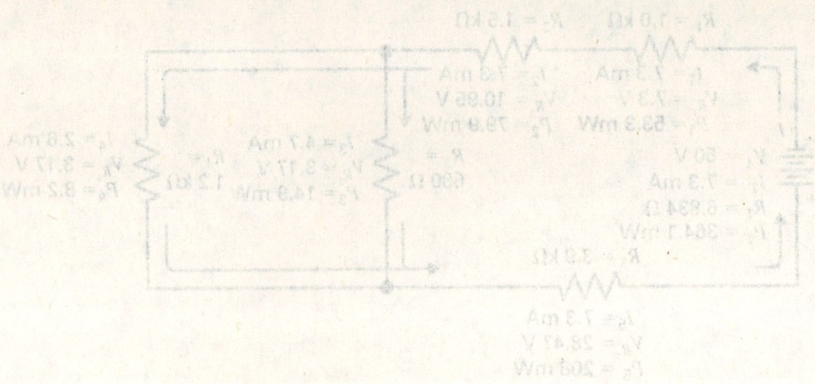

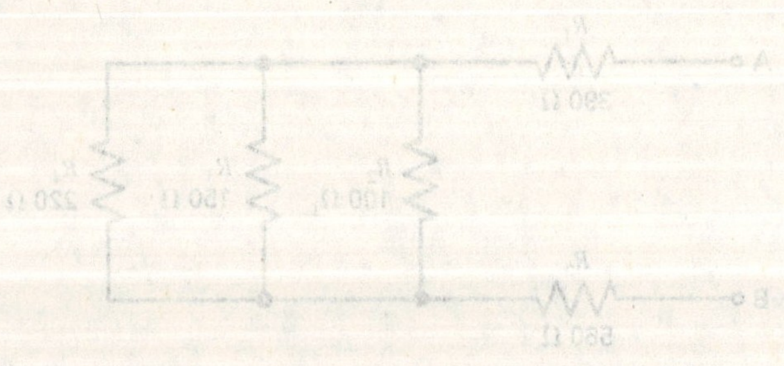

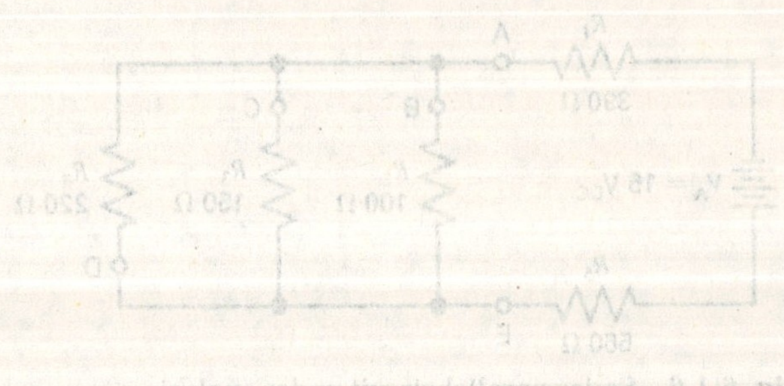

NAME DATE

QUESTIONS FOR EXPERIMENT 6-1

1. Write the formula that correctly determines the total resistance of the circuit in Fig. 6-1.6.

2. With reference to Fig. 6-1.6, which resistors are parallel to, or in series with, R_3? Why?

3. If in Fig. 6-1.1 the voltage across R_1 was determined to be 20 V, what would the circuit supply voltage be? Explain.

4. If in Fig. 6-1.4 the current measured through R_5 was 20 mA, what would be the current through R_1? Why?

5. Write the formula that would give the equivalent resistance of five 22-kΩ resistors connected in parallel. Applying your formula, determine this resistive value.

CRITICAL THINKING QUESTIONS

Note: **The following questions are designed to help you analyze the previous laboratory experiment in a complete and in-depth fashion. To answer these questions, you should review the related material in Grob/Schultz, *Basic Electronics*, twelfth edition.**

1. In analyzing series-parallel circuits, it is believed that with a parallel string across the main line, the branch currents and the total current can be found without knowing the total resistance. Explain how this principle can be true.

2. In analyzing series-parallel circuits, it is believed that when parallel strings have series resistances in the main line, the total resistance must be calculated to find the total current, assuming no branch currents are known. Explain how this principle can be true.

3. In analyzing series-parallel circuits, if the source voltage is applied across the total resistance of the entire circuit, a total current will be produced that will be found in the main line. Explain how this principle can be true.

4. In analyzing series-parallel circuits, it is believed that any individual series resistance has its own voltage drop that is less than the total applied voltage. Explain how this principle can be true.

5. In analyzing series-parallel circuits, it is believed that any individual branch current must be less than the total main-line current. Explain how this principle can be true.

TABLES FOR EXPERIMENT 6-1

TABLE 6-1.1 Individual Resistor Values

	Nominal Resistance	Measured Value
R_1	390 Ω	
R_2	100 Ω	
R_3	150 Ω	
R_4	220 Ω	
R_5	560 Ω	

TABLE 6-1.2 Measurements and Calculations for Resistance, Voltage, and Current

Step		Calculated	Measured	% Error (Step 6)
Step 2	R_T			
Step 4	V_{R_1}			
	V_{R_2}			
	V_{R_3}			
	V_{R_4}			
	V_{R_5}			
Step 5	I_1 (Point A)			
	I_2 (Point B)			
	I_3 (Point C)			
	I_4 (Point D)			
	I_5 (Point E)			

EXPERIMENT RESULTS REPORT FORM

Experiment No.: ______________________ Name: ______________________
Date: ______________________
Experiment Title: ______________________ Class: ______________________
______________________ Instr: ______________________

Explain the purpose of the experiment:

List the first Learning Objective:
OBJECTIVE 1:

After reviewing the results, describe how the objective was validated by this experiment.

List the second Learning Objective:
OBJECTIVE 2:

After reviewing the results, describe how the objective was validated by this experiment.

List the third Learning Objective:
OBJECTIVE 3:

After reviewing the results, describe how the objective was validated by this experiment.

Conclusion:

If required, attach to this form: ☐ Answers to Questions, ☐ Tables, and ☐ Graphs.

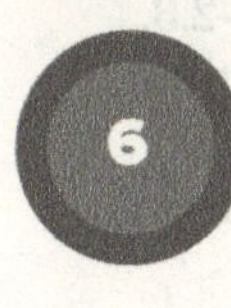

CHAPTER 6

EXPERIMENT 6-2

SERIES-PARALLEL CIRCUITS—NETWORKS

LEARNING OBJECTIVES

At the completion of this experiment, you will be able to:

- Identify a series-parallel circuit.
- Accurately measure voltages and current present in a series-parallel circuit.
- Verify measurements with calculated values.

SUGGESTED READING

Chapter 6, *Basic Electronics*, Grob/Schultz, twelfth edition

INTRODUCTION

Consider the circuit shown in Fig. 6-2.1. Resistors connected together as shown are often called a *resistor network*. In this network, resistors R_2 and R_3 are considered to be in a parallel arrangement. Also, this parallel arrangement is in series with R_1. To calculate the total resistance R_T, the equivalent resistance of R_2 and R_3 must be determined, where

$$R_{\text{eq}} = \frac{R_2 \times R_3}{R_2 + R_3}$$

$$= 150\ \Omega\ \text{(approx.)}$$

Therefore, the equivalent resistance of this parallel arrangement is approximately equal to 150 Ω.

The total resistance is now easily found by adding the two resistances R_1 and R_{eq}, where

$$R_T = R_1 + R_{\text{eq}}$$

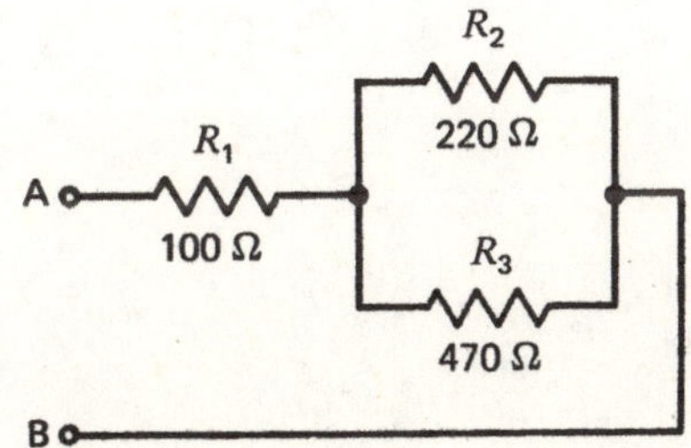

Fig. 6-2.1 Series-parallel circuit.

EQUIPMENT

Ohmmeter	Protoboard or springboard
DC power supply	Leads
DMM	

COMPONENTS

Resistors (all 0.25 W unless indicated otherwise):

(2) 100 Ω	(2) 150 Ω
(1) 120 Ω	(1) 220 Ω
(1) 150 Ω/1 W	(1) 470 Ω

PROCEDURE

1. Measure and record the actual resistance values shown in Table 6-2.1.

2. Connect the circuit shown in Fig. 6-2.2, where $V = 10$ V, $R_1 = 100\ \Omega$, $R_2 = 220\ \Omega$, and $R_3 = 470\ \Omega$.

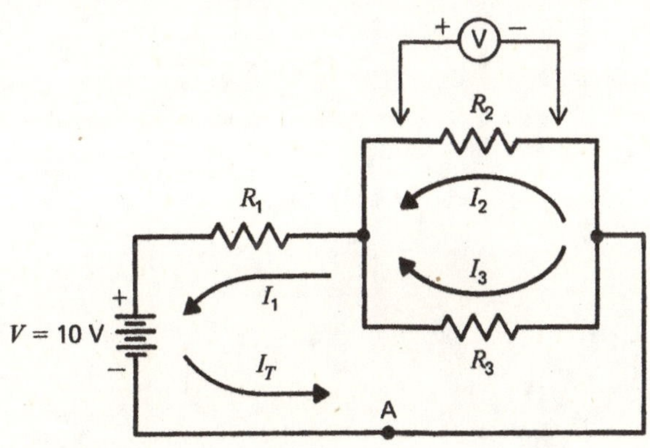

Fig. 6-2.2 Series-parallel circuit.

3. Calculate, measure, and record V_1, V_2, V_3, I_1, I_2, I_3, and I_T (at point A) in Table 6-2.2.

4. Measure and record in Table 6-2.2 the total applied voltage.

5. With the voltage source removed, measure, calculate, and record R_T in Table 6-2.2.

6. Calculate the percentage of error for Table 6-2.2.

7. Construct the circuit shown in Fig. 6-2.3, where

$V_T = 10$ V	$R_3 = 100\ \Omega$
$R_1 = 150\ \Omega/1$ W	$R_4 = 120\ \Omega$
$R_2 = 100\ \Omega$	$R_5 = 150\ \Omega$

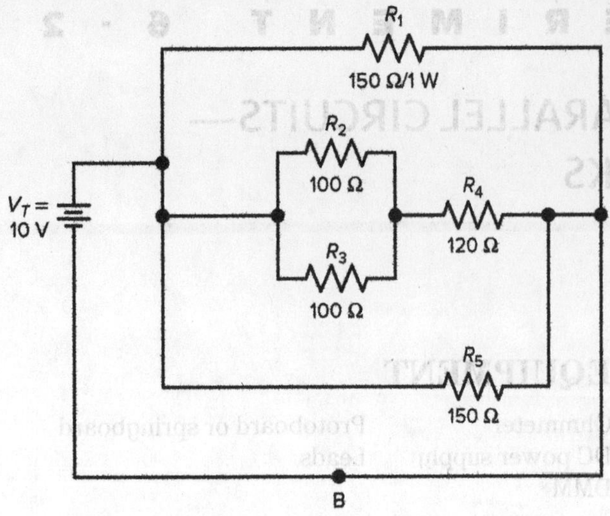

Fig. 6-2.3 Series-parallel circuit.

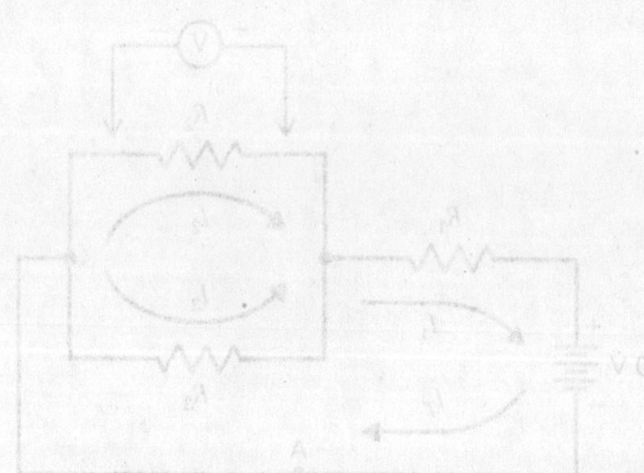

8. Calculate, measure, and record V_1, V_2, V_3, V_4, V_5, I_1, I_2, I_3, I_4, I_5, and I_T (at point B) in Table 6-2.3.
9. Measure and record in Table 6-2.3 the total applied voltage V_T.
10. With the voltage source removed, measure, calculate, and record R_T in Table 6-2.3.
11. Calculate the percentage of error for Table 6-2.3.

NAME DATE

QUESTIONS FOR EXPERIMENT 6-2

1. In your own words, explain what is a series-parallel circuit.

2. In the circuit shown in Fig. 6-2.2, determine the power being dissipated by each resistor. Use the formula $P/I \cdot V =$ power (watts).

3. In Fig. 6-2.2, are the voltages the same across each resistor?

4. In Fig. 6-2.2, are the currents the same through each resistor?

5. Suppose in Fig. 6-2.3 that R_3 developed a short-circuited condition. How would the current flowing through each resistor change? Would the voltage drop across each resistor change? How?

TABLES FOR EXPERIMENT 6-2

TABLE 6-2.1 (values for Fig. 6-2.3)

Nominal Resistance	Measured Resistance
$R_1 = 150\ \Omega$	________
$R_2 = 100\ \Omega$	________
$R_3 = 100\ \Omega$	________
$R_4 = 120\ \Omega$	________
$R_5 = 150\ \Omega$	________

TABLE 6-2.2

	Calculated	Measured	% Error
V_1			
V_2			
V_3			
V_T			
I_1			
I_2			
I_3			
I_T			
R_T			

TABLE 6-2.3

	Calculated	Measured	% Error
V_1			
V_2			
V_3			
V_4			
V_5			
V_T			
I_1			
I_2			
I_3			
I_4			
I_5			
I_T			
R_T			

EXPERIMENT RESULTS REPORT FORM

Experiment No.: ______________________ Name: ______________________

Date: ______________________

Experiment Title: ______________________ Class: ______________________

______________________ Instr: ______________________

Explain the purpose of the experiment:

List the first Learning Objective:
OBJECTIVE 1:

After reviewing the results, describe how the objective was validated by this experiment.

List the second Learning Objective:
OBJECTIVE 2:

After reviewing the results, describe how the objective was validated by this experiment.

List the third Learning Objective:
OBJECTIVE 3:

After reviewing the results, describe how the objective was validated by this experiment.

Conclusion:

If required, attach to this form: ☐ Answers to Questions, ☐ Tables, and ☐ Graphs.

CHAPTER 6

EXPERIMENT 6-3

SERIES-PARALLEL CIRCUITS—ANALYSIS

LEARNING OBJECTIVES

At the completion of this experiment, you will be able to:

- Recognize the basic characteristics of series-parallel resistive banks that are in series with other components.
- Determine that within a series-parallel circuit main-line currents are the same through series resistances.
- Confirm that voltage drops across parallel resistive components in a series-parallel circuit are equivalent.

SUGGESTED READING

Chapter 6, *Basic Electronics*, Grob/Schultz, twelfth edition

INTRODUCTION

Another way to analyze complex series-parallel circuits is to consider resistance banks in series with other resistors. Figure 6-3.1 depicts such a circuit. Here R_1 is in series with a parallel combination of R_2 and R_3. From this point R_4 is seen to be in a series relationship with the parallel combination of R_5 and R_6, leaving R_7 as a series component. If the main-line current from the supply voltage is followed throughout this circuit, we can see that currents are equivalent through series components and that the voltage drops across parallel components are also equivalent. This is demonstrated by the following calculations:

$$\begin{aligned}R_T &= R_1 + R_2 \parallel R_3 + R_4 + R_5 \parallel R_6 + R_7\\ &= 1\text{ k}\Omega + 5.6\text{ k}\Omega \parallel 6.8\text{ k}\Omega + 3.9\text{ k}\Omega + 10\text{ k}\Omega \parallel 2.4\text{ k}\Omega + 2.2\text{ k}\Omega\\ &= 1\text{ k}\Omega + 3070.97\ \Omega + 3.9\text{ k}\Omega + 1935.48\ \Omega + 2.2\text{ k}\Omega\\ &= 12{,}106.45\ \Omega\end{aligned}$$

Figure 6-3.2 shows the equivalent total resistance for Fig. 6-3.1. This total resistance, with an applied voltage of 100 V DC, will create a current of 8.3 mA, as shown below:

$$\begin{aligned}I_T &= V_A/R_T\\ &= 80\text{ V}/12{,}106\ \Omega\\ &= 0.0033\text{ A} = 8.3\text{ mA}\end{aligned}$$

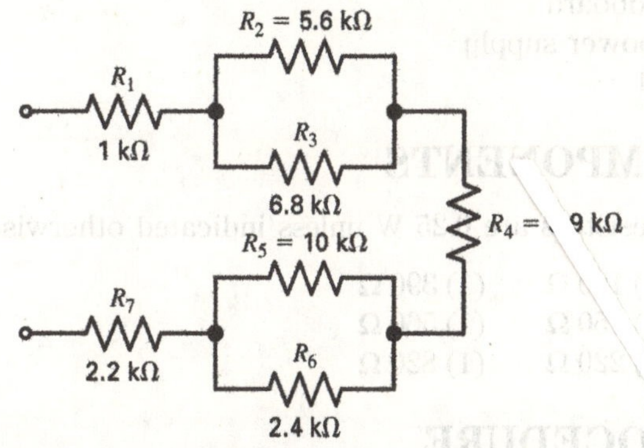

Fig. 6-3.1 Series-parallel circuit for analysis.

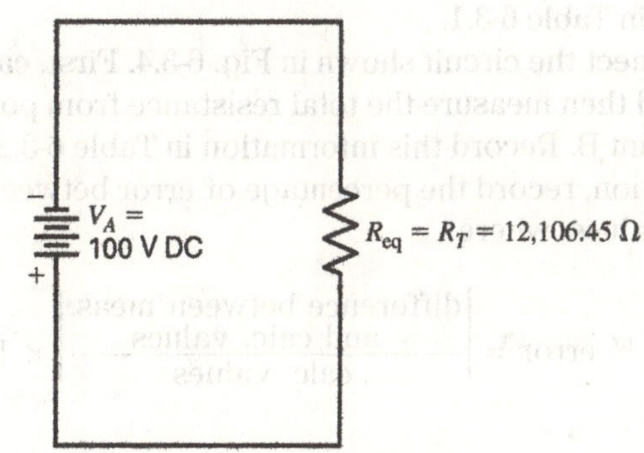

Fig. 6-3.2 The equivalent total resistance.

The values from all calculations for the circuit shown in Fig. 6-3.1 are shown in Fig. 6-3.3.

In summary, there can be more than two parallel resistances in a bank and any number of banks in series. Still, Ohm's law can be applied in the same

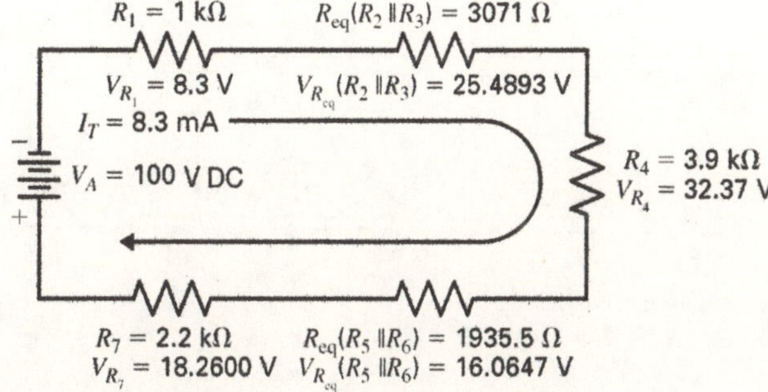

Fig. 6-3.3 Determining the equivalent resistance of a series-parallel circuit.

way to the series and parallel parts of the circuit. The general procedures for circuits of this type is to find the equivalent resistance of each bank and then add all the series resistances.

Again, as in the previous experiment, the most important facts to know are which components are in series with each other and which parts of the circuit are in parallel.

EQUIPMENT

Breadboard
DC power supply
DMM

COMPONENTS

All resistors are 0.25 W unless indicated otherwise:

(1) 100 Ω	(1) 390 Ω
(1) 150 Ω	(1) 560 Ω
(1) 220 Ω	(1) 820 Ω

PROCEDURE

1. Measure and record each resistor value for the resistors required in this experiment. Record the results in Table 6-3.1.

2. Connect the circuit shown in Fig. 6-3.4. First, calculate and then measure the total resistance from point A to point B. Record this information in Table 6-3.2. In addition, record the percentage of error between these values, where

$$\% \text{ error} = \left| \frac{\text{difference between meas. and calc. values}}{\text{calc. values}} \right| \times 100$$

If the error is greater than 10 percent, repeat the calculations and measurements.

3. Connect the circuit shown in Fig. 6-3.5. Adjust the supply voltage to 15 V. Using a voltmeter, take the required measurements to complete Table 6-3.3.

4. With 15 V applied to the circuit, as shown in Fig. 6-3.5, determine with an ammeter the current flowing at each identified point and record this measurement in Table 6-3.3.

5. Perform all the calculations required to complete Table 6-3.3.

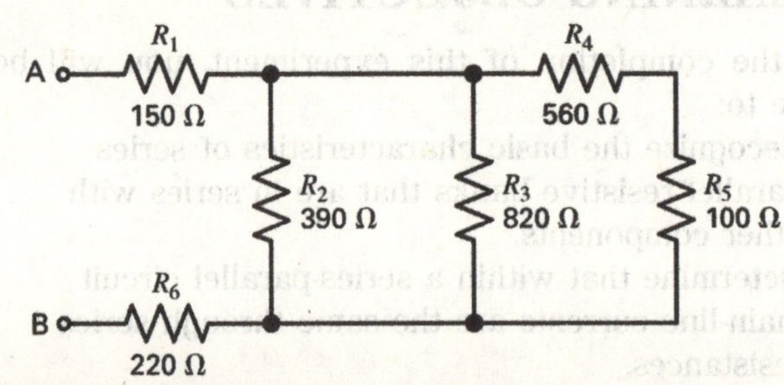

Fig. 6-3.4 Determining the equivalent resistance of the complex circuit.

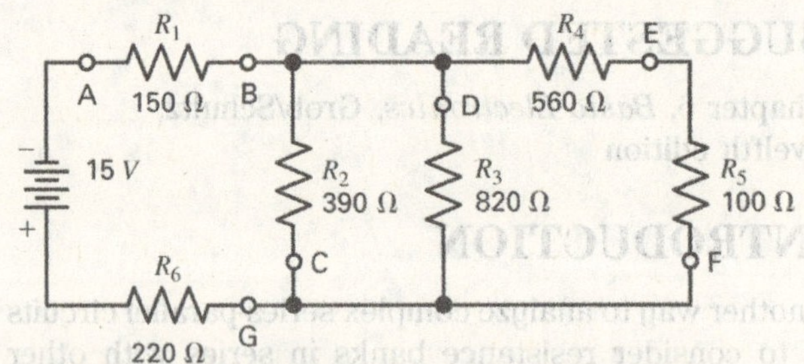

Fig. 6-3.5 Complex series-parallel circuit analysis.

NAME DATE

QUESTIONS FOR EXPERIMENT 6-3

1. Draw a diagram showing three resistors in a bank that is in series with three resistors.

2. From Fig. 6-3.4, describe which resistors are in a parallel circuit arrangement.

3. From your readings in Grob/Schultz, *Basic Electronics*, describe three of the six characteristics of a parallel resistor circuit.

4. From your readings in Grob/Schultz, *Basic Electronics*, describe three of the five characteristics of a series resistor circuit.

5. In Fig. 6-3.5, explain how you would know which resistors are in parallel with each other and which are in series.

CRITICAL THINKING QUESTIONS

Note: The following questions are designed to help you analyze the previous laboratory experiment in a complete and in-depth fashion. To answer these questions, you should review the related material in Grob/Schultz, *Basic Electronics*, twelfth edition.

1. Determine all circuit calculations (voltage drops and currents) for the circuit shown in Fig. 6-3.5, where the applied voltage is changed to 15 V DC. Determine the power dissipation required for each resistor. If the voltage were increased to 25 V DC, would the 0.25-W power rating of the resistors recommended for use in this experiment be adequate?

2. Determine the total resistance R_T for Fig. 6-3.3. Show all calculations.

3. Determine the power dissipation required for each resistor shown in Fig. 6-3.3. Show all calculations.

4. Determine the total resistance R_T for Fig. 6-3.5. Show all calculations.

5. Determine the power dissipation required for each resistor shown in Fig. 6-3.5. Show all calculations.

TABLES FOR EXPERIMENT 6-3

TABLE 6-3.1 Individual Resistor Values

		Nominal Values	Measured Values
Step 1	R_1	150 Ω	______
	R_2	390 Ω	______
	R_3	820 Ω	______
	R_4	560 Ω	______
	R_5	100 Ω	______
	R_6	220 Ω	______

TABLE 6-3.2 Total Resistance R_T and Percentage of Error

		Measured	Calculated	% Error
Step 2	R_T	______	______	______

TABLE 6-3.3 Voltage, Current Measurements, and Percentage of Error

Step		Measured	Calculated	% Error
Step 3	V_{R_1}			
	V_{R_2}			
	V_{R_3}			
	V_{R_4}			
	V_{R_5}			
	V_{R_6}			
Step 4	I_T (Point A)			
	I_{R_1} (Point B)			
	I_{R_2} (Point C)			
	I_{R_3} (Point D)			
	I_{R_4} (Point E)			
	I_{R_5} (Point F)			
	I_{R_6} (Point G)			
				Step 5

EXPERIMENT RESULTS REPORT FORM

Experiment No.: ______________________ Name: ______________________

Date: ______________________

Experiment Title: ______________________ Class: ______________________

______________________ Instr: ______________________

Explain the purpose of the experiment:

List the first Learning Objective:
OBJECTIVE 1:

After reviewing the results, describe how the objective was validated by this experiment.

List the second Learning Objective:
OBJECTIVE 2:

After reviewing the results, describe how the objective was validated by this experiment.

List the third Learning Objective:
OBJECTIVE 3:

After reviewing the results, describe how the objective was validated by this experiment.

Conclusion:

If required, attach to this form: ☐ Answers to Questions, ☐ Tables, and ☐ Graphs.

CHAPTER 6

EXPERIMENT 6-4

SERIES-PARALLEL CIRCUITS—OPENS AND SHORTS

LEARNING OBJECTIVES

At the completion of this experiment, you will be able to:

- Confirm the basic characteristics of series-parallel resistive circuits.
- Identify the characteristics and effects of opens in a series-parallel resistive circuit.
- Identify the characteristics and effects of shorts in a series-parallel resistive circuit.

SUGGESTED READING

Chapter 6, *Basic Electronics*, Grob/Schultz, twelfth edition

INTRODUCTION

The purpose of this experiment is to learn about the characteristics of electrical opens and shorts and how they affect currents and voltages. In this experiment we will see how a series-parallel circuit is affected.

The Short Circuit: A short circuit has practically zero resistance. Therefore, it allows excessive current to flow. An open circuit has the opposite effect, because it has infinitely high resistance with practically zero current. Furthermore, in series-parallel circuits, an open or short circuit in one path changes the circuit for the other paths. For example, in Fig. 6-4.1, the series-parallel circuit becomes a series circuit with only R_1 when there is a short circuit between points A and B.

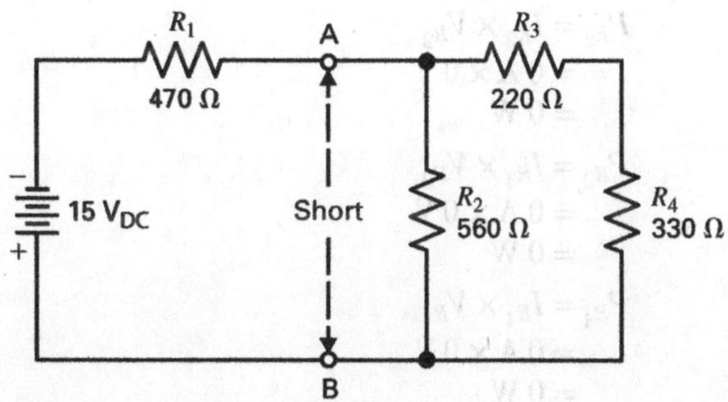

Fig. 6-4.1 Analysis of the short circuit.

Studying the Circuit without a Short: If the circuit shown in Fig. 6-4.1 does not have a short, R_T can be calculated so that R_3 and R_4 are in series and have an equivalent resistance of 550 Ω. This 550-Ω equivalent resistance is in parallel with 560-Ω R_2 and creates an equivalent resistance of 277.48 Ω. This value is now in series with the 470 Ω value of R_1 and provides an overall total equivalent resistance of 747.48 Ω.

Since R_T is known, the main-line current can be determined from Ohm's law as follows:

$$I_T = V_T/R_T$$
$$= 15\text{ V DC}/747.48\ \Omega$$
$$= 0.0201\text{ A} = 20.1\text{ mA}$$

The total power of the circuit can be calculated as follows:

$$P_T = I_T \times V_T$$
$$= 0.0201\text{ A} \times 15\text{ V}$$
$$= 0.3010\text{ W} = 301\text{ mW}$$

Since the total main-line current I_T is known, the associated voltage drops across each of the resistors can be determined. For example,

$$V_{R_1} = I_T \times R_1$$
$$= 0.0201\text{ A} \times 470\ \Omega$$
$$= 9.45\text{ V}$$

Since the applied voltage is 15 V and 9.45 V was dropped across R_1, V_{R2} can be determined to be 5.55 V, as follows:

$$V_{R_2} = V_T - V_{R_1}$$
$$= 15\text{ V} - 9.45\text{ V}$$
$$= 5.55\text{ V}$$

Since 5.55 V is dropped across R_2, the current through R_2 can be determined as follows:

$$I_{R_2} = V_{R_2}/R_2$$
$$= 5.55\text{ V}/560\ \Omega$$
$$= 0.0099\text{ A} = 9.9\text{ mA}$$

As you know, currents divide in parallel branches. The main-line current of 20.1 mA divides between the R_2 branch and the $R_3 + R_4$ branch. If 9.9 mA passes through R_2, and the total current I_T is 20.1 mA, the difference of 10.2 mA will pass through the series combination of the $R_3 + R_4$ branch. Since R_3 and R_4 are in series, this 10.2 mA of current will pass through both series resistors, as follows:

$$I_{R_3} = I_{R_4} = I_T - I_{R_2}$$
$$= 0.02010 \text{ A} - 0.0099 \text{ A}$$
$$= 0.0102 \text{ A} = 10.2 \text{ mA}$$

If the current through R_3 and R_4 is known to be 10.2 mA, the voltage drop across each can be determined:

$$V_{R_3} = I_{R_3} \times R_3$$
$$= 0.0102 \text{ A} \times 220\ \Omega$$
$$= 2.24 \text{ V}$$

and

$$V_{R_4} = I_{R_4} \times R_4$$
$$= 0.0102 \text{ A} \times 330\ \Omega$$
$$= 3.36 \text{ V}$$

Note that the sum of V_{R_3} and V_{R_4} is equivalent to V_{R_2}. This is to be anticipated, since these resistances are in a parallel branch arrangement, and voltage drops are found to be equivalent in parallel branches.

$$V_{R_3} + V_{R_4} = V_{R_2}$$
$$2.24 \text{ V} + 3.36 \text{ V} = 5.60 \text{ V}$$
(approximately equal to 5.55 V)

Power in the circuit can be determined as follows:

$$P_{R_1} = I_{R_1} \times V_{R_1}$$
$$= 0.0201 \text{ A} \times 9.45 \text{ V}$$
$$= 0.1899 \text{ W} = 189.9 \text{ mW}$$

$$P_{R_2} = I_{R_2} \times V_{R_2}$$
$$= 0.0099 \text{ A} \times 5.55 \text{ V}$$
$$= 0.0549 \text{ W} = 54.9 \text{ mW}$$

$$P_{R_3} = I_{R_3} \times V_{R_3}$$
$$= 0.0102 \text{ A} \times 2.24 \text{ V}$$
$$= 0.0228 \text{ W} = 22.8 \text{ mW}$$

$$P_{R_4} = I_{R_4} \times V_{R_4}$$
$$= 0.0102 \text{ A} \times 3.36 \text{ V}$$
$$= 0.0343 \text{ W} = 34.3 \text{ mW}$$

Studying the Same Circuit with a Short: If the circuit shown in Fig. 6-4.1 now has a short between points A and B, R_T can be calculated so that R_3 and R_4 are *still* in series and have an equivalent resistance of 550 Ω. This 550-Ω equivalent resistance is *still* in parallel with the 560-Ω R_2 value and creates an equivalent resistance of 277.48 Ω. However, this equivalent resistance is in parallel with a short, which displays practically no resistance (0 Ω). In this case, the equivalent resistance of the parallel branches is now 0 Ω. This ohmic value is in series with the 470 Ω value of R_1 and provides an overall total equivalent resistance of 470 Ω.

Since R_T is known, the main-line current can be determined from Ohm's law as follows:

$$I_T = V_T/R_T$$
$$= 15 \text{ V DC}/470\ \Omega$$
$$= 0.0319 \text{ A} = 31.9 \text{ mA}$$

The total power of the circuit can be calculated as follows:

$$P_T = I_T \times V_T$$
$$= 0.0319 \text{ A} \times 15 \text{ V}$$
$$= 0.4787 \text{ W} = 478.7 \text{ mW}$$

Since the total main-line current I_T is known, the associated voltage drops across each of the resistors can be determined. For example,

$$V_{R_1} = I_T \times R_1$$
$$= 0.0319 \text{ A} \times 470\ \Omega$$
$$= 14.99 \text{ V (which in practical terms is 15 V)}$$

Since the applied voltage is 15 V and 15 V was dropped across R_1, V_{R_2} can be determined to be 0 V, as follows:

$$V_{R_2} = V_T - V_{R_1}$$
$$= 15 \text{ V} - 15 \text{ V}$$
$$= 0 \text{ V}$$

Since 0 V is dropped across R_2, the current through R_2 can be determined as follows:

$$I_{R_2} = V_{R_2}/R_2$$
$$= 0 \text{ V}/560\ \Omega$$
$$= 0 \text{ A} = 0 \text{ mA}$$
$$I_{R_3} = 0 \text{ A} = 0 \text{ mA}$$
$$I_{R_4} = 0 \text{ A} = 0 \text{ mA}$$
$$V_{R_3} = I_{R_3} \times R_3$$
$$= 0 \text{ A} \times 220\ \Omega$$
$$= 0 \text{ V}$$
$$V_{R_4} = I_{R_4} \times R_4$$
$$= 0 \text{ A} \times 330\ \Omega$$
$$= 0 \text{ V}$$
$$P_{R_1} = I_{R_1} \times V_{R_1}$$
$$= 0.0319 \text{ A} \times 15 \text{ V}$$
$$= 0.4787 \text{ W} = 478.7 \text{ mW}$$

$$P_{R_2} = I_{R_2} \times V_{R_2}$$
$$= 0 \text{ A} \times 0 \text{ V}$$
$$= 0 \text{ W}$$

$$P_{R_3} = I_{R_3} \times V_{R_3}$$
$$= 0 \text{ A} \times 0 \text{ V}$$
$$= 0 \text{ W}$$

$$P_{R_4} = I_{R_4} \times V_{R_4}$$
$$= 0 \text{ A} \times 0 \text{ V}$$
$$= 0 \text{ W}$$

	Normal	Short Circuit
R_T	747.48 Ω	470 Ω
I_T	20.1 mA	31.9 mA
P_T	301 mW	478.7 mW
V_{R_1}	9.45 V	15 V
V_{R_2}	5.55 V	0 V
I_{R_2}	9.9 mA	0 mA
I_{R_3}	10.2 mA	0 mA
I_{R_4}	10.2 mA	0 mA
V_{R_3}	2.24 V	0 V
V_{R_4}	3.36 V	0 V
P_{R_1}	189.9 mW	478.7 mW
P_{R_2}	54.9 mW	0 mW
P_{R_3}	22.8 mW	0 mW
P_{R_4}	34.3 mW	0 mW

Summary Comparison between the Normal and Short Circuit: When a short was created, the overall resistance of the circuit decreased, main-line current increased, and overall power dissipation increased.

The Open Circuit: The circuit in Fig. 6-4.2 becomes a series circuit with just R_1 and R_2 when there is an open between terminals C and D. Since the component values and supply voltages are the same as in Fig. 6-4.1, we will develop the characteristics of the open in this circuit by concentrating on the component calculated values.

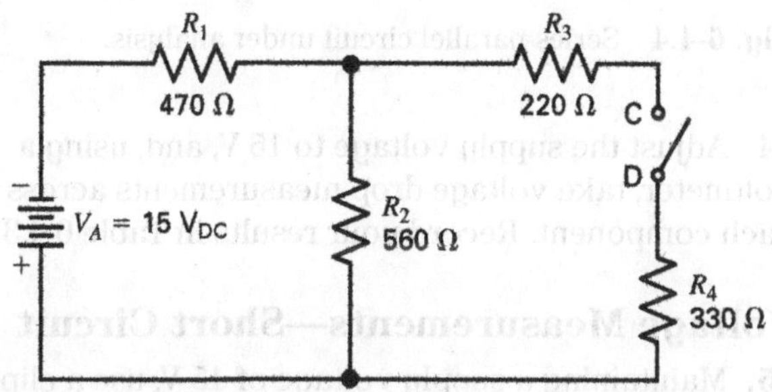

Fig. 6-4.2 Analysis of the open circuit.

When analyzing the effects of this open-circuit problem, remember that, in the normal closed circuit, R_T = 747.48 Ω, I_T = 20.1 mA, P_T = 301 mW, V_{R_1} = 9.45 V, V_{R_2} = 5.55 V, I_{R_2} = 9.9 mA, I_{R_3} = 10.2 mA, I_{R_4} = 10.2 mA, V_{R_3} = 2.24 V, V_{R_4} = 3.36 V, P_{R_1} = 189.9 mW, P_{R_2} = 54.9 mW, P_{R_3} = 22.8 mW, and P_{R_4} = 34.3 mW.

Studying the Effect of an Open Circuit: If the circuit shown in Fig. 6-4.2 is open from points C to D, R_T can be calculated as 560 Ω R_2 in series with the 470 Ω value of R_1. This is a total equivalent resistance of 1030 Ω.

Since R_T is known, the main-line current can be determined from Ohm's law as follows:

$$\begin{aligned} I_T &= V_T/R_T \\ &= 15\text{ V DC}/1030\ \Omega \\ &= 0.0146\text{ A} \approx 14.6\text{ mA} \end{aligned}$$

The total power of the circuit can be calculated as follows:

$$\begin{aligned} P_T &= I_T \times V_T \\ &= 0.0146\text{ A} \times 15\text{ V} \\ &= 0.219\text{ W} = 219\text{ mW} \end{aligned}$$

Since the total main-line current I_T is known, the associated voltage drops across each of the resistors can be determined, for example:

$$\begin{aligned} V_{R_1} &= I_T \times R_1 \\ &= 0.0146\text{ A} \times 470\ \Omega \\ &= 6.8620\text{ V} \end{aligned}$$

Since the applied voltage is 15 V and 6.8620 V was dropped across R_1, V_{R_2} can be determined to be 8.1380 V, as follows:

$$\begin{aligned} V_{R_2} &= V_T - V_{R_1} \\ &= 15\text{ V} - 6.8620\text{ V} \\ &= 8.1380\text{ V} \end{aligned}$$

Since 8.1380 V is dropped across R_2, the current through R_2 can be determined to be

$$\begin{aligned} I_{R_2} &= V_{R_2}/R_2 \\ &= 8.1380\text{ V}/560\ \Omega \\ &= 0.0145\text{ A} = 14.5\text{ mA} \end{aligned}$$

As you know, currents divide in parallel branches. However, the main-line current of 14.6 mA does not divide between the R_2 branch and the $R_3 + R_4$ branch, since the $R_3 + R_4$ branch is open. It would follow therefore that

$$I_{R_3} = I_{R_4} = 0\text{ A}$$

If the current is known to be 0 A through R_3 and R_4, the voltage drop across each can be determined, as follows:

$$\begin{aligned} V_{R_3} &= I_{R_3} \times R_3 \\ &= 0\text{ A} \times 220\ \Omega \\ &= 0\text{ V} \end{aligned}$$

and

$$\begin{aligned} V_{R_4} &= I_{R_4} \times R_4 \\ &= 0\text{ A} \times 330\ \Omega \\ &= 0\text{ V} \end{aligned}$$

Power in the circuit can be determined as follows:

$$\begin{aligned} P_{R_1} &= I_{R_1} \times V_{R_1} \\ &= 0.0146\text{ A} \times 6.8620\text{ V} \\ &= 0.1002\text{ W} = 100.2\text{ mW} \end{aligned}$$

$$P_{R_2} = I_{R_2} \times V_{R_2}$$
$$= 0.0146\ \text{A} \times 8.1380\ \text{V}$$
$$= 0.1188\ \text{W} = 118.8\ \text{mW}$$

$$P_{R_3} = I_{R_3} \times V_{R_3}$$
$$= 0\ \text{A} \times 0\ \text{V}$$
$$= 0\ \text{W}$$

$$P_{R_4} = I_{R_4} \times V_{R_4}$$
$$= 0\ \text{A} \times 0\ \text{V}$$
$$= 0\ \text{W}$$

Summary Comparison between the Normal and Short Circuit: When an open was created, the overall resistance of the circuit increased, main-line current decreased, and overall power dissipation changed markedly.

	Normal	Open Circuit
R_T	747.48 Ω	1030 Ω
I_T	20.1 mA	14.6 mA
P_T	301 mW	218.4 mW
V_{R_1}	9.45 V	6.8620 V
V_{R_2}	5.55 V	8.1380 V
I_{R_2}	9.9 mA	14.5 mA
I_{R_3}	10.2 mA	0 mA
I_{R_4}	10.2 mA	0 mA
V_{R_3}	2.24 V	0 V
V_{R_4}	3.36 V	0 V
P_{R_1}	189.9 mW	100.2 mW
P_{R_2}	54.9 mW	118.8 mW
P_{R_3}	22.8 mW	0 mW
P_{R_4}	34.3 mW	0 mW

EQUIPMENT

Breadboard
DC power supply
DMM

COMPONENTS

All resistors are 0.25 W unless indicated otherwise:

(1) 100 Ω	(1) 560 Ω
(1) 120 Ω	(1) 680 Ω
(1) 470 Ω	(1) 820 Ω

PROCEDURE

1. Measure and record each resistor value for the resistors required in this experiment. Record the results in Table 6-4.1.

2. Connect the circuit shown in Fig. 6-4.3. First, calculate and then measure the total resistance from point A to point B. Then, record this information in

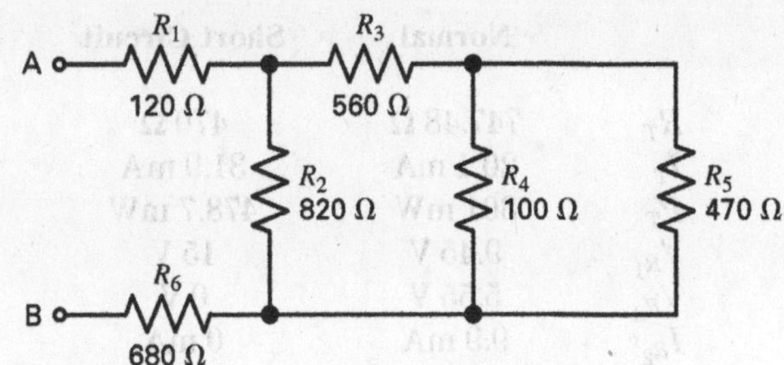

Fig. 6-4.3 Finding the total resistance of a series-parallel circuit.

Table 6-4.2. In addition, record the percentage of error between these values, where

$$\%\ \text{error} = \left| \frac{\text{difference between meas. and calc. values}}{\text{calc. values}} \right| \times 100$$

If the error is greater than 10 percent, repeat the calculations and measurements.

3. Connect the circuit shown in Fig. 6-4.4. Calculate all voltage drops and component currents, and record them in Table 6-4.3.

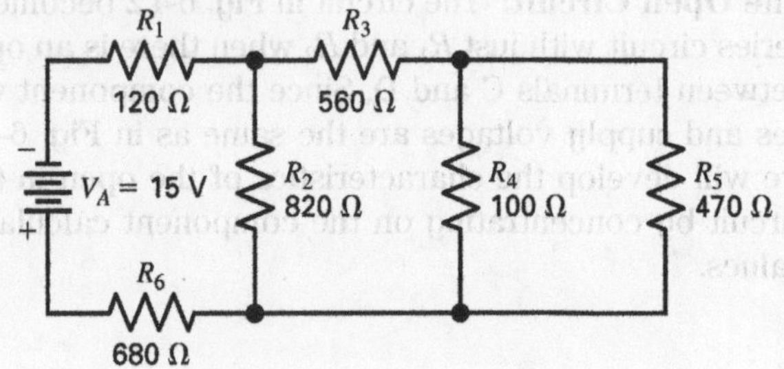

Fig. 6-4.4 Series-parallel circuit under analysis.

4. Adjust the supply voltage to 15 V, and, using a voltmeter, take voltage drop measurements across each component. Record your results in Table 6-4.3.

Voltage Measurements—Short Circuit

5. Maintaining a supply voltage of 15 V, use a clip lead to short R_1. Take voltage drop measurements V_{R_1}, V_{R_2}, V_{R_3}, V_{R_4}, V_{R_5}, and V_{R_6}. Record the results in Table 6-4.3.

6. Remove the clip-lead short across R_1.

7. Maintaining a supply voltage of 15 V, use a clip lead to short R_2. Take voltage drop measurements V_{R_1}, V_{R_2}, V_{R_3}, V_{R_4}, V_{R_5}, and V_{R_6}. Record the results in Table 6-4.3.

8. Remove the clip-lead short across R_2.

9. Maintaining a supply voltage of 15 V, use a clip lead to short R_3. Take voltage drop measurements V_{R_1}, V_{R_2}, V_{R_3}, V_{R_4}, V_{R_5}, and V_{R_6}. Record the results in Table 6-4.3.

10. Remove the clip-lead short across R_3.

11. Maintaining a supply voltage of 15 V, use a clip lead to short R_4. Take voltage drop measurements

V_{R_1}, V_{R_2}, V_{R_3}, V_{R_4}, V_{R_5}, and V_{R_6}. Record the results in Table 6-4.3.

12. Remove the clip-lead short across R_4.

13. Maintaining a supply voltage of 15 V, use a clip lead to short R_5. Take voltage drop measurements V_{R_1}, V_{R_2}, V_{R_3}, V_{R_4}, V_{R_5}, and V_{R_6}. Record the results in Table 6-4.3.

14. Remove the clip-lead short across R_5.

15. Maintaining a supply voltage of 15 V, use a clip lead to short R_6. Take voltage drop measurements V_{R_1}, V_{R_2}, V_{R_3}, V_{R_4}, V_{R_5}, and V_{R_6}. Record the results in Table 6-4.3.

16. Remove the clip-lead short across R_6.

Current Measurements—Short Circuit

17. Maintaining a supply voltage of 15 V, use a clip lead to short R_1 from your circuit. With an ammeter, take current measurements of I_{R_2}, I_{R_3}, I_{R_4}, I_{R_5}, and I_{R_6}. Do not take the current measurement through R_1. Record this information in Table 6-4.3.

18. Remove the clip-lead circuit short across R_1.

19. Maintaining a supply voltage of 15 V, use a clip lead to short R_2 from your circuit. With an ammeter, take current measurements of I_{R_1}, I_{R_3}, I_{R_4}, I_{R_5}, and I_{R_6}. Do not take the current measurement through R_2. Record this information in Table 6-4.3.

20. Remove the clip-lead circuit short across R_2.

21. Maintaining a supply voltage of 15 V, use a clip lead to short R_3 from your circuit. With an ammeter, take current measurements of I_{R_1}, I_{R_2}, I_{R_4}, I_{R_5}, and I_{R_6}. Do not take the current measurement through R_3. Record this information in Table 6-4.3.

22. Remove the clip-lead circuit short across R_3.

23. Maintaining a supply voltage of 15 V, use a clip lead to short R_4 from your circuit. With an ammeter, take current measurements of I_{R_1}, I_{R_2}, I_{R_3}, I_{R_5}, and I_{R_6}. Do not take the current measurement through R_4. Record this information in Table 6-4.3.

24. Remove the clip-lead circuit short across R_4.

25. Maintaining a supply voltage of 15 V, use a clip lead to short R_5 from your circuit. With an ammeter, take current measurements of I_{R_1}, I_{R_2}, I_{R_3}, I_{R_4}, and I_{R_6}. Do not take the current measurement through R_5. Record this information in Table 6-4.3.

26. Remove the clip-lead circuit short across R_5.

27. Maintaining a supply voltage of 15 V, use a clip lead to short R_6 from your circuit. With an ammeter, take current measurements of I_{R_1}, I_{R_2}, I_{R_3}, I_{R_4}, and I_{R_5}. Do not take the current measurement through R_6. Record this information in Table 6-4.3.

28. Remove the clip-lead circuit short across R_6.

Voltage Measurements—Open Circuit

29. Maintaining a supply voltage of 15 V, open R_1 by removing it from the circuit. Take voltage drop measurements V_{R_2}, V_{R_3}, V_{R_4}, V_{R_5}, and V_{R_6}. Do not take the voltage drop measurement across R_1. Record these measurements in Table 6-4.4.

30. Remove the open by replacing R_1 into the circuit.

31. Maintaining a supply voltage of 15 V, open R_2 by removing it from the circuit. Take voltage drop measurements V_{R_1}, V_{R_3}, V_{R_4}, V_{R_5}, and V_{R_6}. Do not take the voltage drop measurement across R_2. Record these measurements in Table 6-4.4.

32. Remove the open by replacing R_2 into the circuit.

33. Maintaining a supply voltage of 15 V, open R_3 by removing it from the circuit. Take voltage drop measurements V_{R_1}, V_{R_2}, V_{R_4}, V_{R_5}, and V_{R_6}. Do not take the voltage drop measurement across R_3. Record these measurements in Table 6-4.4.

34. Remove the open by replacing R_3 into the circuit.

35. Maintaining a supply voltage of 15 V, open R_4 by removing it from the circuit. Take voltage drop measurements V_{R_1}, V_{R_2}, V_{R_3}, V_{R_5}, and V_{R_6}. Do not take the voltage drop measurement across R_4. Record these measurements in Table 6-4.4.

36. Remove the open by replacing R_4 into the circuit.

37. Maintaining a supply voltage of 15 V, open R_5 by removing it from the circuit. Take voltage drop measurements V_{R_1}, V_{R_2}, V_{R_3}, V_{R_4}, and V_{R_6}. Do not take the voltage drop measurement across R_5. Record these measurements in Table 6-4.4.

38. Remove the open by replacing R_5 into the circuit.

39. Maintaining a supply voltage of 15 V, open R_6 by removing it from the circuit. Take voltage drop measurements V_{R_1}, V_{R_2}, V_{R_3}, V_{R_4}, and V_{R_5}. Do not take the voltage drop measurement across R_6. Record these measurements in Table 6-4.4.

40. Remove the open by replacing R_6 into the circuit.

Current Measurements—Open Circuit

41. Maintaining a supply voltage of 15 V, open R_1 by removing it from the circuit. Take current measurements I_{R_2}, I_{R_3}, I_{R_4}, I_{R_5}, and I_{R_6}. Do not take the current measurement through R_1. Record these measurements in Table 6-4.4.

42. Remove the open by replacing R_1 into the circuit.

43. Maintaining a supply voltage of 15 V, open R_2 by removing it from the circuit. Take current measurements I_{R_1}, I_{R_3}, I_{R_4}, I_{R_5}, and I_{R_6}. Do not take the current measurement through R_2. Record these measurements in Table 6-4.4.

44. Remove the open by replacing R_2 into the circuit.

45. Maintaining a supply voltage of 15 V, open R_3 by removing it from the circuit. Take current measurements I_{R_1}, I_{R_2}, I_{R_4}, I_{R_5}, and I_{R_6}. Do not take the current measurement through R_3. Record these measurements in Table 6-4.4.

46. Remove the open by replacing R_3 into the circuit.

47. Maintaining a supply voltage of 15 V, open R_4 by removing it from the circuit. Take current measurements I_{R_1}, I_{R_2}, I_{R_3}, I_{R_5}, and I_{R_6}. Do not take the current measurement through R_4. Record these measurements in Table 6-4.4.

48. Remove the open by replacing R_4 into the circuit.

49. Maintaining a supply voltage of 15 V, open R_5 by removing it from the circuit. Take current measurements I_{R_1}, I_{R_2}, I_{R_3}, I_{R_4}, and I_{R_6}. Do not take the current measurement through R_5. Record these measurements in Table 6-4.4.

50. Remove the open by replacing R_5 into the circuit.

51. Maintaining a supply voltage of 15 V, open R_6 by removing it from the circuit. Take current measurements I_{R_1}, I_{R_2}, I_{R_3}, I_{R_4}, and I_{R_5}. Do not take the current measurement through R_6. Record these measurements in Table 6-4.4.

52. Remove the open by replacing R_6 into the circuit.

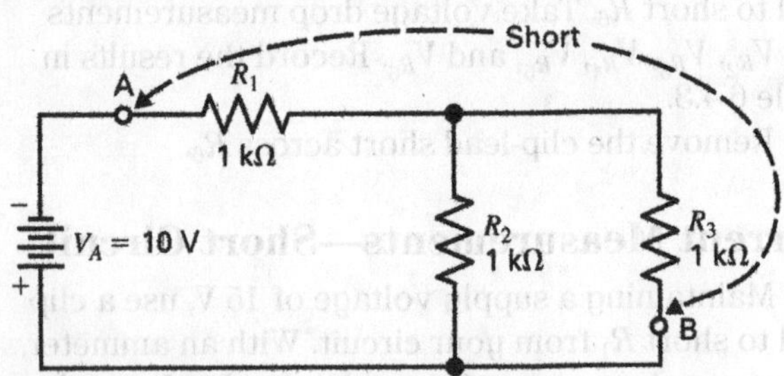

Fig. 6-4.5 Shorted circuit for question 2.

NAME DATE

QUESTIONS FOR EXPERIMENT 6-4

1. What effect would a short across a resistor located in series with a parallel resistive branch have on the level of main-line current flow? Explain.

2. What effect would there be (on individual branch circuit levels) of a short from point A to point B within a circuit (opposite in Fig. 6-4.5)? Explain.

3. If in Fig. 6-4.1, R_2 was shorted, what would happen to the current flow and voltage drop of R_3? What would happen to the voltage and current of R_2? Explain.

4. Calculate the needed level of power dissipation for each resistor shown in Fig. 6-4.4. How does power dissipation change for a shorted component? How would it change across an open component? Explain.

5. According to Grob/Schultz, *Basic Electronics*, twelfth edition, what is the potential difference across an open? Explain.

CRITICAL THINKING QUESTIONS

Note: **The following questions are designed to help you analyze the previous laboratory experiment in a complete and in-depth fashion. To answer these questions, you should review the related material in Grob/Schultz, *Basic Electronics*, twelfth edition.**

1. For the short-circuit current analysis in step 17, the following statement is made: "Maintaining a supply voltage of 15 V, use a clip lead to short R_1 from your circuit. With an ammeter, take current measurements of I_{R_2}, I_{R_3}, I_{R_4}, I_{R_5}, and I_{R_6}. Do not take the current measurement through R_1. Record this information in Table 6-4.3." Why was the current measurement I_{R_1} excluded from the original measurements? What is the anticipated value for I_{R_1}?

2. For the open circuit voltage analysis in step 29, the following statement is made: "Maintaining a supply voltage of 15 V, open R_1 by removing it from the circuit. Take voltage drop measurements V_{R_2}, V_{R_3}, V_{R_4}, V_{R_5}, and V_{R_6}. Do not take the voltage drop measurement across R_1. Record these measurements in Table 6-4.4." Why was the voltage measurement V_{R_1} excluded from the original measurements? What is the anticipated value for V_{R_1}?

3. For the open circuit current analysis in step 41, the following statement is made: "Maintaining a supply voltage of 15 V, open R_1 by removing it from the circuit. Take current measurements I_{R_2}, I_{R_3}, I_{R_4}, I_{R_5}, and I_{R_6}. Do not take the current measurement through R_1. Record these measurements in Table 6-4.4." Why was the current measurement I_{R_1} excluded from the original measurements? What is the anticipated value for I_{R_1}?

TABLES FOR EXPERIMENT 6-4

TABLE 6-4.1 Individual Resistor Values

Resistor	Nominal Value, Ω	Resistive Measurement
R_1	120	______
R_2	820	______
R_3	560	______
R_4	100	______
R_5	470	______
R_6	680	______

TABLE 6-4.2 Total Resistance R_T and Percentage of Error

	Calculated	Measurement	% Error
R_T			

TABLE 6-4.3 Short-Circuit Calculations and Measurements

	Normal Circuit Calculations	Normal Circuit Measurements	Measurements (with Each Resistor *Shorted* One at a Time)					
			R_1	R_2	R_3	R_4	R_5	R_6
V_{R1}								
V_{R_2}								
V_{R_3}								
V_{R_4}								
V_{R_5}								
V_{R_6}								
I_{R_1}								
I_{R_2}								
I_{R_3}								
I_{R_4}								
I_{R_5}								
I_{R_6}								

TABLE 6-4.4 Open-Circuit Calculations and Measurements

	Measurements (with Each Resistor *Opened* One at a Time)					
	R_1	R_2	R_3	R_4	R_5	R_6
V_{R_1}						
V_{R_2}						
V_{R_3}						
V_{R_4}						
V_{R_5}						
V_{R_6}						
I_{R_1}						
I_{R_2}						
I_{R_3}						
I_{R_4}						
I_{R_5}						
I_{R_6}						

EXPERIMENT RESULTS REPORT FORM

Experiment No.: ______________________ Name: ______________________

Date: ______________________

Experiment Title: ______________________ Class: ______________________

______________________ Instr: ______________________

Explain the purpose of the experiment:

List the first Learning Objective:

OBJECTIVE 1:

After reviewing the results, describe how the objective was validated by this experiment.

List the second Learning Objective:

OBJECTIVE 2:

After reviewing the results, describe how the objective was validated by this experiment.

List the third Learning Objective:

OBJECTIVE 3:

After reviewing the results, describe how the objective was validated by this experiment.

Conclusion:

If required, attach to this form: ☐ Answers to Questions, ☐ Tables, and ☐ Graphs.

THE WHEATSTONE BRIDGE

LEARNING OBJECTIVES

At the completion of this experiment, you will be able to:

- Measure current and voltage in a Wheatstone bridge.
- Use a galvanometer.
- Understand how a Wheatstone bridge can be balanced.

SUGGESTED READING

Chapters 5, 6, and 10, *Basic Electronics*, Grob/Schultz, twelfth edition

INTRODUCTION

Many types of bridge circuits are used in electronics. The bridge circuit in this experiment can become a functional ohmmeter as well as a balanced circuit. Here, the Wheatstone bridge has two input terminals, where the battery (or power supply) terminals are connected to the ratio arms. The ratio arms are the key to understanding how the Wheatstone bridge is balanced. The bridge is balanced when there is an equal division of voltages across the bridge output. This output is a current path across the ratio arms, like a bridge between two series-parallel paths. It does not matter what the ratios are for the bridge to be in a balanced state because resistances in parallel (two series resistors in this case) have the same voltage across both branches. However, some ratios may create a more sensitive balance than others, depending upon the total current and total resistance of the ratio arms.

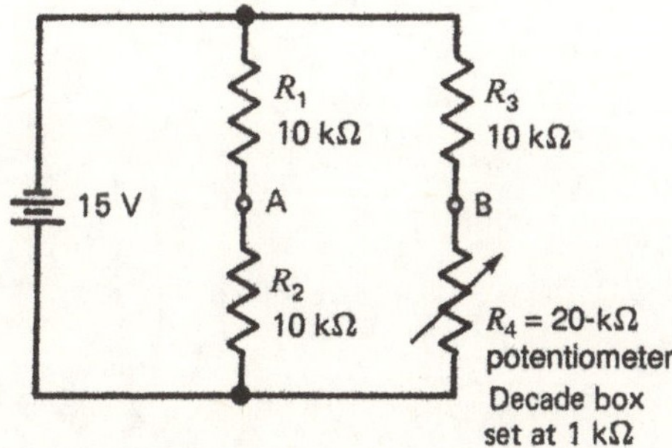

Fig. 6-5.1 Series-parallel circuit for analysis.

If there is an imbalance in the bridge, current will flow through the output path from one ratio arm to the other. In this experiment, a galvanometer is used. It is a dual directional microammeter with the needle zeroed at top dead center.

It is possible to use a VOM. However, extreme caution will be needed to prevent damage to the meter.

EQUIPMENT

DC power supply, 0–10 V
Decade box, or 20-kΩ potentiometer
Galvanometer (analog)
Test leads
DMM

COMPONENTS

(3) 10-kΩ, 0.25-W resistors
Plus other desired resistors (all 0.25 W):

(1) 5.6 kΩ (1) 2.7 kΩ
(1) 1 kΩ (1) 560 Ω
(1) 2.2 kΩ

PROCEDURE

1. Connect the circuit of Fig. 6-5.1.

2. Usually, a galvanometer is placed across the output terminals (A and B) to complete the bridge. However, it is better to study the series-parallel aspects of the Wheatstone bridge first to gain a better understanding of how it works. Thus, measure the voltage across points A and B, and record the results in Table 6-5.1. Put the ground side of the voltmeter on point B. Also, determine the direction of current flow.

3. Increase the value of R_4 to 5 kΩ, measure the voltage across points A and B, and determine current direction. Record the results in Table 6-5.1.

4. Increase the value of R_4 in 1-kΩ steps from 5 to 15 kΩ. Measure the voltage across points A and B at each step, determine current direction, and record the results in Table 6-5.1. This should lead to an understanding of how the value of the ratio arms determines the voltage across the AB output. When finished, turn off the power.

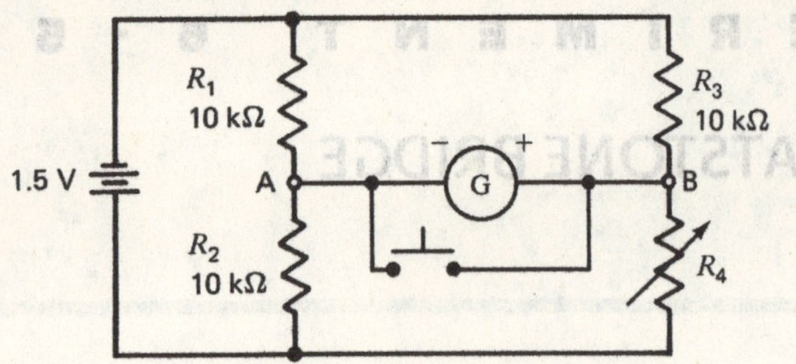

Fig. 6-5.2 Galvanometer circuit.

CAUTION: It will be necessary to reverse the voltmeter leads at some point.

5. Insert a galvanometer across points A and B as shown in Fig. 6-5.2. Be sure R_4 is 10 kΩ. The circuit is now a Wheatstone bridge because there is a true current path across points A and B.

Note: Many galvanometers have a switch (push button or toggle switch on the front of the meter). Until the switch is closed, current flows through a wire (or short circuit) to protect the meter movement. When the switch is closed, current flows through the meter movement. Be careful not to peg the meter.

6. With equal nominal values of R_1 and R_4, turn on the power and finely adjust R_4 so that no current flows across points A and B. This is the same as 0 V across points A and B. Thus, R_4 (decade box) will be finely adjusted and may be 10.5 kΩ or 11.1 kΩ, etc., unless, of course, you are using precision resistors (1 percent tolerance). The important thing is that both voltage and current at points A and B = 0. If so, the bridge is now balanced. Record the exact value of R_4 (decade box) in Table 6-5.2.

7. Now adjust R_4 so that the needle deflects to the right, approximately halfway to full scale, as shown in Fig. 6-5.3. Measure and record in Table 6-5.2 the voltage across R_4 and R_2. Repeat this procedure by adjusting R_4 with the needle deflecting to the left. Record the value of R_4 and the voltage across R_4 and R_2. Keep in mind the following: The bridge was balanced. Then, the bridge was unbalanced on both ratio arms (left and right). This procedure can be analyzed later to determine how the direction of current flow across the bridge indicates the condition of the bridge. Also, record the adjusted value of R_4.

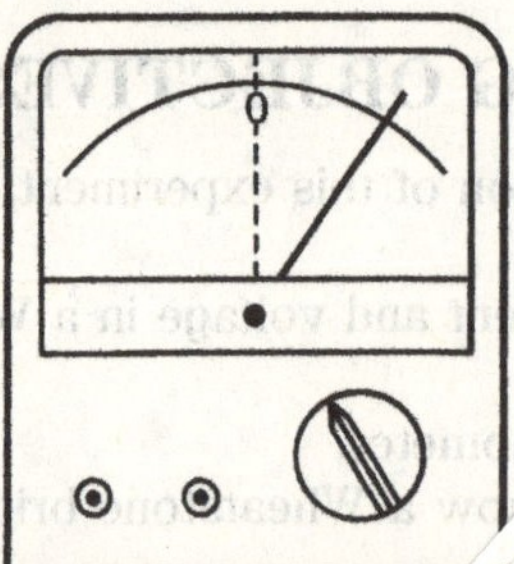

Fig. 6-5.3 Galvanometer.

8. Bring the bridge back to a balanced condition by readjusting R_4 and monitoring the galvanometer. The needle should be dead center—zero current.

9. Turn off the power. Replace R_2 with an unknown resistor with a value between 2 and 8 kΩ. Simply put electrical tape over a resistor. Or, if this is not possible, use a 5.6-kΩ resistor, and pretend you do not know the value.

10. Turn on the power. Adjust R_4 until the bridge is balanced. The value of R_4 (decade box) should be the true value of R_2. Repeat steps 9 and 10 with several resistors (e.g., 1 kΩ, 2.2 kΩ, 2.7 kΩ, and 560 Ω). R_4 should be equal to R_2 (unknown resistor) each time. Every time you adjust R_4, notice the amount of needle deflection. Record the results in a separate table. Be sure to label all values.

NAME DATE

QUESTIONS FOR EXPERIMENT 6-5

1. In procedure steps 9 and 10, the Wheatstone bridge was balanced by matching a decade box to the unknown resistor. Thus, the Wheatstone bridge was used as what kind of meter?

2. Explain the difference between a galvanometer and an ohmmeter.

3. In procedure step 5, suppose the schematic of Fig. 6-5.2 showed R_4 as 50 kΩ, and R_1 to R_3 remained at 10 kΩ. With the power on, would the needle deflect to the right or the left?

4. Which circuit in Fig. 6-5.4 would be more sensitive (greater needle deflection) when attempting to balance the bridge? Why?

5. Explain any differences and/or similarities between the two circuits in Fig. 6-5.5.

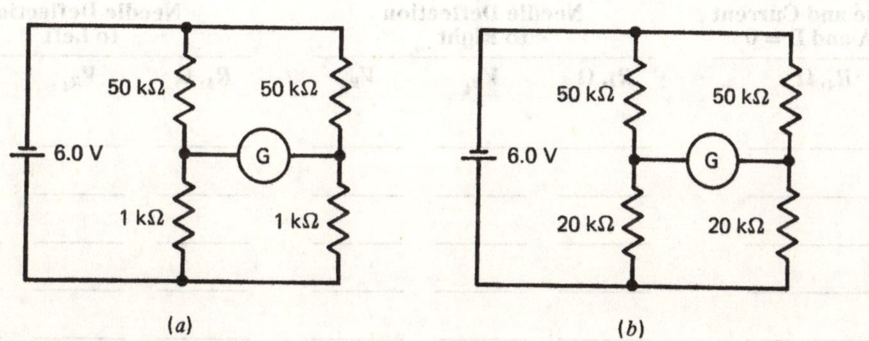

Fig. 6-5.4 Circuits for question 4.

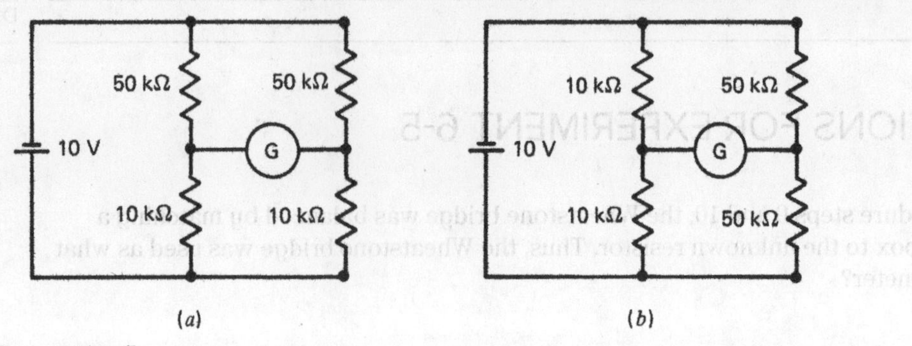

Fig. 6-5.5 Circuits for question 5.

TABLES FOR EXPERIMENT 6-5

TABLE 6-5.1 Data from Fig. 6-5.1

R_4, kΩ	Voltage AB	Current Direction
1		
5		
6		
7		
8		
9		
10		
11		
12		
13		
14		
15		

TABLE 6-5.2 R_1 and R_1 = 10 kΩ, R_4 = Decade Box

	Voltage and Current at A and B = 0	Needle Deflection to Right			Needle Deflection to Left		
	R_4, Ω	R_4, Ω	V_{R_4}	V_{R_2}	R_4, Ω	V_{R_4}	V_{R_2}
R_2 = 10 kΩ							
R_2 = 5.6 kΩ							
R_2 = 1 kΩ							
R_2 = 2.2 kΩ							
R_2 = 2.7 kΩ							
R_2 = 560 Ω							

EXPERIMENT RESULTS REPORT FORM

Experiment No.: ______________________ Name: ______________________

Date: ______________________

Experiment Title: ______________________ Class: ______________________

______________________ Instr: ______________________

Explain the purpose of the experiment:

List the first Learning Objective:

OBJECTIVE 1:

After reviewing the results, describe how the objective was validated by this experiment.

List the second Learning Objective:

OBJECTIVE 2:

After reviewing the results, describe how the objective was validated by this experiment.

List the third Learning Objective:

OBJECTIVE 3:

After reviewing the results, describe how the objective was validated by this experiment.

Conclusion:

If required, attach to this form: ☐ Answers to Questions, ☐ Tables, and ☐ Graphs.

CHAPTER 6

EXPERIMENT 6-6

ADDITIONAL SERIES-PARALLEL CIRCUITS

LEARNING OBJECTIVES

At the completion of this experiment, you will be able to:

- Redraw and simplify a series-parallel circuit.
- Write an equation for the total resistance of a series-parallel circuit.
- Determine which resistances have the greatest effect on R_T.

SUGGESTED READING

Chapter 6, *Basic Electronics*, Grob/Schultz, twelfth edition

INTRODUCTION

The following points summarize the concepts of a series-parallel circuit:

- The R_T of two parallel branches equals the product divided by the sum.

$$R_T = \frac{R_1 \times R_2}{R_1 + R_2}$$

- Adding resistance in series increases the total resistance and decreases the total current.
- Adding resistance in parallel increases the total current and decreases the total resistance.
- The equivalent resistance R_{eq} of a series-parallel network must be determined before the total resistance R_T can be found.

Consider the circuit shown in Fig. 6-6.1. This resistive network has a total resistance of

$$R_T = R_1 \| R_{eq}$$

The two vertical lines in the equation are used as a symbol to indicate resistances in parallel. Thus, $R_T = R_1 \| R_{eq}$ can be literally taken to mean

$$R_T = R_1 \text{ in parallel with } R_{eq}$$

To determine R_T, the circuit must be reduced or redrawn so that all the resistors, except R_1, can be combined into one equivalent resistance R_{eq}. This is because R_1 is in parallel with the equivalent total resistance of R_2 through R_5.

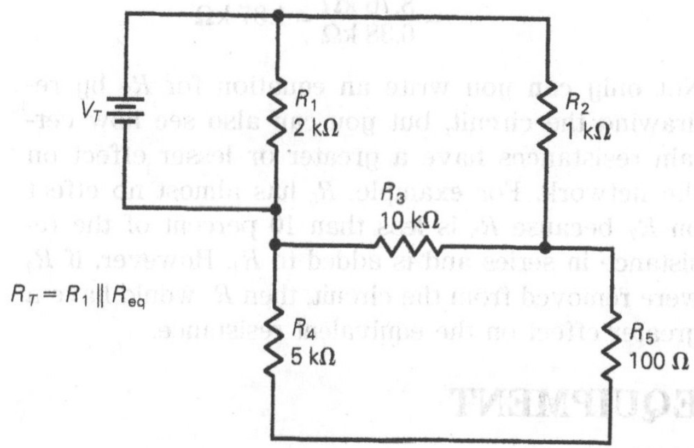

Fig. 6-6.1 Resistive network.

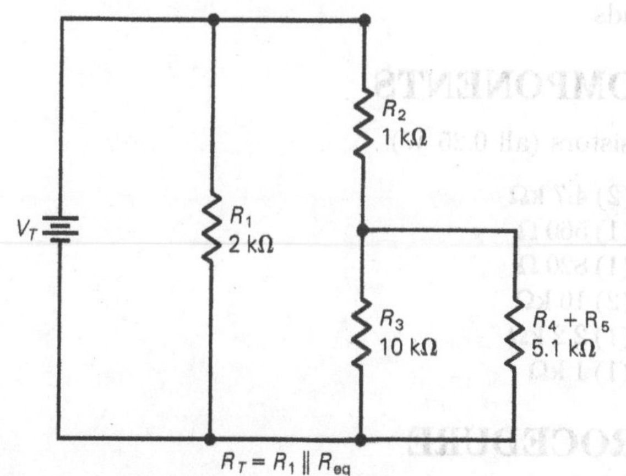

Fig. 6-6.2 Redrawn resistive network.

Although the circuit can be redrawn in more than one way, there is a method that can be used easily: Look for the resistors farthest from the power source, and combine them, working backward toward the source. The result of this method is

$$R_{eq} = [(R_4 + R_5) \| R_3] + R_2$$

The network can be redrawn as shown in Fig. 6-6.2. Notice that the redrawn figure makes it easier to write the equation. In this case, you can begin with the sum of $R_5 + R_4$ in parallel with R_3. This would be

$$R_4 + R_5 = 5.1\ \text{k}\Omega$$

Then, by using the product-over-the-sum method

$$\frac{R_3 \times 5.1\ \text{k}\Omega}{R_3 + 5.1\ \text{k}\Omega} = \frac{10\ \text{k}\Omega \times 5.1\ \text{k}\Omega}{10\ \text{k}\Omega + 5.1\ \text{k}\Omega}$$

$$= \frac{51.0\ \text{k}\Omega}{15.1\ \text{k}\Omega} = 3.38\ \text{k}\Omega$$

Next, add 3.38 kΩ to 1 kΩ R_2 so that $R_{eq} = 4.38$ kΩ. Finally,

$$R_T = R_1 \| R_{eq} = \frac{2\ \text{k}\Omega \times 4.38\ \text{k}\Omega}{2\ \text{k}\Omega + 4.38\ \text{k}\Omega}$$

$$= \frac{8.76\ \text{k}\Omega}{6.38\ \text{k}\Omega} = 1.37\ \text{k}\Omega$$

Not only can you write an equation for R_T by redrawing the circuit, but you can also see how certain resistances have a greater or lesser effect on the network. For example, R_5 has almost no effect on R_T because R_5 is less than 10 percent of the resistance in series and is added to R_4. However, if R_4 were removed from the circuit, then R_5 would have a greater effect on the equivalent resistance.

EQUIPMENT

DMM
DC power supply
Protoboard or springboard
Leads

COMPONENTS

Resistors (all 0.25 W):

- (2) 4.7 kΩ
- (1) 560 Ω
- (1) 820 Ω
- (2) 10 kΩ
- (1) 2.2 kΩ
- (1) 1 kΩ

PROCEDURE

1. Measure and record the resistor values shown in Table 6-6.1.
2. Connect the circuit shown in Fig. 6-6.3.

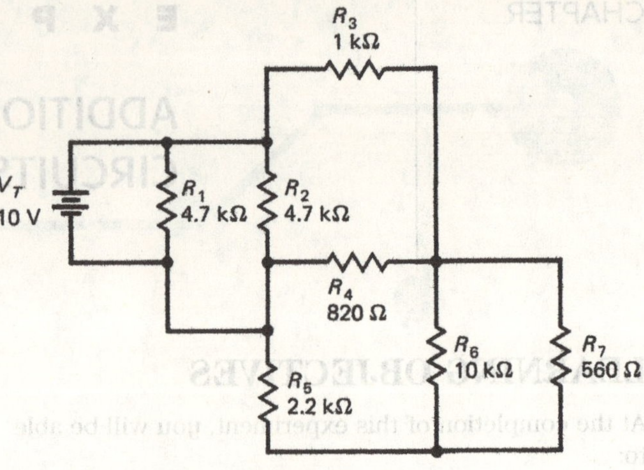

Fig. 6-6.3 Resistive network.

3. Measure and record the voltages around the circuit as shown in Table 6-6.1.
4. Disconnect the power supply V_T, and measure the total circuit resistance. Record the results in Table 6-6.1.
5. Disconnect R_1 and V_T, and measure and record the resistance R_{eq}.
6. Calculate the current through each resistor, using Ohm's law and the measured voltage drops. Record the results in Table 6-6.1 and include I_T, the total circuit current.
7. Remove R_6 from the circuit and measure the total circuit resistance. Record the results in Table 6-6.1.
8. Replace R_2 with a 10-kΩ resistor, and measure the total circuit resistance. Record the results in Table 6-6.1. (R_6 is still removed.)
9. On a separate sheet of 8 × 11 in. paper, redraw the circuit so that only four resistances represent the simplified circuit. Label all resistances so that the combined resistances are easy to identify. For example, one resistance might be $(R_3 + R_4) \| R_2$, etc.
10. Write an equation for the total resistance, and show how to calculate R_T by using R_{eq} and the product-over-the-sum method. Use the same sheet of paper as in step 9 above.

NAME ______________________________ DATE ______________

QUESTIONS FOR EXPERIMENT 6-6

1. Which resistor in the circuit of Fig. 6-6.2 has the least effect on R_T and why?

2. What would happen to the circuit of Fig. 6-6.2 if R_1 were decreased to 10 Ω?

3. What would happen to the circuit of Fig. 6-6.2 if R_1 were increased to 10 MΩ?

4. Which resistor in the circuit of Fig. 6-6.2 has the least effect on I_T?

5. Which resistor in the circuit of Fig. 6-6.2 has the greatest effect on I_T?

TABLES FOR EXPERIMENT 6-6

TABLE 6-6.1

Resistance Values Nominal or Calculated	Ω Measured	*V* Measured	*I* Calculated
$R_1 = 4.7$ kΩ	______	______	______
$R_2 = 4.7$ kΩ	______	______	______
$R_3 = 1$ kΩ	______	______	______
$R_4 = 820$ Ω	______	______	______
$R_5 = 2.2$ kΩ	______	______	______
$R_6 = 10$ kΩ	______	______	______
$R_7 = 560$ Ω	______	______	______
$R_T =$ ______	______		$I_T =$ ______
$R_{eq} =$ ______	______		
R_T with R_6 removed	______		
R_T with $R_2 = 10$ kΩ	______		

EXPERIMENT RESULTS REPORT FORM

Experiment No.: ____________________ **Name:** ____________________

Date: ____________________

Experiment Title: ____________________ **Class:** ____________________

____________________ **Instr:** ____________________

Explain the purpose of the experiment:

List the first Learning Objective:

OBJECTIVE 1:

After reviewing the results, describe how the objective was validated by this experiment.

List the second Learning Objective:

OBJECTIVE 2:

After reviewing the results, describe how the objective was validated by this experiment.

List the third Learning Objective:

OBJECTIVE 3:

After reviewing the results, describe how the objective was validated by this experiment.

Conclusion:

If required, attach to this form: ☐ Answers to Questions, ☐ Tables, and ☐ Graphs.

CHAPTER 6

EXPERIMENT 6-7

ADDITIONAL SERIES-PARALLEL OPENS AND SHORTS

LEARNING OBJECTIVES

At the completion of this experiment, you will be able to:

- Determine the changes in circuit current and voltage drops resulting from a short circuit.
- Determine the changes in circuit current and voltage drops resulting from an open circuit.
- Predict changes in circuit current and voltage drops resulting from open and short circuits.

SUGGESTED READING

Chapter 6, *Basic Electronics*, Grob/Schultz, twelfth edition

INTRODUCTION

A short circuit has practically zero resistance. Its effect, therefore, is to allow excessive current to flow in a circuit, although this is not usually intentional. An open circuit has the opposite effect because an open circuit has infinitely high resistance with practically zero current.

Therefore, if one path in a circuit changes (becomes open or short), the circuit's voltage, resistance, and current in the other paths change as well. For example, the series-parallel circuit shown in Fig. 6-7.1 becomes a series circuit when there is a short across circuit points A and B.

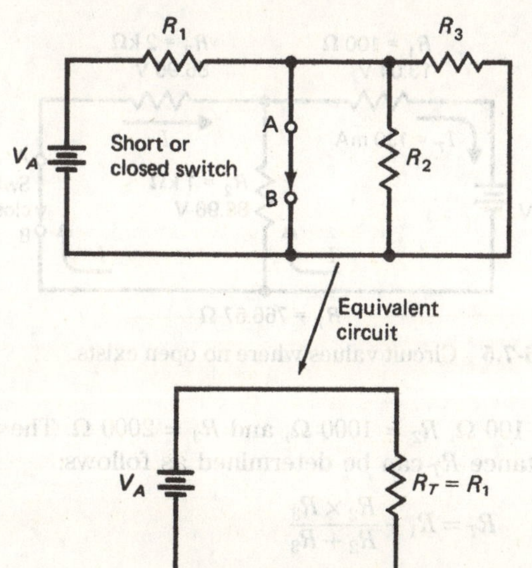

Fig. 6-7.1 Effect of a short in a series-parallel circuit.

As an example of an open circuit, the series-parallel circuit in Fig. 6-7.2 becomes a series circuit with just R_1 and R_2 when there is an open between points A and B.

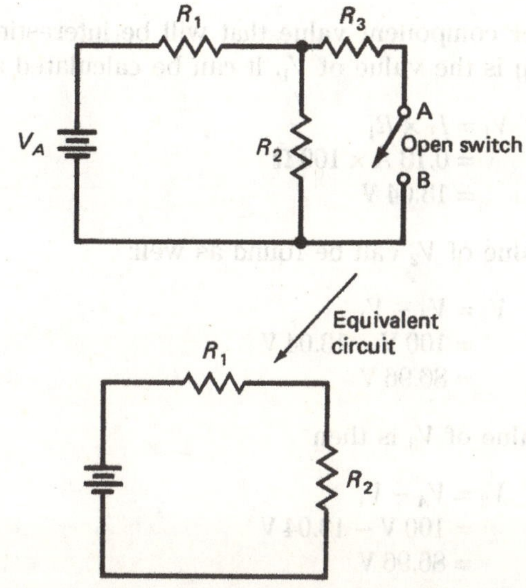

Fig. 6-7.2 Effects of an open in a series-parallel circuit.

The Short Circuit

You can determine the effect of a short in a series-parallel circuit. For example, in the circuit of Fig. 6-7.1, a switch is shown between points A and B. This switch represents a possible flaw in construction or operation of an actual circuit.

Refer to Fig. 6-7.3, where the circuit does not have a short. Specific component values have been assigned

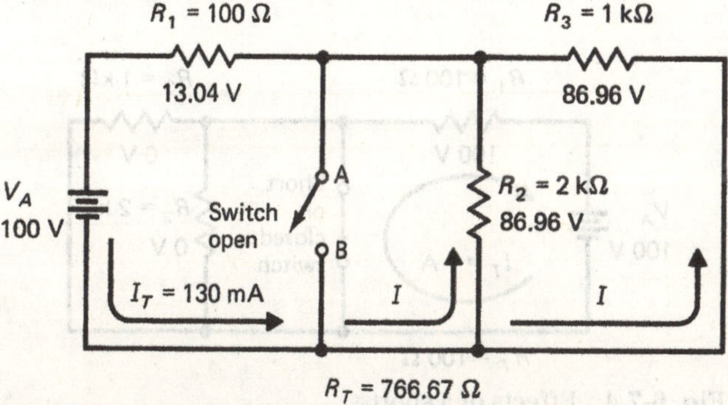

Fig. 6-7.3 Circuit values where no short exists.

to the same circuit as in Fig. 6-7.1. As shown in Fig. 6-7.3, R_1 = 100 Ω, R_2 = 1000 Ω, and R_3 = 2000 Ω. The total resistance R_T can be determined as follows:

$$R_T = R_1 + \frac{R_2 \times R_3}{R_2 + R_3}$$
$$= 766.67\ \Omega$$

Knowing the total resistance is essential in determining the total circuit current. If the applied voltage V_A is 100 V, then

$$I_T = \frac{V_A}{R_T}$$
$$= \frac{100\text{ V}}{766.67\ \Omega}$$
$$= 0.13\text{ A} \quad (\text{or } 130\text{ mA})$$

Another component value that will be interesting to identify is the value of V_1. It can be calculated as

$$V_1 = I_T \times R_1$$
$$= 0.13\text{ A} \times 100\ \Omega$$
$$= 13.04\text{ V}$$

The value of V_2 can be found as well:

$$V_2 = V_A - V_1$$
$$= 100\text{ V} - 13.04\text{ V}$$
$$= 86.96\text{ V}$$

The value of V_3 is then

$$V_3 = V_A - V_1$$
$$= 100\text{ V} - 13.04\text{ V}$$
$$= 86.96\text{ V}$$

The voltage values of V_2 and V_3 should be equivalent since R_2 and R_3 form a parallel circuit.

Figure 6-7.3 shows the calculated effect of a short with a series-parallel circuit. If the circuit is shorted from point A to B, then the effects on resistors R_2 and R_3 are eliminated. The elimination of R_2 and R_3 creates predictable changes in circuit current and the voltage drop of R_1. The circuit of Fig. 6-7.4 shows the electrical effects. Since R_2 and R_3 are eliminated, the only effective resistance left in the circuit is R_1.

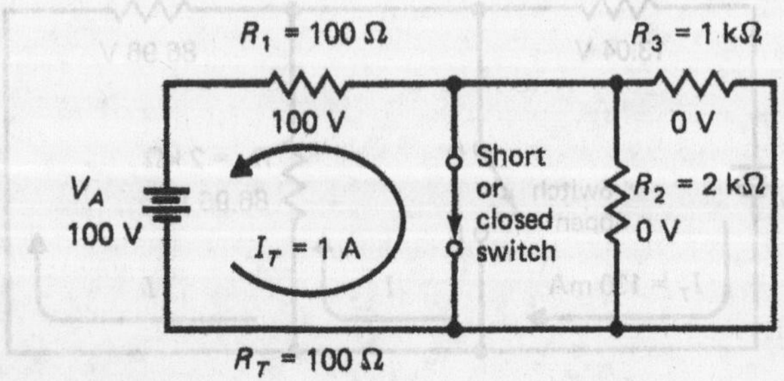

Fig. 6-7.4 Effects of a short.

The total circuit current I_T can then be calculated as

$$I_T = \frac{V_A}{R_1}$$
$$= \frac{100\text{ V}}{100\ \Omega}$$
$$= 1\text{ A}$$

The voltage drop across R_1 is then

$$V_1 = I_T \times R_1$$
$$= 1\text{ A} \times 100\ \Omega$$
$$= 100\text{ V}$$

The voltage drop across R_2 is thus

$$V_2 = V_A - V_1$$
$$= 100\text{ V} - 100\text{ V}$$
$$= 0\text{ V}$$

The voltage drop across R_3 is

$$V_3 = V_A - V_1$$
$$= 100\text{ V} - 100\text{ V}$$
$$= 0\text{ V}$$

The voltage drops across R_2 and R_3 should equal 0 V since their resistance values are reduced to 0 Ω because of the short.

In summary, these circuit changes occur in two ways. First, the circuit currents increase significantly. Second, the voltage drop from the increase in current flow also increases.

The Open Circuit

An open circuit provides practically infinite resistance to the applied voltage V_A. Its overall effect on the circuit would be zero or minimal current flow. You can determine the effect of an open path in a series-parallel circuit. For example, in the circuit of Fig. 6-7.5, component values have been assigned as

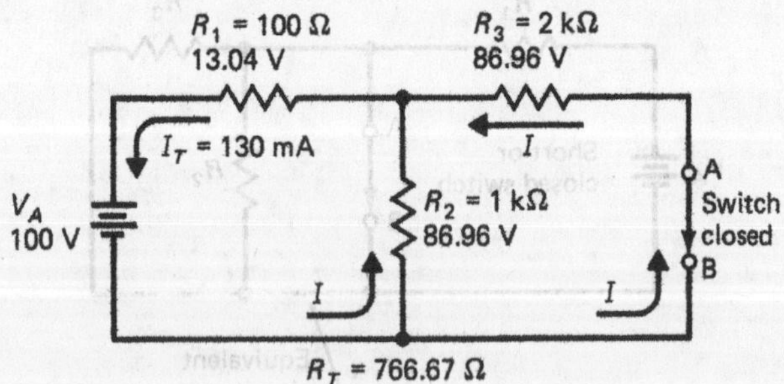

Fig. 6-7.5 Circuit values where no open exists.

R_1 = 100 Ω, R_2 = 1000 Ω, and R_3 = 2000 Ω. The total resistance R_T can be determined as follows:

$$R_T = R_1 + \frac{R_2 \times R_3}{R_2 + R_3}$$
$$= 766.67\ \Omega$$

Knowing the total resistance is essential in determining the total circuit current. If the applied voltage V_A is 100 V, then

$$I_T = \frac{V_A}{R_T}$$
$$= \frac{100\ \text{V}}{766.67\ \Omega}$$
$$= 0.13\ \text{A} \quad (\text{or } 130\ \text{mA})$$

Another component value that will be important to identify is the voltage value of V_1. It can be calculated as

$$V_1 = I_T \times R_1$$
$$= 0.13\ \text{A} \times 100\ \Omega$$
$$= 13.04\ \text{V}$$

The voltage of V_2 is

$$V_2 = V_A - V_1$$
$$= 100\ \text{V} - 13.04\ \text{V}$$
$$= 86.96\ \text{V}$$

The voltage of V_3 is

$$V_3 = V_A - V_1$$
$$= 100\ \text{V} - 13.04\ \text{V}$$
$$= 86.96\ \text{V}$$

The values are shown in Fig. 6-7.5.

If the circuit is open between points A and B, then R_3 is no longer part of the circuit. The removal of R_3 creates a predictable change in the circuit current and voltage drops of R_1 and R_2.

The circuit of Fig. 6-7.6 shows the overall electrical effects when the switch is opened between points A and B. The circuit current I_T can be calculated after the total resistance R_T is found:

$$R_T = R_1 + R_2$$
$$= 100\ \Omega + 1000\ \Omega$$
$$= 1100\ \Omega$$

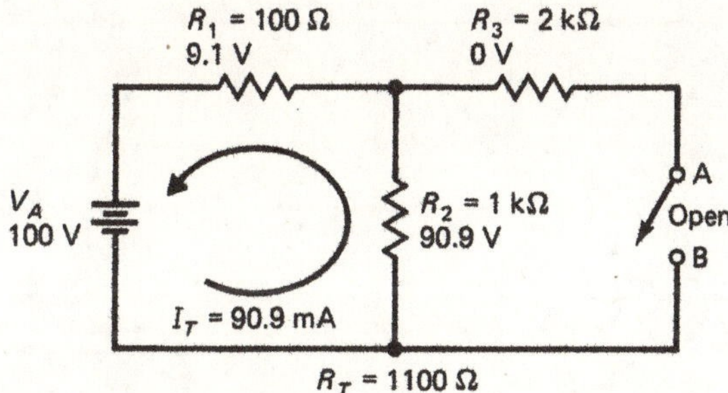

Fig. 6-7.6 Effects of an open.

Then I_T can be calculated as

$$I_T = \frac{V_A}{R_1}$$
$$= \frac{100\ \text{V}}{1100\ \Omega}$$
$$= 0.0909\ \text{A} \quad (\text{or } 90.9\ \text{mA})$$

The voltage drop across R_1 is then

$$V_1 = I_T \times R_1$$
$$= 0.0909\ \text{A} \times 100\ \Omega$$
$$= 9.09 \text{ or } 9.1\ \text{V}$$

The voltage drop across R_2 is then

$$V_2 = I_T \times R_2$$
$$= 0.0909\ \text{A} \times 1000\ \Omega$$
$$= 90.9\ \text{V}$$

The voltage value of R_3 is 0 V due to the open circuit.

EQUIPMENT

DMM
Power supply
Protoboard

COMPONENTS

(1) 150-Ω, 1-W resistor
(2) 150-Ω, 0.25-W resistors
(1) SPST switch

PROCEDURE

1. Measure and record in Table 6-7.1 the values of the three resistors used in this experiment.

2. Construct the circuit of Fig. 6-7.7, and adjust the voltage of the power supply to 10 V.

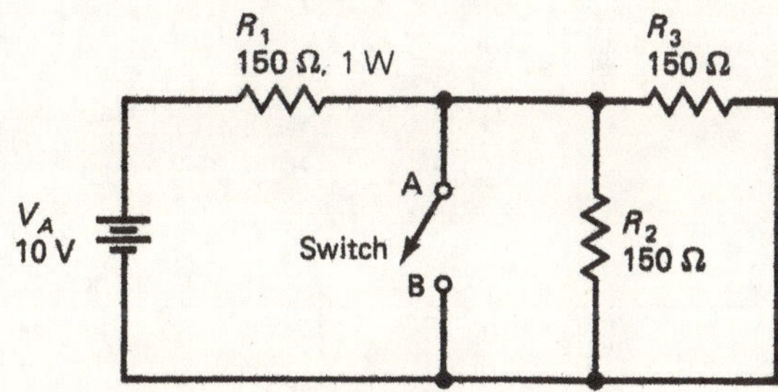

Fig. 6-7.7 Series-parallel circuit.

3. Calculate and record in Table 6-7.1, for the circuit of Fig. 6-7.7, the values of the total resistance R_T; the currents I_{R_1}, I_{R_2}, and I_{R_3}; and the voltages V_{R_1}, V_{R_2}, and V_{R_3}.

4. Measure and record in Table 6-7.1 the measured values of I_{R_1}, I_{R_2}, I_{R_3}, V_{R_1}, V_{R_2}, and V_{R_3}.

5. Connect points A and B.

6. Calculate and record in Table 6-7.1 the values of R_T, I_{R_1}, I_{R_2}, I_{R_3}, V_{R_1}, V_{R_2}, and V_{R_3}.

7. Measure and record in Table 6-7.1 the values of I_{R_1}, I_{R_2}, I_{R_3}, V_{R_1}, V_{R_2}, and V_{R_3}.

8. Turn off the power supply, and disconnect the circuit.

9. Measure and record in Table 6-7.2 the values of the three resistors used in this experiment.

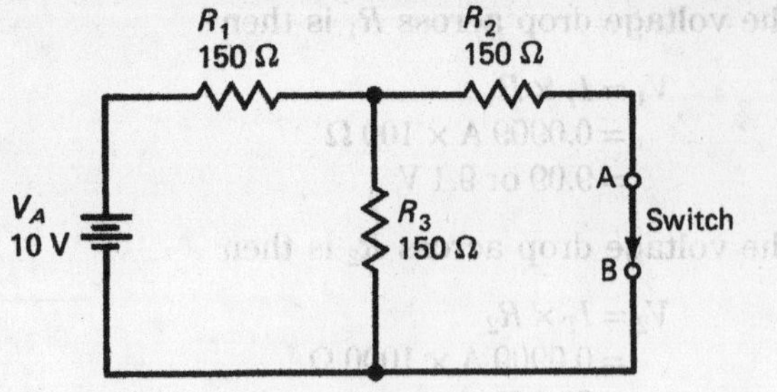

Fig. 6-7.8 Series-parallel circuit.

10. Construct the circuit of Fig. 6-7.8, and adjust the voltage of the power supply to 10 V.

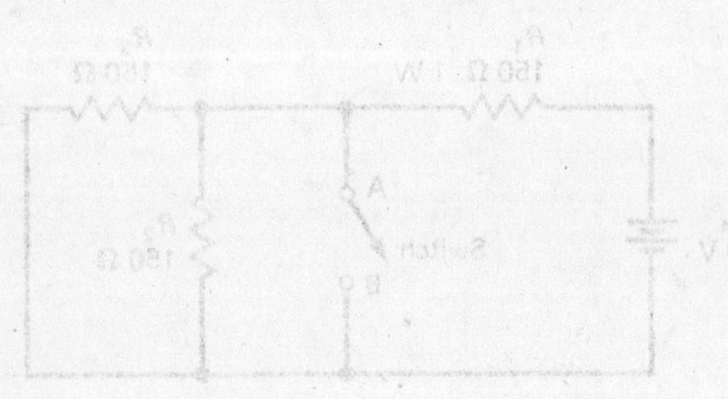

11. Calculate and record in Table 6-7.2, for the circuit of Fig. 6-7.8, the values of the total resistance R_T; currents I_{R_1}, I_{R_2}, and I_{R_3}; and voltages V_{R_1}, V_{R_2}, and V_{R_3}.

12. Measure and record in Table 6-7.2 the measured values of I_{R_1}, I_{R_2}, I_{R_3}, V_{R_1}, V_{R_2}, and V_{R_3}.

13. Disconnect or open points A and B.

14. Calculate and record in Table 6-7.2 the values of R_T, I_{R_1}, I_{R_2}, I_{R_3}, V_{R_1}, V_{R_2}, and V_{R_3}.

15. Measure and record in Table 6-7.2 the values of I_{R_1}, I_{R_2}, I_{R_3}, V_{R_1}, V_{R_2}, and V_{R_3}.

16. Turn off the power supply and disconnect the circuit.

NAME DATE

QUESTIONS FOR EXPERIMENT 6-7

1. What are the characteristics of a shorted circuit?

2. What are the characteristics of an open circuit?

3. Compare Tables 6-7.1 and 6-7.2. What are the differences and similarities?

TABLES FOR EXPERIMENT 6-7

TABLE 6-7.1 ($V_A = 10$ V)

Unshorted Circuit, Fig. 6-7.7		
Component	**Measured**	**Calculated**
R_1	______	
R_2	______	
R_3	______	
R_T		______
I_{R_1}	______	______
I_{R_2}	______	______
I_{R_3}	______	______
V_{R_1}	______	______
V_{R_2}	______	______
V_{R_3}	______	______
Shorted Circuit, Fig. 6-7.7		
R_T		______
I_{R_1}	______	______
I_{R_2}	______	______
I_{R_3}	______	______
V_{R_1}	______	______
V_{R_2}	______	______
V_{R_3}	______	______

TABLE 6-7.2 ($V_A = 10$ V)

Shorted Circuit, Fig. 6-7.8		
Component	**Measured**	**Calculated**
R_1	______	
R_2	______	
R_3	______	
R_T		______
I_{R_1}	______	______
I_{R_2}	______	______
I_{R_3}	______	______
V_{R_1}	______	______
V_{R_2}	______	______
V_{R_3}	______	______
Opened Circuit, Fig. 6-7.8		
R_T		______
I_{R_1}	______	______
I_{R_2}	______	______
I_{R_3}	______	______
V_{R_1}	______	______
V_{R_2}	______	______
V_{R_3}	______	______

EXPERIMENT RESULTS REPORT FORM

Experiment No.: ______________________ Name: ______________________

Date: ______________________

Experiment Title: ______________________ Class: ______________________

______________________ Instr: ______________________

Explain the purpose of the experiment:

List the first Learning Objective:

OBJECTIVE 1:

After reviewing the results, describe how the objective was validated by this experiment.

List the second Learning Objective:

OBJECTIVE 2:

After reviewing the results, describe how the objective was validated by this experiment.

List the third Learning Objective:

OBJECTIVE 3:

After reviewing the results, describe how the objective was validated by this experiment.

Conclusion:

If required, attach to this form: ☐ Answers to Questions, ☐ Tables, and ☐ Graphs.

CHAPTER 7

EXPERIMENT 7-1

VOLTAGE DIVIDERS WITH LOADS

LEARNING OBJECTIVES

At the completion of this experiment, you will be able to:

- Identify a voltage divider circuit.
- Define the purpose for voltage divider circuits.
- Describe voltage divider loading effects.

SUGGESTED READING

Chapter 7, *Basic Electronics*, Grob/Schultz, twelfth edition

INTRODUCTION

A series circuit is also a voltage divider. That is, the total voltage applied to a series circuit is divided among the series resistors. Because the current is the same value in all parts of a series circuit, voltage is divided among the series resistances in direct proportion to the value of resistance.

For example, imagine four resistors in series. Each resistor is equal in value. Therefore, any one resistor will receive one-fourth of the total applied voltage. In other words, the total voltage is divided by 4 across each resistor.

Another way to determine the voltage across any resistor in series, without using a meter, is to add the total resistance, divide that value into any single resistor in series, and multiply by the total voltage. This is the proportional method. For example, see Fig. 7-1.1. Therefore,

$$V_{R_2} = \frac{R_2}{R_T} \times V_T = \frac{2\text{ k}\Omega}{6\text{ k}\Omega} \times 6\text{ V} = 2\text{ V}$$

R_1 1 kΩ
$V_T = 6$ V
R_2 2 kΩ
R_3 3 kΩ

Fig. 7-1.1 Proportional method circuit.

Although these simple voltage dividers are limited in their use, when a load is placed across any series resistance, the voltage divider is then extremely useful as a voltage tap. For example, the circuit in Fig. 7-1.2 shows a parallel load current through R_{load}.

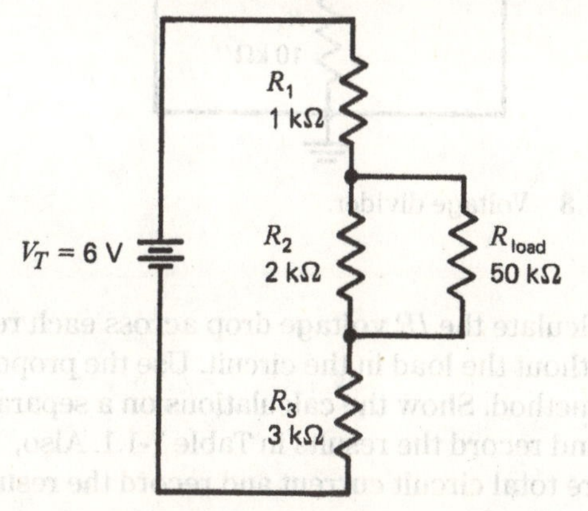

Fig. 7-1.2 Simple voltage divider with parallel load.

This parallel load, if its resistance was more than 10 times the value of R_2, would actually be sharing the *IR* voltage drops across R_2 without changing the division of voltage considerably. For example, if $R_{\text{load}} = 50\text{ k}\Omega$,

$$\frac{R_2 \times R_{\text{load}}}{R_2 + R_{\text{load}}} = \frac{100 \times 10^6}{52 \times 10^3} = 2\text{ k}\Omega\text{ (approx.)}$$

Thus, R_{load} could share the same approximate voltage as R_2, but it would have its own current.

While voltage dividers are mainly used to tap off part of a total voltage, it is necessary to remember that the addition of a load will always have some effect upon the circuit current and, many times, the proportional *IR* voltage drops.

EQUIPMENT

DC power supply, 0–10 V
DMM
Test leads

COMPONENTS

Resistors (all 0.25 W):

(3) 1 kΩ (1) 47 kΩ
(3) 10 kΩ (1) 4.7 kΩ
(1) 470 Ω (1) 100 kΩ

PROCEDURE

1. Connect the circuit of Fig. 7-1.3. Do not connect R_{load}.

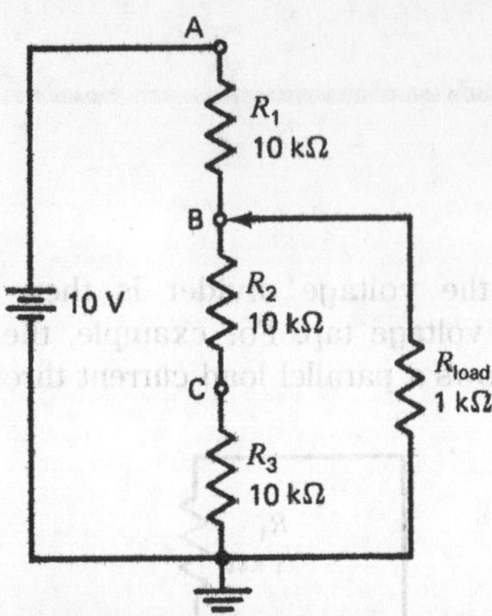

Fig. 7-1.3 Voltage divider.

2. Calculate the *IR* voltage drop across each resistor without the load in the circuit. Use the proportional method. Show the calculations on a separate sheet and record the results in Table 7-1.1. Also, measure total circuit current and record the results in Table 7-1.1.

3. With the load connected to point B, measure the total circuit current, load current, and bleeder current. The bleeder current is the steady drain on the source, the current through R_3. Record the values in Table 7-1.2.

4. Measure the voltages across R_1, R_2, R_3, and R_{load}. Record these values in Table 7-1.2.

5. Connect R_{load} to point A. Repeat steps 3 and 4. Record the results.

6. Connect R_{load} to point C. Repeat steps 3 and 4. Record the results.

7. Change the value of R_{load} to 100 kΩ and repeat steps 3 to 6.

8. Connect the circuit of Fig. 7-1.4. This is a voltage divider with loads.

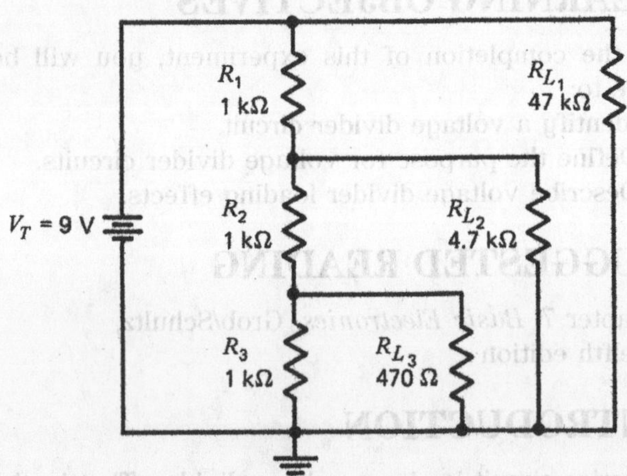

Fig. 7-1.4 Complex voltage divider.

9. Measure and record the *IR* voltage drops across R_1, R_2, R_3, R_{L_1}, R_{L_2}, and R_{L_3} in Table 7-1.3.

10. Calculate the current through each resistor and the total circuit current. Record the results in Table 7-1.3.

NAME DATE

QUESTIONS FOR EXPERIMENT 7-1

1. Explain what is meant by the term *voltage divider.*

2. Refer to the circuit of Fig. 7-1.4 (loaded voltage divider). Explain what would happen to the total circuit current, voltage, and resistance if R_{L_2} and R_{L_3} were removed.

3. If Fig. 7-1.4 were used to tap the total voltage of 9 V into three equal parts (without the loads), explain why V_{R_1}, V_{R_2}, and V_{R_3} (with respect to ground) would not be 3 V each. Does the value of any one load greatly affect the original unloaded divider?

4. Redraw Fig. 7-1.4 and show the path of current flow by drawing arrows where necessary.

5. What effect would reversing the battery polarity have on the circuit of Fig. 7-1.4?

TABLES FOR EXPERIMENT 7-1

TABLE 7-1.1 No-Load Values for Fig. 7-1.3

	Calculated *IR* Drop	Measured Current
R_1		
R_2		
R_3		
I_T		

TABLE 7-1.2 Measured Values for Fig. 7-1.3, Circuit under Load

$R_L =$ 1 kΩ	Load at Point A *IR* Drops	Load at Point A Current	Load at Point B *IR* Drops	Load at Point B Current	Load at Point C *IR* Drops	Load at Point C Current
R_1						
R_2						
R_3						
R_{load}						
I_T						
I_B						
$R_L =$ 100 kΩ						
R_1						
R_2						
R_3						
R_{load}						
I_T						
I_B						

TABLE 7-1.3 Circuit Values for Fig. 7-1.4

	IR Drop, Measured	Current, Calculated*
R_1		
R_2		
R_3		
R_{L_1}		
R_{L_2}		
R_{L_3}		
R_T		

*V (*IR* drop measured)/*R* nominal = calculated current.

EXPERIMENT RESULTS REPORT FORM

Experiment No.: ______________________ Name: ______________________

Date: ______________________

Experiment Title: ______________________ Class: ______________________

______________________ Instr: ______________________

Explain the purpose of the experiment:

List the first Learning Objective:

OBJECTIVE 1:

After reviewing the results, describe how the objective was validated by this experiment.

List the second Learning Objective:

OBJECTIVE 2:

After reviewing the results, describe how the objective was validated by this experiment.

List the third Learning Objective:

OBJECTIVE 3:

After reviewing the results, describe how the objective was validated by this experiment.

Conclusion:

If required, attach to this form: ☐ Answers to Questions, ☐ Tables, and ☐ Graphs.

CHAPTER 7

EXPERIMENT 7-2

CURRENT DIVIDERS

LEARNING OBJECTIVES

At the completion of this experiment, you will be able to:

- Understand how parallel circuits act as current dividers.
- Explain the proportional method for solving branch currents.
- Understand the parallel conductance method for solving branch currents.

SUGGESTED READING

Chapter 7, *Basic Electronics*, Grob/Schultz, twelfth edition

INTRODUCTION

In the same way that series circuits are also voltage dividers, parallel circuits are also current dividers. That is, the total current is divided among the parallel branches in inverse proportion to the resistance in any branch. Therefore, main-line current increases as branches are added.

To find the branch currents without knowing the total voltage across the bank, a formula can be used. This formula is a proportional method for solving unknown branch currents. For example, the branch currents for Fig. 7-2.1 can be found by using this proportional method. It is based on the fact that currents divide in inverse proportion to their resistances.

Notice two things in Fig. 7-2.1. First, the numerator for each branch resistance is the value of the "opposite" branch resistance. Second, it is only necessary to calculate one branch current and subtract it from the total current. The remainder will be the other branch current.

$I_T = 30$ A

$I_{R_1} = \frac{40}{60} \times I_T = 20$ A

V_{unknown}

R_1 20 Ω

R_2 40 Ω

I_T

$I_{R_2} = \frac{20}{60} \times I_T = 10$ A

Note: $I_{R_x} = \frac{R_{\text{opposite }x}}{R_1 + R_2} \times I_T$

Fig. 7-2.1 Two-branch current divider.

Another method for determining branch current division is the parallel conductance method. Remember that conductance $G = 1/R$. Note that conductance and current are directly proportional. This is true because the greater the resistance, the less will be the current. With any number of parallel branches, each branch current can be calculated without knowing the voltage across the bank. The formula is

$$I_x = \frac{G_x}{G_T} \times I_T$$

For example, the branch current I_{R_1} in Fig. 7-2.2 can be found as follows:

$$I_{R_1} = \frac{G_1}{G_T} \times I_T$$

$$G_1 = \frac{1}{R_1} = \frac{1}{10\ \Omega} = 0.1\ \text{S}$$

Note: Siemens (S) is the reciprocal of ohms.

$$R_T = \frac{1}{1/R_1 + 1/R_2 + 1/R_3}$$

$$= 6.25\ \Omega$$

$$G_T = \frac{1}{R_T} = \frac{1}{6.25\ \Omega} = 0.16\ \text{S}$$

$$I_{R_1} = \frac{0.1}{0.16} \times 40\ \text{mA}$$

$$= 25\ \text{mA}$$

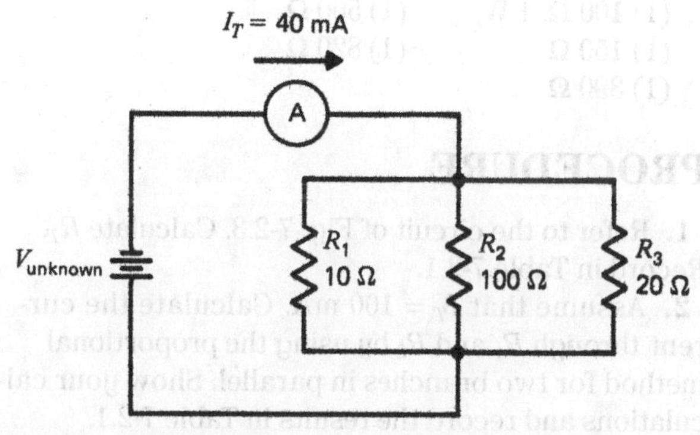

Fig. 7-2.2 Three-branch current divider.

Because $I_{R_1} = 25$ mA, it is easy to see that the remaining currents will be

$$I_{R_2} + I_{R_3} = +40 \text{ mA} - 25 \text{ mA} = 15 \text{ mA}$$

Thus, the previous formula (proportional method) can be used to solve for the remaining two branch currents.

Another method is to find the total conductance and use the following formula to solve for each branch current in Fig. 7-2.2:

$$G_1 = \frac{1}{R_1} = \frac{1}{10\ \Omega} = 0.10 \text{ S}$$

$$G_2 = \frac{1}{R_2} = \frac{1}{100\ \Omega} = 0.01 \text{ S}$$

$$G_3 = \frac{1}{R_3} = \frac{1}{20\ \Omega} = 0.05 \text{ S}$$

Therefore, the total conductance is

$$G_1 + G_2 + G_3 = 0.1 + 0.01 + 0.05 = 0.16 \text{ S}$$

To calculate branch currents, use the following formula:

$$I_{R_x} = \frac{G_{R_x}}{G_T} \times I_T$$

In the case of Fig. 7-2.2,

$$I_{R_1} = \frac{0.10}{0.16} \times 40 \text{ mA} = 25 \text{ mA}$$

$$I_{R_2} = \frac{0.01}{0.16} \times 40 \text{ mA} = 2.5 \text{ mA}$$

$$I_{R_3} = \frac{0.05}{0.16} \times 40 \text{ mA} = 12.5 \text{ mA}$$

$$I_T = 40 \text{ mA}$$

EQUIPMENT

DC power supply
Ammeter
Protoboard or springboard
VTVM/VOM
Leads

COMPONENTS

Resistors (all 0.25 W unless indicated otherwise):

(1) 100 Ω, 1 W
(1) 150 Ω
(1) 390 Ω
(1) 560 Ω
(1) 820 Ω

PROCEDURE

1. Refer to the circuit of Fig. 7-2.3. Calculate R_T. Record in Table 7-2.1.
2. Assume that $I_T = 100$ mA. Calculate the current through R_1 and R_2 by using the proportional method for two branches in parallel. Show your calculations and record the results in Table 7-2.1.

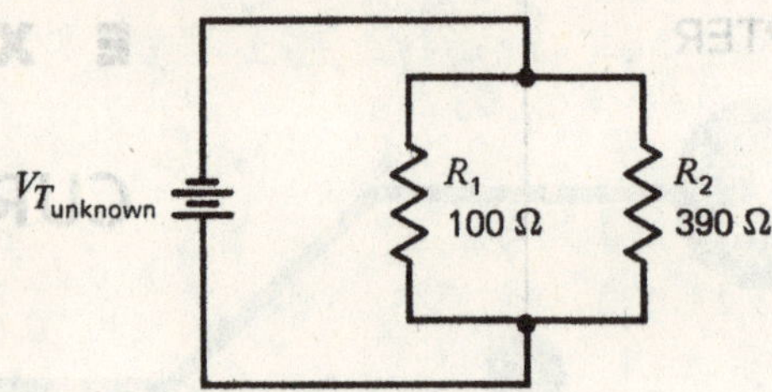

Fig. 7-2.3 Two-branch current divider.

3. Using the values you calculated for I_{R_1} and I_{R_2}, calculate the total voltage across the parallel bank (use Ohm's law). Show your calculations on a separate sheet, and record the value of V_T in Table 7-2.1.
4. Connect the circuit of Fig. 7-2.3. Calculate R_T.
5. Add another resistor, $R_3 = 820\ \Omega$, to the parallel bank, and apply the total voltage you calculated in step 3 to the circuit.
6. Measure the total current and the current through each branch. Record the results in Table 7-2.2.
7. Connect the circuit of Fig. 7-2.4.
8. Adjust the voltage so that total current = 30 mA.
9. Use the parallel conductance formula to determine the branch conductances G_{R_2} to G_{R_5}. Also,

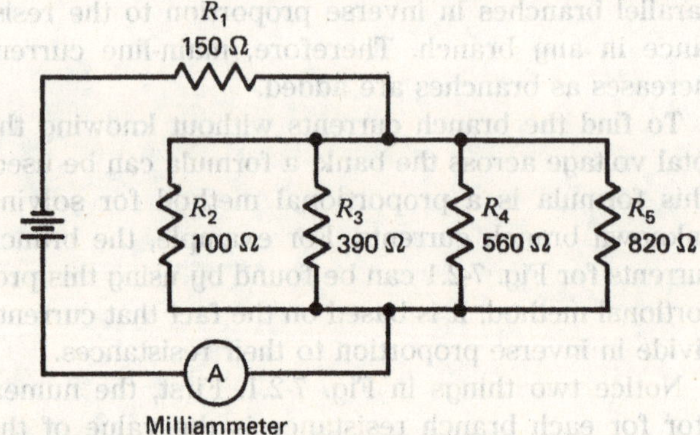

Fig. 7-2.4 Four-branch current divider.

measure the current through each branch and record the results in Table 7-2.3. Show all your calculations and record the results in Table 7-2.3.

Note: Do not measure *any* voltages. Doing so will defeat the purpose of this experiment.

10. Using the values you have determined for the circuit of Fig. 7-2.4, calculate the voltages across R_1 and the bank. Remember, do not measure these voltages. How does the calculated voltage compare to the applied voltage? (Answer in your report.)

NAME ______ DATE ______

QUESTIONS FOR EXPERIMENT 7-2

Answer true (T) or false (F) to the following:

_____ **1.** Series circuits divide current, and parallel circuits divide voltages.

_____ **2.** Conductance G is the reciprocal of branch current.

_____ **3.** Refer to Fig. 7-2.4. If another resistor, $R_6 = 100\ \Omega$, were added in parallel to the bank, the voltage across the bank would increase.

_____ **4.** Refer to Fig. 7-2.4. If the series resistor R_1 were short-circuited, the total current would decrease.

_____ **5.** Refer to Fig. 7-2.4. If R_3 and R_4 were opened, the voltage across the series resistor R_1 would decrease.

_____ **6.** Refer to Fig. 7-2.4. If the total voltage were halved and the total circuit resistance doubled, there would be no effect upon total current.

_____ **7.** Refer to Fig. 7-2.4. If R_1 were opened, the total circuit current would increase.

_____ **8.** The voltage across any parallel bank is increased as the total conductance of the bank is increased.

_____ **9.** The total current in a parallel bank is inversely proportional to its conductance.

_____ **10.** The total current in a parallel bank is directly proportional to its total resistance.

TABLES FOR EXPERIMENT 7-2

TABLE 7-2.1

	Nominal R, Ω	Calculated I, mA	Total Calculated V, V
R_1	100		
R_2	390		
R_T			

TABLE 7-2.2

	Nominal R, Ω	Measured I, mA	From Table 7-2.1 V, V
R_1	100		
R_2	390		
R_3	820		
R_T			

TABLE 7-2.3

	Nominal R, Ω	Calculated G, S	Calculated I, mA	Calculated V, V*	Measured I, mA
R_1	150		30.0		30.0
R_2	100				
R_3	390				
R_4	560				
R_5	820				
R_T			30.0		3.0

Note: Calculate R_T.
$^*V_x = I$ calculated $\times R$ nominal.

Calculations for Step 2

EXPERIMENT RESULTS REPORT FORM

Experiment No.: ______________________ Name: ______________________

Date: ______________________

Experiment Title: ______________________ Class: ______________________

______________________ Instr: ______________________

Explain the purpose of the experiment:

List the first Learning Objective:

OBJECTIVE 1:

After reviewing the results, describe how the objective was validated by this experiment.

List the second Learning Objective:

OBJECTIVE 2:

After reviewing the results, describe how the objective was validated by this experiment.

List the third Learning Objective:

OBJECTIVE 3:

After reviewing the results, describe how the objective was validated by this experiment.

Conclusion:

If required, attach to this form: ☐ Answers to Questions, ☐ Tables, and ☐ Graphs.

CHAPTER E X P E R I M E N T 7 - 3

POTENTIOMETERS AND RHEOSTATS AS DIVIDERS

LEARNING OBJECTIVES

At the completion of this experiment, you will be able to:

- Identify the circuit configuration of a potentiometer.
- Identify the circuit configuration of a rheostat.
- Contrast the differences and similarities of potentiometers and rheostats used in divider circuits.

SUGGESTED READING

Chapters 2 and 7, *Basic Electronics*, Grob/Schultz, twelfth edition

INTRODUCTION

Potentiometers and rheostats are variable resistances and are used to vary voltage and current in a circuit. A rheostat is a two-terminal device. The potentiometer is a three-terminal device, as shown in Fig. 7-3.1.

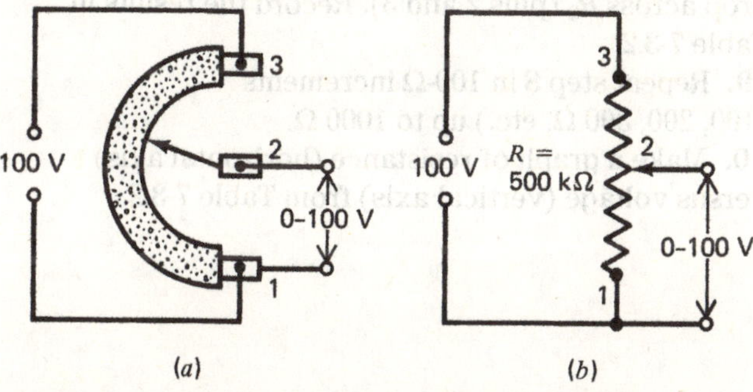

Fig. 7-3.1 Potentiometer connected across voltage source to function as a voltage divider. (*a*) Wiring diagram. (*b*) Schematic diagram.

The maximum resistance is seen between the two end terminals. The middle terminal mechanically adjusts and taps a proportion of this total resistance. A potentiometer can be used as a rheostat by connecting one end terminal to the other, as shown in Fig. 7-3.2.

The primary purpose of a potentiometer (pot) is to tap off a variable voltage from a voltage source, as shown in Fig. 7-3.3.

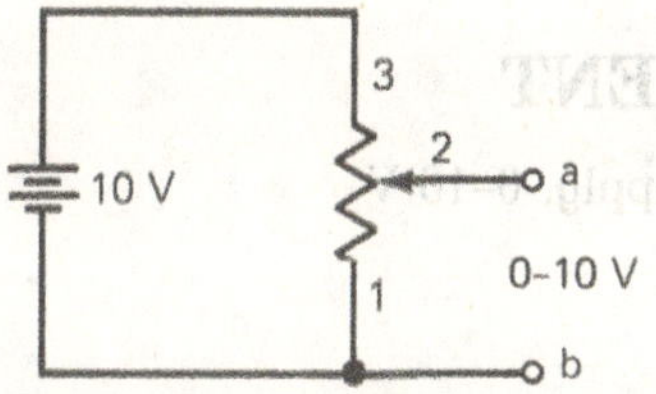

Fig. 7-3.3 Variable voltage source.

As pin 2 is rotated up toward pin 3, the voltage at ab increases until a 10-V level is achieved. If pin 2 is rotated downward toward pin 1, then the voltage present at ab decreases to zero, or approximately zero. The variance of voltages may appear to be presented in a linear or nonlinear fashion, depending upon the manufacturer's type of potentiometer.

The primary purpose of a rheostat is to vary current though a load. This is accomplished by locating the rheostat in series with the load and source voltage. In this way, the total resistance R_T can be varied and indirectly vary the total current. This circuit configuration is shown in Fig. 7-3.4.

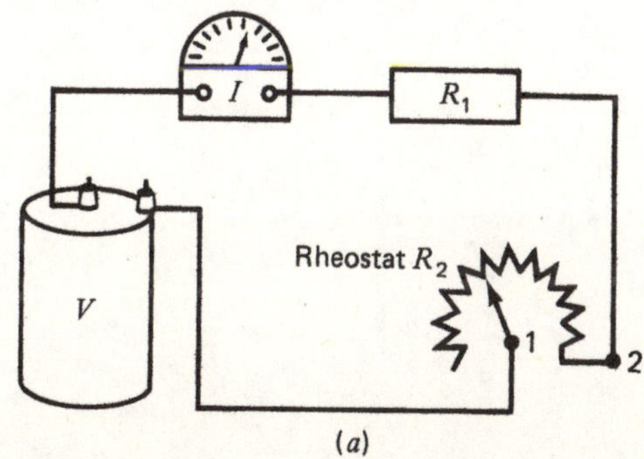

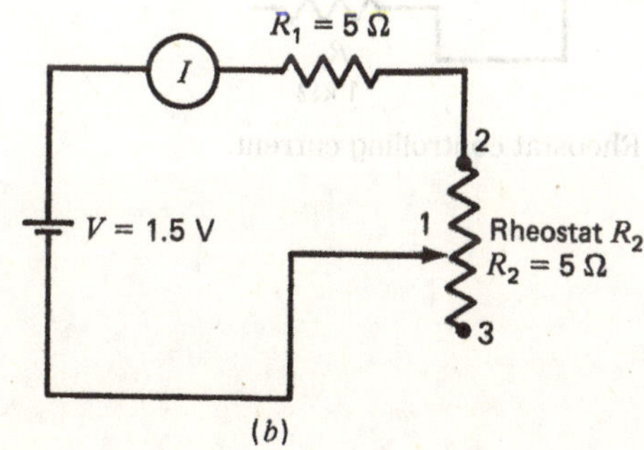

Fig. 7-3.2 Rheostat connected in a series circuit to vary the current. (*a*) Wiring diagram with ammeter to measure *I*. (*b*) Schematic diagram.

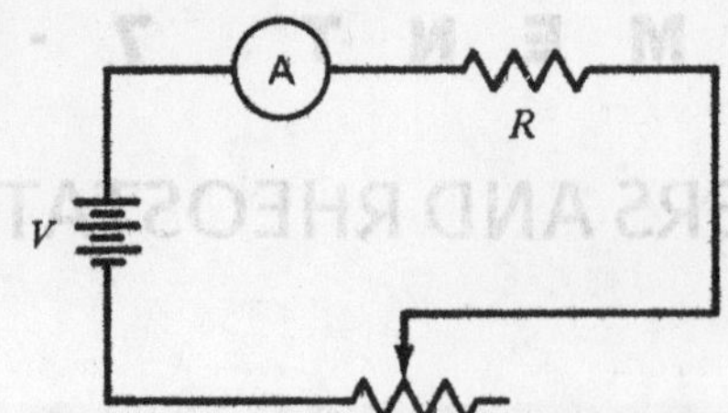

Fig. 7-3.4 Rheostat circuit to vary current.

In summary, rheostats are:

Two-terminal devices.
Found in series with loads and voltage sources.
Used to vary total current.

Potentiometers are:

Three-terminal devices.
Found to have end terminals connected across voltage sources.
Used to tap off part of the voltage source.

EQUIPMENT

DC power supply, 0–10 V
DMM
Protoboard
Test leads

COMPONENTS

(1) 100-Ω, 1-W resistor
(1) 1-kΩ, 1-W potentiometer, linear taper

PROCEDURE

1. Connect the circuit shown in Fig. 7-3.5, where $V = 10$ V, $R_1 = 100$ Ω, and R_2 (pot) – 1 kΩ.

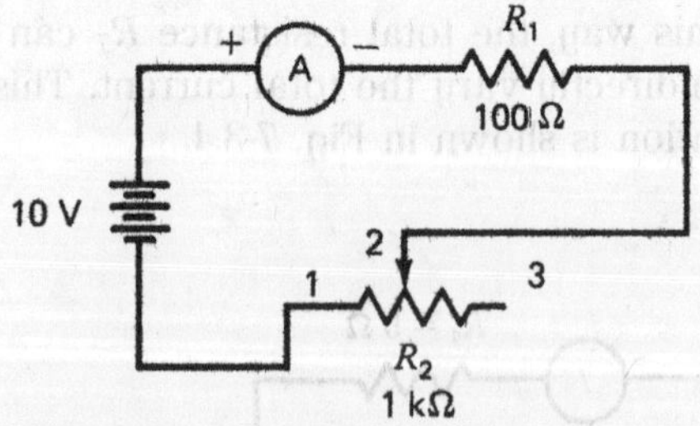

Fig. 7-3.5 Rheostat controlling current.

2. Turn on the power supply, and adjust R_2 for a minimum resistance value (maximum I).

3. Remove R_2 from the circuit and connect it to an ohmmeter (points 1 and 2). Set R_2 to 100 Ω and reconnect R_2 to the circuit. Measure the current flow and record in Table 7-3.1.

4. Repeat step 3 in 100-Ω increments (100, 200, 300 Ω, etc.) up to 1000 Ω.

5. Make a graph of resistance (horizontal axis) versus current (vertical axis) from Table 7-3.1.

6. Connect the circuit shown in Fig. 7-3.6, where $V = 10$ V and R_2 (pot) = 1 kΩ.

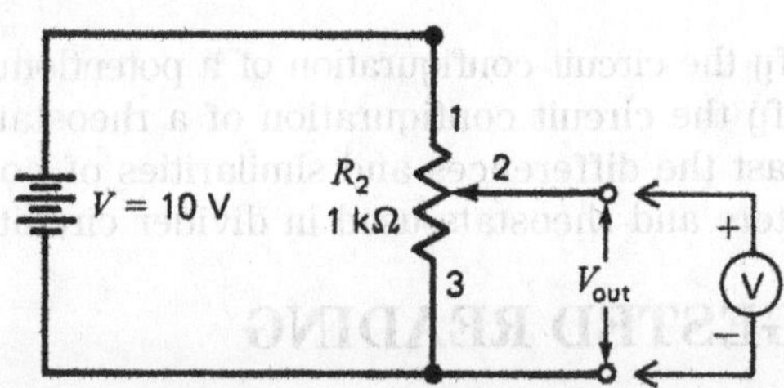

Fig. 7-3.6 Potentiometer voltage divider.

7. Turn on the power supply, and adjust R_2 for a minimum resistance value (minimum V).

8. Remove R_2 from the circuit and connect it to an ohmmeter (points 2 and 3). Set R_2 to 100 Ω, and reconnect R_2 to the circuit. Measure the voltage drop across R_2 (pins 2 and 3). Record the results in Table 7-3.2.

9. Repeat step 8 in 100-Ω increments (100, 200, 300 Ω, etc.) up to 1000 Ω.

10. Make a graph of resistance (horizontal axis) versus voltage (vertical axis) from Table 7-3.2.

NAME ______________________ DATE ____________

QUESTIONS FOR EXPERIMENT 7-3

1. How many circuit connections to a potentiometer are needed?

2. How many circuit connections to a rheostat are needed?

3. Determine maximum power consumption from the graphs you completed in steps 5 and 10. What are the actual necessary wattages of R_1 and R_2?

TABLES FOR EXPERIMENT 7-3

TABLE 7-3.1

Resistance, Ω	Measured Current
100	______
200	______
300	______
400	______
500	______
600	______
700	______
800	______
900	______
1000	______

TABLE 7-3.2

Resistance, Ω	Measured Voltage
100	______
200	______
300	______
400	______
500	______
600	______
700	______
800	______
900	______
1000	______

GRAPH FOR EXPERIMENT 7-3

Graph 7-3.1 Graphing results from Table 7-3.2 where resistance is located on the *horizontal axis* versus voltage on the *vertical axis.*

EXPERIMENT RESULTS REPORT FORM

Experiment No.: ______________________ Name: ______________________

Date: ______________________

Experiment Title: ______________________ Class: ______________________

______________________ Instr: ______________________

Explain the purpose of the experiment:

List the first Learning Objective:

OBJECTIVE 1:

After reviewing the results, describe how the objective was validated by this experiment.

List the second Learning Objective:

OBJECTIVE 2:

After reviewing the results, describe how the objective was validated by this experiment.

List the third Learning Objective:

OBJECTIVE 3:

After reviewing the results, describe how the objective was validated by this experiment.

Conclusion:

If required, attach to this form: ☐ Answers to Questions, ☐ Tables, and ☐ Graphs.

CHAPTER 7

EXPERIMENT 7-4

VOLTAGE DIVIDER DESIGN

LEARNING OBJECTIVES

At the completion of this experiment, you will be able to:

- Design a voltage divider for given load requirements.
- Understand the concept of negative voltage.
- Understand the concepts of common ground and bleeder current.

SUGGESTED READING

Chapter 7, *Basic Electronics*, Grob/Schultz, twelfth edition

INTRODUCTION

This experiment requires that you design and test a loaded voltage divider circuit. You can imagine that you are actually designing this divider for the output of a power supply in any piece of electronic equipment. Often, designing a circuit teaches you more than simply measuring and analyzing data.

In previous experiments, you became familiar with the concepts of voltage and current dividers. You may recall that the loads on a voltage divider are also current dividers because they are parallel circuits.

To review, consider the circuit of Fig. 7-4.1.

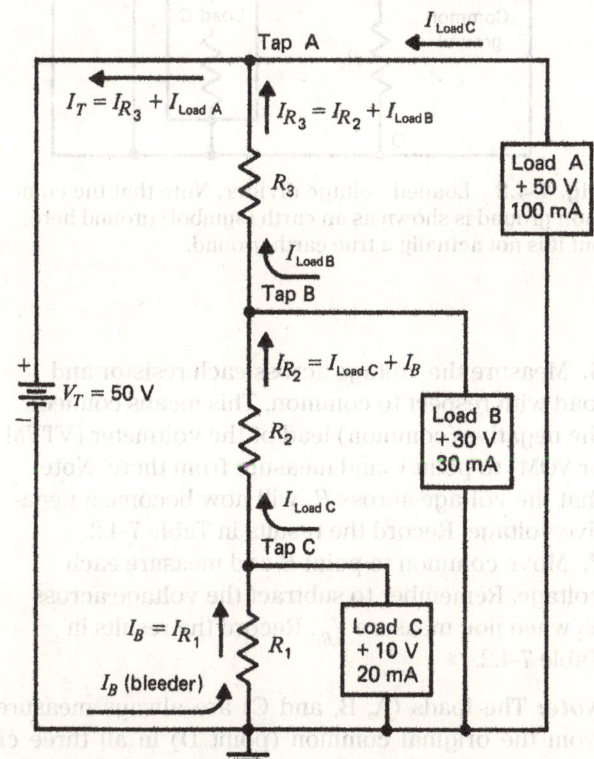

Fig. 7-4.1 Loaded voltage divider.

1. *Loads:* The loads are resistances that require specific voltages and, often, specific currents. Here, imagine each load is a separate circuit board that requires certain voltages and currents in order to properly perform a function such as amplification.
2. I_B: This is known as *bleeder current*. It is a steady drain on the source and has no-load currents passing through it. Typically, bleeder currents are calculated to be 10 percent of the total current.
3. V_T: Total voltage is fixed. It is like a budget that the designer is allowed to work with.
4. *Taps:* These are the places where the loads are connected in order to tap off their specified voltages.
5. R_1, R_2, and R_3: These are the voltage divider resistances that you, the designer, will be determining by your calculations. In this case, Fig. 7-4.1 has a given budget of 50 V (total). Your calculations will divide this voltage into the necessary values to supply the loads.
6. *Ground:* This symbol shows a common or earth ground that is connected to one side of the power supply. Remember that ground is a reference point, like zero.

How to Calculate the Voltage Divider?

Use a table, as shown in Table 7-4.1, in order to keep your values organized. Find the current in each R.

$$I_{R_1} = I_B = \text{approx. 10 percent of } I_T \text{ loads}$$

Thus

$$I_B = 150 \text{ mA} \times 0.1 = 15 \text{ mA}$$
$$I_{\text{Load A}} = 100 \text{ mA}$$
$$I_{\text{Load B}} = 30 \text{ mA}$$
$$I_{\text{Load C}} = 20 \text{ mA}$$
$$I_T = 150 \text{ mA}$$

Knowing the value of $I_B = I_{R_1}$, I_{R_2} and I_{R_3} are calculated as follows (compare to Fig. 7-4.1):

$$I_{R_2} = I_B + I_{\text{Load C}} = 15 \text{ mA} + 20 \text{ mA} = 35 \text{ mA}$$
$$I_{R_3} = I_{R_2} + I_{\text{Load B}} = 35 \text{ mA} + 30 \text{ mA} = 65 \text{ mA}$$

How to Calculate the Voltage across Each Resistor?

The voltages at the taps are also the voltages across the loads with respect to ground. However, the voltages across R_2 and R_3 are not in parallel only with these loads. Therefore, only the voltage across R_1 is the same as $V_{\text{Load C}} = 10$ V. V_{R_2} and V_{R_3} are calculated as follows (compare to Fig. 7-4.1):

$$V_{R_1} = V_{\text{Tap C}} = 10 \text{ V}$$
$$V_{R_2} = V_{\text{Tap B}} - V_{\text{Tap C}} = 30 \text{ V} - 10 \text{ V} = 20 \text{ V}$$
$$V_{R_3} = V_{\text{Tap A}} - V_{\text{Tap B}} = 50 \text{ V} - 30 \text{ V} = 20 \text{ V}$$

How to Calculate Each Resistor?

Now that the voltages and currents for each resistor have been determined, it is easy to use Ohm's law to calculate the value of each resistor:

$$R_1 = \frac{V_{R_1}}{I_{R_1}} = \frac{10 \text{ V}}{15 \text{ mA}} = 666.7 \ \Omega$$

$$R_2 = \frac{V_{R_2}}{I_{R_2}} = \frac{20 \text{ V}}{35 \text{ mA}} = 571.4 \ \Omega$$

$$R_3 = \frac{V_{R_3}}{I_{R_3}} = \frac{20 \text{ V}}{65 \text{ mA}} = 307.7 \ \Omega$$

These values of voltage divider resistance should provide the specified load requirements.

EQUIPMENT

DC power supply
DMM
Protoboard or springboard
Leads

COMPONENTS

Resistors (all 0.25 W):

(2) 100 Ω
(1) 120 Ω
(1) 150 Ω
(4) 470 Ω
(1) 560 Ω

PROCEDURE

1. Design a voltage divider, similar to Fig. 7-4.1, with the following specifications:

$V_T = 20$ V
Load A = 20 V, 40 mA
Load B = 15 V, 25 mA
Load C = 5 V, 10 mA

2. On a separate sheet, draw the circuit similar to Fig. 7-4.1 and fill in Table 7-4.1 (columns 1, 2, and 4).

3. Verify that your circuit works by connecting the circuit and measuring the voltages across R_1, R_2, R_3, and each load. Use resistors within 10 percent of the load values. Record the voltages (measured) in Table 7-4.1 (column 3). If any value of voltage falls outside plus or minus 20 percent of your calculated values, redesign the circuit.

4. With your circuit connected, remove earth ground (if attached).

5. Using the circuit of Fig. 7-4.2 as an example, you will move the ground point (common: without an earth ground connection) to point C. Although no actual connection is yet made, point C now becomes the zero reference point.

Note: $V_{R_3} = V_{\text{Point A}} - V_{R_2}$.

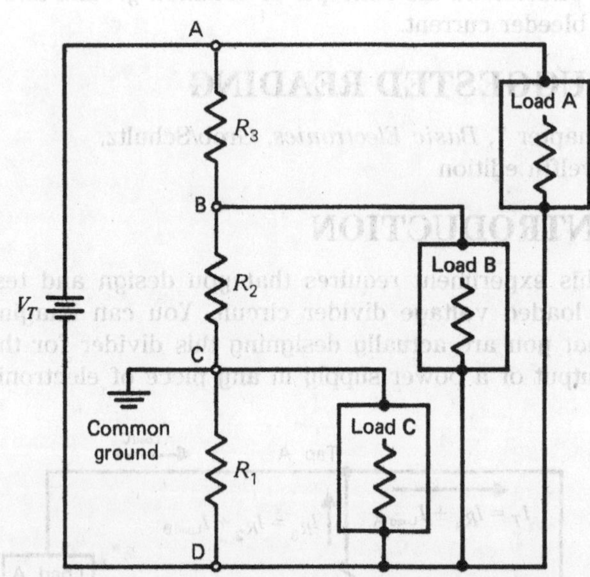

Fig. 7-4.2 Loaded voltage divider. Note that the common ground is shown as an earth (symbol) ground here, but it is not actually a true earth ground.

6. Measure the voltage across each resistor and load with respect to common. This means connect the negative (common) lead of the voltmeter (VTVM or VOM) to point C and measure from there. Note that the voltage across R_1 will now become a negative voltage. Record the results in Table 7-4.2.

7. Move common to point B and measure each voltage. Remember to subtract the voltage across R_2 when you measure V_{R_1}. Record the results in Table 7-4.2.

Note: The loads (A, B, and C) are always measured from the original common (point D) in all three circuits. But the divider voltages are measured from point B, C, or D, depending on the configurations.

NAME ______________________ DATE __________

QUESTIONS FOR EXPERIMENT 7-4

1. **Explain the difference between earth ground and a common reference point ground.**

2. **Explain how a negative voltage can be obtained from a voltage divider. Explain the effects upon *I*, *V*, and *R*, if any.**

3. **Explain, in your own words, what is meant by *bleeder current*.**

4. **Explain the difference, if any, between a loaded voltage divider and a series-parallel circuit.**

5. **Explain how the circuit of Fig. 7-4.1 would be affected if:**
   A. R_2 **were short-circuited** **B.** R_2 **were opened**

TABLES FOR EXPERIMENT 7-4

TABLE 7-4.1

Divider	(1) Calculated R Ω	(2) Actual R Value, Ω	(3) Measured V, V	(4) Calculated I, mA
Load A				
Load B				
Load C				
R_1				
R_2				
R_3				

TABLE 7-4.2

Divider	Measured V, V Common = Point C	Measured V, V Common = Point B
Load A		
Load B		
Load C		
R_1		
R_2		
R_3		

EXPERIMENT RESULTS REPORT FORM

Experiment No.: ____________________ Name: ____________________
Date: ____________________
Experiment Title: ____________________ Class: ____________________
____________________ Instr: ____________________

Explain the purpose of the experiment:

List the first Learning Objective:
OBJECTIVE 1:

After reviewing the results, describe how the objective was validated by this experiment.

List the second Learning Objective:
OBJECTIVE 2:

After reviewing the results, describe how the objective was validated by this experiment.

List the third Learning Objective:
OBJECTIVE 3:

After reviewing the results, describe how the objective was validated by this experiment.

Conclusion:

If required, attach to this form: ☐ Answers to Questions, ☐ Tables, and ☐ Graphs.

CHAPTER 8

EXPERIMENT 8-1

ANALOG AMMETER DESIGN

LEARNING OBJECTIVES

At the completion of this experiment, you will be able to:

- Determine the internal resistance of a basic D'Arsonval meter movement.
- Design an ammeter circuit from this meter movement.
- Understand how shunts are used with coils.

SUGGESTED READING

Chapter 8, *Basic Electronics*, Grob/Schultz, twelfth edition

INTRODUCTION

Range of an Ammeter

The small size of the wire with which an ammeter's movable coil is wound places severe limits on the current that may be passed through the coil. Consequently, the basic D'Arsonval movement may be used to indicate or measure only very small currents—for example, microamperes or milliamperes, depending on meter sensitivity.

To measure a larger current, a shunt must be used with the meter. A shunt is a heavy, low-resistance conductor connected across the meter terminals to carry most of the load current. This shunt has the correct amount of resistance to cause only a small part of the total circuit current to flow through the meter coil. The meter current is proportional to the load current. If the shunt is of such a value that the meter is calibrated in milliamperes, the instrument is called a *milliammeter*. If the shunt is of such a value that the meter is calibrated in amperes, it is called an *ammeter*.

A single type of standard meter movement is generally used in all ammeters, no matter what the range of a particular meter is. For example, meters with working ranges of 0 to 10 A, 0 to 5 A, or 0 to 1 A all use the same galvanometer movement. The designer of the ammeter calculates the correct shunt resistance required to extend the range of the meter movement to measure any desired amount of current. This shunt is then connected across the meter terminals. Shunts may be located inside the meter case (internal shunt) or somewhere away from the meter (external shunt), with leads going to the meter.

Extending the Range by Use of Shunts

For limited current ranges (below 50 A), internal shunts are most often employed. In this manner, the range of the meter may be easily changed by selecting the correct internal shunt having the necessary current rating. Before the required resistance of the shunt for each range can be calculated, the resistance of the meter movement must be known.

For example, suppose it is desired to use a 100-μA D'Arsonval meter having a resistance of 100 Ω to measure line currents up to 1 A. The meter deflects full scale when the current through the 100-Ω coil is 100 μA. Therefore, the voltage drop across the meter coil is *IR*, or

$$0.0001 \times 100 = 0.01 \text{ V}$$

Because the shunt and coil are in parallel, the shunt must also have a voltage drop of 0.01 V. The current that flows through the shunt is the difference between the full-scale meter current and the line current. In this case, the meter current is 0.0001 A. This current is negligible compared with the line (shunt) current, so the shunt current is approximately 1 A. The resistance R_S of the shunt is therefore,

$$R_S = \frac{V}{I} = \frac{0.01}{1} = 0.01\ \Omega \text{ (approx.)}$$

and the range of the 100-μA meter has been increased to 1 A by paralleling it with the 0.01-Ω shunt.

The 100-μA instrument may also be converted to a 10-A meter by the use of a proper shunt. For full-scale deflection of the meter, the voltage drop V across the shunt (and across the meter) is still 0.01 V. The meter current is again considered negligible, and the shunt current is now approximately 10 A. The resistance R_S of the shunt is therefore,

$$R_S = \frac{V}{I} = \frac{0.01}{10} = 0.001\ \Omega$$

The same instrument may likewise be converted to a 50-A meter by the use of the proper type of shunt. The current I_S through the shunt is approximately 50 A, and the resistance R_S of the shunt is

$$R_S = \frac{V}{I_S} = \frac{0.01}{50} = 0.0002\ \Omega$$

EQUIPMENT

DC power supply
Protoboard or springboard
Leads
DMM

COMPONENTS

(1) 0- to 1-mA meter movement

Resistors:

(1) 150 kΩ, 0.25 W   Other resistors as calculated.
(1) 470 Ω, 0.25 W

Potentiometers:

(1) 5 kΩ   (1) 100 kΩ

(1) SPST switch
(1) SPDT switch

PROCEDURE

1. Measure the internal resistance of the meter movement by connecting the circuit shown in Fig. 8-1.1.

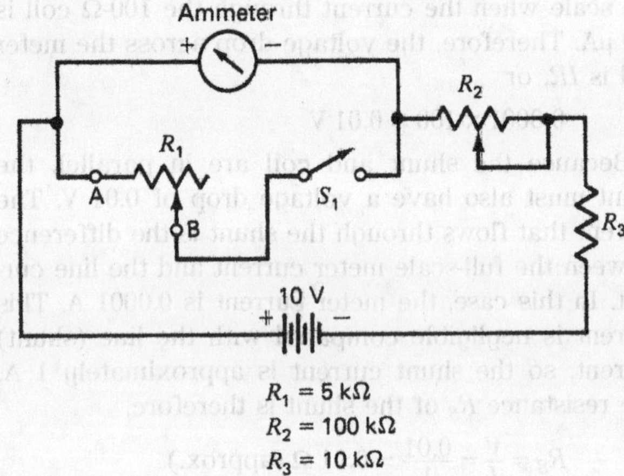

Fig. 8-1.1 Internal resistance measurement.

2. With S_1 open, turn on the power supply and adjust it for 10 V.

3. Adjust R_2 so that the scale upon the meter movement reads at full-scale deflection.

4. Close S_1 and adjust R_1 so that the scale upon the meter movement reads at half-scale deflection. The currents will evenly divide between R_1 and the internal resistance r_m of the meter movement when $R_1 = r_m$.

5. Measure and record in Table 8-1.1 the voltage dropped across V_m.

6. Measure and record r_m in Table 8-1.1 by turning off the power supply, disconnecting R_1 from the circuit, and measuring from point A to B. At this point, $r_m = R_1$.

7. Calculate and record I_m in Table 8-1.1, where

$$I_m = \frac{V_m}{r_m}$$

8. Record in Table 8-1.2 the value of I_m for full-scale deflection, the r_m (meter movement's internal resistance) for the meter movement, and the value V_m needed for full-scale deflection.

9. Construct the following dual-range ammeter in Fig. 8-1.2. Range 1 will measure 30 mA full scale, and range 2 will measure 100 mA full scale.

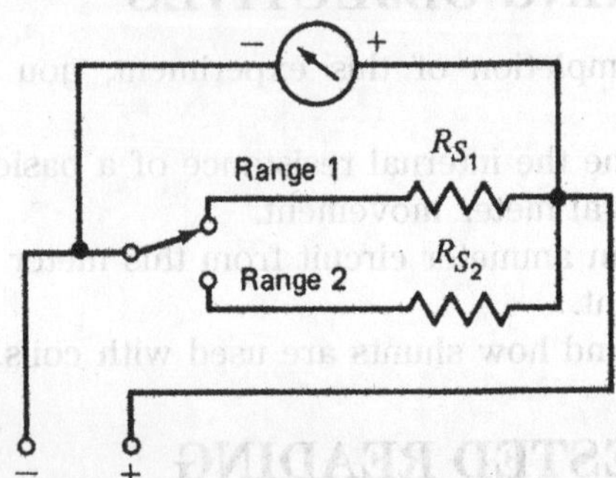

Fig. 8-1.2 Ammeter circuit.

10. Using the following formula, determine the multiplier resistors R_{S_1} and R_{S_2}:

$$R_S = (I_m \times r_m)I_S$$

For a 30-mA full-scale deflection,

$$R_{S_1} = (I_m \times r_m)/30\text{ mA}$$

Record this value in Table 8-1.2.

For a 100-mA full-scale deflection,

$$R_{S_2} = (I_m \times r_m)/100\text{ mA}$$

Record this value in Table 8-1.2.

11. Connect your ammeter into the circuit configuration shown in Fig. 8-1.3, where I_2 is an ammeter of known accuracy and I_1 is your ammeter design on another meter movement. Complete Table 8-1.3 by turning on and adjusting the power supply in accordance with Tablc 8 1.3. Record the values of I_1 and I_2 for each power supply setting.

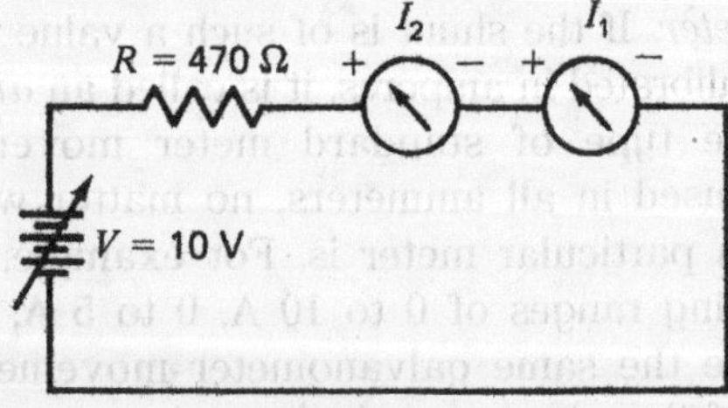

Fig. 8-1.3 Ammeter test setup.

12. Determine the percentage of accuracy for Table 8-1.3.

NAME _______________ DATE _______________

QUESTIONS FOR EXPERIMENT 8-1

1. Describe how ammeters are connected in a circuit to measure current.

2. Design an ammeter circuit that will measure 1.5 A with a 0- to 100-mA full-scale deflection meter movement.

REPORT

Write a complete report. Discuss the measured and calculated results. Also, discuss the three most significant aspects of the experiment, and write a conclusion.

TABLES FOR EXPERIMENT 8-1

TABLE 8-1.1

r_m	I_m	V_m

TABLE 8-1.2 Meter Movement

I_m, A	r_m, Ω	V_m, V
Shunt R_{S_1}	**Shunt R_{S_2}**	

TABLE 8-1.3

Range, mA	Voltage Setting	I_1	I_2	% Accuracy
0–30				
0–100				

EXPERIMENT RESULTS REPORT FORM

Experiment No.: ______________________ Name: ______________________

Date: ______________________

Experiment Title: ______________________ Class: ______________________

______________________ Instr: ______________________

Explain the purpose of the experiment:

List the first Learning Objective:

OBJECTIVE 1:

After reviewing the results, describe how the objective was validated by this experiment.

List the second Learning Objective:

OBJECTIVE 2:

After reviewing the results, describe how the objective was validated by this experiment.

List the third Learning Objective:

OBJECTIVE 3:

After reviewing the results, describe how the objective was validated by this experiment.

Conclusion:

If required, attach to this form: ☐ Answers to Questions, ☐ Tables, and ☐ Graphs.

EXPERIMENT 8-2

ANALOG VOLTMETER DESIGN

LEARNING OBJECTIVES

At the completion of this experiment, you will be able to:

- Determine the internal resistance of a basic D'Arsonval movement.
- Design a voltmeter from this meter movement.
- Understand coil resistance.

SUGGESTED READING

Chapter 8, *Basic Electronics*, Grob/Schultz, twelfth edition

INTRODUCTION

D'Arsonval Meter

The stationary permanent-magnet moving-coil meter is the basic movement used in most measuring instruments for servicing electric equipment. This type of movement is commonly called the D'Arsonval movement because it was first employed by the Frenchman D'Arsonval in making electrical measurements.

The basic D'Arsonval movement consists of a stationary permanent magnet and a movable coil. When current flows through the coil, the resulting magnetic field reacts with the magnetic field of the permanent magnet and causes the coil to rotate. The greater the amount of current flow through the coil, the stronger the magnetic field produced; the stronger this field, the greater the rotation of the coil. To determine the amount of current flow, a means must be provided to indicate the amount of coil rotation.

Voltmeter

The 100-μA D'Arsonval meter used as the basic meter for the ammeter may also be used to measure voltage if a high resistance is placed in series with the moving coil of the meter. When this is done, the unit containing the resistance is commonly called a *multiplier*. A simplified diagram of a voltmeter is shown in Fig. 8-2.1.

Extending the Range

The value of the necessary series resistance is determined by the current required for full-scale deflection of the meter and by the range of voltage to be measured. Because the current through the meter circuit is directly proportional to the applied voltage, the meter scale can be calibrated directly in volts for a fixed series resistance.

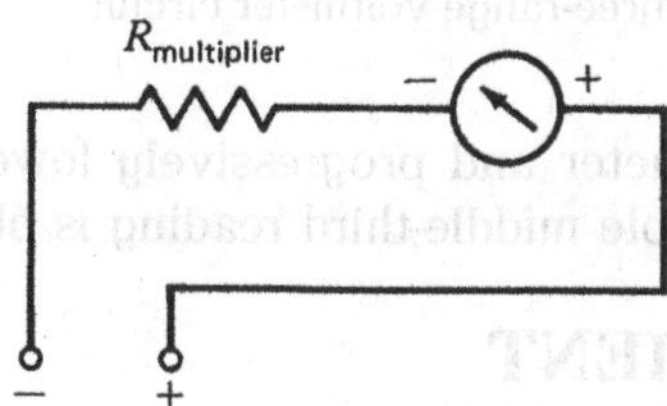

Fig. 8-2.1 Basic voltmeter circuit with $R_{multiplier}$.

For example, assume that the basic meter (microammeter) is to be made into a voltmeter with a full-scale reading of 1 V. The coil resistance of the basic meter is 100 Ω, and 0.0001 A causes a full-scale deflection. The total resistance R of the meter coil and the series resistance is

$$r_m = \frac{V_m}{I_m}$$

$$= \frac{1}{100\ \mu A}$$

$$= 10{,}000\ \Omega$$

and the series resistance alone is

$$R_1 = 10{,}000 - 100$$
$$= 9900\ \Omega$$

Multirange voltmeters utilize 1 meter movement with a convenient switching arrangement. A multirange voltmeter with three ranges is shown in Fig. 8-2.2. The total circuit resistance for each of the three ranges, beginning with the 1-V range, is

$$R_1 = \frac{V_m}{I_m} = \frac{1}{100} = 0.01\ M\Omega$$

$$R_2 = \frac{V_m}{I_m} = \frac{100}{100} = 1\ M\Omega$$

$$R_3 = \frac{V_m}{I_m} = \frac{1000}{100} = 10\ M\Omega$$

Voltage-measuring instruments are always connected across (in parallel with) a circuit. If the approximate value of the voltage to be measured is not known, it is best to start with the highest range

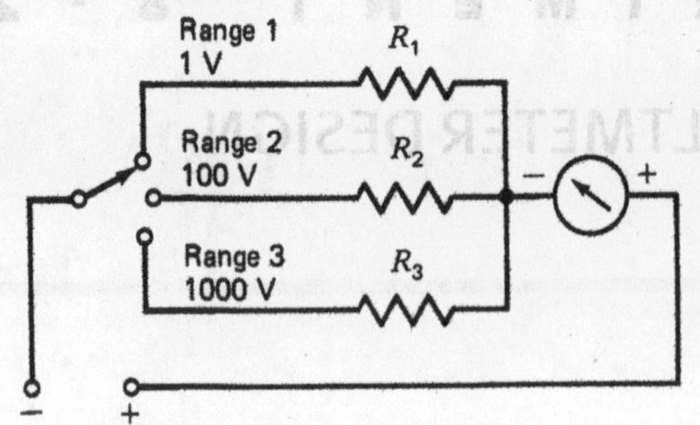

Fig. 8-2.2 Three-range voltmeter circuit.

of the voltmeter and progressively lower the range until a suitable middle-third reading is obtained.

EQUIPMENT

DC power supply, 10 V
Voltmeter of known accuracy
Protoboard or springboard
Test leads
DMM

COMPONENTS

Resistors:

(1) 150 kΩ, 0.25 W Other resistors as calculated.

Potentiometers (linear taper):

(1) 100 kΩ (1) 5 kΩ

(1) SPST switch
(1) SPDT switch
(1) 50-μA meter movement of unknown resistance

PROCEDURE

1. Measure the internal resistance of the meter movement by connecting the circuit shown in Fig. 8-2.3.

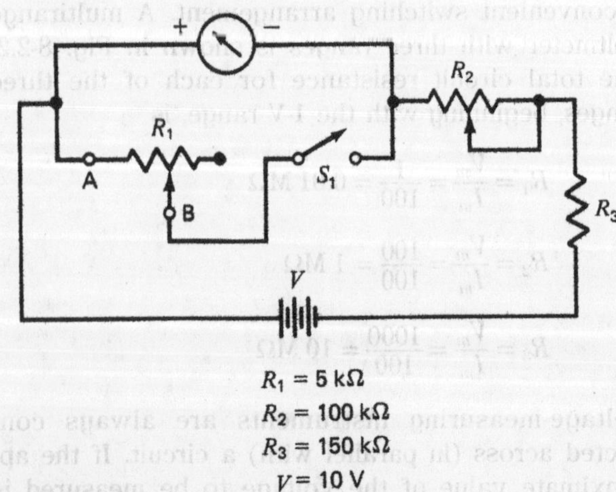

Fig. 8-2.3 Measuring internal resistance.

2. With S_1 open, turn on the power supply and adjust it for 10 V.

3. Adjust R_2 so that the scale of the meter movement is at full-scale deflection.

4. Close S_1 and adjust R_1 so that the scale upon the movement is at half-scale deflection. The currents will evenly divide between R_1 and the internal resistance r_m of the meter movement when $R_1 = r_m$.

5. Measure and record in Table 8-2.1 the voltage dropped across V_m.

6. Measure and record r_m in Table 8-2.1 by turning off the power supply, disconnecting R_1 from the circuit, and measuring from points A to B. At this point, $r_m = R_1$.

7. Calculate and record I_m in Table 8-2.1, where

$$I_m = \frac{V_m}{r_m}$$

8. Record in Table 8-2.2 the I_m for full-scale deflection, the r_m (meter movement's internal resistance) for the meter movement, and the value of V_m for full-scale deflection.

9. Construct the dual-range voltmeter in Fig. 8-2.4 so that range 1 will measure 5 V full scale and range 2 will measure 10 V full scale. Use the following formula to determine the multiplier resistors R_1 and R_2:

$$R_{\text{multiplier}} = \frac{V_{\text{FS}}}{I_{\text{FS}}} - r_m$$

$$= \frac{V_{\text{intended}}}{I_m} - r_m$$

Note: FS means full scale.

For a 5-V full-scale deflection,

$$R_1 = \frac{5\ \text{V}}{I_m} - r_m$$

For a 10-V full-scale deflection,

$$R_2 = \frac{10\ \text{V}}{I_m} - r_m$$

Record these calculated values in Table 8-2.2.

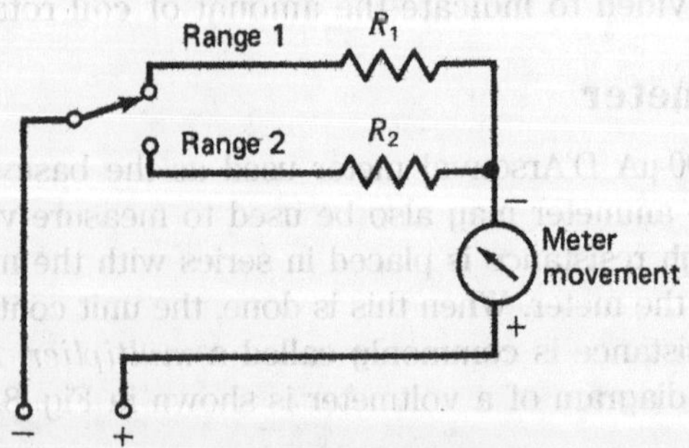

Fig. 8-2.4 Voltmeter dual-range circuit.

10. Connect your voltmeter into the circuit configuration of Fig. 8-2.5, where V_2 is a voltmeter of known accuracy and V_1 is your voltmeter design.

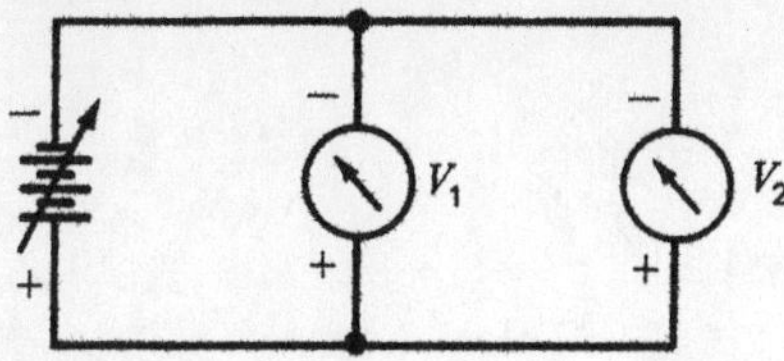

Fig. 8-2.5 Measuring voltages.

11. Complete Table 8-2.3 by turning on and adjusting the power supply in accordance with Table 8-2.3. Record the values of V_1 and V_2 for each power supply setting.

12. Determine the percentage of accuracy for Table 8-2.3.

NAME ______________________ DATE ______________

QUESTION FOR EXPERIMENT 8-2

1. **Design a voltmeter that will measure 0 to 30 V DC by using a 100-mA meter movement.**

TABLES FOR EXPERIMENT 8-2

TABLE 8-2.1

0–50 µA
V_m = ______
r_m = ______
I_m = ______

TABLE 8-2.2 Meter Movement

I_m = ______

r_m = ______

V_m = ______

R_1 = ______

R_2 = ______

TABLE 8-2.3

Power Supply Voltages	V_1	V_2	% Accuracy
Range 1: 1–5 V			
1 V	______	______	______
2 V	______	______	______
3 V	______	______	______
4 V	______	______	______
5 V	______	______	______
Range 2: 1–10 V			
1 V	______	______	______
2 V	______	______	______
3 V	______	______	______
4 V	______	______	______
5 V	______	______	______
6 V	______	______	______
7 V	______	______	______
8 V	______	______	______
9 V	______	______	______
10 V	______	______	______

EXPERIMENT RESULTS REPORT FORM

Experiment No.: ______________________ Name: ______________________

Date: ______________________

Experiment Title: ______________________ Class: ______________________

______________________ Instr: ______________________

Explain the purpose of the experiment:

List the first Learning Objective:
OBJECTIVE 1:

After reviewing the results, describe how the objective was validated by this experiment.

List the second Learning Objective:
OBJECTIVE 2:

After reviewing the results, describe how the objective was validated by this experiment.

List the third Learning Objective:
OBJECTIVE 3:

After reviewing the results, describe how the objective was validated by this experiment.

Conclusion:

If required, attach to this form: ☐ Answers to Questions, ☐ Tables, and ☐ Graphs.

CHAPTER 8

EXPERIMENT 8-3

ANALOG OHMMETER DESIGN

LEARNING OBJECTIVES

At the completion of this experiment, you will be able to:

- Determine the internal resistance of a basic D'Arsonval meter movement.
- Design an ohmmeter from this meter movement.
- Understand how current works in a coil.

SUGGESTED READING

Chapter 8, *Basic Electronics*, Grob/Schultz, twelfth edition

INTRODUCTION

The ohmmeter consists of a DC milliammeter, with a few added features. The added features are:

1. A DC source of potential
2. One or more resistors (one of which is variable)

A simple ohmmeter circuit is shown in Fig. 8-3.1.

The ohmmeter's pointer deflection is controlled by the amount of battery current passing through the moving coil. Before measuring the resistance of an unknown resistor or electric circuit, the test leads of the ohmmeter are first short-circuited together, as shown in Fig. 8-3.1. With the leads short-circuited, the meter is calibrated for proper operation on the selected range. (While the leads are short-circuited, meter current is maximum and the pointer deflects a maximum amount, somewhere near the zero position on the ohms scale.) When the variable resistor is adjusted properly, with the leads short-circuited, the meter pointer will come to rest exactly on the zero graduation. This indicates *zero resistance* between the test leads, which in fact are short-circuited together. The zero readings of series-type ohmmeters are sometimes on the right-hand side of the scale, whereas the zero reading for ammeters and voltmeters is generally to the left-hand side of the scale. When the test leads of an ohmmeter are separated, the meter pointer will return to the left side of the scale, due to the interruption of current and the spring tension acting on the movable-coil assembly.

After the ohmmeter is adjusted for zero reading, it is ready to be connected in a circuit to measure resistance. A typical circuit and ohmmeter arrangement is shown in Fig. 8-3.2.

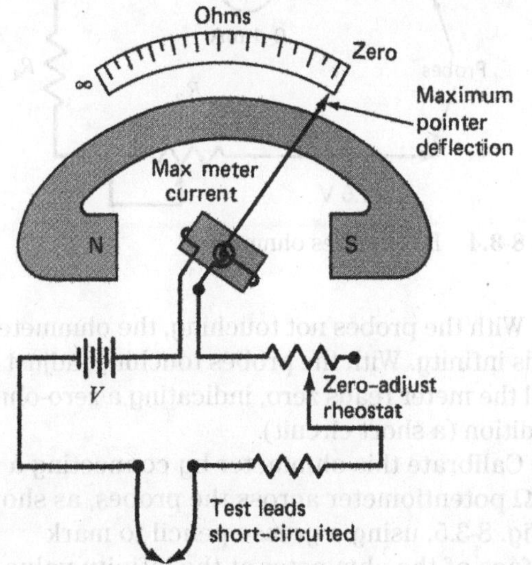

Fig. 8-3.1 Simple series ohmmeter.

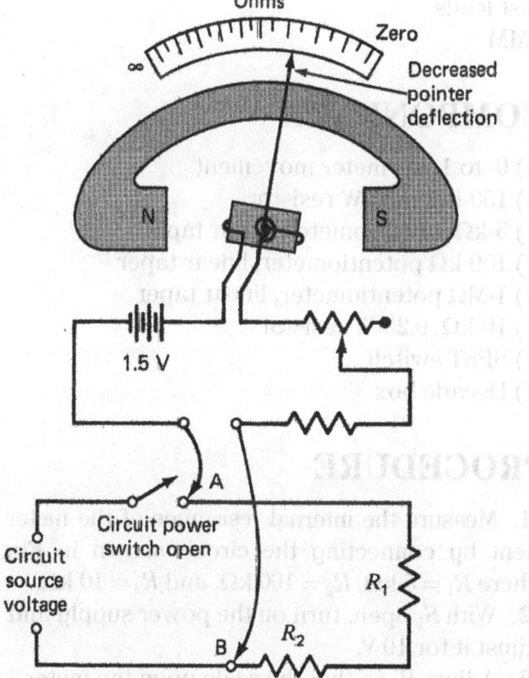

Fig. 8-3.2 Typical ohmmeter arrangement.

The power switch of the circuit to be measured should always be in the OFF position. This prevents the circuit's source voltage from being applied across the meter, which could cause damage to the meter movement.

The test leads of the ohmmeter are connected across (in parallel with) the circuit to be measured (see Fig. 8-3.2). This causes the current produced by the meter's internal battery to flow through the circuit being tested. Assume that the meter test leads are connected at points A and B of Fig. 8-3.2. The amount of current that flows through the meter coils will depend on the resistance of resistors R_1 and R_2, plus the resistance of the meter. Since the meter has been preadjusted (zeroed), the amount of coil movement now depends solely upon the resistance of R_1 and R_2. The inclusion of R_1 and R_2 raised the total series resistance, decreased the current, and thus decreased the pointer deflection. The pointer will now come to rest at a scale figure indicating the combined resistance of R_1 and R_2. If R_1 or R_2, or both, were replaced with a resistor(s) having a larger ohmic value, the current flow in the moving coil of the meter would be decreased still more. The deflection would also be further decreased, and the scale indication would read a still higher circuit resistance. Movement of the moving coil is proportional to the amount of current flow. The scale reading of the meter, in ohms, is inversely proportional to current flow in the moving coil.

EQUIPMENT

DC power supply, 0–10 V
Test leads
DMM

COMPONENTS

(1) 0- to 1-mA meter movement
(1) 150-kΩ, 0.25-W resistor
(1) 5-kΩ potentiometer, linear taper
(1) 100-kΩ potentiometer, linear taper
(1) 1-MΩ potentiometer, linear taper
(1) 10-kΩ, 0.25-W resistor
(1) SPST switch
(1) Decade box

PROCEDURE

1. Measure the internal resistance of the meter movement by connecting the circuit shown in Fig. 8-3.3, where $R_1 = 5$ kΩ, $R_2 = 100$ kΩ, and $R_3 = 10$ kΩ.

2. With S_1 open, turn on the power supply and adjust it for 10 V.

3. Adjust R_2 so that the scale upon the meter movement reads at full-scale deflection.

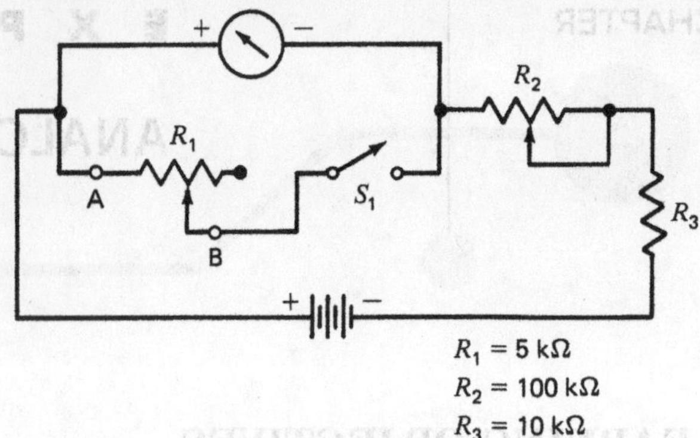

Fig. 8-3.3 Measuring internal resistance.

4. Close S_1 and adjust R_1 so that the scale upon the movement is at half-scale deflection. The currents will evenly divide between R_1 and the internal r_m of the meter movement when $R_1 = r_m$.

5. Measure and record in Table 8-3.1 the voltage dropped across V_m.

6. Measure and record r_m in Table 8-3.1 by turning the power supply off, disconnecting R_1 from the circuit, and measuring from points A to B. At this point, $r_m = R_1$.

7. Calculate and record I_m in Table 8-3.1, where

$$I_m = \frac{V_m}{r_m}$$

8. Record in Table 8-3.1 the r_m for full-scale deflection of a 0- to 1-mA meter movement.

9. Construct the series-type ohmmeter shown in Fig. 8-3.4, where $R_1 = 1.5\ \text{k}\Omega - r_m$. (The resistance value of R_1 may have to be created by using a resistance decade box.) R_2 will be used to set the ohmmeter to zero ohms.

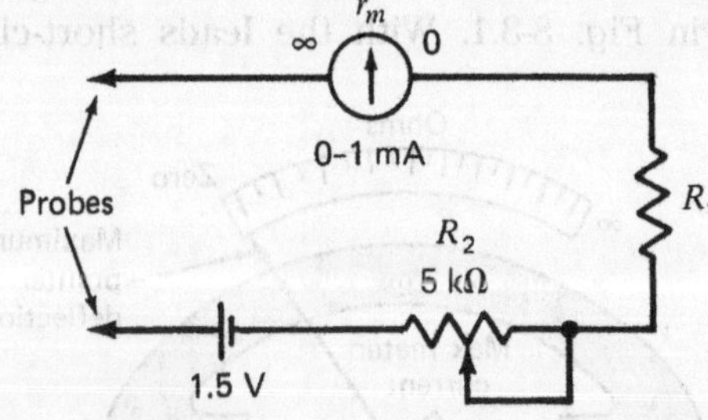

Fig. 8-3.4 Basic series ohmmeter.

10. With the probes not touching, the ohmmeter reads infinity. With the probes touching, adjust R_2 until the meter reads zero, indicating a zero-ohm condition (a short circuit).

11. Calibrate this ohmmeter by connecting a 1-MΩ potentiometer across the probes, as shown in Fig. 8-3.5, using a grease pencil to mark the face of the ohmmeter at the infinity value. Complete Table 8-3.2.

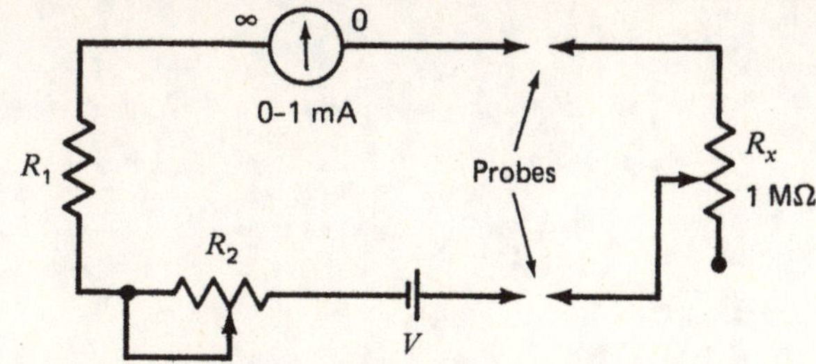

Fig. 8-3.5 Ohmmeter test circuit.

12. After calibrating the ohmmeter, measure the several resistances shown in Table 8-3.2 with an ohmmeter of known accuracy and your ohmmeter design.

13. Complete Table 8-3.3, and determine the percentage of accuracy. You will measure and record the resistance values in that table.

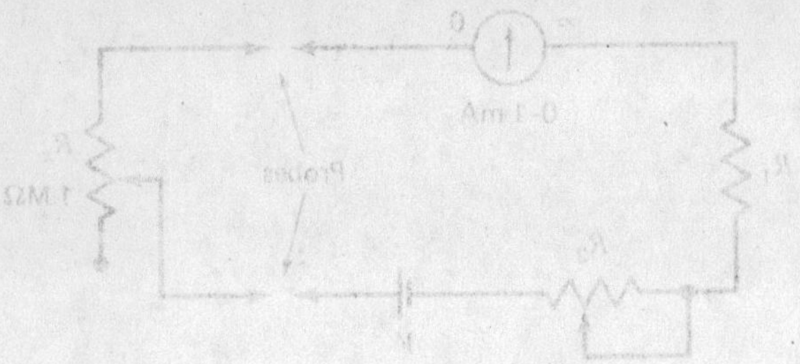

NAME ______________________ DATE ____________

QUESTIONS FOR EXPERIMENT 8-3

Answer true (T) or false (F) to the following:

_____ **1.** An ohmmeter is used to measure voltage and current.

_____ **2.** An ohmmeter has an internal battery.

_____ **3.** The infinity symbol (∞) on an ohmmeter indicates a short circuit.

_____ **4.** The ohmmeter's leads are placed across the resistance to be measured.

_____ **5.** When the ohmmeter leads are short-circuited, the needle will probably indicate zero.

_____ **6.** Ohmmeters do not require internal current-limiting resistances or shunt paths.

_____ **7.** The ohms or resistance scale that reads from left to right is called a *back-off scale.*

_____ **8.** The zero-ohm adjustment should not be used when changing ranges.

_____ **9.** For greater values of resistance, a less sensitive meter is required to read lesser values of current.

_____ **10.** An ohmmeter can be destroyed or have its fuse blown if it is used to measure resistance in a circuit where power is applied.

TABLES FOR EXPERIMENT 8-3

TABLE 8-3.1

Steps 5, 6, and 7
V_m = ______
r_m = ______
I_m = ______
Step 8
r_m = ______

TABLE 8-3.2

External R_x, Ω	Amount of Deflection	Scale Reading
0		
750		
1500		
3000		
150,000		
500,000		

TABLE 8-3.3

R	Known Meter	Design Meter	% Accuracy
100 Ω			
1 kΩ			
4.7 kΩ			
22 kΩ			
100 kΩ			
1 MΩ			

EXPERIMENT RESULTS REPORT FORM

Experiment No.: ____________________ Name: ____________________

Date: ____________________

Experiment Title: ____________________ Class: ____________________

____________________ Instr: ____________________

Explain the purpose of the experiment:

List the first Learning Objective:

OBJECTIVE 1:

After reviewing the results, describe how the objective was validated by this experiment.

List the second Learning Objective:

OBJECTIVE 2:

After reviewing the results, describe how the objective was validated by this experiment.

List the third Learning Objective:

OBJECTIVE 3:

After reviewing the results, describe how the objective was validated by this experiment.

Conclusion:

If required, attach to this form: ☐ Answers to Questions, ☐ Tables, and ☐ Graphs.

CHAPTER 9

EXPERIMENT 9-1

KIRCHHOFF'S LAWS

LEARNING OBJECTIVES

At the completion of this experiment, you will be able to:

- Validate Kirchhoff's current and voltage laws.
- Gain proficiency with lab equipment and technique.
- Predict the path of current flow using Kirchhoff's current law.

SUGGESTED READING

Chapter 9, *Basic Electronics*, Grob/Schultz, twelfth edition

INTRODUCTION

In 1847, Gustav R. Kirchhoff formulated two laws that have become fundamental to the study of electronics. They are as follows:

1. The algebraic sum of the currents into and out of any point must be equal to zero.
2. The algebraic sum of the applied source voltages and the *IR* voltage drops on any closed path must be equal to zero.

At first, Kirchhoff's laws seem obvious. That is, it seems obvious to state that whatever goes into a circuit must also equal what comes out of it. However, Kirchhoff's laws are used to analyze circuits that are not simple series or parallel or series-parallel circuits. For example, circuits that contain more than one voltage source and that contain hundreds of transistors cannot be understood without using Kirchhoff's laws to analyze those circuits. In its simplest form, Kirchhoff's current law could be used to analyze the circuit in Fig. 9-1.1.

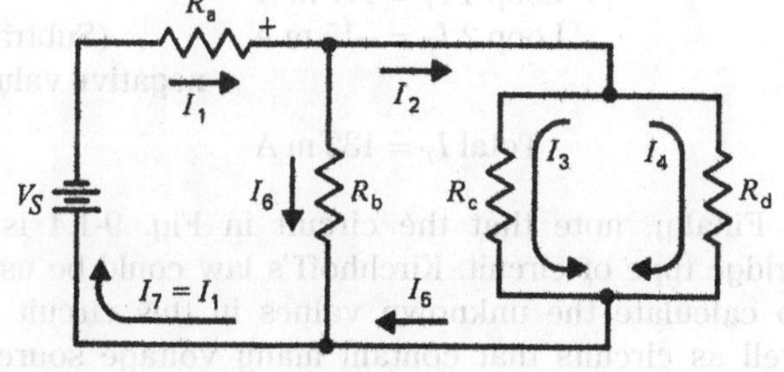

Fig. 9-1.1 Series-parallel circuit.

Simplified version:

$I_1 = I_2 + I_6$ $\quad I_2 = I_3 + I_4$ $\quad I_5 = I_2$

$I_6 = I_1 - I_2$ $\quad I_3 = I_2 - I_4$ $\quad I_7 = I_6 + I_5$

$I_2 = I_1 - I_6$ $\quad I_4 = I_5 - I_3$ $\quad I_7 = I_1$

Algebraic version:

$+I_1 - I_6 - I_2 = 0$ $\quad +I_5 + I_6 - I_7 = 0$

$+I_2 - I_3 - I_4 = 0$ $\quad I_T = I_1 = I_7$

Note: For the circuit of Fig. 9-1.1, if V_S = 20 V, the sum of the *IR* voltage drops across the series-parallel combination of R_a, R_b, R_c, and R_d will also equal 20 V.

Be sure that you understand the algebraic version of the circuit. It has become standard practice to assign positive and negative values to the currents as follows: The current into any point is positive (+), and the current out of any point is negative (–). Also, this would be true if there were more than one path for the current to enter or leave. Do not confuse this with electron flow.

Although the circuit of Fig. 9-1.1 is a series-parallel circuit, it was used to demonstrate how Kirchhoff's laws operate. Now consider the circuit of Fig. 9-1.2. This circuit could not be solved for its currents and *IR* voltage drops without using Kirchhoff's laws because of the two source voltages.

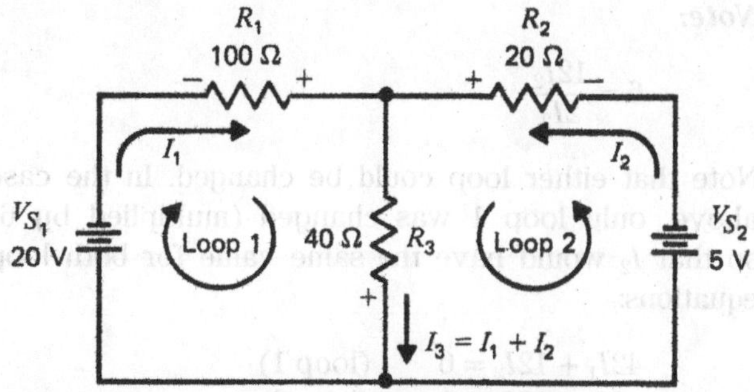

Fig. 9-1.2 Circuit with two voltage sources.

By using the loop method, Kirchhoff's laws can be used to determine the currents and *IR* voltages in the circuit by calculation. For example,

$$V_{S1} = 20\text{ V} - V_{R_1} - V_{R_3} = 0 \qquad \text{(loop 1)}$$

$$V_{S_2} = 5\text{ V} - V_{R_2} - V_{R_3} = 0 \qquad \text{(loop 2)}$$

By following the loop around its path, Kirchhoff's law provides a method. Take the source as positive, and take the individual path resistance as negative. The sum is equal to zero. This means that circuits will be divided into separate loops as if the other voltage source and its corresponding path did not exist. The IR voltage drops would be calculated as follows:

$$V_{R_1} = I_1R_1 = I_1 \times 100\ \Omega$$
$$V_{R_2} = I_2R_2 = I_2 \times 20\ \Omega$$
$$V_{R_3} = I_3R_3 = (I_1 + I_2)R_3 = (I_1 + I_2) \times 40\ \Omega$$

Refer to the equations for loops 1 and 2. The IR voltages would now be replaced in the formulas as follows:

For loop 1,

$$20\text{ V} - 100(I_1) - 40(I_1 + I_2) = 0$$
$$-140I_1 - 40I_2 = -20\text{ V}$$

and for loop 2,

$$5\text{ V} - 20(I_2) - 40(I_1 + I_2) = 0$$
$$-60I_2 - 40I_1 = -5\text{ V}$$

Note that the final equations are transposed versions that can be simplified further by division:

$$\frac{-140I_1 - 40I_2}{-20\text{ V}} = \frac{-20\text{ V}}{-20\text{ V}} \text{ or } 7I_1 + 2I_2 = 1 \quad \text{(loop 1)}$$

$$\frac{-60I_2 - 40I_1}{-5\text{ V}} = \frac{-5\text{ V}}{-5\text{ V}} \text{ or } 12I_2 + 8I_1 = 1 \quad \text{(loop 2)}$$

To solve for the currents, isolate the I_2 currents by making them the same value. Multiply the loop 1 equation by 6:

$$42I_1 + 12I_2 = 6 \quad \text{(loop 1)}$$

Note:

$$6 = \frac{12I_2}{2I_2}$$

Note that either loop could be changed. In the case above, only loop 1 was changed (multiplied by 6) so that I_2 would have the same value for both loop equations:

$$42I_1 + 12I_2 = 6 \quad \text{(loop 1)}$$
$$8I_1 + 12I_2 = 1 \quad \text{(loop 2)}$$

Subtracting the two equations, term by term, will eliminate I_2 because $12I_2 - 12I_2 = 0$. Therefore,

$$34I_1 = 5$$
$$I_1 = 147\text{ m A}$$

By using the loop equation method, Kirchhoff's law, the current through the resistance of R_1 is determined. Also, the direction of current flow is correct as assumed because the answer was a positive value.

To calculate I_2, substitute 147 mA for I_1 in either loop equation. Substituting in loop 2,

$$8(147\text{ mA}) + 12I_2 = 1$$
$$1.18\text{ A} + 12I_2 = 1$$
$$12I_2 = 1 - 1.18\text{ A} = -0.18\text{ A}$$
$$I_2 = -0.015\text{ A} = -15\text{ mA}$$

Because the negative sign appears in the solution for the current I_2, it follows that the current through I_2 is opposite in direction from the polarity shown in the circuit of Fig. 9-1.2, the assumed direction using the loop method.

Now that Kirchhoff's law has been used to determine the currents through the circuit of Fig. 9-1.2, the circuit can be redrawn to reflect the correct values, as shown in Fig. 9-1.3.

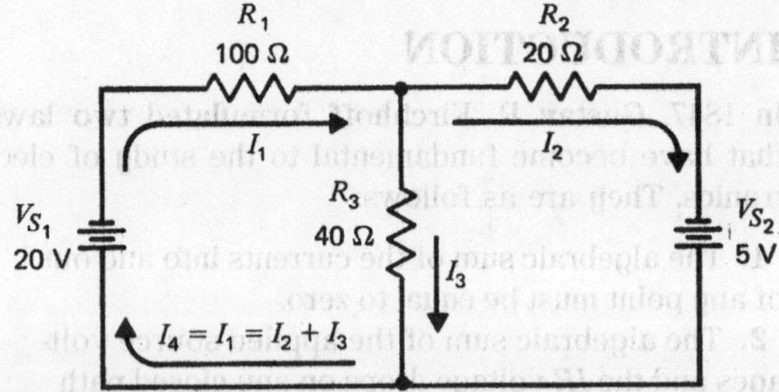

Fig. 9-1.3 Redrawn version of the circuit in Fig. 9-1.2.

In the case of the circuit in Fig. 9-1.3, the 20-V DC source actually overrides the 5-V DC source. This series opposition results in approximately 15 V DC.

Because R_1 is 100 Ω, it is easy to verify the Kirchhoff's law loop equation solution as follows:

$$\frac{15\text{ V}}{100\ \Omega + (20\ \Omega \parallel 40\ \Omega)} = \frac{15\text{ V}}{113.3\ \Omega} = 132\text{ m A}$$

$$\text{Loop 1 } I_T = 147\text{ m A}$$
$$\text{Loop 2 } I_T = -15\text{ m A} \quad \text{(Subtract negative value)}$$

$$\text{Total } I_T = 132\text{ m A}$$

Finally, note that the circuit in Fig. 9-1.4 is a bridge type of circuit. Kirchhoff's law could be used to calculate the unknown values in this circuit as well as circuits that contain many voltage sources and many current paths.

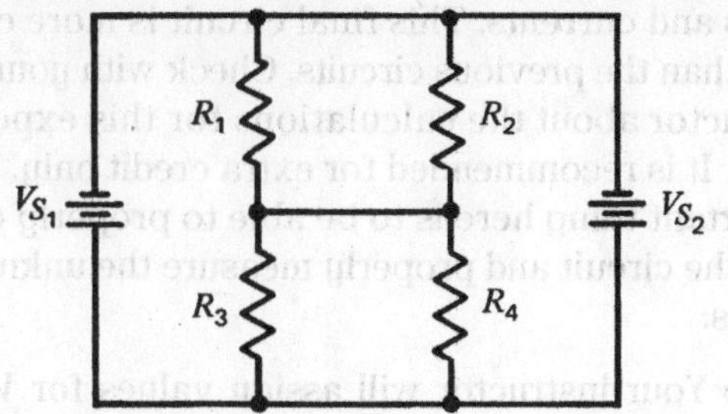

Fig. 9-1.4 Bridge-type circuit.

EQUIPMENT

DC power supply, 0–10 V
DMM
Springboard or protoboard
Connecting leads

COMPONENTS

Resistors (all 0.25 W):

(1) 100 Ω	(1) 390 Ω
(1) 220 Ω	(1) 470 Ω
(1) 330 Ω	

(2) Dry cells (flashlight batteries, 1.5 V)

PROCEDURE

1. Measure and record the values of voltage for each dry cell you are using. It may be necessary to solder connecting wires onto the ends of the batteries. Record the results in Table 9-1.1. Together, series-aiding, the two batteries should be approximately 3.0 V.

2. For each of the three circuits in Figs. 9-1.5 to 9-1.7, use batteries for voltages of 1.5 V and use the power supply for the other source voltage. If both 3.0 V and 1.5 V are used, connect the two 1.5-V batteries in series for 3.0 V. Measure and record the values of current in Table 9-1.2. Also, measure and record the value of the *IR* voltage drop across each resistance. Finally, use arrows to indicate the direction of current flow.

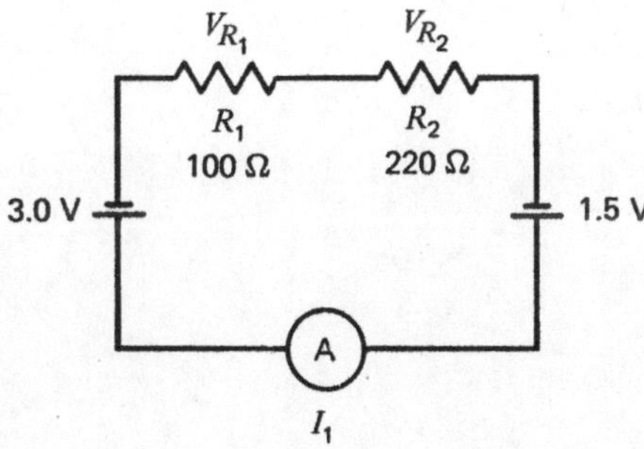

Fig. 9-1.5 Series-opposing voltage circuit.

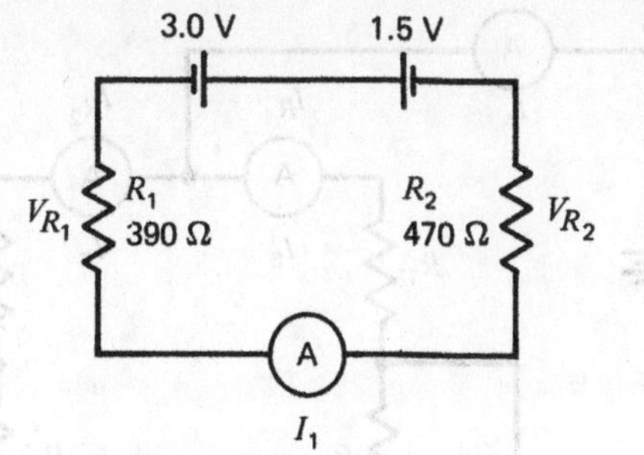

Fig. 9-1.6 Series-opposing voltage circuit.

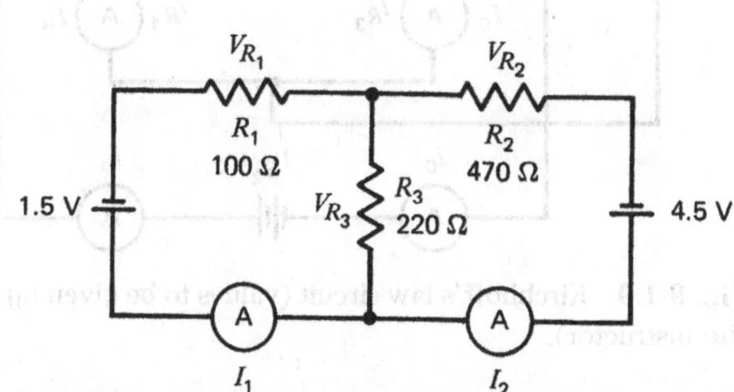

Fig. 9-1.7 Kirchhoff's circuit for analysis.

3. For the circuit of Fig. 9-1.8, refer to the procedure outlined in the introduction to this experiment. Preferably, use the loop method of Kirchhoff's law to determine the values and polarities of both current and voltage for the circuit. Then use the lab to verify the values (within 10 to 15 percent). Record all the calculated values, and then measure and record in Table 9-1.3 all the values determined in the lab for the circuit. Be sure to include arrows alongside or directly above the meters marked A. Include Fig. 9-1.8 in your report. Note that Fig. 9-1.8 will be used to verify Kirchhoff's law. That is, you will compare your calculations to your measurements. Do not forget to include the values of *IR* drops across each resistance.

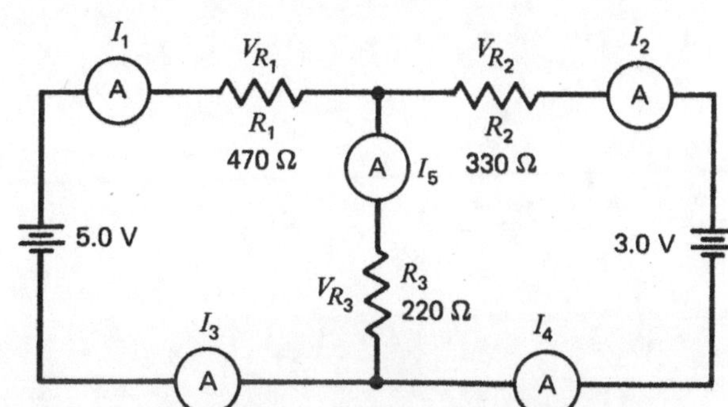

Fig. 9-1.8 Kirchhoff's circuit. Use loop method to determine the values and polarities of current and voltage.

4. Connect the circuit of Fig. 9-1.9. Measure and record in Table 9-1.4 all the values of *IR* voltage

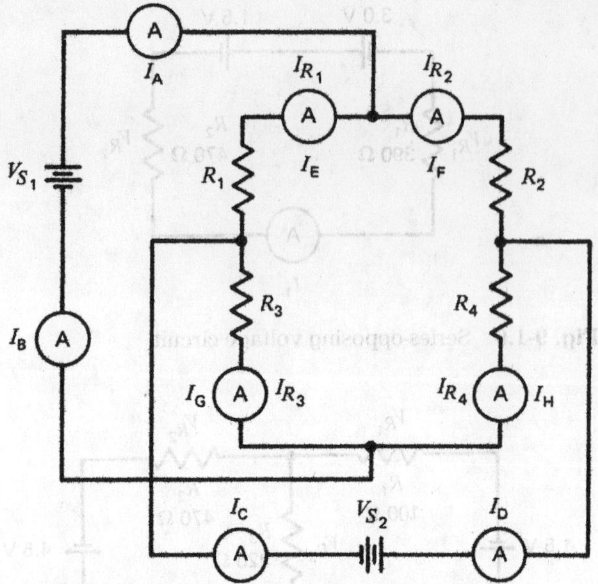

Fig. 9-1.9 Kirchhoff's law circuit (values to be given by the instructor).

drops and currents. This final circuit is more complex than the previous circuits. Check with your instructor about the calculations for this experiment. It is recommended for extra credit only. The important thing here is to be able to properly connect the circuit and properly measure the unknown values.

Note: **Your instructor will assign values for V_{S_1}, V_{S_2}, R_1, R_2, R_3, and R_4. This ensures that the experiment cannot be precalculated.**

NAME DATE

QUESTIONS FOR EXPERIMENT 9-1

1. State Kirchhoff's current law (KCL).

2. State Kirchhoff's voltage law (KVL).

TABLES FOR EXPERIMENT 9-1

TABLE 9-1.1 Dry Cell Voltages

Battery	Measured Voltage
1	______
2	______

TABLE 9-1.2

Circuit	V_{R_1}	V_{R_2}	V_{R_3}	I_1	I_2
Fig. 9-1.5	______	______		______	
Fig. 9-1.6	______	______		______	
Fig. 9-1.7	______	______	______	______	______

TABLE 9-1.3

	Calculated	Measured
I_1		
I_2		
I_3		
I_4		
I_5		
V_{R_1}		
V_{R_2}		
V_{R_3}		

TABLE 9-1.4

	Current	Voltage
R_1		
R_2		
R_3		
R_4		
I_A		
I_B		
I_C		
I_D		
I_E		
I_F		
I_G		
I_H		

EXPERIMENT RESULTS REPORT FORM

Experiment No.: ______________________ Name: ______________________

Date: ______________________

Experiment Title: ______________________ Class: ______________________

______________________ Instr: ______________________

Explain the purpose of the experiment:

List the first Learning Objective:

OBJECTIVE 1:

After reviewing the results, describe how the objective was validated by this experiment.

List the second Learning Objective:

OBJECTIVE 2:

After reviewing the results, describe how the objective was validated by this experiment.

List the third Learning Objective:

OBJECTIVE 3:

After reviewing the results, describe how the objective was validated by this experiment.

Conclusion:

If required, attach to this form: ☐ Answers to Questions, ☐ Tables, and ☐ Graphs.

CHAPTER 10

EXPERIMENT 10-1

NETWORK THEOREMS

LEARNING OBJECTIVES

At the completion of this experiment, you will be able to:

- Thevenize a circuit.
- Nortonize a circuit.
- Identify the difference between a Thevenin circuit and a Norton circuit.

SUGGESTED READING

Chapter 10, *Basic Electronics*, Grob/Schultz, twelfth edition

INTRODUCTION

The analyses of Ohm's law and Kirchhoff's laws have been of primary use in the solution of relatively simple solutions of DC circuits. In the analysis of relatively complex circuits, a more powerful method is required. In the case of simplifying complex circuits, Thevenin's and Norton's theorems are used. This technique involves reducing a complex network to a simple circuit, which acts like the original circuit. In general, any circuit with many voltage sources and components, with no regard made to interconnection, can be represented by an equivalent circuit with respect to a pair of terminals in the equivalent circuit.

Thevenin's theorem states that a circuit can be replaced by a single voltage source V_{Th} in series with a single resistance R_{Th} connected to two terminals. This is shown in Fig. 10-1.1.

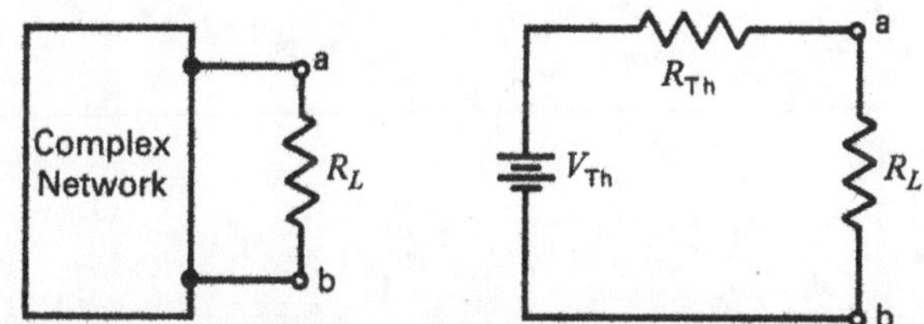

Fig. 10-1.1 Thevenin's circuit.

Norton's analysis is used to simplify a circuit in terms of currents rather than voltage, as is done in Thevenin circuits. Norton's circuit can be used to reduce a complex network into a simple parallel circuit that consists of a current source I_N and a parallel resistance to R_N. An example of this is shown in Fig. 10-1.2.

In the procedure that follows, the techniques of thevenizing and nortonizing simple voltage source–resistor networks will be developed as the procedure is completed.

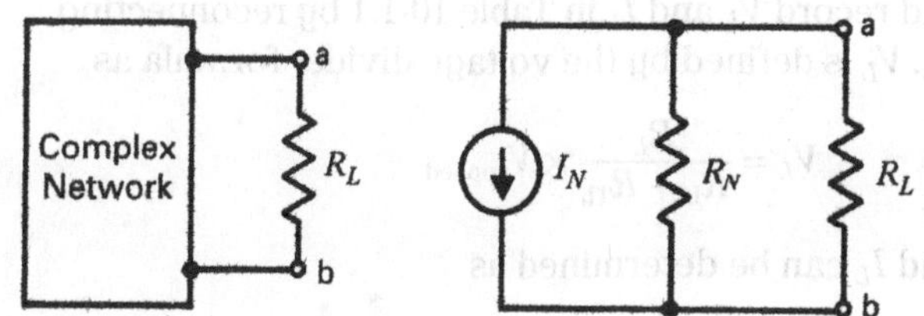

Fig. 10-1.2 Norton's circuit.

EQUIPMENT

DC power supply
DMM
Protoboard or springboard
Test leads

COMPONENTS

Resistors:

(1) 100 Ω, 1 W
(1) 220 Ω, 0.25 W
(1) 270 Ω, 0.25 W

PROCEDURE

Thevenizing

1. Construct the circuit shown in Fig. 10-1.3, where $R_1 = 100\ \Omega$, $R_2 = 270\ \Omega$, $R_L = 220\ \Omega$, and V is adjusted to 10 V.

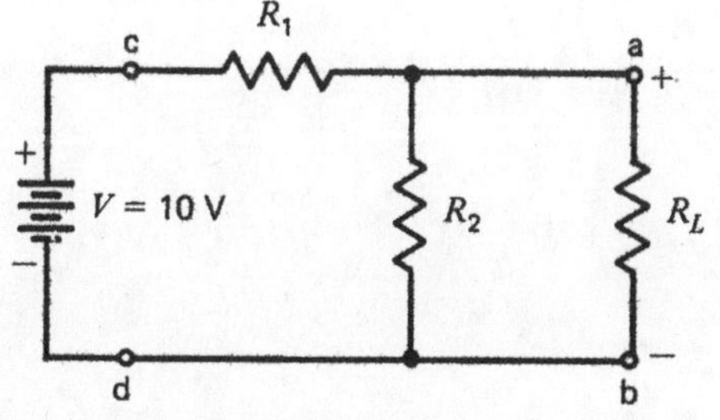

Fig. 10-1.3 Thevenizing a circuit.

2. Open the circuit at points a and b by disconnecting R_L from the circuit. The remainder of the circuit connected to a and b will be thevenized. Calculate, measure, and record in Table 10-1.1 the voltage across points a and b. Note that $V_{ab} = V_{Th}$.

3. Turn off the power supply and completely remove it from the circuit.

4. With the power supply removed, connect points c and d.

5. Calculate, measure, and record in Table 10-1.1 the value of R_{ab}. Note that $R_{ab} = R_{Th}$.

V_{Th} has now been determined and is found to be in series with R_{Th}. Since R_L was disconnected, this Thevenin equivalent can be applied to any value of R_L.

6. Reconnect the circuit shown in Fig. 10-1.3. Calculate (using the voltage divider formula), measure, and record V_L and I_L in Table 10-1.1 by reconnecting R_L. V_L is defined by the voltage divider formula as

$$V_L = \frac{R_L}{R_L + R_{Th}} \times V_{applied}$$

and I_L can be determined as

$$I_L = \frac{V_L}{R_L}$$

The same answers could be determined by using Ohm's law. The advantage of thevenizing the circuit is that the effect of R_L can be calculated easily for different values.

7. Complete Table 10-1.1 for the percentage of accuracy.

Nortonizing

8. Construct the circuit shown in Fig. 10-1.4, where $R_1 = 100\ \Omega$, $R_2 = 270\ \Omega$, $R_L = 220\ \Omega$, and V is adjusted to 10 V.

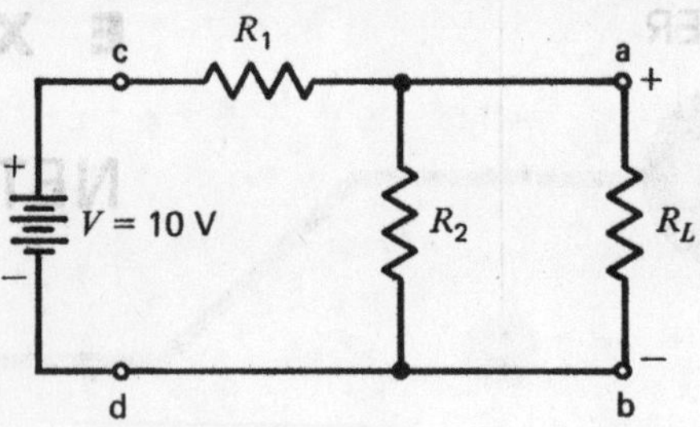

Fig. 10-1.4 Nortonizing a circuit.

9. Short-circuit points a and b together. This will also short-circuit R_2, and this will create a circuit condition in which resistor R_1 is in series with the power supply. Calculate, measure, and record in Table 10-1.2 the current flowing through R_1.

Note: $I_{R_1} = I_N$ (Norton).

10. Determine R_N by removing the short circuit, and remove R_L. This will leave points a and b unconnected to R_L.

11. Turn off the power supply and completely remove it from the circuit.

12. With the power supply removed, connect point c to d.

13. Calculate, measure, and record in Table 10-1.2 the value R_{ab}. Note that $R_{ab} = R_N$.

14. Reconnect the circuit shown in Fig. 10-1.4. Calculate, measure, and record I_L and V_L in Table 10-1.2 by reconnecting R_L.

15. Complete Table 10-1.2 for the percentage of accuracy.

NAME DATE

QUESTIONS FOR EXPERIMENT 10-1

1. What are the primary use and importance of thevenizing a circuit?

2. What are the primary use and importance of nortonizing a circuit?

3. Draw a Thevenin equivalent of the circuit shown in Fig. 10-1.3.

4. Draw a Norton equivalent of Fig. 10-1.4.

5. Is the statement made at the end of procedure step 6 that reads, "The advantage of thevenizing the circuit is that the effect of R_L can be calculated easily for different values," valid? Explain and prove by example.

TABLES FOR EXPERIMENT 10-1

TABLE 10-1.1

	Calculated	Measured	% Accuracy
V_{Th}			
R_{Th}			
V_L			
I_L			

TABLE 10-1.2

	Calculated	Measured	% Accuracy
I_N			
R_N			
I_L			
V_L			

EXPERIMENT RESULTS REPORT FORM

Experiment No.: ______________________ Name: ______________________

Date:

Experiment Title: ______________________ Class: ______________________

______________________ Instr: ______________________

Explain the purpose of the experiment:

List the first Learning Objective:
OBJECTIVE 1:

After reviewing the results, describe how the objective was validated by this experiment.

List the second Learning Objective:
OBJECTIVE 2:

After reviewing the results, describe how the objective was validated by this experiment.

List the third Learning Objective:
OBJECTIVE 3:

After reviewing the results, describe how the objective was validated by this experiment.

Conclusion:

If required, attach to this form: ☐ Answers to Questions, ☐ Tables, and ☐ Graphs.

CHAPTER 11

EXPERIMENT 11-1

CONDUCTORS AND INSULATORS

LEARNING OBJECTIVES

At the completion of this experiment, you will be able to:

- Identify the characteristics of conductors and insulators.
- Identify and validate the pole and throw of a switch.
- Describe the operating characteristics of a fuse.

SUGGESTED READING

Chapter 11, *Basic Electronics*, Grob/Schultz, twelfth edition

INTRODUCTION

This experiment investigates the electrical characteristics of components and materials commonly found within the electronics industry and used by electronic technicians. Understanding these electrical characteristics promotes a general understanding of electronics while at the same time protecting the safety of the electronic technician within the workplace. This experiment will investigate four areas of interest: conductors and insulators, body resistance and safety, electrical switches, and electrical fuses.

Conductors, Insulators, and Semiconductors Electrical materials are generally classified into one of three areas: conductors, insulators, or semiconductors. When electrons can easily move from one atom to another in a material, it is considered a conductor. In general, metals are a good example of conductors. In the electronics industry, typical conductors are copper, aluminum, silver, and gold.

A material with atoms in which the electrons tend to stay in their own orbits, and cannot easily move from one atom to another, is considered an insulator because insulators cannot easily conduct electricity. Insulators are a category of material that does not permit current to flow when voltage is applied because of its high electrical resistance. However, insulators are able to hold or store electricity better than conductors are. Examples of typical insulators are rubber, certain plastics, paper, glass, and air.

Semiconductors are materials that will conduct less than metal conductors and more than insulators. Examples of materials that are classified as semiconductors are carbon, germanium, and silicon.

Body Resistance and Safety The human body contains various levels of electrical resistance. However, consider that a current of 100 mA through the heart will probably kill most people. The concept of maintaining electrical safety while working with electrical circuits is important. It is crucial to respect the potential harm of electrical circuits and test equipment. Body resistance can be easily and safely measured by the use of an ohmmeter. Skin resistance is typically 500 kΩ, whereas the resistance of blood may be less than 100 Ω.

One of the "rules of thumb" that is often used is the 1-10-100 rule of current. This rule states that 1 mA of current through the human body can be felt, 10 mA of current is sufficient to make muscles contract to the point where you cannot let go of a power source, and 100 mA is sufficient to stop the heart. According to Ohm's law,

$$V = IR$$

where V is voltage in volts, I is current in amps, and R is resistance in ohms.

Using a 9-V battery and assuming that 500 kΩ is a typical level of body resistance in the equation, we determine that 18 μA of current will flow. The level of 18 μA is below what is determined to be our "feel" threshold of 1 mA. However, if you were to have a cut finger, your resistance would be lower because blood electrolyte is a good conductor. At around 100 Ω, in fact, a current of 90 mA results—sufficient to stop the human heart, resulting in death. Therefore, it is for safety purposes that students are encouraged to increase bodily resistance while working on electrical circuitry. This can be accomplished by the use of insulated hand tools, gloves, and rubber-soled shoes.

Electrical Switches

A switch is a component that allows for the control of the flow of electrical current. The switch operates within two states, closed or open. When a switch is closed, it is considered in the ON position. When a switch is opened, it is considered in the OFF position. A closed switch exhibits practically zero resistance, and an open switch is considered to be practically an infinite amount of resistance.

Toggle-type switches are usually described as having a certain number of poles and throws. The number of poles and throws that they contain generally classifies switches.

The most basic type of switch is a single-pole, single-throw (SPST) switch. Other popular switch types include the single-pole, double-throw (SPDT); double-pole, single-throw (DPST); and the double-pole, double-throw (DPDT). The schematics for each type of switch are shown in Fig. 11-1.1.

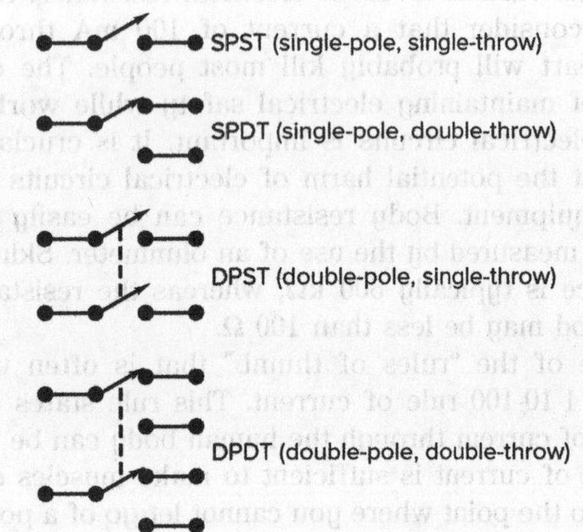

Fig. 11-1.1 Various toggle switch schematics.

Electrical Fuses

Typically, electrical circuits use a fuse in series with a circuit as a protection against an electrical overload resulting from a short circuit. This is accomplished when excessive current melts the fuse element, blowing the fuse and opening the series circuit. The purpose is to let the fuse blow before the components are damaged.

A complete (closed) circuit or path for direct current consists of a voltage source such as a battery or power supply, closed switch, a load such as an electric lamp, and connecting conductors, such as copper wire (see Fig. 11-1.2). In such a closed circuit there is electric current, and if sufficient current flows, the lamp will light. If rubber cords instead of copper wires were used to tie the battery to the lamp, the lamp would not light, because rubber is an insulator and does not permit current to flow through it in the circuit.

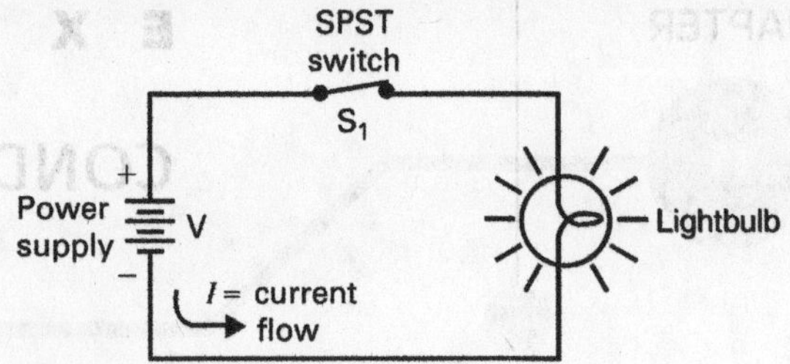

Fig. 11-1.2 Closed circuit.

The tungsten filament in the electric lamp is a conductor, but it is not as good a conductor as the connecting copper wires. Tungsten is used for the filament because its properties cause it to glow and give off light when an electric current heats it.

It is apparent that not all materials are equally good conductors of electricity. Specific materials are used for electrical components because of their unique characteristics as conductors and nonconductors.

EQUIPMENT

DMM
Glass cartridge fuse holder
DC power supply

COMPONENTS

SPST toggle switch
SPDT toggle switch
DPST toggle switch
DPDT toggle switch
Operating glass cartridge fuse (0.25 A)
Inoperative glass cartridge fuse (0.25 A)
47 Ω, 2 W
56 Ω, 2 W

SPECIAL MATERIALS

A beaker of water
A beaker of saltwater
10-kΩ resistor
Capacitor (disc)
Inductor
Diode (1N4004)
Nichrome wire
Copper wire
Wood dowel
Plastic rod
Glass rod
Paper
Fuse (operational)
Fuse (nonoperational)

PROCEDURE

Part A: Investigating the Resistive Characteristics of Insulators and Conductors

1. Note for Multimeters: When using either digital or analog multimeters to measure resistance, it is important to hold the meter probes by their insulation, not their metal tips. If you come into contact with the metal probes, the meter will attempt to measure your body resistance, which can interfere with accurate resistance measurements of the object you are measuring.

2. Measure and record the resistance of each of the 15 items listed in Table 11-1.1.

Note: It may be necessary to adjust your meter's range switch (RX scale) to make accurate measurements.

3. Reverse the meter's polarity by interchanging the test leads of the meter. Again, measure and record the resistance of each of the 14 items listed in Table 11-1.1.

Part B: Investigating the Body Resistance

4. Set your ohmmeter to the highest resistance measurement range of the meter.

5. Under the supervision of your instructor, measure your body resistance by loosely holding the red (positive) and black (negative) test probe tip between the thumb and forefinger of each hand. Measure the ohmmeter's value and record your body resistance in Table 11-1.2.

6. Reverse the ohmmeter leads, read the ohmmeter's value, and record your body resistance in Table 11-1.2.

7. Increase you grip by squeezing the probes (but be careful not to break the skin), measure the ohmmeter's value, and record your body resistance in Table 11-1.2.

8. Wet your thumb and finger, and repeat the measurement. Read the ohmmeter's value, and record your body resistance in Table 11-1.2.

Part C: The Switch

9. Through the use of an ohmmeter, identify each of the switches. Classify each of the switches as to SPST, SPDT, DPST, and DPDT. Draw their respective schematic diagram in Table 11-1.3, and identify their poles and throws.

Part D: The Fuse

10. Measure and record in Table 11-1.4, the resistance values of R_1 (47 Ω, 2 W) and R_2 (56 Ω, 2 W).

11. Using an *operating* glass cartridge fuse and holder, construct the circuit as shown in Fig. 11-1.3.

Note: Do not turn on the power supply yet.

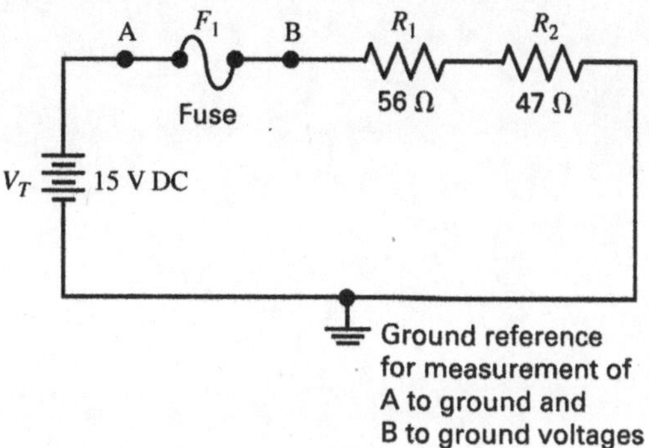

Fig. 11-1.3 Fused series circuit.

12. If the internal resistance of the fuse is assumed to be zero, calculate (from the measured values) the total series resistance R_T of the circuit, as shown in Fig. 11-1.4, and record this information in Table 11-1.4.

13. If it would be unacceptable to exceed the current-carrying capability of the circuit fuse, then calculate the maximum voltage, V_{max}, that could be applied to this circuit without damaging the fuse. Record this information in Table 11-1.4.

14. Turn on the power supply and adjust the voltage to 10 V DC. Using Ohm's law, calculate the total expected series circuit current I_T and record in Table 11-1.4.

15. Calculate the expected voltage from point A to ground, and record in Table 11-1.4.

16. Measure the voltage from point A to ground, and record in Table 11-1.4.

17. Calculate the expected voltage from point B to ground, and record in Table 11-1.4.

18. Measure the voltage from point B to ground, and record in Table 11-1.4.

19. Calculate the expected voltage across R_1, and record in Table 11-1.4.

20. Measure the voltage across R_1, and record in Table 11-1.4.

21. Calculate the expected voltage across R_2, and record this information in Table 11-1.4.

22. Measure the voltage across R_2, and record in Table 11-1.4.

23. Calculate the expected voltage across points A and B, and record in Table 11-1.4.

24. Measure the voltage across the fuse, or from points A to B, and record in Table 11-1.4.

25. Turn off the power supply, and replace the *operable* fuse with an *inoperable* glass cartridge fuse. Repeat procedure steps 15 through 24.

NAME DATE

QUESTIONS FOR EXPERIMENT 11-1

1. Considering that a current of 100 mA through your heart will almost certainly kill you, how much voltage across your hands would be lethal if you have a body resistance of 250 kΩ?

2. How many connecting terminals does the SPST switch have? The SPDT? The DPST? The DPDT?

3. What is the voltage drop across a closed switch? What is the voltage drop across "an open switch"?

4. The double-pole, double-throw switch can be used to reverse the polarity of voltage across the terminals of a DC motor. From reading Chapter 11, *Basic Electronics*, Grob/Schultz, twelfth edition, draw the schematic diagram that would accomplish this task.

5. When an electrical open appears with a component found within a series circuit, such as a fuse or resistor, what can the assumed voltage be across the circuit's open? Is this always true? Explain what the implication is for electrical safety.

6. Explain your results from Table 11-1.1 concerning the two fuses used in this experiment.

7. Explain your results from Table 11-1.2. Using this information, how could you improve your safety while working with electrical equipment?

8. Explain your results from procedural steps 23 and 24 in Table 11-1.4. Using this information, how could you improve your safety while working with electrical equipment?

TABLES FOR EXPERIMENT 11-1

TABLE 11-1.1 Conductive Materials

Materials to Be Measured	Resistance in Ω	Resistance in Ω
Water		
Saltwater		
10-kΩ resistor		
Disc capacitor		
Inductor		
1N4004 diode		
Nichrome wire		
Copper wire		
Wood dowel		
Plastic rod		
Glass rod		
Paper		
Operational fuse		
Nonoperational fuse		

TABLE 11-1.2 Body Resistance

Procedural Step and Measurement		Resistance Value
5	Loosely holding	Ω
6	Reverse ohmmeter measurement	Ω
7	Increased grip	Ω
8	Wet measurement	Ω

TABLE 11-1.3 Switches

Schematic Diagram Including Pole and Throw Identifications	Switch Classification
	SPST
	SPDT
	DPST
	DPDT

TABLE 11-1.4 Fused Series Circuits

Procedural Step	Resistor	Measured Value
	Analysis of Fused (Operable) and Unfused (Inoperable) Series Circuits	
10	R_1	________ Ω
10	R_2	________ Ω

Procedural Step	Maximum Parameters	Calculated/Measured Value with Appropriate Unit
	Establishing Maximum Circuit Parameters	
12	R_T	________
13	V_{max}	________
14	I_T	________

Procedural Step	Location	Calculated/Measured Value with Appropriate Unit
	Analysis with the Operable Fuse Circuit	
15	Calculated voltage A to ground	________
16	Measured voltage A to ground	________
17	Calculated voltage B to ground	________
18	Measured voltage B to ground	________
19	Calculated voltage R_1 to ground	________
20	Measured voltage R_1 to ground	________
21	Calculated voltage R_2 to ground	________
22	Measured voltage R_2 to ground	________
23	Calculated voltage A to B	________
24	Measured voltage A to B	________
	Analysis with the Inoperable Fuse Circuit	
15-Repeated	Calculated voltage A to ground	________
16-Repeated	Measured voltage A to ground	________
17-Repeated	Calculated voltage B to ground	________
18-Repeated	Measured voltage B to ground	________
19-Repeated	Calculated voltage R_1 to ground	________
20-Repeated	Measured voltage R_1 to ground	________
21-Repeated	Calculated voltage R_2 to ground	________
22-Repeated	Measured voltage R_2 to ground	________
23-Repeated	Calculated voltage A to B	________
24-Repeated	Measured voltage A to B	________

EXPERIMENT RESULTS REPORT FORM

Experiment No.: ______________________ Name: ______________________

Date: ______________________

Experiment Title: ______________________ Class: ______________________

______________________ Instr: ______________________

Explain the purpose of the experiment:

List the first Learning Objective:

OBJECTIVE 1:

After reviewing the results, describe how the objective was validated by this experiment.

List the second Learning Objective:

OBJECTIVE 2:

After reviewing the results, describe how the objective was validated by this experiment.

List the third Learning Objective:

OBJECTIVE 3:

After reviewing the results, describe how the objective was validated by this experiment.

Conclusion:

If required, attach to this form: ☐ Answers to Questions, ☐ Tables, and ☐ Graphs.

CHAPTER 12

EXPERIMENT 12-1

BATTERY INTERNAL RESISTANCE

LEARNING OBJECTIVES

At the completion of this experiment, you will be able to:

- Validate the concept of internal resistance in a power source.
- Determine the internal resistance of a dry cell battery and a DC power supply (generator).
- Graph or plot decreasing terminal voltage versus load current.

SUGGESTED READING

Chapter 12, *Basic Electronics,* Grob/Schultz, twelfth edition

INTRODUCTION

Any source of electric power that produces a continuous output voltage can be called a *generator.* All generators have some internal resistance, labeled r_i. This internal resistance has its own *IR* voltage drop, because it is in series with any load connected to the generator. In other words, the internal resistance of a source subtracts from the generated voltage, resulting in a decreased voltage across the output terminals. In a battery, r_i is due to the chemical makeup inside; in a power supply, r_i is due to the internal circuitry of the supply.

For example, Fig. 12-1.1 is a schematic representation of a 9-V battery with 100 Ω of internal resistance.

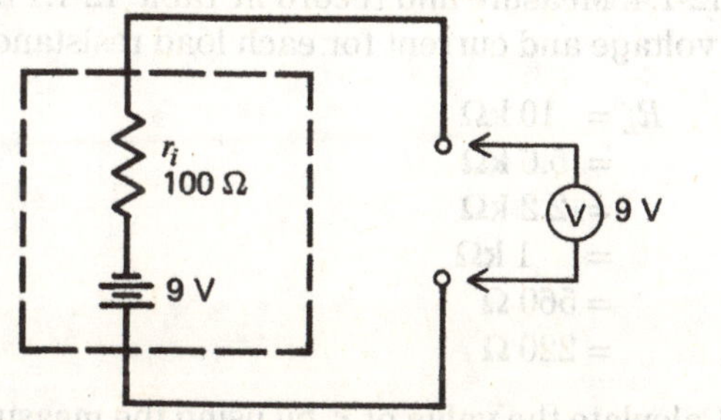

Fig. 12-1.1 A 9-V battery with r_i = 100 Ω.

Notice that the dotted line indicates that r_i is actually inside the battery. This battery has 9 V across its output terminals when it is measured with a voltmeter. If r_i were equal to 100 kΩ, the voltmeter would still measure 9 V across the output terminals. Thus, the value of r_i does not affect the output voltage. However, if a load is connected across the output terminals, then the value of r_i becomes significant. In any case, Fig. 12-1.2 shows that the battery's internal resistance now becomes a resistance in series with the load.

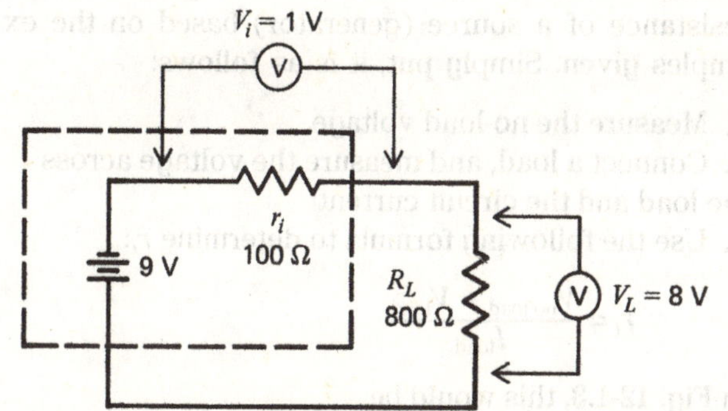

Fig. 12-1.2 A 9-V battery under load.

With a load resistance of 800 Ω connected across the output terminals, the voltmeter will now measure 8 V instead of 9 V. The other 1 V is now across the internal resistance of the battery. If r_i were equal to 100 kΩ, for example, the voltage across the output terminals would be almost 0 V due to the excessive value of r_i. In that case, the battery would be worn out or depleted.

Most bench power supplies have a fixed value of internal resistance that does not vary, regardless of the load value. Remember, without the load connected, the circuit is an open load. Therefore, the voltage drop across r_i equals zero. In this case, the total voltage is still available across the output, and it is called *open-circuit voltage* or *no-load voltage.*

In the example of Fig. 12-1.2, the total circuit current I_T is equal to

$$I_T = \frac{V_L}{R_L} = \frac{8\ \text{V}}{800\ \Omega} = 0.01\ \text{A}$$

As the load resistance decreases, more circuit current will flow. If R_L decreases to 350 Ω, the current will increase and the load will require more current. Also, the voltage drop across the load will decrease and the voltage drop across V_i will increase.

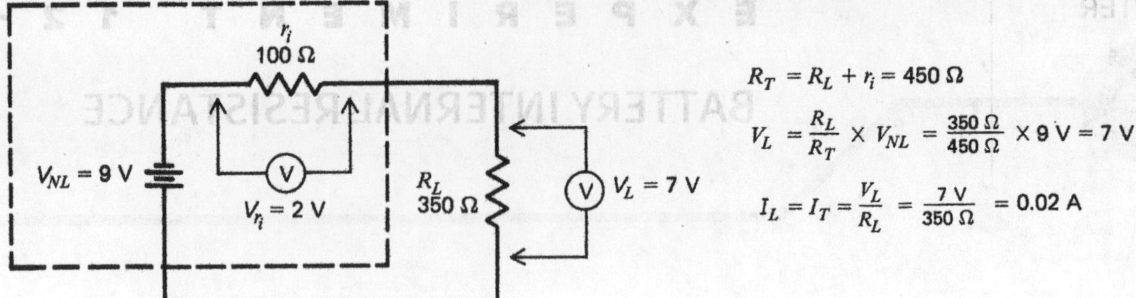

Fig. 12-1.3 Determining r_i.

Notice that as the load resistance decreased, the circuit current increased. Thus, the terminal voltage (the same thing as the load IR voltage) decreased. Therefore, the terminal voltage drops with more load current.

There is a method for determining the internal resistance of a source (generator) based on the examples given. Simply put, it is as follows:

1. Measure the no-load voltage.
2. Connect a load, and measure the voltage across the load and the circuit current.
3. Use the following formula to determine r_i:

$$r_i = \frac{V_{\text{no load}} - V_{\text{load}}}{I_{\text{load}}}$$

In Fig. 12-1.3, this would be

$$r_i = \frac{V_{NL} - V_L}{I_L}$$

$$= \frac{9\ V - 7\ V}{0.02\ A}$$

$$= \frac{2\ V}{0.02\ A}$$

$$= 100\ \Omega$$

Finally, in general, if a generator has a very low internal resistance in relation to load resistance, it is considered a constant voltage source because the voltage across r_i will subtract very little from the load voltage. If the value of r_i is very great in relation to load resistance, the generator is considered a constant current source because the load resistance will have little effect upon the total resistance ($r_i + R_L$) and the total circuit current.

EQUIPMENT

DC power supply
DMM
1.5-V battery
Protoboard or springboard
Test leads

COMPONENTS

Resistors (all 0.25 W unless indicated otherwise):

(1) 220 Ω (1) 2.2 kΩ
(2) 560 Ω (1) 5.6 kΩ
(2) 1 kΩ (1) 10 kΩ

PROCEDURE

1. Connect the circuit of Fig. 12-1.4. Do not connect the load yet. Measure and record in Table 12-1.1 the no-load voltage (V_{NL}) across the output terminals.

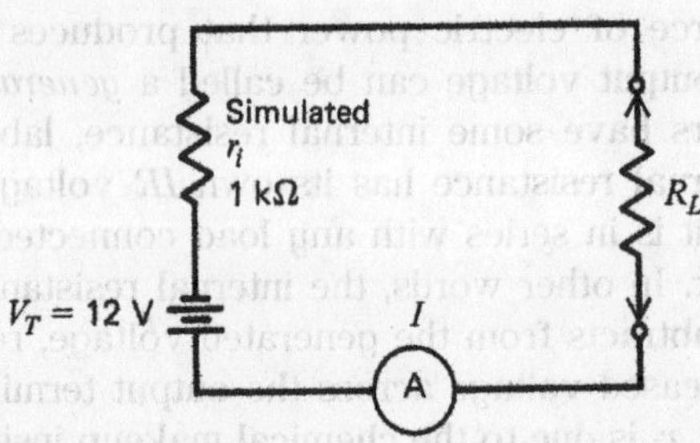

Fig. 12-1.4 Power supply with simulated r_i. Note that r_i is a 1-kΩ resistor in series.

2. Connect the following loads to the circuit of Fig. 12-1.4. Measure and record in Table 12-1.1 the load voltage and current for each load resistance.

R_L = 10 kΩ
= 5.6 kΩ
= 2.2 kΩ
= 1 kΩ
= 560 Ω
= 220 Ω

3. Calculate the value of r_i by using the measured values of load voltage and current for each load resistance. Show your calculations on a separate sheet of paper, and record the results in Table 12-1.1.
4. Change the value of r_i to 560 Ω and repeat steps 1 to 3 above. Record the results in Table 12-1.2.

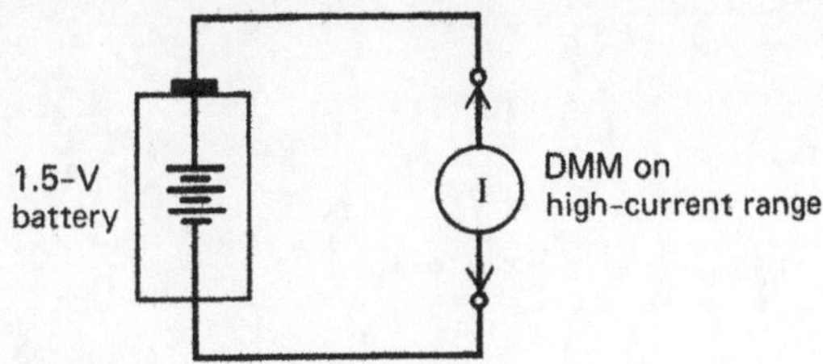

Fig. 12-1.5 Short-circuit method.

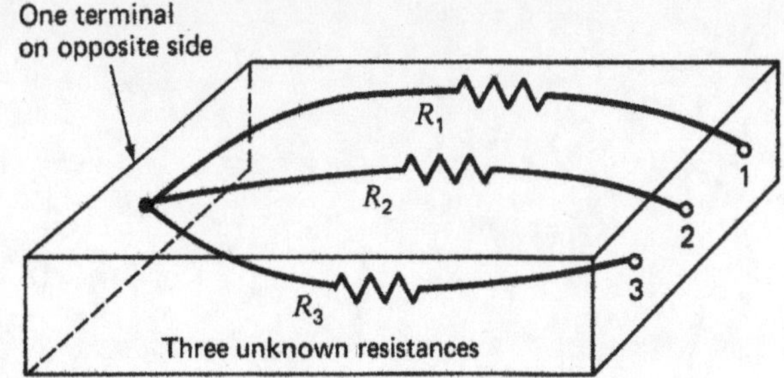

Fig. 12-1.6 Resistance box containing three unknown resistors.

5. Change the value of V_T to 6 V (using $r_i = 560\ \Omega$), and repeat steps 1 to 3. Record the results in Table 12-1.3.

6. Measure the voltage across a 1.5-V battery, and record the value in Table 12-1.4.

7. Measure the short-circuit current of a 1.5-V battery by placing an ammeter across the output terminals of the battery for no longer than approximately 5 s, as shown in Fig. 12-1.5. Record the value in Table 12-1.4.

CAUTION: Use the 10- or 12-A scale range. Some meters have special input jacks for this purpose.

8. Calculate r_i by using Ohm's law. If available, repeat with a 22 V or any other size battery that will not damage the meter.

Note: You can do this for any value of battery, provided a meter of large enough current capacity is used.

9. Using the method for determining r_i (steps 1 to 3), use an unknown value of r_i (three times) and use a 1-kΩ load resistor. This can be done by disguising the value of a resistance with black electrical tape or by placing the resistance inside a chassis, as illustrated in Fig. 12-1.6. Use the circuit in Fig. 12-1.7 for this step. Record the results in Table 12-1.5.

10. *OPTIONAL:* Plot the results of steps 1 to 3 (Table 12-1.1). For example, V_L versus I_L. Use regular graph paper, *not* semilog graph paper. See Fig. 12-1.8. See Appendix D for suggestions on how to make graphs.

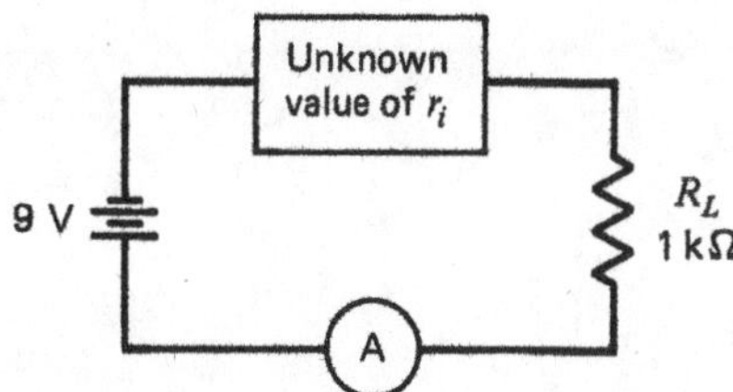

Fig. 12-1.7 Circuit for determining the internal resistance.

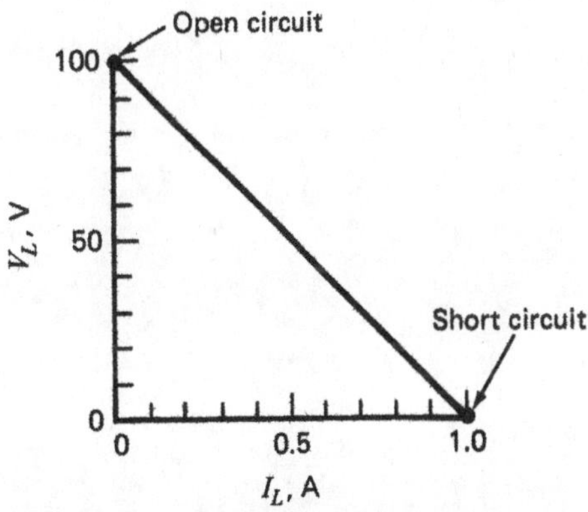

Fig. 12-1.8 How terminal voltage V_L drops with more load current I_L.

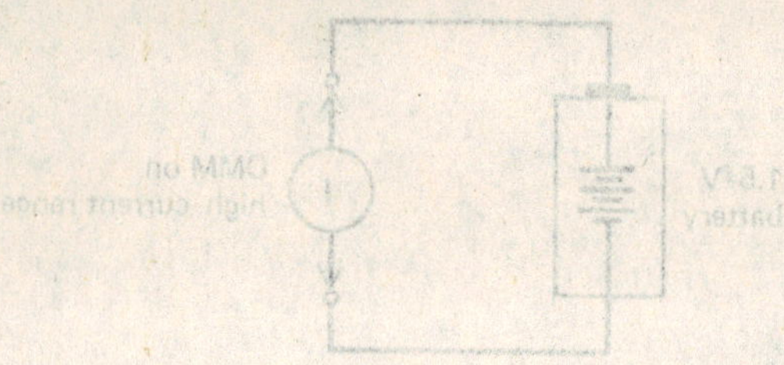

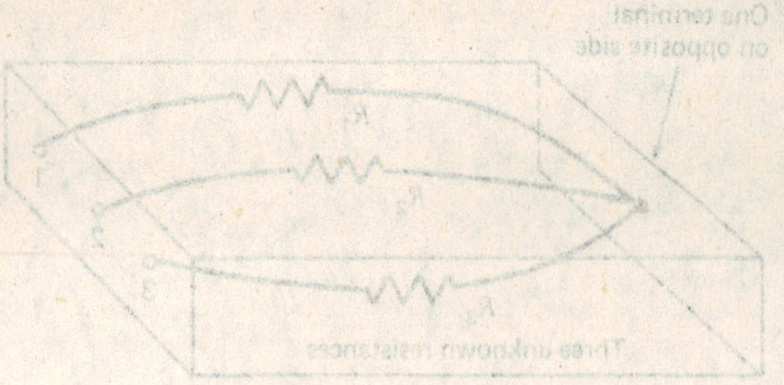

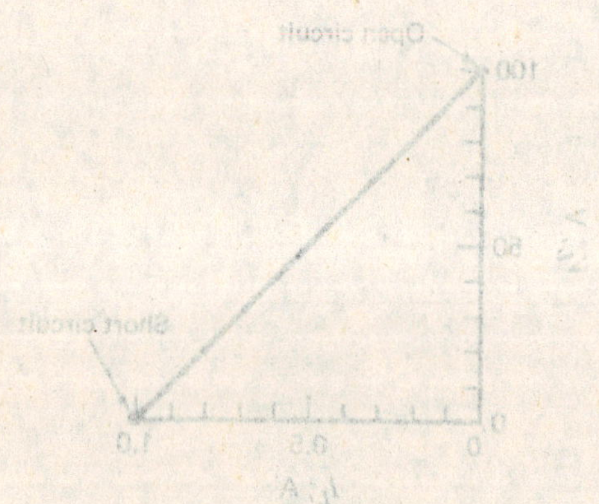

NAME ______________________________ DATE ______________

QUESTIONS FOR EXPERIMENT 12-1

Answer true (T) or false (F) to the following:

_____ **1.** Batteries have internal resistance, but DC power supplies do not.

_____ **2.** As load resistance increases, the terminal voltage decreases.

_____ **3.** As load current increases, terminal voltage increases.

_____ **4.** Connecting four batteries in parallel, each with $r_i = 100\ \Omega$, would increase the total r_i four times.

_____ **5.** Connecting four batteries in series, each with $r_i = 100\ \Omega$, would increase the total r_i four times.

_____ **6.** The internal resistance of a generator is always in parallel with a load.

_____ **7.** Subtracting the load voltage from the no-load voltage gives a remainder that is equal to the *IR* voltage drop in the internal resistance of the source.

_____ **8.** Internal resistance is in series with the load resistance.

_____ **9.** Short-circuiting a battery will not drain the battery.

_____ **10.** A 1.5-V dry cell battery with 1 Ω of internal resistance is probably a depleted battery.

TABLES FOR EXPERIMENT 12-1

TABLE 12-1.1 $r_i = 1\ \text{k}\Omega$ (Steps 1–3)

R_L	Measured V_L, V	Measured I_L, A	Calculated r_i, Ω
10 kΩ			
5.6 kΩ			
2.2 kΩ			
1 kΩ			
560 Ω			
200 Ω			

V_{NL} = __________

TABLE 12-1.2 $r_i = 560\ \Omega$ (Step 4)

R_L	Measured V_L, V	Measured I_L, A	Calculated r_i, Ω
10 kΩ			
5.6 kΩ			
2.2 kΩ			
1 kΩ			
560 Ω			
220 Ω			

TABLE 12-1.3 $V_T = 6$ V; $r_i = 560\ \Omega$ (Step 5)

R_L	Measured V_L, V	Measured I_L, A	Calculated r_i, Ω
10 kΩ			
5.6 kΩ			
2.2 kΩ			
1 kΩ			
560 Ω			
220 Ω			

TABLE 12-1.4 Steps 6–8

	V_{NL}, V	Short-circuit I, A	Calculated r_i, Ω
1.5-V battery			
Additional ____ -V battery			

Note: Only the instructor will know the value of the three unknown values of r_i. In this way, your lab techniques and your ability to follow procedures will be tested.

TABLE 12-1.5 r_i Unknown; $r_L = 1$ kΩ (Step 9)

	Measured V_L, V	Measured I_L, A	Calculated r_i, Ω
Measured V_{NL} = ____			
r_i No. 1			
r_i No. 2			
r_i No. 3			

Chassis or box number (if applicable): ____________

EXPERIMENT RESULTS REPORT FORM

Experiment No.: ______________________ Name: ______________________

Date: ______________________

Experiment Title: ______________________ Class: ______________________

______________________ Instr: ______________________

Explain the purpose of the experiment:

List the first Learning Objective:

OBJECTIVE 1:

After reviewing the results, describe how the objective was validated by this experiment.

List the second Learning Objective:

OBJECTIVE 2:

After reviewing the results, describe how the objective was validated by this experiment.

List the third Learning Objective:

OBJECTIVE 3:

After reviewing the results, describe how the objective was validated by this experiment.

Conclusion:

If required, attach to this form: ☐ Answers to Questions, ☐ Tables, and ☐ Graphs.

CHAPTER 12

EXPERIMENT 12-2

LOAD MATCH AND MAXIMUM POWER

LEARNING OBJECTIVES

At the completion of this experiment, you will be able to:

- Validate that maximum source power is transferred to a load when the value of source $r_i = R_L$.
- Plot a graph of load power for differing values of load resistance.
- Understand the concept of maximum efficiency versus maximum power.

SUGGESTED READING

Chapter 12, *Basic Electronics*, Grob/Schultz, twelfth edition

INTRODUCTION

When the internal resistance of a generator is equal to the load resistance, the load is considered matched to the source. The matching of load to source resistance is significant because the source can then transfer maximum power to the load.

Whenever $R_L = r_i$, maximum power is transferred to the load. When load resistance is more than r_i, the output voltage is more but the circuit current is less. When the load resistance is less than r_i, the output voltage is less but the circuit current is more. This experiment will provide data that you can analyze and thus prove that these concepts are valid.

The circuit of Fig. 12-2.1 and the accompanying graph of Fig. 12-2.2 illustrate the concept of matching a load to an internal source resistance to obtain maximum power transfer.

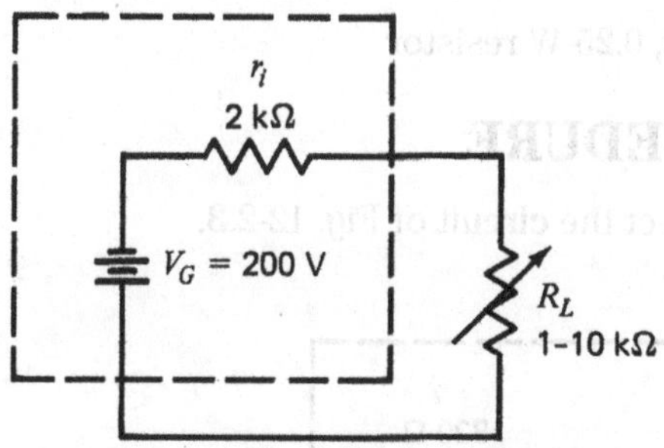

Fig. 12-2.1 Maximum power transfer circuit for analysis.

Because of the voltage divider formed by r_i and R_L, there is an equal voltage division: half of V_G is across r_i and half of V_G is across R_L. Under these circumstances, the load develops the maximum power that is possible using the particular source.

Referring to Fig. 12-2.1, as R_L increases, current decreases, resulting in less power dissipated in r_i. This results in more circuit efficiency because less power is lost across r_i. However, when $r_i = R_L$, the circuit efficiency is 50 percent.

$$\frac{P_L}{P_T} \times 100 = \text{circuit efficiency}$$

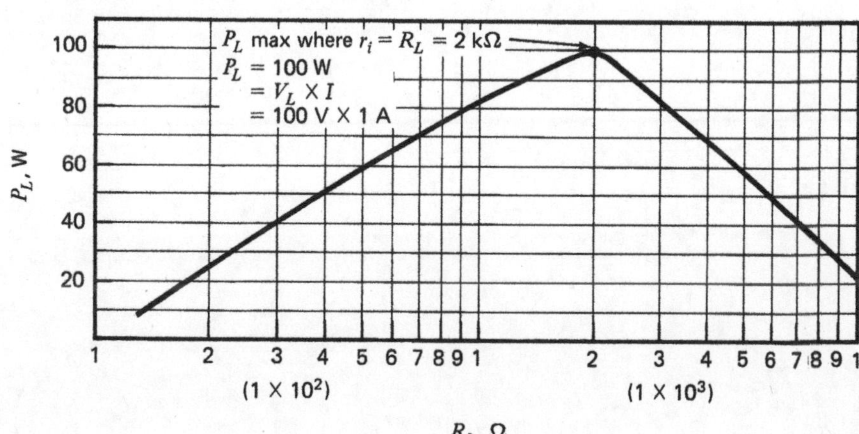

Fig. 12-2.2 Semilog graph of P_L versus R_L.

where P_T is the total power dissipated by the circuit, or

$$P_T = P_L + P_{r_i}$$

By this definition, 100 percent circuit efficiency means that absolutely no power is being dissipated.

EQUIPMENT

DMM
DC power supply
Decade box
Protoboard or springboard
Leads

COMPONENTS

(1) 820-Ω, 0.25-W resistor

PROCEDURE

1. Connect the circuit of Fig. 12-2.3.

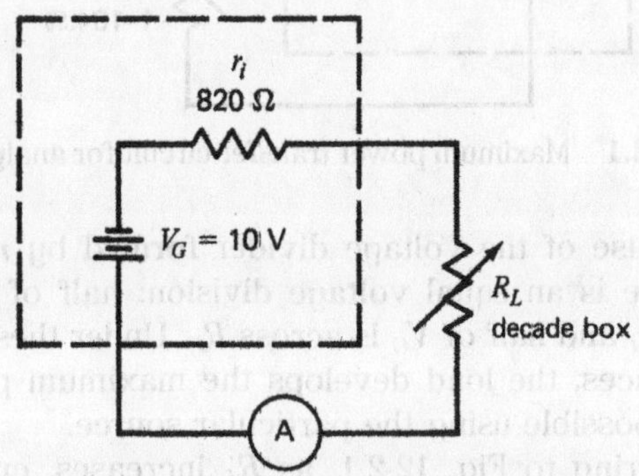

Fig. 12-2.3 Maximum power transfer circuit.

2. Increase the load resistance in 100-Ω steps from 100 Ω to 1 kΩ. Measure the voltage across R_L and the current at each step. Record the results in Table 12-2.1 for each step.
3. Increase the load resistance from 1 kΩ to 10 kΩ in 1-kΩ steps. Measure the voltage across R_L and the current at each step. Record the results in Table 12-2.1 for each step.
4. Calculate the IR voltage drop across r_i at each step, as $V_{r_i} = V_G - V_{R_1}$, and record the results in Table 12-2.1.
5. Calculate the load power dissipated at each step of R_L, as

$$P_L = V_L \times I$$

Record in Table 12-2.1.

6. Calculate the power dissipated across r_i at each step of R_L as

$$P_{r_i} = V_{r_i} \times I$$

Record in Table 12-2.1.

7. Calculate the total power dissipated in the circuit for each step of R_L as

$$P_T = V_G \times I$$

Note that

$$V_G = V_L + V_{r_i}$$

Record in Table 12-2.1.

8. Calculate circuit efficiency for each step of R_L as

$$\frac{P_L}{P_T} \times 100$$

expressed as a percentage.

9. Plot a graph of load resistance versus load power, using your data. Use two-cycle semilog paper. Prepare this graph as if it were to be used in a professional situation. It should be neat and well organized, it should include a title and all the possible values, and critical parameters should be labeled.

NAME DATE

QUESTIONS FOR EXPERIMENT 12-2

1. Explain the difference between circuit efficiencies of 1, 50, and 100 percent. In other words, explain what is meant by *circuit efficiency* as it relates to transfer of maximum power.

2. Explain what would happen if the circuit of Fig. 12-2.3 had an internal resistance of 100 kΩ.

3. Explain what would happen if the circuit of Fig. 12-2.3 had an internal resistance of 0.001 Ω.

4. Explain why semilog paper is used to graph the data.

5. Explain how you could get maximum power transferred to a 15-kΩ load if the internal resistance of your source were 10 kΩ.

TABLE FOR EXPERIMENT 12-2

TABLE 12-2.1 Data for Circuit Fig. 12-2.3

R_L	Measured V_L, V	Measured I, A	Calculated V_{r_i}, V	Calculated P_L, W	Calculated P_{r_i}, W	Calculated P_T, W	% Efficiency
100 Ω							
200 Ω							
300 Ω							
400 Ω							
500 Ω							
600 Ω							
700 Ω							
800 Ω							
900 Ω							
1 kΩ							
2 kΩ							
3 kΩ							
4 kΩ							
5 kΩ							
6 kΩ							
7 kΩ							
8 kΩ							
9 kΩ							
10 kΩ							

EXPERIMENT RESULTS REPORT FORM

Experiment No.: ______________________ Name: ______________________

Date: ______________________

Experiment Title: ______________________ Class: ______________________

______________________ Instr: ______________________

Explain the purpose of the experiment:

List the first Learning Objective:

OBJECTIVE 1:

After reviewing the results, describe how the objective was validated by this experiment.

List the second Learning Objective:

OBJECTIVE 2:

After reviewing the results, describe how the objective was validated by this experiment.

List the third Learning Objective:

OBJECTIVE 3:

After reviewing the results, describe how the objective was validated by this experiment.

Conclusion:

If required, attach to this form: ☐ Answers to Questions, ☐ Tables, and ☐ Graphs.

CHAPTER 13

EXPERIMENT 13-1

MAGNETISM

LEARNING OBJECTIVES

At the completion of this experiment, you will be able to:

- Validate that current in a conductor has an associated magnetic field.
- Understand the concept of shielding.
- Examine the left-hand rule to determine magnetic polarity.

SUGGESTED READING

Chapter 13, *Basic Electronics*, Grob/Schultz, twelfth edition

INTRODUCTION

Any electric current has an associated magnetic field that can do the work of attraction or repulsion. Not only is the magnetic field useful for doing work, it is also the cause of unwanted attraction and repulsion. Thus, it is often necessary to shield particular circuits to prevent one component from affecting another.

The most common example of magnetic force is that produced by a magnet. The magnet, with its north and south poles, acts as a generator that produces an external magnetic field provided by the opposite magnetic poles of the magnet. The idea is like the two opposite terminals of a battery that have opposite charges. Also, the earth itself is a huge natural magnet, having both north and south poles. Thus, the needle of a compass (also a magnet) is attracted to the north pole because the atoms that make up the needle have been aligned in such a way that their magnetic field is attracted to the magnetic field of the earth's north pole.

Fig. 13-1.1 Magnetic field of force around a bar magnet. (*a*) Field outlined by iron filings. (*b*) Field indicated by lines of force.

It is these magnetic fields that are the subject of electromagnetism. These fields are thought of as lines of force, called *magnetic flux*, as shown in Fig. 13-1.1.

If current is flowing in a conductor, there is a similar magnetic field that can be used in conjunction with the fields of a magnet. For example, PM (permanent magnet) loudspeakers found in most radios, televisions, and public address systems all use the principles of magnetism to produce the audible sound we listen to.

Finally, the opposite effect of current moving through a conductor is a magnetic field in motion, forcing electrons to move. This action is called *induction*. Inductance is produced by the motion of magnetic lines of flux cutting across a conductor, thus forcing free electrons in the conductor to move, as shown in Fig. 13-1.2.

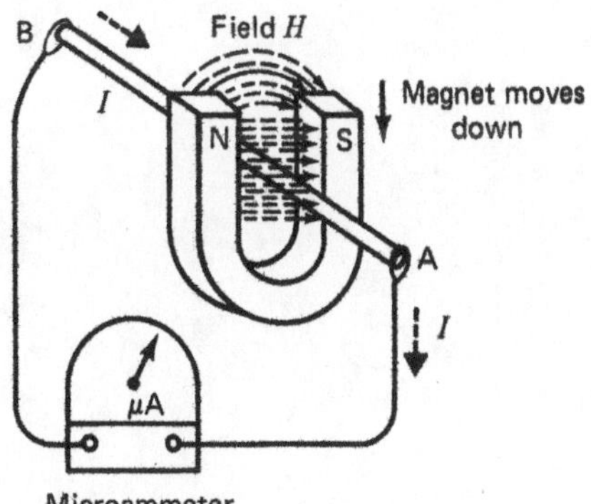

Fig. 13-1.2 Magnetically induced current in a conductor. Current *I* is electron flow.

EQUIPMENT

DC power supply
Galvanometer or microammeter
Heavy-duty horseshoe magnet (>20-lb pull)
Magnetic compass
Shield (6 × 6 in.) conductance sheet metal

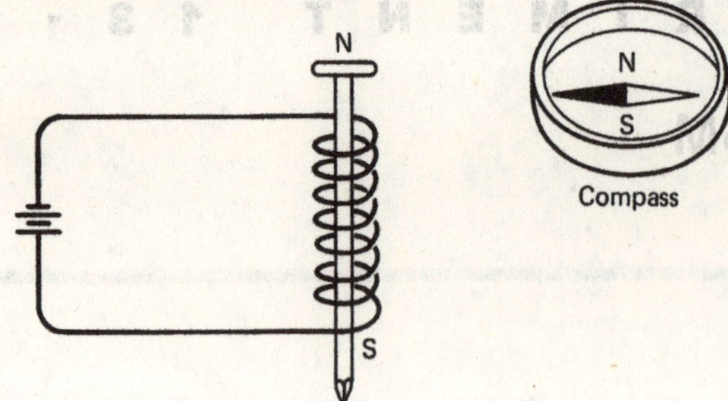

Fig. 13-1.3 Electromagnetic circuit for step 1.

COMPONENTS

2 to 3 ft of thin insulated wire
(1) No. 18 iron nail
Iron filings

PROCEDURE

1. Connect the circuit of Fig. 13-1.3. Have the compass and the iron filings nearby. Wrap the insulated wire evenly around the nail (about 10 to 15 turns).
2. The left-hand rule states that if a coil is grasped with the fingers of the left hand curled around the coil in the direction of electron flow, the thumb (extended) points to the north pole of the coil. Imagine your left hand grasped around the coil of wire wound around the nail. Determine which end is the north pole.
3. Turn on the power, and slowly move the compass close to both ends of the needle. Determine which end of the nail is north and which is south. Compare the results to step 2 above.
4. Place the shield between the compass and the nail in the circuit, and repeat step 3. Note the results.
5. Turn off the power. Remove the nail and pass it through the iron filings. Note the results. Disconnect the circuit.
6. Connect the circuit of Fig. 13-1.4. Using the same wire as in steps 1 to 4 above, loop the wire around the magnet many times, making sure that the galvanometer is on the lowest range.
7. Move the horseshoe magnet up and down, as necessary, and note the amount of current produced. Try moving the magnet more rapidly, then slowly.
8. Disconnect the circuit.

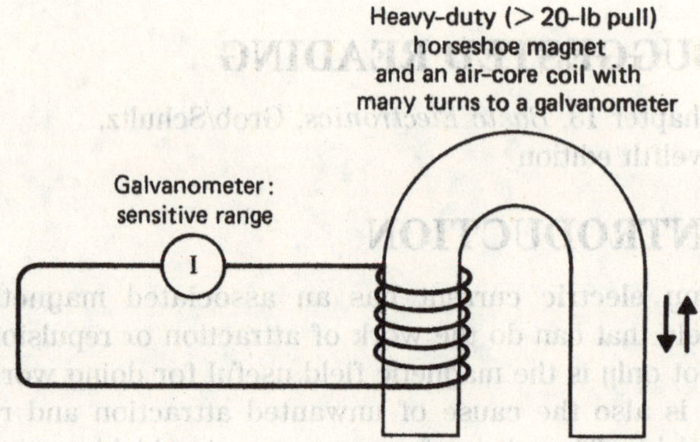

Fig. 13-1.4 Electromagnetic generator circuit for step 6.

NAME DATE

QUESTIONS FOR EXPERIMENT 13-1

1. Explain what is meant by shielding.

2. Discuss the results of moving the magnet (in step 7) faster or slower.

3. Explain which end of the nail attracted the iron filings, and why.

4. Discuss any differences between the results of steps 1 to 4 and steps 6 to 8.

5. Explain the left-hand rule as it was applied in this experiment.

EXPERIMENT RESULTS REPORT FORM

Experiment No.: ______________________ Name: ______________________

Date: ______________________

Experiment Title: ______________________ Class: ______________________

______________________ Instr: ______________________

Explain the purpose of the experiment:

List the first Learning Objective:

OBJECTIVE 1:

After reviewing the results, describe how the objective was validated by this experiment.

List the second Learning Objective:

OBJECTIVE 2:

After reviewing the results, describe how the objective was validated by this experiment.

List the third Learning Objective:

OBJECTIVE 3:

After reviewing the results, describe how the objective was validated by this experiment.

Conclusion:

If required, attach to this form: ☐ Answers to Questions, ☐ Tables, and ☐ Graphs.

CHAPTER 13

EXPERIMENT 13-2

ELECTROMAGNETISM AND COILS

LEARNING OBJECTIVES

At the completion of this experiment you will be able to:

- Identify the physical characteristics of a solenoid coil.
- Understand what circuit characteristics influence magnetomotive force.
- Compare the DC characteristics of a solenoid coil.

SUGGESTED READING

Chapter 13, *Basic Electronics*, Grob/Schultz, twelfth edition

INTRODUCTION

Electromagnetic fields are always associated with the flow of current in an electric circuit. From this current flow, magnetic units are defined. A coil of wire is also known as a *solenoid*, and as current flows through the coil, magnetic relationships are also defined.

The movement of current through a coil provides a magnetizing force known as *magnetomotive force*, or *mmf*. This magnetomotive force is directly dependent upon the amount of current flow. Also, magnetomotive force creates a magnetic field density (*H*) that will decrease with the length of the solenoid. In addition, the field density (*H*) creates flux density (*B*) that is directly related to the permeability (μ) of the core of the solenoid.

An ideal solenoid is one that enhances the ability to create magnetic flux density (*H*). This is accomplished by creating a solenoid that has a length much greater than its diameter. Like the magnetic field that is created by passing a current through a single wire loop, the solenoid concentrates these magnetic field inside the coil and provides opposing magnetic poles at the respective ends. These effects are enhanced by the number of wire turns around the coils, the length of the coil, and the type of core material used within the coil.

In addition, the strength of the magnetic field depends upon the level of current that flows in the coil. This can be summarized as:

- The *larger* current, the *stronger* the magnetic field.
- The *greater* the number of wire turns in a specific length of solenoid, the *greater* the concentration of the field.

This experiment analyzes the importance of the size (number of turns) and the type of core material used within a coil.

EQUIPMENT

Breadboard
24 V DC power supply
DMM

COMPONENTS

Resistor: 33 Ω 2 W
Normally open push-button switch

SPECIAL MATERIALS

12-inch wooden or plastic ruler
Solenoid coil (150 turns)
Solenoid coil (300 turns)
5-inch length of ¼-inch diameter plastic rod
5-inch length of ¼-inch diameter iron rod
5-inch length of ¼-inch diameter wood rod
5-inch length of ¼-inch diameter brass rod
5-inch length of ¼-inch diameter aluminum rod
5-inch length of ¼-inch diameter glass rod
Compass

PROCEDURE

1. Construct the circuit from the pictorial diagram as shown in Fig. 13-2.1.

2. Position the magnetic compass approximately 2 inches from the 150-turn coil. Turn on the power supply, and adjust the supply's voltage until the ammeter reads 200 mA. Move the magnetic compass in such a way that there is just a slight needle deflection noted.

3. Insert the plastic rod into the center of the solenoid coil, and note any changes in the position of the magnetic compass. Is there any change in the circuit's series current? Record your observation in Table 13-2.1.

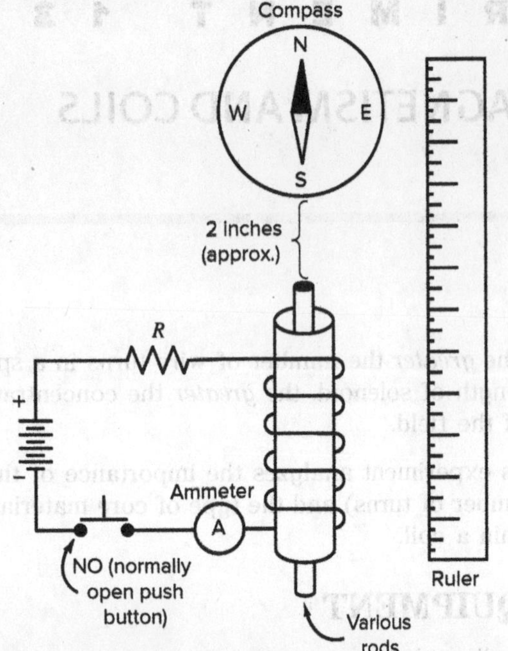

Fig. 13-2.1 Solenoid coil apparatus.

4. Insert the iron rod into the center of the solenoid coil, and note any changes in the position of the magnetic compass. Is there any change in the circuit's series current? Record your observation in Table 13-2.1.

5. Insert the wood dowel into the center of the solenoid coil, and note any changes in the position of the magnetic compass. Is there any change in the circuit's series current? Record your observation in Table 13-2.1.

6. Insert the brass rod into the center of the solenoid coil, and note any changes in the position of the magnetic compass. Is there any change in the circuit's series current? Record your observation in Table 13-2.1.

7. Insert the aluminum rod into the center of the solenoid coil, and note any changes in the position of the magnetic compass. Is there any change in the circuit's series current? Record your observation in Table 13-2.1.

8. Insert the glass rod into the center of the solenoid coil, and note any changes in the position of the magnetic compass. Is there any change in the circuit's series current? Record your observation in Table 13-2.1.

9. Reverse the polarity of the power supply, and repeat procedure steps 2 through 8. Record your observations in Table 13-2.2.

10. Reverse the polarity of the power supply, substitute the 300-turn coil for the 150-turn coil, and repeat procedure steps 2 through 8. Record your observations in Table 13-2.3.

11. Reverse the polarity of the power supply, and repeat procedure steps 2 through 8. Record your observations in Table 13-2.4.

NAME ______________________ DATE ________

QUESTIONS FOR EXPERIMENT 13-2

1. Which of the materials inserted into the solenoid coil had the greatest effect on the compass? Which of the materials had the least effect?

2. Explain why some of the materials inserted into the solenoid had a greater effect than other materials.

3. What were the observable changes, if any, when the power supply's polarity was reversed?

4. Explain any observable changes when a 300-turn coil replaced the 150-turn coil.

5. What circuit characteristics influence the level of magnetomotive force?

TABLES FOR EXPERIMENT 13-2

TABLE 13-2.1 Circuit Characteristics of the 150-Turn Coil

Procedural Step	Record the Observed Change in Compass Deflection	Record the Observed Change in Measured Circuit Current
3	________	________
4	________	________
5	________	________
6	________	________
7	________	________
8	________	________

TABLE 13-2.2 150-Turn Coil with a Reversed Supply Polarity

Procedural Step	Record the Observed Change in Compass Deflection	Record the Observed Change in Measured Circuit Current
3		
4		
5		
6		
7		
8		

TABLE 13-2.3 Circuit Characteristics of the 300-Turn Coil

Procedural Step	Record the Observed Change in Compass Deflection	Record the Observed Change in Measured Circuit Current
3		
4		
5		
6		
7		
8		

TABLE 13-2.4 300-Turn Coil with a Reversed Supply Polarity

Procedural Step	Record the Observed Change in Compass Deflection	Record the Observed Change in Measured Circuit Current
3		
4		
5		
6		
7		
8		

EXPERIMENT RESULTS REPORT FORM

Experiment No.: ______________________ Name: ______________________

Date: ______________________

Experiment Title: ______________________ Class: ______________________

______________________ Instr: ______________________

Explain the purpose of the experiment:

List the first Learning Objective:

OBJECTIVE 1:

After reviewing the results, describe how the objective was validated by this experiment.

List the second Learning Objective:

OBJECTIVE 2:

After reviewing the results, describe how the objective was validated by this experiment.

List the third Learning Objective:

OBJECTIVE 3:

After reviewing the results, describe how the objective was validated by this experiment.

Conclusion:

If required, attach to this form: ☐ Answers to Questions, ☐ Tables, and ☐ Graphs.

CHAPTER 14

EXPERIMENT 14-1

RELAYS

LEARNING OBJECTIVES

At the completion of this experiment, you will be able to:

- Identify a relay.
- Describe the function of a magnetic field associated with the flow of electric current.
- List and explain significant relay ratings.

SUGGESTED READING

Chapter 14, *Basic Electronics*, Grob/Schultz, twelfth edition

INTRODUCTION

Current moving through a conductor can move the needle of a compass if it is nearby. Also, a magnetic field in motion can move current through a nearby conductor.

Electromagnetism combines the characteristics of an electric current and magnetism. Electrons in motion always have an associated magnetic field. These electromagnetic effects have many practical applications that form the basis of motors, generators, solenoids, and relays.

A relay is an electromechanical device that uses either an AC or DC actuated electromagnet to open or close one or more sets of contacts. Relay contacts that are open when the relay is not energized are called *normally open (NO)* contacts, and relay contacts that are closed when the relay is not energized are called *normally closed (NC)* contacts. Relay contacts are held in their resting position either by a spring or by some other mechanism.

Figure 14-1.1 presents the basic relay and schematic symbols that are used to represent relay contacts. Similar to mechanical switches, the switching contacts of a relay can have any number of poles and throws. A relay will always have two electrical connections for the coil or electromagnet. Depending upon the type or configuration, the relay can have any number of poles and throws. For example, Fig. 14-1.1 depicts an SPDT (single-pole, double-throw) relay.

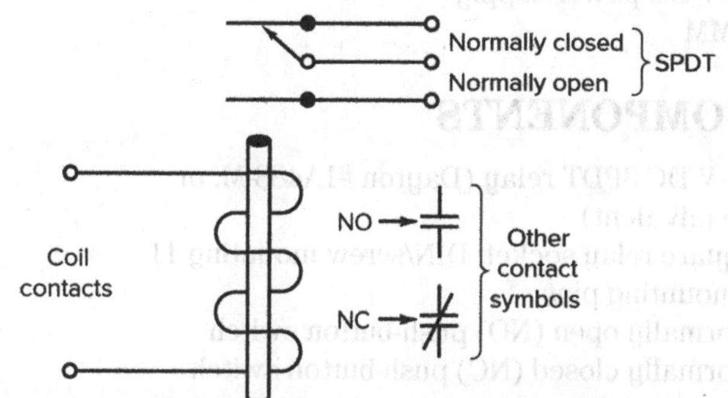

Fig. 14-1.1 Basic SPDT relay and related schematic symbols.

When terminals 1 and 2 are connected to a source voltage, current will flow through the relay coil and an electromagnet is formed. If there is sufficient current in the relay coil, the normally open (NO) contacts 3 and 4 close, and the normally closed (NC) contacts 4 and 5 open.

Relay Specifications

Manufacturers of electromechanical relays always supply a specification sheet for each of their relays. The specification sheet contains voltage and current ratings for both the relay coil and its switching contacts. The specification sheet also includes information regarding the location of the relay coil and switching contact terminals.

The following is an explanation of a relay's most important ratings.

Pickup current: The minimum amount of relay coil current necessary to energize or operate the relay.

Holding current: The minimum amount of current required keeping a relay energized or operating. (The holding current is less than the pickup current.)

Dropout voltage: The maximum relay coil voltage at which the relay is no longer energized.

Contact voltage rating: The maximum voltage the relay contacts are capable of switching safely.

Contact current rating: The maximum current the relay contacts are capable of switching safely.

Contact voltage drop: The voltage drop across the closed contacts of a relay when operating.

Insulation resistance: The resistance measured across the relay contacts in the open position.

EQUIPMENT

24-V DC power supply
DMM

COMPONENTS

24-V DC 3PDT relay (Dayton #1A488-M, or equivalent)
Square relay socket, DIN/screw mounting 11 mounting pins
Normally open (NO) push-button switch
Normally closed (NC) push-button switch

PROCEDURE

1. Obtain and inspect a 24-V DC 3PDT relay.
2. Obtain and inspect a compatible relay socket.
3. Align and plug the relay into the relay socket. Observe, inspect, and determine the terminals associated with the relay's coil (both normally open and closed contacts). On the schematic diagram shown in Table 14-1.1, identify and record the pin numbers for the coil, poles, and normally open (NO)/ normally closed (NC) contacts for the type of relay you are using for this experiment. Also in this table, associate the coil, poles, NO, and NC contacts for this relay to the relay's pin numbers.

OPTIONAL:

4. Obtain the manufacturer's specification sheet for the type of relay that you will be using for this experiment. Determine from the specification sheet the operating voltage, current, and other electrical characteristics of the relay. Record this information in Table 14-1.2.
5. Connect the schematic diagram shown in Fig. 14-1.2.
6. Turn on the power supply, and increase the voltage level until the relay's coil just becomes energized. You will be able to determine this condition by hearing the relay "click" or by observing the relay's internal electrical contacts moving. When a relay is energized, it is referred to as being "tripped." Note the power supply's voltage and circuit current level at this juncture, and record your results in Table 14-1.3. These measurements are known as the *pickup voltage* and *pickup current*.

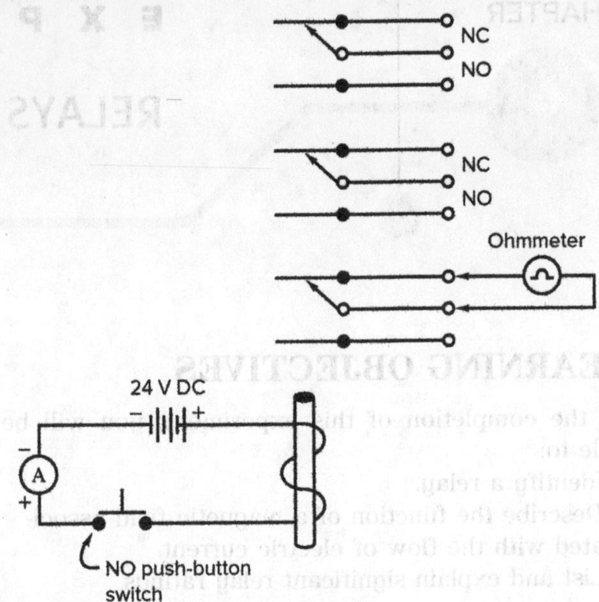

Fig. 14-1.2 Relay schematic diagram.

7. Reduce the power supply voltage until the relay's coil is de-energized. Note the power supply's voltage at this point, and record this value in Table 14-1.3. This measurement is known as the *dropout voltage*. Through experimentation, by increasing and decreasing the power supply's voltage, determine the minimum amount of current necessary to keep the relay energized. Note this current level and record its value in Table 14-1.3. This measurement is known as the *holding current*.
8. Construct the circuit as shown in Fig. 14-1.3.
9. Turn on and adjust power supply to 24 V DC. Press the NO push-button switch A, and note the circuit action.
10. Press the NC push-button switch B, and note the circuit action.
11. Repeat procedural steps 9 and 10 several times.
12. Describe the observed circuit operation in Table 14-1.4.

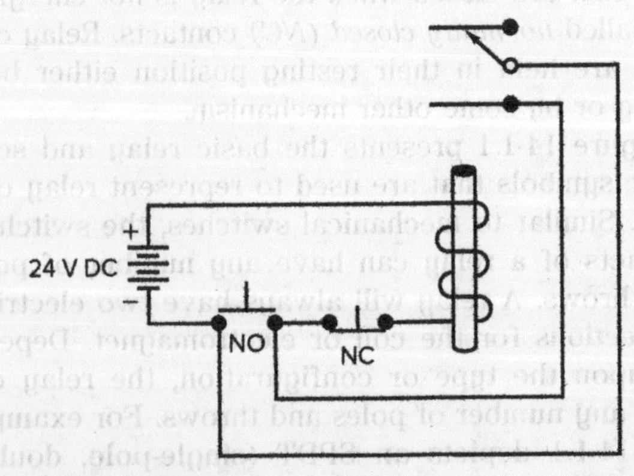

Fig. 14-1.3 Latching relay circuit.

NAME DATE

QUESTIONS FOR EXPERIMENT 14-1

1. Describe how an ohmmeter could be used to determine if the relay's coil had opened.

2. Understanding that the relay is an electromechanical device, what electrical problems could you see developing when using the device over a long period of time?

3. If a set of NC relay contacts is closed when the relay's coil is energized, is the relay operating correctly?

4. It has been stated that one of the main advantages of using a relay is its ability to control high-power loads with a low amount of input power. Could you describe why this is an advantage?

5. What is the advantage of using a relay over a mechanical switch in remote control applications?

6. What is the purpose (and/or probable use of) the schematic circuit shown in Fig. 14-1.3?

TABLES FOR EXPERIMENT 14-1

TABLE 14-1.1 Identifying the Pins and Functions of a Relay

Using the pin numbers on the schematic diagram shown below, identify the function of each pin. If you are using a different type of relay, other than a 3PDT, create your own table.	Pin numbers and relay function: Use the terms coil, pole, NO, and NC in completing this section of the table.	
*	Pin Number	Relay Term*
	1	______
	2	______
	3	______
	4	______
	5	______
	6	______
	7	
	8	______
	9	______
	10	
	11	

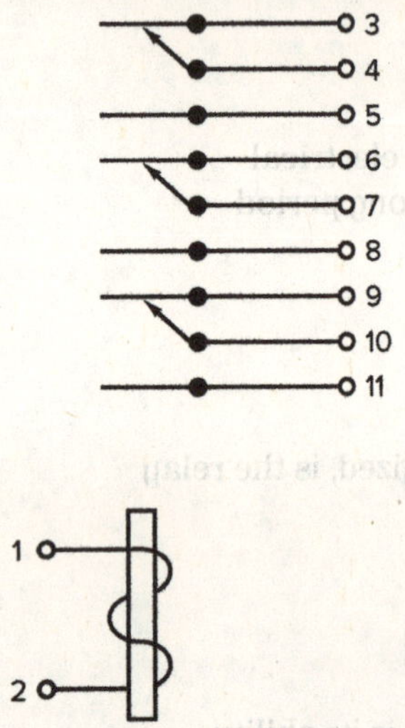

*Answers can vary.

TABLE 14-1.2 Optional: Operational Characteristics

Electrical Specification	Electrical Rating	Short Written Decription That Describes the Electrical Specification
Pickup voltage	______ volts	
Pickup current	______ amps	
Holding current	______ amps	
Dropout voltage	______ volts	
Contact voltage rating	______ volts	
Contact current rating	______ amps	
Contact voltage drop	______ volts	
Insulation resistance	______ ohms	

TABLE 14-1.3 Measured Operational Characteristics

Procedural Step	Operational Characteristic	Measured Value
6	___________ Pickup voltage	___________ volts
6	___________ Pickup current	___________ amps
7	___________ Dropout voltage	___________ volts
7	___________ Holding current	___________ amps

TABLE 14-1.4 Measured Operational Characteristics

Procedural Step 12	For the circuit constructed from the latching relay schematic shown in Fig. 14-1.3, describe below the observed circuit action.

I observed ____

____

____

____

____

____

EXPERIMENT RESULTS REPORT FORM

Experiment No.: ______________________ Name: ______________________

Date: ______________________

Experiment Title: ______________________ Class: ______________________

______________________ Instr: ______________________

Explain the purpose of the experiment:

List the first Learning Objective:

OBJECTIVE 1:

After reviewing the results, describe how the objective was validated by this experiment.

List the second Learning Objective:

OBJECTIVE 2:

After reviewing the results, describe how the objective was validated by this experiment.

List the third Learning Objective:

OBJECTIVE 3:

After reviewing the results, describe how the objective was validated by this experiment.

Conclusion:

If required, attach to this form: ☐ Answers to Questions, ☐ Tables, and ☐ Graphs.

CHAPTER 15

EXPERIMENT 15-1

AC VOLTAGE AND OHM'S LAW

LEARNING OBJECTIVES

At the completion of this experiment, you will be able to:

- Validate the Ohm's law expression for alternating current, where

$$V = I \times R$$

$$I = \frac{V}{R}$$

$$R = \frac{V}{I}$$

- Determine the power dissipation in an alternating current circuit.
- Operate an AC oscillator or signal generator.

SUGGESTED READING

Chapter 15, *Basic Electronics*, Grob/Schultz, twelfth edition

INTRODUCTION

This experiment is designed to introduce you to alternating current (AC) voltages and to validate that Ohm's law can still be used to determine current, voltage, or resistance if two of the three terms of an AC circuit are known. The amount of electric power, measured in watts, can be determined for AC circuits by using Ohm's law in the same way as for direct current (DC) circuits.

Alternating current is electron flow in two directions: positive and negative. As shown in Fig. 15-1.1, two batteries can be switched on and off alternately to achieve the effect of AC voltage. Notice that, depending on which battery is switched on, the direction of current flow through the resistor will alternate. Now, imagine that you could move the rocker switch back and forth 100 times in 1 s. If you could do it, you would create an alternating current of 100 cycles/s, expressed as 100 hertz (Hz). Note that 1 Hz = 1 cycle/s, named after the 19th-century scientist Hertz.

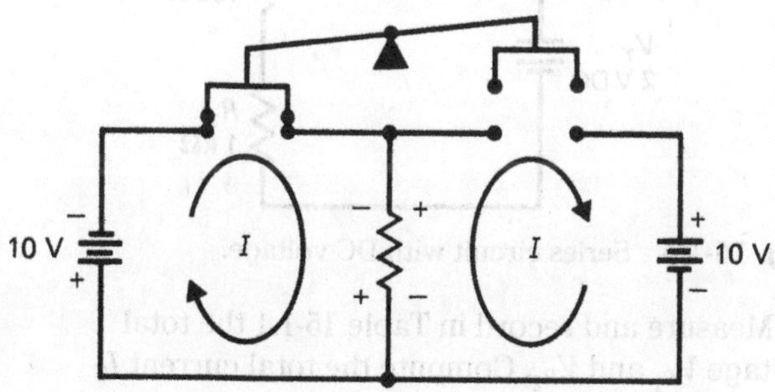

Fig. 15-1.1 Rocker switch alternates current flow.

The concept of alternating current is described not only by the frequency (in hertz) at which it alternates but also by the amount of voltage that is being alternated. In the circuit of Fig. 15-1.1, the two batteries supply a voltage to the resistor at 10 V in each polarity or direction. Therefore, it is valid to say that the voltage across the resistor has +10 or −10 V across it at a particular time. In fact, this is a total potential of 20 V from one peak value (+10 or −10 V) to another.

Because it takes some amount of time for the voltage to reach the resistor, depending on the frequency of the alternating current, each potential voltage rises to its peak value (+10 or −10 V) and returns to zero before rising to the alternate peak value. And because this alternation of current flow occurs in a back-and-forth or oscillating manner, it is represented by a sine-wave symbol. Notice that Fig. 15-1.2 shows the number line of Fig. 15-1.3 combined with a sine wave to illustrate how AC voltage is represented.

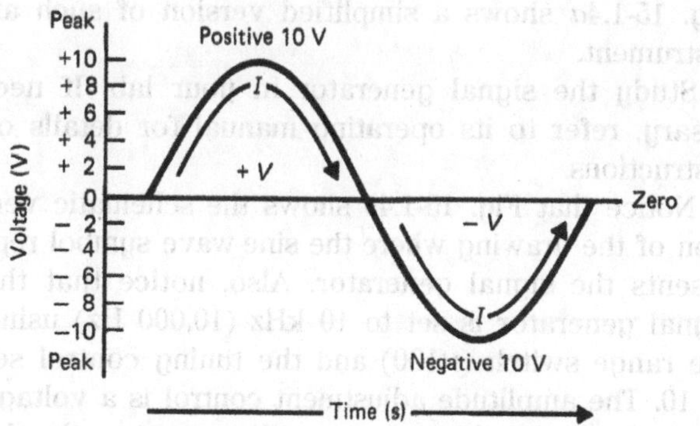

Fig. 15-1.2 Sine-wave illustration of AC voltage: peak to peak.

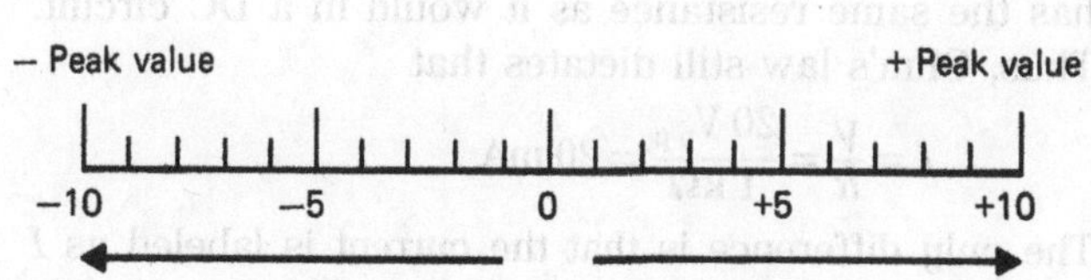

Fig. 15-1.3 Peak values on a number line.

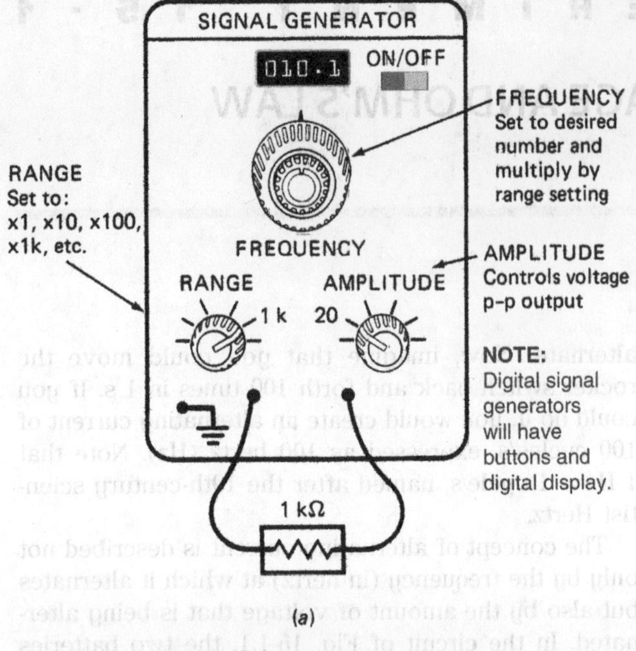

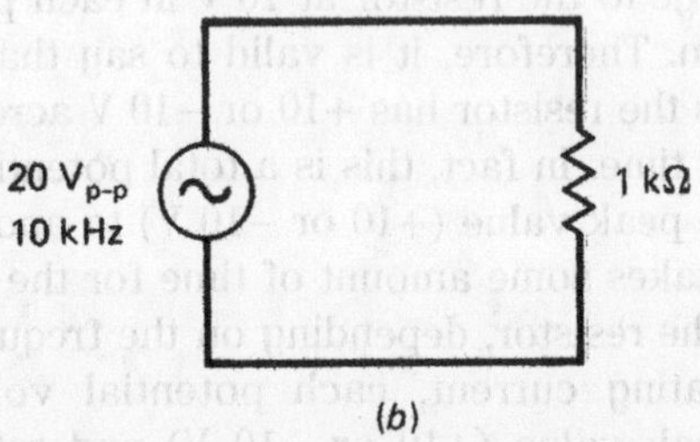

Fig. 15-1.4 (*a*) Simplified signal generator connected to a 1-kΩ resistor. (*b*) Schematic.

The most common form of applying AC voltage to a circuit is by using an oscillator or signal generator. This instrument can be simple or complex, depending on its style and manufacture. However, Fig. 15-1.4*a* shows a simplified version of such an instrument.

Study the signal generator in your lab. If necessary, refer to its operating manual for details or instructions.

Notice that Fig. 15-1.4*b* shows the schematic version of the drawing where the sine-wave symbol represents the signal generator. Also, notice that the signal generator is set to 10 kHz (10,000 Hz) using the range switch (×100) and the tuning control set to 10. The amplitude adjustment control is a voltage adjust knob that is used to set the output to the desired voltage (p-p) level.

In the AC circuit of Fig. 15-1.4, the 1-kΩ resistor has the same resistance as it would in a DC circuit. Thus, Ohm's law still dictates that

$$I = \frac{V}{R} = \frac{20\ V_{\text{p-p}}}{1\ \text{k}\Omega} = 20\ \text{mA}$$

The only difference is that the current is labeled as I (p-p) to show that it is the result of an AC voltage (p-p).

In AC circuits, an ammeter is not used to measure current because it is only built to measure current flow in one direction (direct current). Do not put a DC ammeter in an AC circuit.

Finally, power in an AC circuit is calculated slightly differently from that in a DC circuit. To calculate the power dissipation, the p-p voltage must be converted to its DC equivalent value. This is also called the *rms* (root-mean-square) *value* and is equal to 70.7 percent of the peak voltage. For example, a 20-V p-p voltage will produce the same heating or lighting power (in watts) as a 7-V DC battery. Thus, AC power is calculated as:

$$\frac{V_{\text{p-p}}}{2 \times 0.707} = V_{\text{DC}} \text{ or } V_{\text{rms}} \text{ for use in the power equation}$$

$$\text{Power (watts)} = \frac{V^2}{R}$$

$$\text{where } V = \frac{V_{\text{p-p}}}{2} \times 0.707$$

In summary, AC voltage can be used in any Ohm's law equation where I or V is labeled as a p-p value and converted to its DC equivalent (rms) for power calculations. Although AC voltages are usually measured on an oscilloscope, the following procedure will use a voltmeter (DMM) to validate Ohm's law.

EQUIPMENT

Signal generator DMM (AC voltage)
DC power supply
Springboard
Leads

COMPONENTS

Resistors (all 0.25 W):

(1) 100 Ω
(1) 1 kΩ

PROCEDURE

1. Connect the circuit of Fig. 15-1.5.

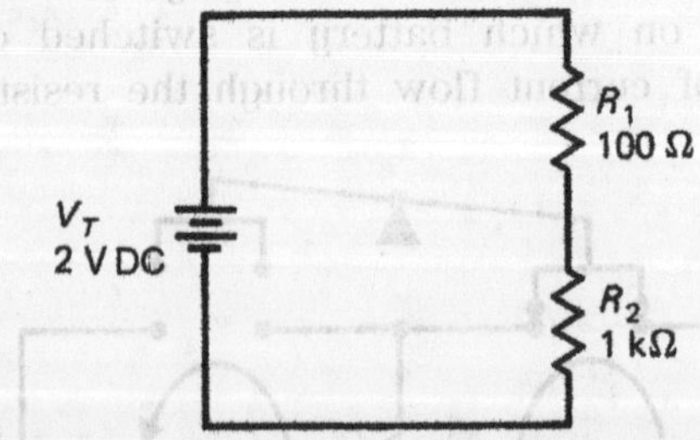

Fig. 15-1.5 Series circuit with DC voltage.

2. Measure and record in Table 15-1.1 the total voltage V_{R_1} and V_{R_2}. Compute the total current I_T and I_{R_1} and I_{R_2} by using Ohm's law.

3. Connect the circuit of Fig. 15-1.6.

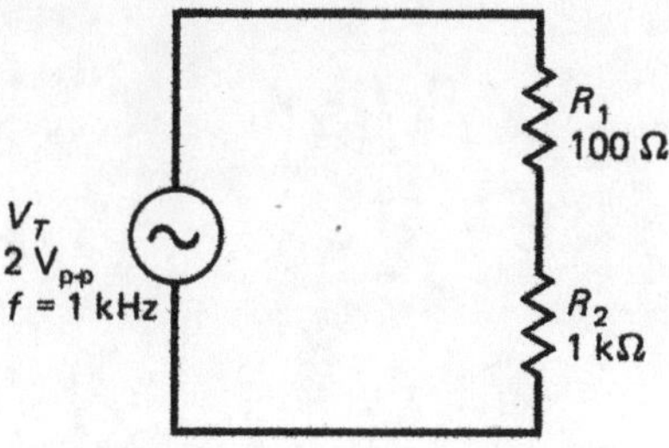

Fig. 15-1.6 Series circuit with AC voltage.

Note: Connect the entire circuit, and then measure V_T to be sure it is 2 $V_{p\text{-}p}$.

4. Repeat step 2 for the circuit of Fig. 15-1.6. Current should be calculated as:

$$I_{p\text{-}p} = \frac{V_{p\text{-}p}}{R}$$

5. Change the frequency to 5 kHz and repeat step 2 for Fig. 15-1.6.

6. Calculate the power dissipated by each circuit as shown in Table 15-1.1.

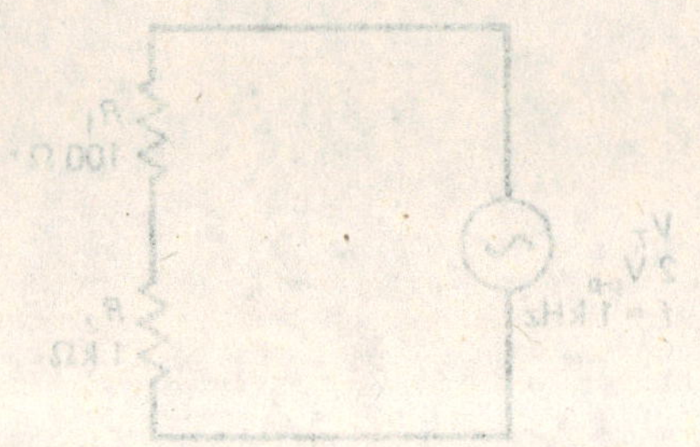

NAME _______________ DATE _______________

QUESTIONS FOR EXPERIMENT 15-1

Answer true (T) or false (F) to the following:

_____ **1.** Resistors do not function the same in AC circuits as in DC circuits.

_____ **2.** A 10-V battery and a 7-V (p-p) signal will produce the same amount of power across a 2-kΩ resistor.

_____ **3.** Ohm's law can be used only to find current in DC circuits.

_____ **4.** Current in an AC circuit can be measured with an ammeter just as a DC circuit can.

_____ **5.** A signal generator does not produce any current in an AC circuit.

TABLES FOR EXPERIMENT 15-1

TABLE 15-1.1

Resistance	Nominal Value	Volts Measured	Current $I = V/R$	Power, Watts, V^2/R
DC CIRCUIT—Fig. 15-1.5				
R_1	100 Ω	_____	_____	_____
R_2	1 kΩ	_____	_____	_____
R_T	1.1 kΩ	_____	_____	_____
AC CIRCUIT—Fig. 15-1.6* f = 1 kHz				
R_1	100 Ω	_____	_____	_____
R_2	1 kΩ	_____	_____	_____
R_T	1.1 kΩ	_____	_____	_____
AC CIRCUIT—Fig. 15-1.6* f = 5 kHz				
R_1	100 Ω	_____	_____	_____
R_2	1 kΩ	_____	_____	_____
R_T	1.1 kΩ	_____	_____	_____

*Note: Use p-p values. For example, $10\ V_{p\text{-}p} \div 100\ \Omega = 100$ mA, 1 p-p.

Also, power = $\left(\frac{V_{p\text{-}p}}{2} \times 0.707\right)^2 \div R$ for AC circuits.

EXPERIMENT RESULTS REPORT FORM

Experiment No.: ______________________ Name: ______________________

Date: ______________________

Experiment Title: ______________________ Class: ______________________

______________________ Instr: ______________________

Explain the purpose of the experiment:

List the first Learning Objective:
OBJECTIVE 1:

After reviewing the results, describe how the objective was validated by this experiment.

List the second Learning Objective:
OBJECTIVE 2:

After reviewing the results, describe how the objective was validated by this experiment.

List the third Learning Objective:
OBJECTIVE 3:

After reviewing the results, describe how the objective was validated by this experiment.

Conclusion:

If required, attach to this form: ☐ Answers to Questions, ☐ Tables, and ☐ Graphs.

CHAPTER 15

EXPERIMENT 15-2

BASIC OSCILLOSCOPE MEASUREMENTS

LEARNING OBJECTIVES

At the completion of this experiment, you will be able to:

- Operate an oscilloscope.
- Measure DC voltages.
- Measure AC voltages.

SUGGESTED READING

Chapter 15, *Basic Electronics*, Grob/Schultz, twelfth edition; the operating instructions for your oscilloscope; and Appendix F of this lab manual.

INTRODUCTION

This experiment continues the AC portion of your laboratory studies. The same rules for series and parallel circuits and Ohm's law still apply to AC circuits as they did for DC circuits. However, there is a difference. As the name *alternating current* implies, current is alternating its direction. That is, electron flow reverses its direction, and therefore positive and negative polarities will alternate every time electron flow changes direction.

Using the Oscilloscope

To study AC voltages in the lab, it is necessary to learn how to operate an oscilloscope. Although the oscilloscope is a complex instrument, it is basically a voltmeter. It measures and displays AC voltages as well as DC voltages.

There are some aspects of AC voltage that must be understood before attempting to measure them. AC voltages have frequency; that is, one alternation occurs over some period of time. For example, the AC wall outlet has a frequency of 60 hertz (Hz). This means that 60 complete cycles of electron flow, from negative to positive and back again, occur during 1 s. A frequency of 20 kHz means that 20,000 cycles occur during 1 s.

DC voltage does not have frequency. It is a steady, or direct, current flow that does not change over time. This difference in frequency is a major factor in understanding AC voltages.

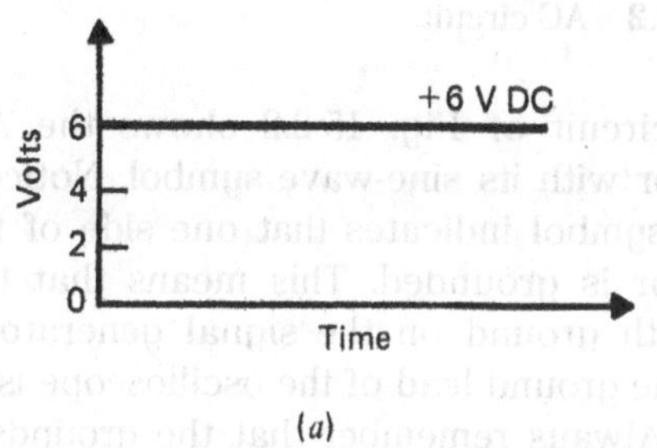

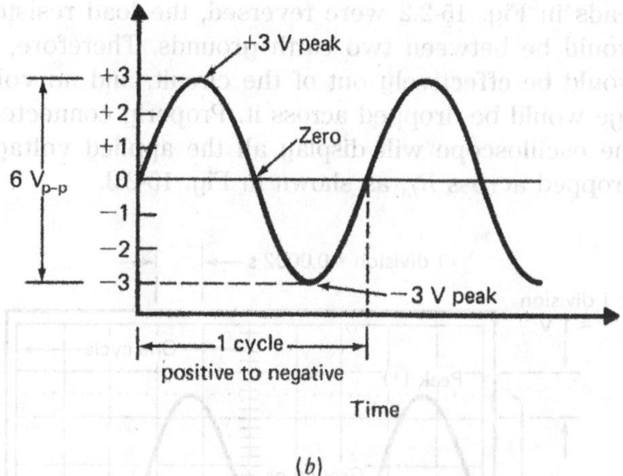

Fig. 15-2.1 DC versus AC voltages. (*a*) DC voltage plot. (*b*) AC voltage plot.

The other important difference between AC and DC voltage is the way that magnitude is measured. DC voltage is usually steady and constant. AC voltage, because of its alternating character, takes some time to reach a peak value, and then it reverses direction and becomes a zero value before it reaches a peak value in the other direction. Figure 15-2.1 illustrates how this works. Notice how the DC and AC voltages differ. They differ in magnitude and frequency. The +6 VDC is referenced from zero. The 6-V peak-to-peak (p-p) AC voltage is referenced from zero but is described by its peak-to-peak relationship. Zero volts still exist, but the AC voltage goes above it and below it. If, for example, the cycle of Fig. 15-2.1*b* occurred during 1 ms, the frequency would be

$$f = \frac{1}{t} = \frac{1}{0.001\ \text{s}} = 1000\ \text{Hz or 1 kHz}$$

where f = frequency

t = time it takes for a complete alternation

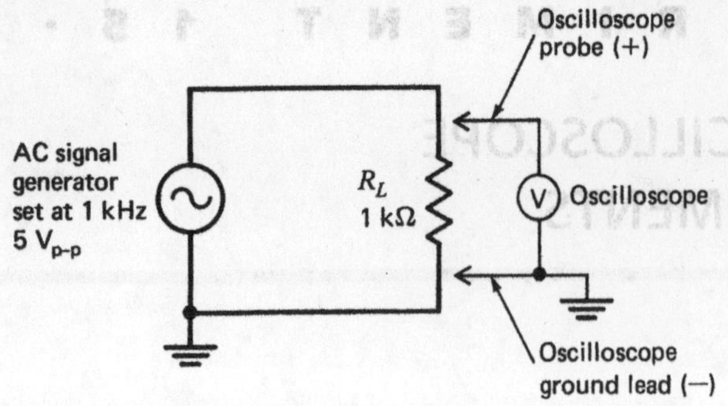

Fig. 15-2.2 AC circuit.

The circuit of Fig. 15-2.2 shows the AC signal generator with its sine-wave symbol. Notice that the ground symbol indicates that one side of the signal generator is grounded. This means that there is a true earth ground on the signal generator. This is where the ground lead of the oscilloscope is also connected. Always remember that the grounds must be connected together. If, for example, the oscilloscope leads in Fig. 15-2.2 were reversed, the load resistor would be between two earth grounds. Therefore, it would be effectively out of the circuit, and no voltage would be dropped across it. Properly connected, the oscilloscope will display all the applied voltage dropped across R_L, as shown in Fig. 15-2.3.

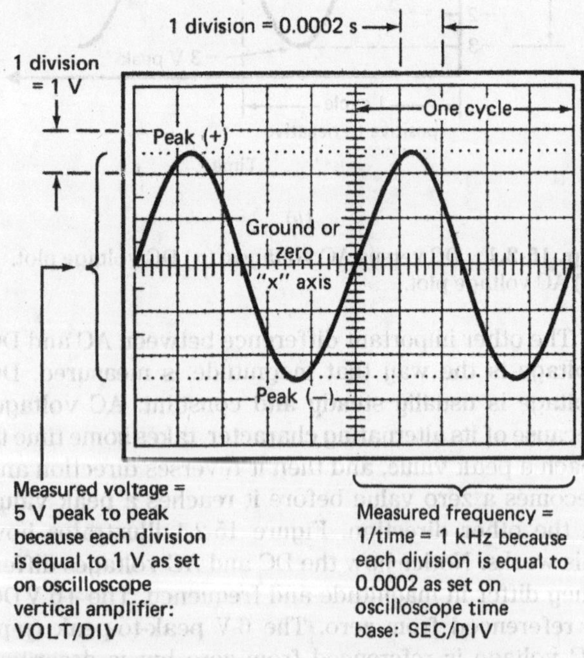

Fig. 15-2.3 Typical oscilloscope display for circuit shown in Fig. 15-2.2: 5 V_{p-p} sine wave at 1 kHz. Note that $f = 1/t$. Here, five divisions = 5 × 0.0002 s = 0.001. Thus $f = 1/0.001 = 1$ kHz.

If a DC voltage were measured, the oscilloscope display would be different. Figure 15-2.4*b* shows how the oscilloscope would display the measured voltage. Notice that the DC measurement shows a

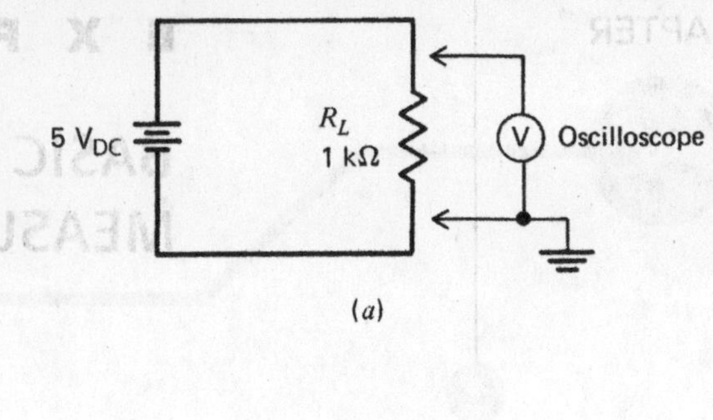

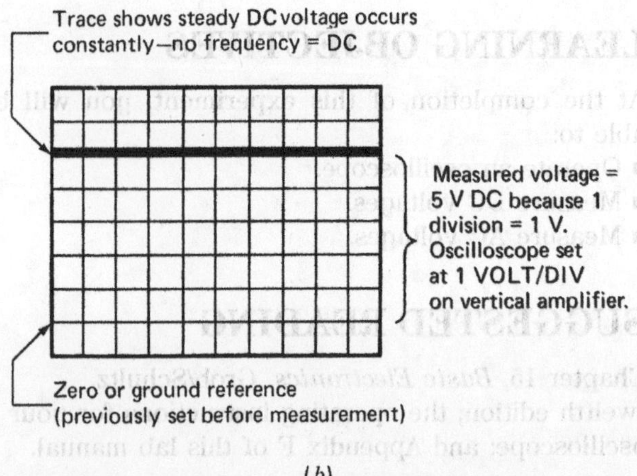

Fig. 15-2.4 (*a*) Oscilloscope measurement of DC voltage. (*b*) Oscilloscope display of DC measurement.

trace that is a straight line. No frequency and no peak-to-peak value are displayed because there is no AC voltage, only a steady DC voltage.

For both AC and DC measurements, the oscilloscope must be adjusted prior to reading the trace from the display. In the same way that it may have taken you several attempts before you could adjust an ohmmeter and properly measure a resistor, it may take several attempts before you will be able to measure voltages by using the oscilloscope. The following procedures are meant to instruct you how to use a scope. However, because there are so many different brands of oscilloscopes, you may receive some preliminary instructions from your laboratory instructor, or you can read the operator's instructions for your oscilloscope. Most oscilloscope manuals have a section that is intended to get you started. It is a good idea to spend some time examining your oscilloscope before you start.

Keep in mind the following main points when you start the following procedures:

1. An oscilloscope is a voltmeter.
2. It measures magnitude and frequency of a signal.
3. The display is like a television display: You adjust it for the best picture. If a sine wave appears larger or smaller, it is due to the operator's adjustment.

4. The magnitude and frequency of the input signal are adjusted by using the signal generator controls.
5. Avoid ground loops by keeping all ground connections at the same point when measuring AC voltages.

Digital vs. Analog Oscilloscopes: Analog scopes are older and larger in size (weight) than digital scopes because of their older circuitry and because they use a cathode ray tube (CRT) for the display. When a signal is measured by an analog oscilloscope, the signal appears to glow on the display just like an old television set. Because of the CRT, the analog oscilloscope has a limited range of frequencies it can display clearly. At low frequencies, the signal can appear as a single dot moving slowly across the screen, instead of the wave you are expecting. At higher frequencies, close to 1 GHz, the CRT cannot react fast enough and the trace may become too dim to see. In general, analog scopes are slowly disappearing for many reasons. But they are still valid and you can use them for the exercises in this lab manual.

The Digital oscilloscope (also known as DSO or Digitizing Storage Oscilloscope) has an LCD display and is lightweight, portable, and uses some type of microprocessor and analog to digital converter. Digital oscilloscopes can react quickly to changing signals and they can also store the waveforms in their memory. The stored waveforms can then be printed or sent to a computer. The DSO creates a digital value from the analog signal that is being measured and then it displays it. Most digital oscilloscopes can also measure very high frequencies and for that reason alone, they are replacing analog oscilloscopes. In addition, hand-held oscilloscopes are now also incorporated into DMMs so that simply pressing a button can turn the DMM into an oscilloscope.

Inputs, Traces, Probes: Most oscilloscopes allow two signals to be input and displayed and are called Dual Trace scopes. With two inputs at the same time, you can compare one signal to another—such as comparing a circuit's input to output signals, or comparing one digital (square) wave to another where one is a reference. Also, most bench-top oscilloscopes have specialized probes that are shielded (coaxial cable) against interfering signals. The connector on the scope is usually a BNC connector (The probes also have polarity like the DMM: one positive and one negative (ground). Hand-held DMM scopes may not have high-quality shielding like a specialized probe but are usually OK. In addition, many scopes have multiple inputs or channels to measure multiple signals at the same time. An example would be the input and the output of a filter or amplifier displayed on the scope to compare them. For this experiment, you will only be using one channel input.

Basic Oscilloscope Controls: Both analog and digital oscilloscopes have controls for the vertical or amplitude (voltage) display, and the horizontal or time display. These two controls define the grid values on the display. So, the oscilloscope is simply measuring the signal you input and you control the way it is displayed. All the other knobs and buttons on the oscilloscope perform secondary functions so keep that in mind if the controls seem confusing. Here are the most important controls:

- **Vertical: volts per division:** This defines the number of volts each vertical (Y axis) grid line represents on the display. For example, a 1 volt wave would be too small to see at 10 V/div; it would only take up 1/10 of the space between two grid lines. However, you can make it appear larger by decreasing the number of volts per division. At 1 V/div, the amplitude of the wave would appear over 5 grid lines and be easy to see, especially because most scopes have 8 divisions on the vertical or Y axis.
- **Horizontal: sec per division:** This defines the number of seconds each horizontal (X axis) grid line represents on the display. For example, at 1 sec/div, a 1 KHz signal (1,000 cycles per second) would appear 1,000 times between each grid line. This would make it impossible to see any one wave. However, at 0.1 m sec/div (1/10,000 second per division), a single 1 KHz wave would appear over 10 horizontal grid lines. This would cover the entire display if there are 10 divisions on the horizontal or X axis.
- **Position controls:** Use the vertical and horizontal position adjustments to center the signal as needed. This moves the trace up/down or left/right as needed.
- **DC, AC, and GND coupling:** These are used to control what signals are input. Most of the time you use DC coupling so that all signals, including DC, are input to the channel. The AC position uses a capacitor to block DC so that only AC is input. The GND position sets the input to zero volts (ground) and is usually used to check where 0 V appears on the display—this can also verify the scope is working.
- **Trigger:** For the trace to be stable, this setting tells the scope each time the sweep reaches the extreme right side of the screen before drawing the next trace. The trigger is usually set to AUTO. But it can be set to another signal such as the other input, the line voltage (60 Hz), or even an external signal you connect. For this course, the trigger should be left in the auto position which is the same as the input channel you are using for your measurement.
- **Other controls:** Again, don't be concerned with the other controls such as Chop, Invert, Step,

Hold and many others, especially on digital scopes. You can learn about them later in your training because they have special uses.

Note: When you first turn on power, the oscilloscope should be in the default or factory settings. Digital oscilloscopes often have a memory which allows custom preset conditions at power on—so check with your instructor or the manual to be sure you are not in a custom preset mode.

Signal Generators Supply the Current and Voltage

A signal generator is like a power supply because it is a source of power—voltage and current—but it supplies AC voltages instead of DC voltages. The voltages are in the lower rf (radio frequency) range, usually less than 1 GHz, with very safe or low values of voltage and current. Some signal generators supply only sine waves and are often called oscillators, but many are capable of supplying many other signals, including nonsinusoidal waveforms such as square waves and sawtooth waves. These instruments are called *function generators*. Regardless of the type you have, you will need to learn how to use its front panel to supply AC voltages. If it is extremely complex, you may need to ask for help or refer to the operating manual.

EQUIPMENT

Oscilloscope (preferably, a late-model, solid-state, auto-triggering type, including an operator's manual)
Signal generator or function generator
VTVM/VOM/DMM
Voltmeter
DC power supply
Springboard
Leads

COMPONENTS

Resistors (all 0.25 W):

(1) 2.2 kΩ (1) 10 kΩ
(1) 4.7 kΩ

PROCEDURE

On your oscilloscope, with power off, identify these controls or inputs—do not set them yet.

1. Probe and the input connector for the channel you will use (usually A or 1). Some scopes require you push a button for the channel you will be using.
2. Vertical (Y axis or amplitude) control: volts/div for the channel you will be using (A or 1). Also, the position control for the horizontal
3. Horizontal (X axis or time base) control: sec/div (usually applies to all channels).
4. Coupling switch: AC, DC, and GND.
5. Trigger: Usually set to INT (internal).

Analog Oscilloscopes only: Before you begin, set the following controls: (1) Intensity knobs control the brightness off the beam for the CRT and you need to adjust it so the beam is not bright—this will avoid damage to the CRT. (2) The Beam Finder, when pressed, verifies the instrument is probably working if you see a trace (beam). Press this button and then adjust the Vertical and Horizontal positions up or down so trace is in the middle center of the display. (3) Focus—use this to sharpen the trace so you see it clearly. (4) The CAL knob allows the vertical or horizontal sections to be offset or calibrated. Turn the CAL knob fully clockwise.

After identifying these controls, you are ready to begin.

Measuring DC Voltages

1. Set up the oscilloscope; locate the oscilloscope probe. It should be connected to the channel 1 (or A) input for dual-channel scopes. The probe should have a positive and negative lead. Be sure the scope is plugged into the AC power line. Do not turn on the power yet. Set the front panel controls as follows:

Set the INTENSITY to midrange.
For dual-trace scopes, set the VERTICAL MODE (trigger) to channel 1 or channel A.
Set the VOLTS/DIV (vertical amplifier, channel 1) to 1 V per division.
Set the AC-DC-GROUND switch (input coupling) to the ground position (GND).
Set the SEC/DIV (time base) to 0.2 s.
Set the TRIGGERING for auto, or for normal (manual) if no auto trigger exists.
Also, be sure the trigger control is set to the INTERNAL mode, not to external or line mode.

Now, turn on the power.

Note: The remaining front panel controls will vary, depending upon the model. Therefore, be sure to consult your scope manual or your instructor if you have any concerns.

2. The trace should appear as a straight line across the display graticule. Note that the graticule is 8 divisions vertically and 10 divisions horizontally. If the trace does not appear, ask for assistance. Next, use the following controls to adjust the trace so that a sharply defined trace is displayed at the middle of the *y* axis, as shown in Fig. 15-2.5.

VERTICAL POSITION control adjusts trace up and down.

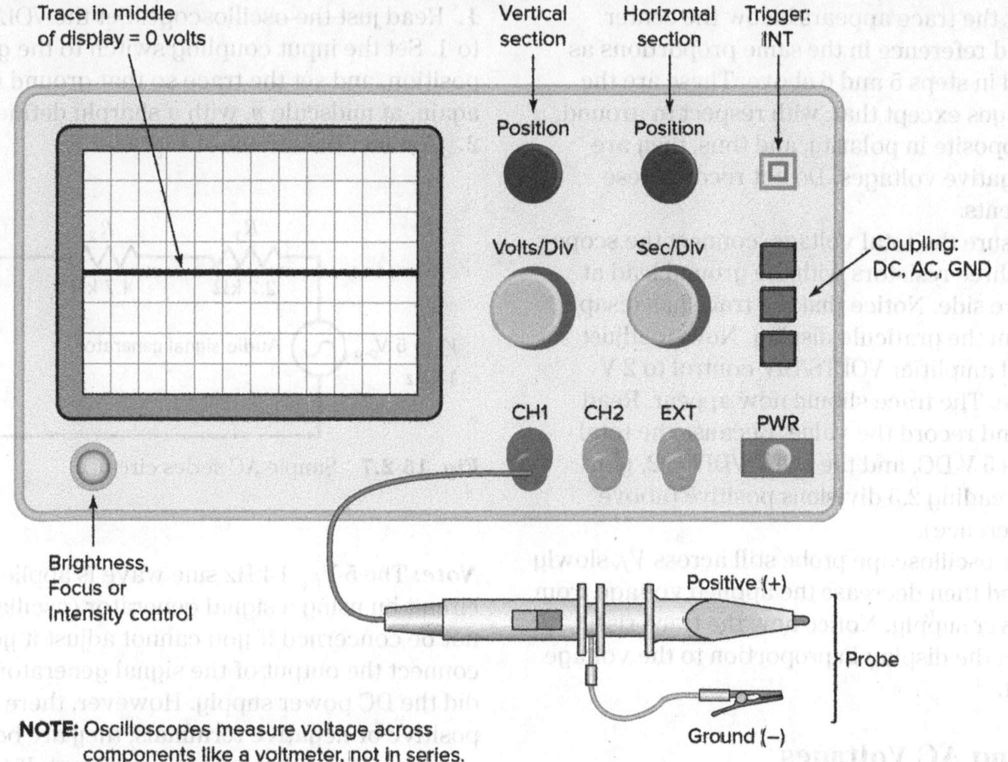

Fig. 15-2.5 Generic oscilloscope front panel.

FOCUS control or brightness sharpens the trace.
HORIZONTAL POSITION control adjusts trace right to left.

Here, in Fig. 15-2.5, the trace is a sharply defined straight line. Because the input coupling switch is set to GROUND, this trace means that ground, or zero volts, is at the division line. Thus, each division line above the ground line equals +1 V, and each division line below equals −1 V. This is similar to zeroing a VTVM.

If your trace looks like the one in Fig. 15-2.5, then you are doing fine. If not, repeat steps 1 and 2 or ask for assistance.

3. Leave the scope as it is. Connect the circuit of Fig. 15-2.6.

Note: Use the DC power supply and a voltmeter.

4. Using the voltmeter, measure and record all the voltages for the circuit of Fig. 15-2.6. Refer to Table 15-2.1. These voltmeter-measured DC

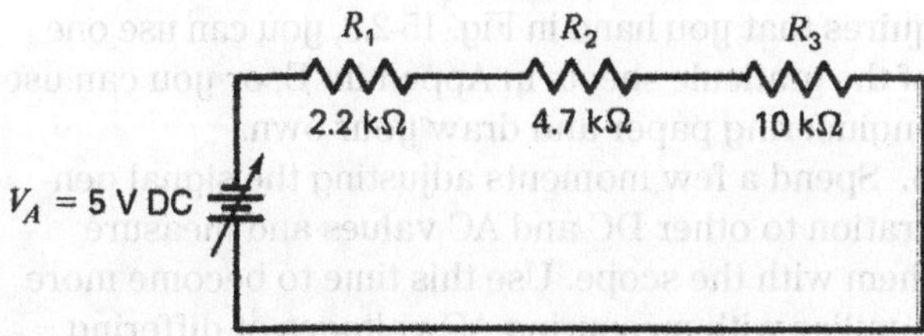

Fig. 15-2.6 Simple DC series circuit.

voltages will be compared to the same voltages measured with the scope.

5. Now, use the oscilloscope to measure the same DC voltages. Connect the ×1 (times 1) scope probe across R_1 (2.2 kΩ). Move the resistors around to avoid group loops if necessary. Usually, for this type of experiment it is not necessary.

With the scope leads across R_1, you should still see no change in the scope display because the input coupling is set to ground. Therefore, now switch this control to the DC position. The trace should now move slightly higher on the graticule display. Here is how you would read it.

Because the channel 1 VOLTS/DIV control is set at 1.0 V per division, each major vertical division on the graticule is equal to 1.0 V. Therefore, if the trace appears at 0.7 division aboveground, the DC measured voltage is equal to 0.7 division multiplied by 1.0 V (0.7 × 1.0 = 0.7 V). Multiply the VOLTS/DIV value by the number of divisions. Using 0.1 V per division is more accurate.

Record the voltage in Table 15-2.1.

6. Go ahead and measure the remaining DC voltages for R_2 and R_3 of Fig. 15-2.6, using the oscilloscope. These values should be the same as those measured with the voltmeter, plus or minus 15 percent. Do not measure V_T yet.

7. Reverse the power supply leads on the circuit of Fig. 15-2.6, and remeasure each resistor voltage.

Notice that the trace appears below the center zero/ground reference in the same proportions as it appeared in steps 5 and 6 above. These are the same voltages except that, with respect to ground, they are opposite in polarity, and thus, they are read as negative voltages. Do not record these measurements.

8. To measure the total voltage, connect the scope across all three resistors with the ground lead at the negative side. Notice that the trace has disappeared from the graticule display. Now, readjust the vertical amplifier VOLTS/DIV control to 2 V per division. The trace should now appear. Read the trace and record the value. Because the total voltage is +5 V DC, and the VOLTS/DIV = 2, you should be reading 2.5 divisions positive (above ground reference).

9. With the oscilloscope probe still across V_T, slowly increase and then decrease the applied voltage from the DC power supply. Notice how the trace rises and falls on the display in proportion to the voltage adjustment.

Measuring AC Voltages

The oscilloscope will display AC voltages so that their waveshapes can be observed. In this way, the amplitude and the frequency can be measured, referenced to the y or x axis: The amplitude is read by analyzing the vertical y axis; the frequency is read by analyzing the horizontal x axis.

Notice in Fig. 15-2.3, in the introduction that the sine-wave amplitude is five vertical divisions from peak to peak. Because the VOLTS/DIV control is set at 1 V per division, each peak is about 2.5 divisions with respect to ground, which equals 5 V peak to peak. From the peak-to-peak value, the following values, typical for describing AC voltages, can be determined:

$$V_{peak} \text{ or } V_{max} = \frac{V_{p\text{-}p}}{2}$$

$$V_{av} = \frac{V_{p\text{-}p}}{2} \times 0.636$$

$$V_{rms} = \frac{V_{p\text{-}p}}{2} \times 0.707$$

Frequency is read as the number of cycles that occur during 1 s (cycles per second = hertz). Notice in Fig. 15-2.3 that one cycle occurs over five horizontal divisions. This is measured from zero to zero on the x axis as long as ground (zero reference) is set at a desired place and noted. Here, time base control (sometimes called the *horizontal amplifier*) is set for 0.2 ms per division. Thus, one cycle occurs over five divisions, or 5 × 0.2 ms = 0.001 s (1 ms). Because $f = 1/t$, the frequency equals 1 kHz, or 1,000 cycles per second.

1. Read just the oscilloscope. Set the VOLTS/DIV to 1. Set the input coupling switch to the ground position, and set the trace so that ground is, once again, at midscale y, with a sharply defined trace.

2. Connect the circuit of Fig. 15-2.7.

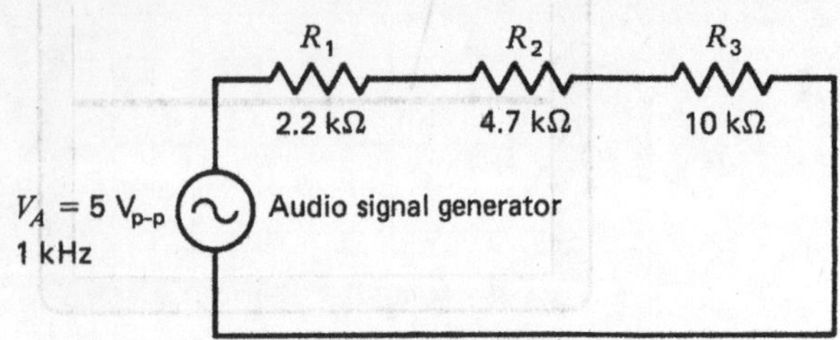

Fig. 15-2.7 Simple AC series circuit.

Note: The 5-$V_{p\text{-}p}$ 1-kHz sine wave is applied to the circuit by using a signal generator (oscillator). Do not be concerned if you cannot adjust it yet. Simply connect the output of the signal generator as you did the DC power supply. However, there are no positive or negative terminals; they are both the same because it is alternating current. If there is a grounded terminal, avoid using it if possible.

3. Connect the scope leads across the entire circuit. Then, set the input coupling for AC. The display should be a sine wave. Adjust the amplitude and frequency controls of the signal generator so that the trace is approximately 5 $V_{p\text{-}p}$ at 1 kHz. The display should be very similar to the one shown in Fig. 15-2.3.

4. Go ahead and measure all the voltages in the circuit—R_1, R_2, R_3—and of course V_T that is now being displayed. Record the values in Table 15-2.2. Change the VOLTS/DIV, or any other setting, so that you get the biggest display without going off the screen. Record all the values in Table 15-2.2, and calculate the remaining values.

5. Reconnect the scope leads across the entire circuit and adjust the signal generator so that 4 $V_{p\text{-}p}$ at 20 kHz is the applied voltage. Then, adjust the scope for the biggest amplitude possible without going off the display. Also, adjust it so that only two complete cycles appear on the screen. Draw a picture of the trace on the graticule of Fig. 15-2.8, and record the VOLTS/DIV and SEC/DIV setting you used to get this display. If your lab instructor requires that you hand in Fig. 15-2.8, you can use one of the graticule sheets in Appendix E, or you can use engineering paper and draw your own.

6. Spend a few moments adjusting the signal generation to other DC and AC values and measure them with the scope. Use this time to become more familiar with measuring AC voltages of differing frequencies and waveshapes.

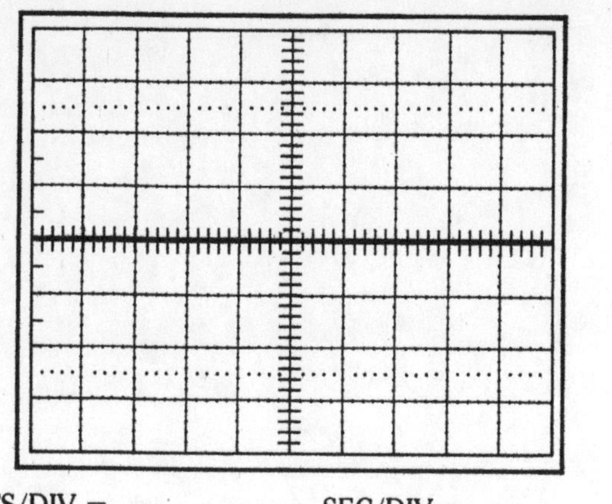

VOLTS/DIV = __________ SEC/DIV = __________

Fig. 15-2.8 Graticule. Displayed waveform =4 $V_{p\text{-}p}$ at 20 kHz.

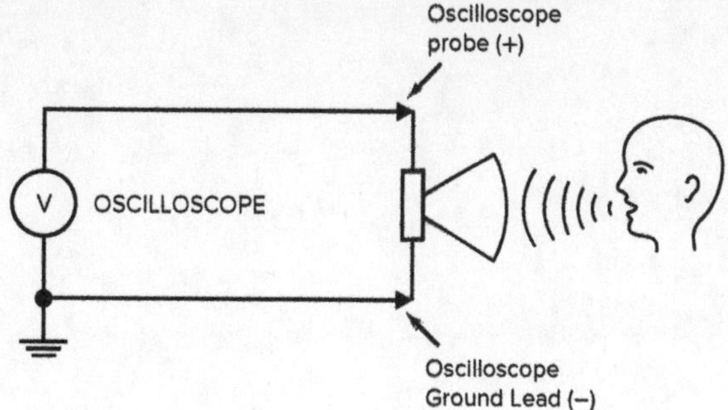

Fig. 15-2.9 Oscilloscope connected to a permanent magnet speaker.

FURTHER EXPLORATION

1. Disconnect the above circuit. Connect a permanent magnet speaker to the input leads of the oscilloscope, as shown in Fig. 15-2.9.
2. Set the input coupling switch to alternating current.
3. While whistling into the speaker, adjust the VOLTS/DIV, or any other oscilloscope setting, so that you get the largest display without going off the screen. As you review the circuit action, determine the operating nature of the speaker. For example, is the speaker producing direct or alternating current? How can the frequency of the signal be changed? Predict how the display will change if the oscilloscope has its input coupling switch changed from AC to DC coupling.

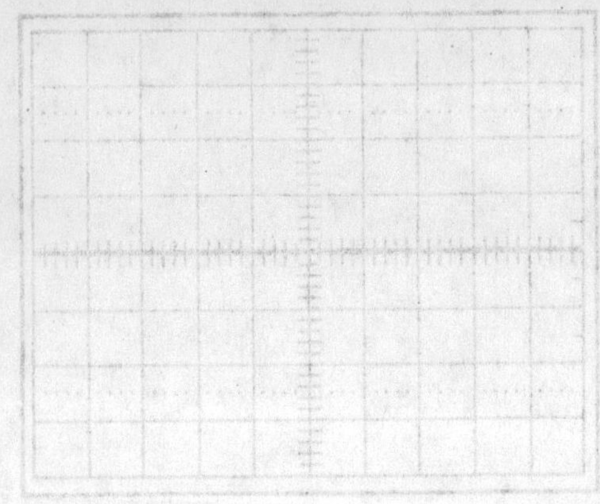

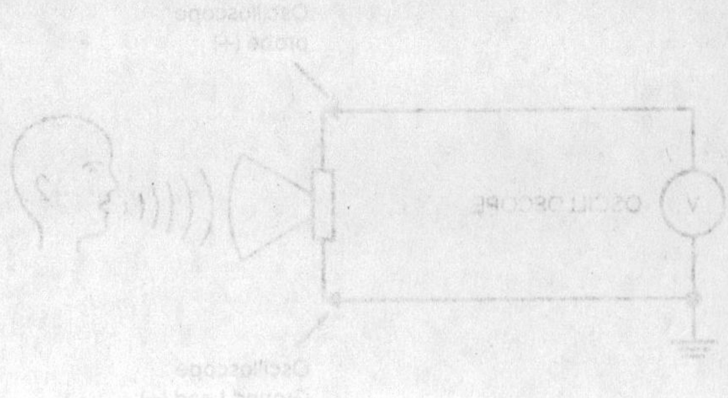

NAME DATE

QUESTIONS FOR EXPERIMENT 15-2

Answer true (T) or false (F) to the following:

_____ **1.** An oscilloscope can measure both AC and DC voltages.

_____ **2.** The horizontal amplifier section controls the amplitude of the display.

_____ **3.** The vertical amplifier section can be adjusted to increase or decrease the peak-to-peak voltage display.

_____ **4.** The oscilloscope probe does not have a grounded side.

_____ **5.** The oscilloscope input coupling switch is usually set to GND, AC, or DC, depending upon the desired usage.

TABLES FOR EXPERIMENT 15-2

TABLE 15-2.1 Measured DC Voltages for Fig. 15-2.6

	Voltmeter	Scope
V_{R_1}		
V_{R_2}		
V_{R_3}		
V_T		

TABLE 15-2.2 AC Voltages for Fig. 15-2.7

	Measured $V_{p\text{-}p}$	Calculated V_{peak}	Calculated V_{av}	Calculated V_{rms}
$R_1 = 2.2\ k\Omega$				
$R_2 = 4.7\ k\Omega$				
$R_3 = 10\ k\Omega$				
V_T				

EXPERIMENT RESULTS REPORT FORM

Experiment No.: ______________________ **Name:** ______________________

Date: ______________________

Experiment Title: ______________________ **Class:** ______________________

______________________ **Instr:** ______________________

Explain the purpose of the experiment:

List the first Learning Objective:
OBJECTIVE 1:

After reviewing the results, describe how the objective was validated by this experiment.

List the second Learning Objective:
OBJECTIVE 2:

After reviewing the results, describe how the objective was validated by this experiment.

List the third Learning Objective:
OBJECTIVE 3:

After reviewing the results, describe how the objective was validated by this experiment.

Conclusion:

If required, attach to this form: ☐ Answers to Questions, ☐ Tables, and ☐ Graphs.

CHAPTER 15

EXPERIMENT 15-3

ADDITIONAL AC OSCILLOSCOPE MEASUREMENTS

LEARNING OBJECTIVES

At the completion of this experiment, you will be able to:

- Improve your ability to operate a signal generator.
- Operate various oscilloscope front panel controls.
- Measure various waveshapes and calculate values.

SUGGESTED READING

Chapter 15, *Basic Electronics*, Grob/Schultz, twelfth edition

INTRODUCTION

An oscilloscope, commonly called a *scope*, is a sophisticated voltmeter that operates in the time domain. It measures and displays both AC and DC signals as they occur over time. The previous experiment introduced you to basic DC and AC measurements in a series circuit. You measured amplitude (voltage) and time, computing frequency (Hz) from the time measurement and calculating peak, rms, and average values from the p-p voltage measurements. You learned the basic operation, including how to set the trace to ground (0 V) for both DC and AC measurements. This experiment will allow you to practice the skills introduced in the previous experiment, and it will introduce new ones that will prepare you for future labs.

Understanding the Oscilloscope System

Every scope has four basic systems: display, horizontal, vertical, and trigger. Each of the four systems is controlled or set up by front panel inputs, switches, and adjustments. Knowing what these systems do will help you understand how to use the front panel. If you have not done so already, read your oscilloscope manual to gain a general idea of how those systems work.

The Oscilloscope Probe

When you use an oscilloscope to measure higher frequencies, the probe prevents other signals such as radio signals and 60-Hz power-line radiated signals from getting into the measurement. In addition, the oscilloscope does not have the high resistance (impedance) of a DC voltmeter, so it could allow also too much circuit current to flow through it in some cases. All of these effects would *load down* the circuit, which means that the oscilloscope would, in effect, become part of the circuit rather than an isolated measuring device. Your measurements would therefore be inaccurate. To avoid this, the probe is specially designed to block out unwanted signals and increase the resistance. Resistive probes are sometimes called *attenuator probes* because they decrease the signal amplitude by 10 times or more (×10 probe). Other probes are also available for specific measurements, usually of higher frequencies or lower circuit impedances (50 Ω).

Signal Generators

You cannot learn to use an oscilloscope without using a signal generator to supply signals to the circuit to be measured. A signal generator is like a power supply because it is a source of power—voltage and current—but it supplies AC voltages instead of DC voltages. The voltages are in the lower rf (radio frequency) range, usually less than 1 GHz, with very safe or low values of voltage and current. As mentioned previously, some signal generators supply only sine waves and are often called oscillators, but many are capable of supplying many other signals, including nonsinusoidal waveforms such as square waves and sawtooth waves. These instruments are called *function generators,* and they are invaluable for testing digital circuits.

EQUIPMENT

Signal generator (function generator)
Oscilloscope: auto-triggering, properly grounded
Probe: × 1

COMPONENTS

Resistors: 2.2 kΩ, 4.7 kΩ, 10 kΩ

PROCEDURE

Front Panel Identification

1. Identify the front panel elements of your scope by writing down which section they belong to as follows. On a separate sheet of paper, make four columns and label them as DISPLAY, VERTICAL, HORIZONTAL, and TRIGGER. In each column, list the switches, buttons, knobs, input connections, etc., that are in that system.

Setup for a Free-Running Display

A free-running display has no signal applied, and the trace is not triggered (it does not start at any particular time). Setting up the scope in this manner, every time you use it, will help you learn its operation, provide a consistent starting point or reference, and help you avoid incorrect measurements.

2. With the scope power off, set the following controls:

INTENSITY: mid range
VOLTS/DIV: largest number possible (least sensitivity)
INPUT COUPLING (AC-DC-GROUND): (ground) GND
SEC/DIV (time base): 1 ms
TRIGGERING: set to INTERNAL (source) and AUTO (mode)
SLOPE: (+)

3. Turn on the power.

4. Use the POSITION knob to center the beam and adjust for a sharp display in the center of the screen. You now have a starting point to begin making measurements.

5. Turn off the scope and change all the settings randomly. Repeat steps 2, 3, and 4.

Measuring the AC Signals (Fields) in the Air

This simple measurement should allow you to see the 60-Hz signals that are present in the air because they are radiated everywhere around you. By using your body as a receiving antenna, the scope can measure the 60-Hz signals present in the air.

6. With the probe and lead connected to the vertical input, set the following:

INPUT COUPLING: AC
SEC/DIV: 5 ms

7. Hold the end of the probe tip and adjust the VOLTS/DIV until a signal appears. If the trace is not stable, switch the trigger mode from AUTO to NORMAL and set the SLOPE to (+).

Observations: Note the waveform. Is it 60 Hz? If so, what is the amplitude? Switch the INPUT COUPLING to the DC position and notice any change.

8. Return the scope to the free-running setup.

Using Front Panel Controls

9. Measure the resistors used in this experiment and replace any that are not within tolerance. Then connect the circuit of Fig. 15-3.1.

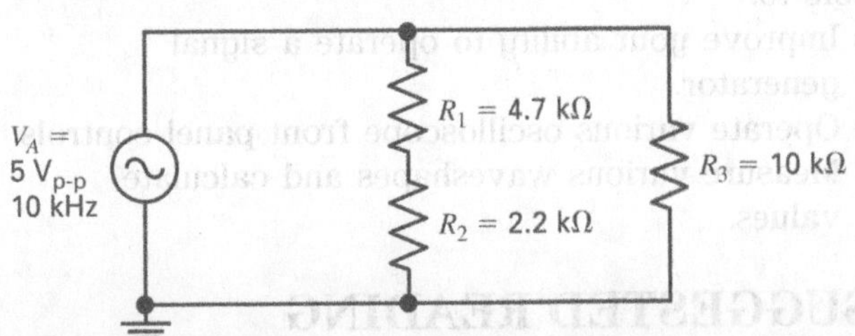

Fig. 15-3.1 Sine-wave series-parallel circuit.

10. Adjust the oscilloscope as necessary to measure the p-p voltages across each resistor. Record the values in Table 15-3.1.

11. SLOPE and LEVEL—This trigger control determines where the trace starts. With the scope leads across R_1, adjust the slope LEVEL back and forth to see what effect it has on the display. Then set the SLOPE button to (−) or opposite of what it was. Measure V_{p-p} across R_1, R_2, and R_3 and record the results.

Measuring Square Waves or Rectangular Waves

Square waves are displayed just like sine waves except that the measurements are different. Both have p-p, peak, and Hz, but the square wave does not have rms or average voltage values. Instead, square-wave measurements have the values shown in Fig. 15-3.2.

12. Return the scope to the free-running setup and connect the circuit of Fig. 15-3.3, where the signal generator (function generator) now supplies a square wave or a rectangular wave. Note the type of wave in Table 15-3.2.

13. For the circuit of Fig. 15-3.3, measure the following values across R_1, R_2, and R_3: V_{p-p}, period (s), pulse width (s), and rise time. Rise time is the time it takes to go from 10 to 90 percent of the magnitude (see Fig. 15-3.2). Record the values in Table 15-3.2.

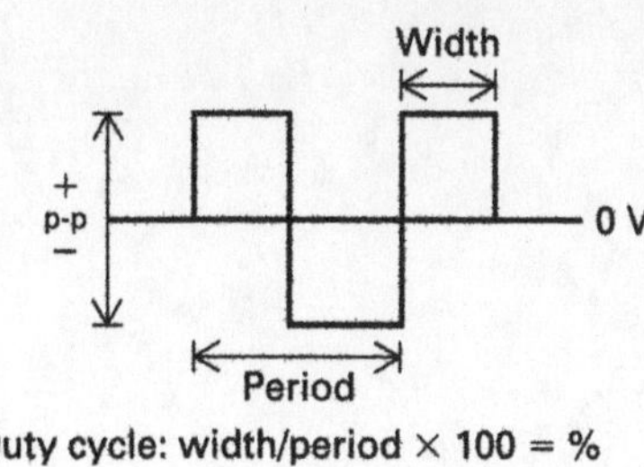

(a)

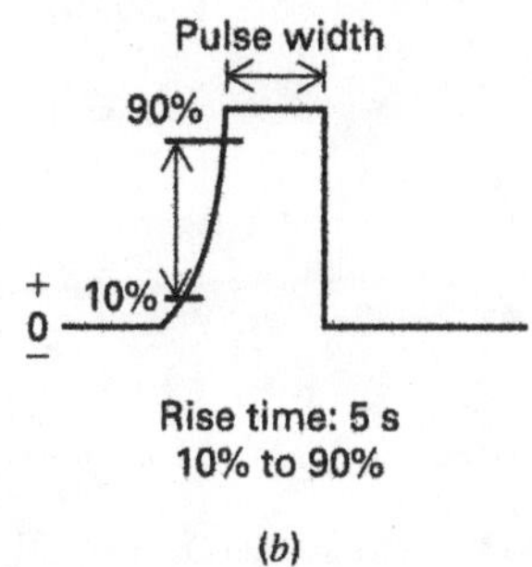

(b)

Fig. 15-3.2 Square- or rectangular-wave measurements. (*a*) Square or rectangular wave. (*b*) Pulse.

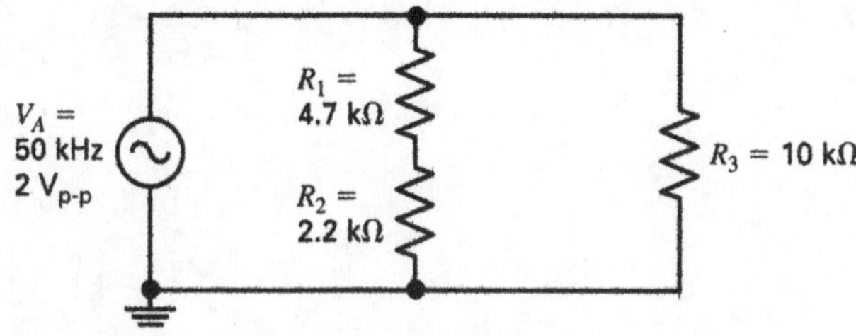

Fig. 15-3.3 Square-wave measurement circuit.

Measuring Sawtooth Waves or Triangular Waves

14. Return the scope to the free-running setup, and connect the circuit of Fig. 15-3.4, where the signal generator (function generator) now supplies a sawtooth wave or triangular wave.

15. Measure the $V_{p\text{-}p}$ value across R_1, R_2, and R_3. Record the results in Table 15-3.2.

16. PROBE ADJUST SIGNAL—This control is used to compensate attenuator probes to make accurate measurements of digital circuits. Locate this control on the front panel, and try to measure the signal by connecting the tip of the scope probe (+) to it and the negative side to any ground connection. Observe the signal—it is used for adjusting ×10 probes. If you are using a ×1 probe, no adjustment is required. If you have a ×10 probe, refer to the scope manual for information on how to adjust a ×10 probe. In Table 15-3.2, record the values of the signal (×1 probe) as you did for the square-wave measurements.

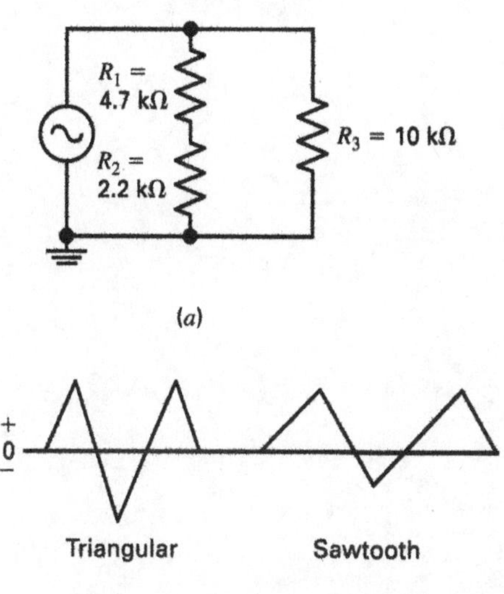

(b)

Fig. 15-3.4 Sawtooth- or triangular-wave circuit and waveshape. (*a*) Circuit. (*b*) Waveshape.

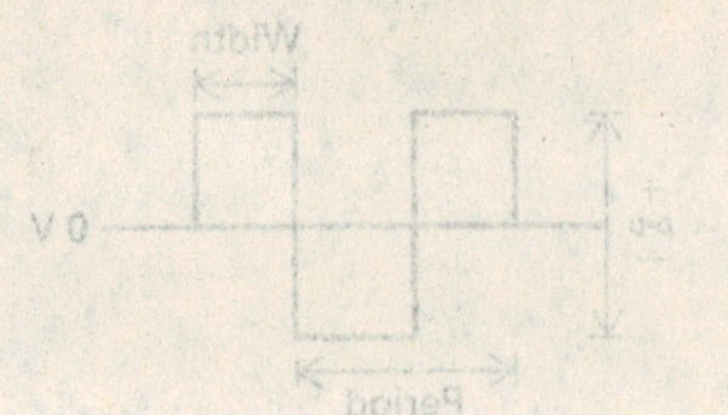

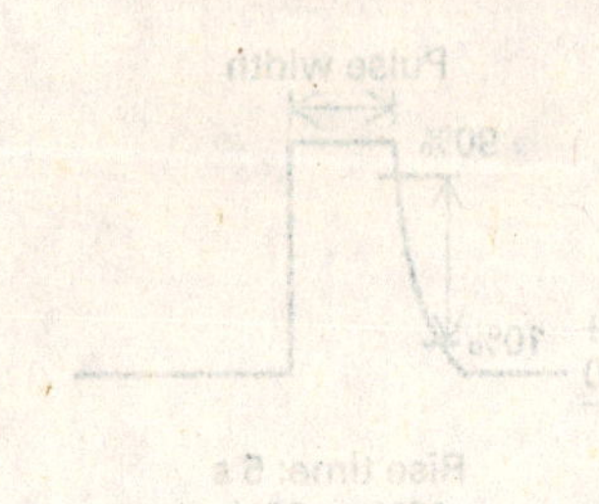

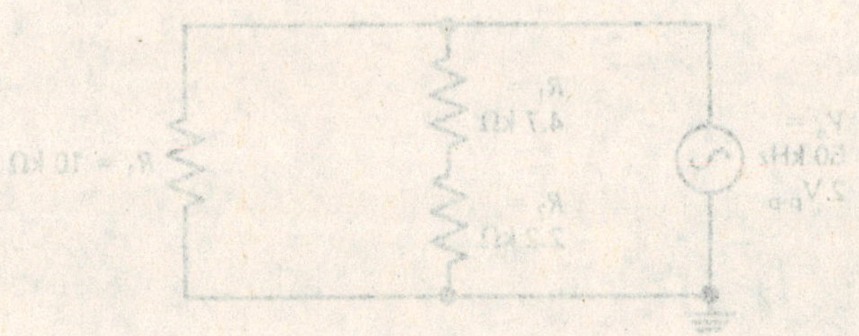

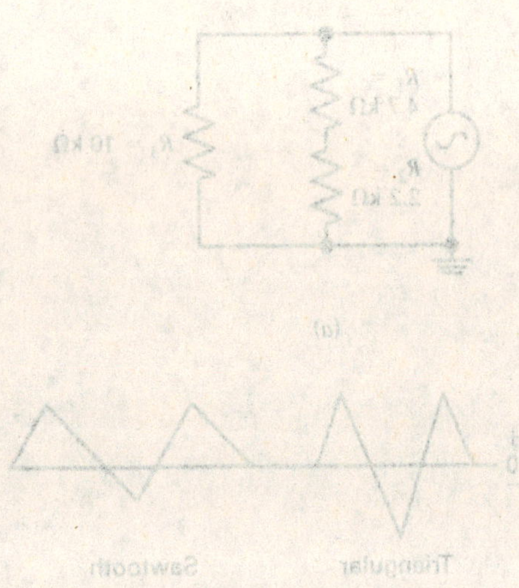

NAME ____ DATE ______________

QUESTIONS FOR EXPERIMENT 15-3

Provide a short answer for each of these questions, referring to the measurements.

1. What is the purpose of starting with a free-running setup?

2. What do the SLOPE and LEVEL adjusted?

3. What observations did you make in steps 7 and 8?

4. What is rise time?

5. Does a square wave have an rms value? Why?

TABLES FOR EXPERIMENT 15-3

TABLE 15-3.1 Sine-Wave Measurements

$V_A = 5\ V_{p\text{-}p}$ $f = 10$ kHz	$V_{p\text{-}p}$	SLOPE — $V_{p\text{-}p}$
R_1		
R_2		
R_3		

TABLE 15-3.2 Wave Type=____________ (Square or Rectangular)

$V_A = 2\ V_{p\text{-}p}$ $f = 50$ kHz	$V_{p\text{-}p}$	Period (s)	Pulse Width (s)	Rise Time 10% to 90% (s)
R_1				
R_2				
R_3				

Wave Type=____________ (Sawtooth or Triangular)

	$V_{p\text{-}p}$
R_1	
R_2	
R_3	

PROBE ADJ Signal

$V_{p\text{-}p}$	Period	Pulse Width	Rise Time

EXPERIMENT RESULTS REPORT FORM

Experiment No.: ______________________ Name: ______________________

Date: ______________________

Experiment Title: ______________________ Class: ______________________

______________________ Instr: ______________________

Explain the purpose of the experiment:

List the first Learning Objective:

OBJECTIVE 1:

After reviewing the results, describe how the objective was validated by this experiment.

List the second Learning Objective:

OBJECTIVE 2:

After reviewing the results, describe how the objective was validated by this experiment.

List the third Learning Objective:

OBJECTIVE 3:

After reviewing the results, describe how the objective was validated by this experiment.

Conclusion:

If required, attach to this form: ☐ Answers to Questions, ☐ Tables, and ☐ Graphs.

CHAPTER 15

EXPERIMENT 15-4

OSCILLOSCOPE MEASUREMENTS: SUPERPOSING AC ON DC

LEARNING OBJECTIVES

At the completion of this experiment, you will be able to:

- Investigate the effective AC resistance of a power supply.
- Study the function of a bypass capacitor and a coupling capacitor.
- Develop a method for measuring both AC and DC components in the same circuit.

SUGGESTED READING

Chapter 15, *Basic Electronics*, Grob/Schultz, twelfth edition

INTRODUCTION

Applied voltage will be dropped proportionally across individual resistances, depending upon their ohmic value. The voltage distribution can be solved, as shown in Fig. 15-4.1*a* and *b*.

If a circuit has both an AC and a DC voltage source, as shown in Fig. 15-4.2, there will be both an AC and a DC voltage distribution. The voltage distribution is solved in the same manner as in Fig. 15-4.1.

If a bypass capacitor is connected across one of the series resistors, the capacitor can act as a short circuit to the AC component. Only a negligible amount of reactance will be created as a result of the chosen capacitor size and the frequency of the AC signal. The capacitor will become a parallel path for the AC component while blocking the DC component, as shown in Fig. 15-4.3. Note that the DC component is restricted to R_3 (4.7kΩ) for a current path. The AC signal will make use of the capacitive path, because the charging and discharging action of the capacitor, in this case, creates only 3.39 Ω of reactance.

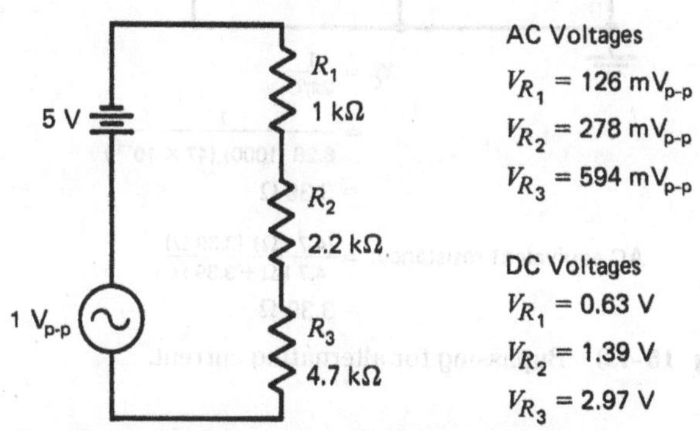

Fig. 15-4.2 AC and DC resistive circuit.

Another method of superposing an AC signal onto a DC voltage is through capacitive coupling. In Fig. 15-4.4, DC voltages are developed across R_1, R_2, and R_3 but are isolated from the signal generator by the coupling capacitor C_1. The DC resistance of the power supply can be considered approximately 0 Ω. The AC signal is applied to the parallel combination of $R_2 + R_3$ and R_1.

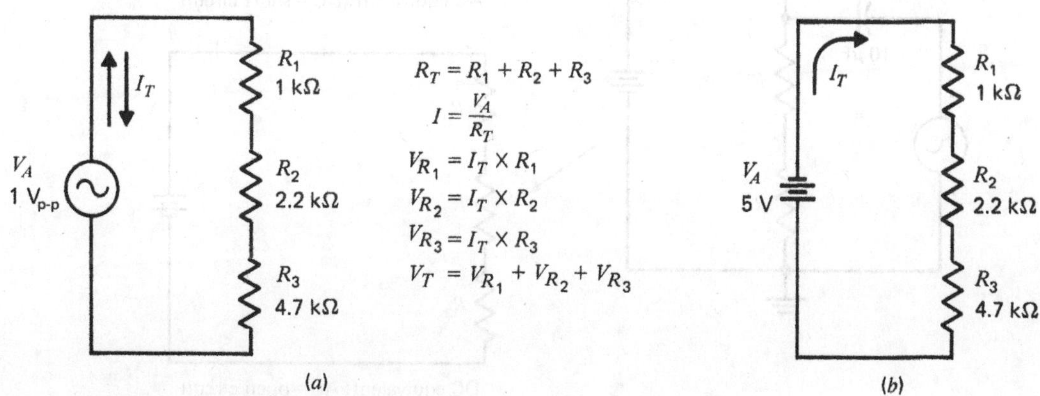

Fig. 15-4.1 (*a*) AC resistive circuit. (*b*) DC resistive circuit.

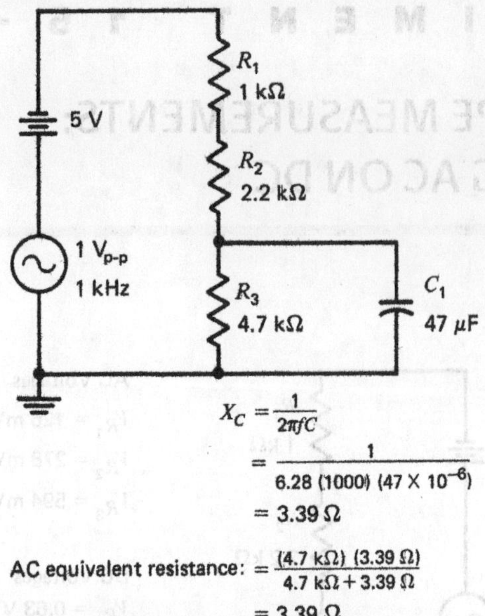

Fig. 15-4.3 Bypassing for alternating current.

Note: When measuring a DC component while an AC component is present, the oscilloscope will display the AC component at the level of the DC component. This means that a sine wave will appear above or below the zero reference when the input coupling is switched to measure direct current. It will be necessary to develop a technique to read the DC component. Therefore, the sine-wave zero (x axis) value must be determined in relation to the DC zero reference. This will require adjusting the oscilloscope for each reading. Be sure that you develop this technique. Refer to Fig. 15-4.5.

EQUIPMENT

Oscilloscope
Audio-frequency signal generator
DMM
Protoboard or springboard
Test leads

COMPONENTS

Resistors (all 0.25 W):

(1) 1 kΩ (1) 4.7 kΩ
(1) 2.2 kΩ

(1) 47-μF electrolytic capacitor

PROCEDURE

1. Connect the circuit of Fig. 15-4.1*a*. Measure and record the voltages across each resistance as indicated in Table 15-4.1.

2. Connect the circuit of Fig. 15-4.1*b*. Measure and record the voltages across each resistance as indicated in Table 15-4.1.

3. Repeat procedures 1 and 2 for the circuit shown in Fig. 15-4.2, and complete the information requested in Table 15-4.2.

4. After recording the measured voltages, calculate the same voltages in accordance with the techniques reviewed in the introduction and record these in Tables 15-4.1 and 15-4.2. Compare the measured and calculated values.

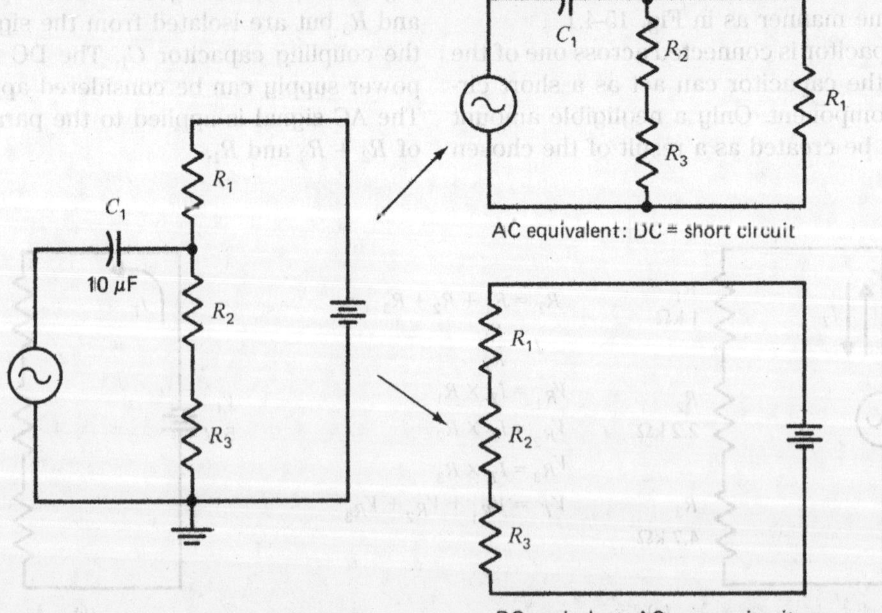

Fig. 15-4.4 Superposing an AC signal on a DC voltage by capacitive coupling with equivalent circuits shown.

REMINDER: Be sure to refer to the note (previous page) at this point in the procedure.

Optional Procedure Steps

Note: Be sure to avoid ground loops.

5. Connect the circuit of Fig. 15-4.3. Measure and record all AC and DC voltages. Prepare your own data table, similar to Table 15-4.2, and label it Table 15-4.3.

6. Connect the circuit of Fig. 15-4.4. Measure and record all AC and DC voltages. Prepare your own data table, and label it Table 15-4.4.

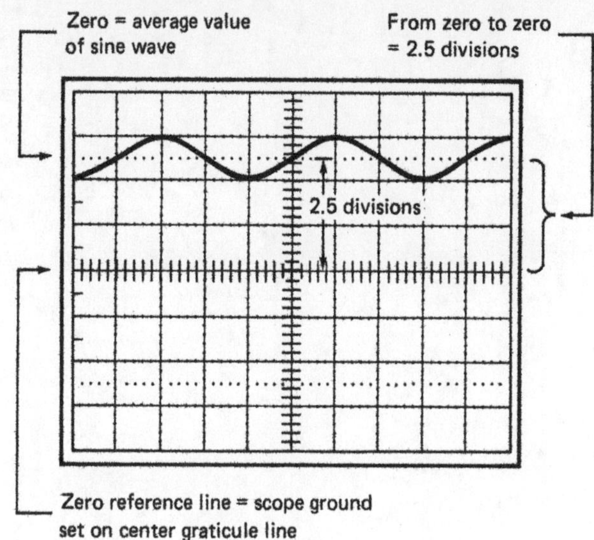

Fig. 15-4.5 Oscilloscope graticule display.

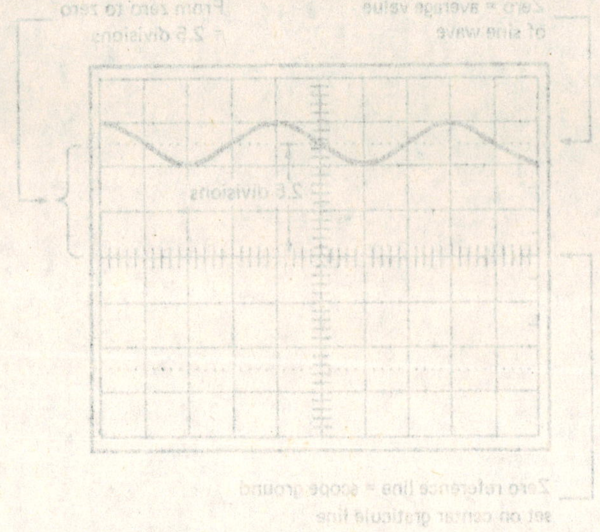

NAME ______ DATE ______

QUESTIONS FOR EXPERIMENT 15-4

1. What function does the coupling capacitor perform?

2. What function does the bypassing capacitor perform?

3. Was the effective AC resistance of the power supply equal to 0 Ω?

4. Draw the AC and DC equivalent circuit diagrams for the circuit shown in Fig. 15-4.2.

5. Is AC resistance the same as DC resistance? Explain.

TABLES FOR EXPERIMENT 15-4

TABLE 15-4.1 Measured and Calculated Voltages for Fig. 15-4.1*a* and *b*

Procedure Step		V_{R_1}	V_{R_2}	V_{R_3}	V_T
1	AC measured				
2	DC measured				
4	AC calculated				
	DC calculated				

TABLE 15-4.2 Measured and Calculated Voltages for Fig. 15-4.2

Procedure Step		V_{R_1}	V_{R_2}	V_{R_3}	V_T
3	AC measured				
	DC measured				
4	AC calculated				
	DC calculated				

EXPERIMENT RESULTS REPORT FORM

Experiment No.: ______________________ Name: ______________________

Date: ______________________

Experiment Title: ______________________ Class: ______________________

______________________ Instr: ______________________

Explain the purpose of the experiment:

List the first Learning Objective:

OBJECTIVE 1:

After reviewing the results, describe how the objective was validated by this experiment.

List the second Learning Objective:

OBJECTIVE 2:

After reviewing the results, describe how the objective was validated by this experiment.

List the third Learning Objective:

OBJECTIVE 3:

After reviewing the results, describe how the objective was validated by this experiment.

Conclusion:

If required, attach to this form: ☐ Answers to Questions, ☐ Tables, and ☐ Graphs.

CHAPTER 16

EXPERIMENT 16-1

CAPACITORS

LEARNING OBJECTIVES

At the completion of this experiment, you will be able to:

- Troubleshoot a capacitor to verify that it is functional.
- Verify the DC and AC response of a capacitor.
- Use a capacitor to route or remove high frequencies.

SUGGESTED READING

Chapter 16, *Basic Electronics*, Grob/Schultz, twelfth edition

INTRODUCTION

A capacitor is made of two conductors separated by an insulator. Most capacitors are made of two metal plates with some type of insulating material such as air, ceramic, mica, paper, or an electrolyte. The insulating material is called a *dielectric*. Capacitors are like batteries because, at low frequencies (DC), they can store a charge. Therefore, it is always best to be cautious when working with larger capacitors because they can be a hazard and cause shock. To avoid this potential hazard, be sure to discharge capacitors safely, by shorting them out while you remain well insulated.

Like resistors, capacitors have a code which shows their value, but the code is different from resistor color codes. Capacitors are rated in units called *farads* (F), which is a measure of how much charge exists in the insulator's electrical field between the two plates. Some manufacturers label capacitors by printing the numerical value directly on the capacitor. Other capacitors are labeled with specific color codes, which may include temperature ratings. A special type of capacitor, the electrolytic capacitor, is polarity sensitive and has a minus sign (negative) printed on it to show which way to connect it with respect to the DC source. Be careful: **If it is connected backward, it can explode** if enough voltage is applied. Figure 16-1.1 shows some typical capacitors and their codes.

The two plates of a capacitor have opposite charges that create a potential difference (voltage) when a DC source is applied. If you remove the capacitor from the source of the charge, it will still hold the charge. However, it will have very little current unless it is an extremely large capacitor, in which case you should be careful not to discharge it with your hands or other parts of your body. Figure 16-1.2 is a circuit diagram of a capacitor storing a DC charge.

Capacitors are an open circuit (large resistance) to direct current. The DC voltage is stored on the plates, but no current passes through from one plate to another because of the insulation. Therefore, there is no current. When some capacitors fail, however, they can allow current to flow. Such capacitors are often called *leaky*.

You can use an ohmmeter to verify the operation of a capacitor. The resistance reading will be several million ohms or more (an open circuit). In the case of electrolytic capacitors, you may find one that is leaky. Other types of capacitors can also be damaged and will then give a measurement less than infinite ohms. Most good capacitors will measure at least 10 MΩ. Figure 16-1.3 shows a circuit for checking a capacitor with an ohmmeter.

If a capacitor is connected in series with an AC signal source, the capacitor will lose its insulating quality and may become a short circuit to the AC signal. The higher the frequency, the more a capacitor will act like a short circuit. Its plates will charge and discharge at the same rate as the AC signal source. The greater the value of the capacitor, the more easily it will charge and discharge.

The reason the capacitor acts like a short is that the plates charge and discharge as the polarity of the source alternates between positive and negative. No current ever actually crosses the dielectric, but the effect is the same as if it did. This concept of the capacitor reacting to AC frequency will be examined more thoroughly in a future experiment.

Figure 16-1.4 shows a capacitor acting like a short in an AC circuit where the voltmeter reads 0 V, there is no resistance, and there is no *IR* voltage drop across the capacitor.

A capacitor can block DC but allows AC to pass when both AC and DC are present in a circuit. Therefore, a capacitor can be used to block DC from entering a circuit where only the AC signal

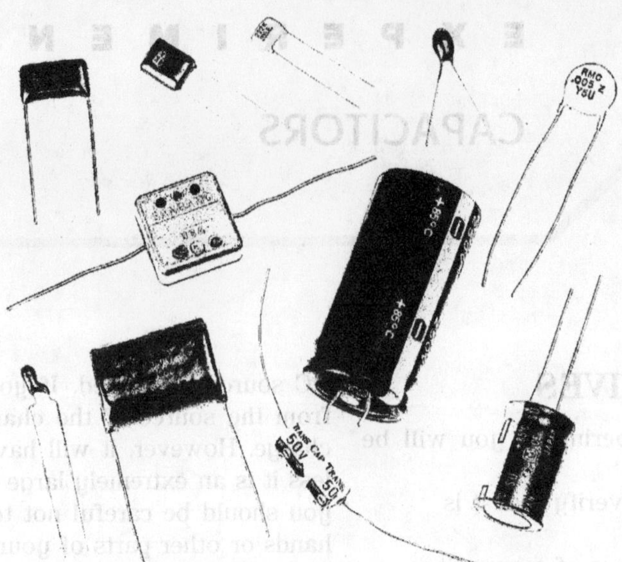

Type	Symbol	Uses
Fixed	C	Stores charge in dielectric; passes AC voltage but blocks DC voltage
Electrolytic	C +	Fixed value with large *C* but has polarity
Variable	C	Variable capacitor; used for tuning

Color Codes for Tubular and Disc Ceramic Capacitors

BAND OR DOT COLOR	CAPACITANCE IN PICOFARADS: SIGNIFICANT DIGITS 1st	SIGNIFICANT DIGITS 2nd	CAPACITANCE MULTIPLIER	TOLERANCE ≤10 pF	TOLERANCE >10 pF	TEMPERATURE COEFFICIENT ppm°C (5-DOT SYSTEM)	TEMPERATURE COEFFICIENT 6-DOT SYSTEM SIG. FIG.	TEMPERATURE COEFFICIENT MULTIPLIER
Black	0	0	1	±2.0 pF	±20%	0	0.0	−1
Brown	1	1	10	±0.1 pF	±1%	−33		−10
Red	2	2	100		±2%	−75	1.0	−100
Orange	3	3	1000		±3%	−150	1.5	−1000
Yellow	4	4				−230	2.0	−10000
Green	5	5		±0.5 pF	±5%	−330	3.3	+1
Blue	6	6				−470	4.7	+10
Violet	7	7				−750	7.5	+100
Gray	8	8	0.01	±0.25 pF		+150 to −1500		+1000
White	9	9	0.1	±1.0 pF	±10%	+100 to −75		+10000
Silver	—	—						
Gold	—	—						

Fig. 16-1.1 Capacitors and codes.

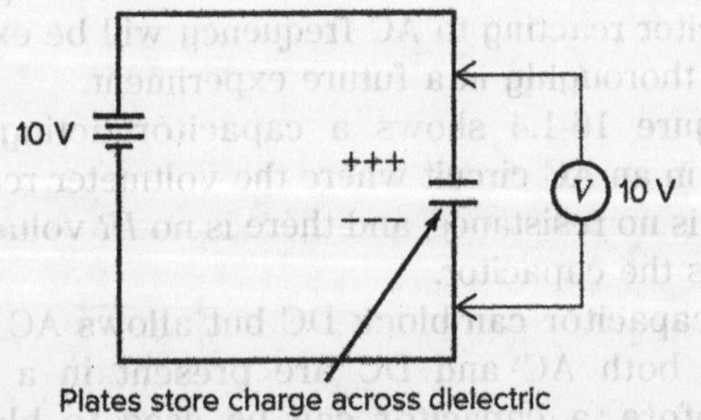

Fig. 16-1.2 Capacitor storing a DC charge.

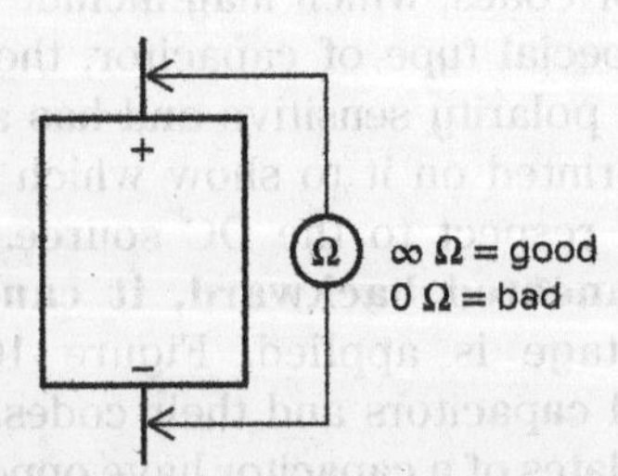

Fig. 16-1.3 Checking a capacitor with an ohmmeter.

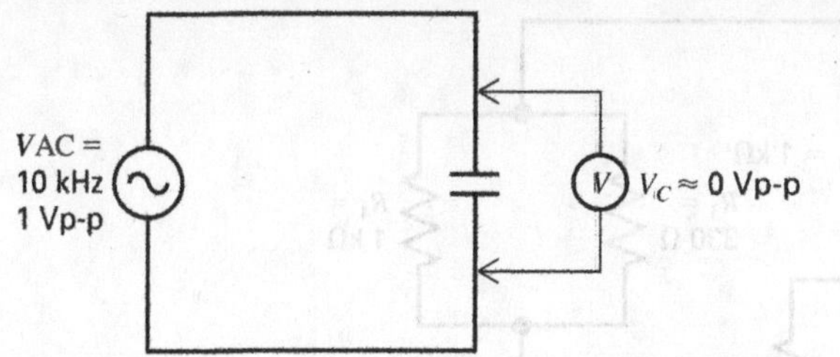

Fig. 16-1.4 Capacitor is a short to alternating current at high frequencies.

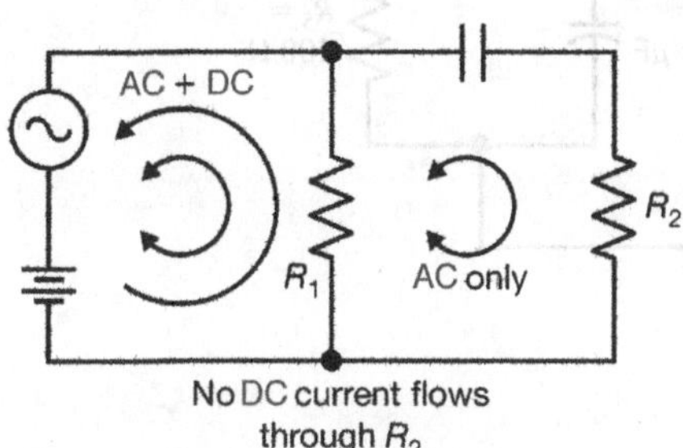

Fig. 16-1.5 Capacitor blocking direct current in a circuit.

is desired. Figure 16-1.5 shows the DC voltage blocked from the other part of the circuit where alternating current is present.

EQUIPMENT

Protoboard
DC power supply
AC signal generator
DMM
Oscilloscope

COMPONENTS

All resistors 0.25 W unless indicated otherwise:

(2) 100-Ω resistor
(1) 330-Ω resistor
(2) 1-kΩ resistor
(1) 0.01-μF capacitor
(1) 0.068-μF capacitor
(1) 0.1-μF capacitor
(1) 10-μF capacitor
(1) 100-μF capacitor

PROCEDURE

1. Verify the operation of each capacitor listed in Table 16-1.1 with an ohmmeter to be sure the capacitor is an open to direct current. Put a check mark next to each value in Table 16-1.1 to show that the capacitor is good. Be sure to replace any bad components.

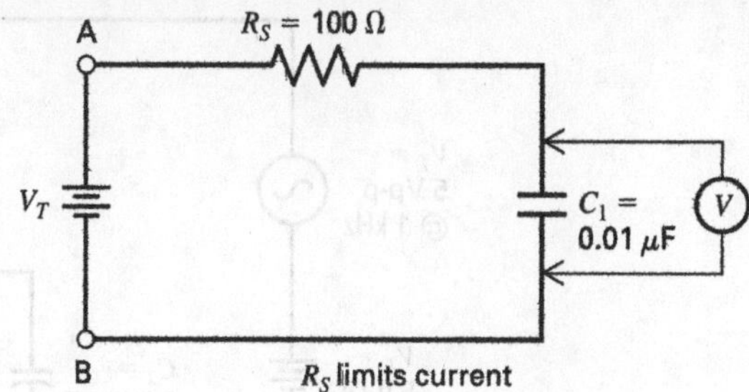

Fig. 16-1.6 A DC circuit with a capacitor in series.

2. Connect the circuit of Fig. 16-1.6, but do not connect the capacitor.

3. Apply 1 V across A–B and then turn off the power supply. Insert the capacitor in the circuit and turn on the power supply.

4. Measure the voltage across the capacitor, and record the voltage in Table 16-1.1.

5. Increase the power supply to 5 V (across A–B), and measure the voltage across the capacitor. Record the value in Table 16-1.1.

6. Go to the next value of capacitor in Table 16-1.1. Measure the voltage across the capacitor at 1 V, and at 5 V as in the steps above and record the results in the table.

7. Repeat step 6 for all the remaining capacitors.

8. Connect the circuit of Fig. 16-1.7, but not connect the capacitor. Adjust the signal generator to 1 $V_{p\text{-}p}$ at 1 kHz across A–B.

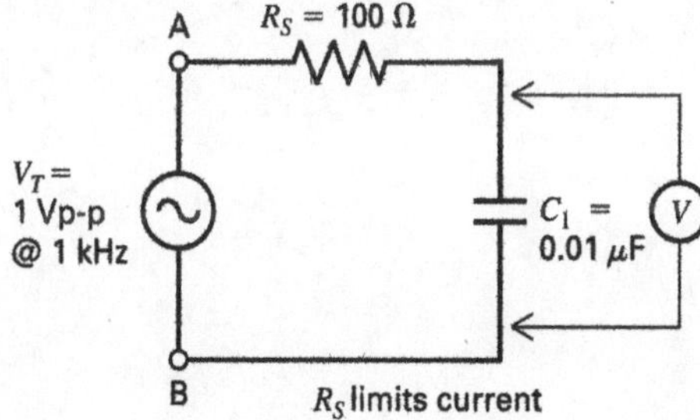

Fig. 16-1.7 An AC circuit with a capacitor in series.

9. Insert the first capacitor and measure the p-p voltage across it. Record the value in Table 16-1.1.

10. Increase the voltage to 5 $V_{p\text{-}p}$ (at 1 kHz). Measure and record the voltage across the capacitor.

11. Increase the frequency to 100 kHz. Repeat the measurements at 1 $V_{p\text{-}p}$ and 5 $V_{p\text{-}p}$ at this new frequency, and record the results in Table 16-1.1. Be sure to monitor the voltage as you adjust the frequency.

12. Carefully replace the capacitor with the next value of capacitor, and repeat all the measurements at 1 $V_{p\text{-}p}$ and at 5 $V_{p\text{-}p}$ for frequencies of 1 and 100 kHz, being sure to record the values in Table 16-1.1. Do this

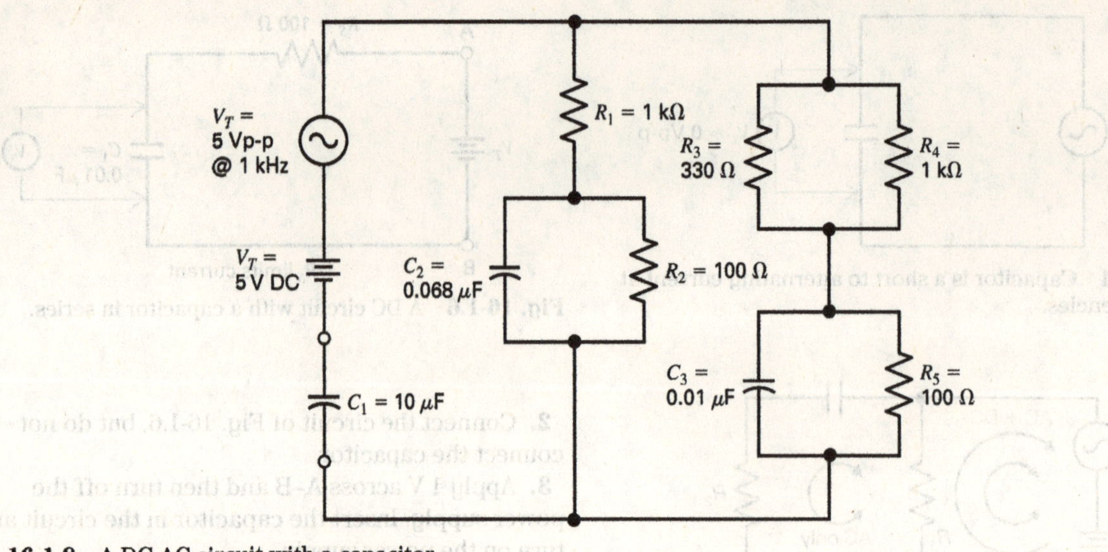

Fig. 16-1.8 A DC-AC circuit with a capacitor.

for all the remaining capacitors so that you have all the required data.

13. **Connect the circuit of Fig. 16-1.8.**

14. **Measure the DC and AC voltages across the resistors. Record the results in Table 16-1.2. Measure the capacitor voltage and note it separately.**

15. **Carefully remove capacitor C_1 from the circuit, and reconnect the circuit.**

16. **Measure all the DC and AC voltages across the resistors again. Record the results in Table 16-1.2.**

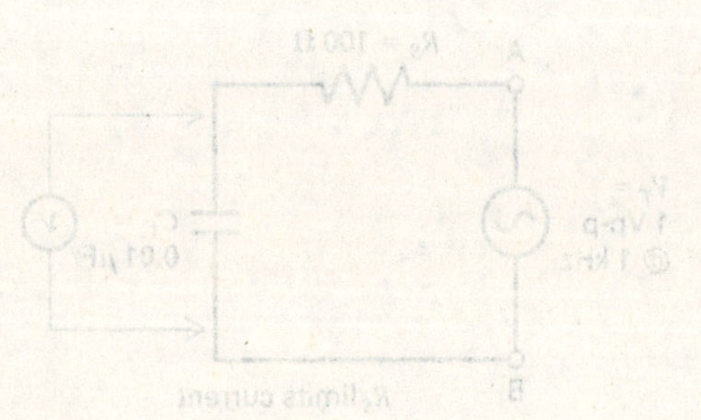

NAME DATE

QUESTIONS FOR EXPERIMENT 16-1

1. Why is a capacitor able to block direct current?

2. When is a capacitor like a 100-MΩ resistor?

3. When is a capacitor like a 0.001-Ω resistor?

4. Why is a capacitor able to pass high frequencies?

5. How can you verify the operation of a capacitor?

6. Is it possible for a capacitor to be leaky?

7. What type of capacitor needs to be connected with respect to polarity?

8. What precautions should you take to be sure a capacitor cannot be a hazard?

9. What differences, if any, did you notice between the capacitor voltage at 1 kHz and at 100 kHz in the circuit?

10. Describe the effects of removing the blocking capacitor C_1 from the last circuit, including any measurements that were significant.

TABLES FOR EXPERIMENT 16-1

TABLE 16-1.1

C (µF) (✓)	$V_T = 1$ V	$V_T = 5$ V	$f = 1$ kHz		$f = 100$ kHz	
	V_C	V_C	$V_{AC} = 1\ V_{p\text{-}p}$	$V_{AC} = 5\ V_{p\text{-}p}$	$V_{AC} = 1\ V_{p\text{-}p}$	$V_{AC} = 5\ V_{p\text{-}p}$
0.01						
0.068						
0.1						
10						
100						

TABLE 16-1.2 V_{C_1} = ________

R	With C_1		Without C_1	
	V_{DC}	V_{AC}	V_{DC}	V_{AC}
$R_1 = 1$ kΩ				
$R_2 = 100$ Ω				
$R_3 = 330$ Ω				
$R_4 = 1$ kΩ				
$R_5 = 100$ Ω				

EXPERIMENT RESULTS REPORT FORM

Experiment No.: ____________________ Name: ____________________
Date: ____________________
Experiment Title: ____________________ Class: ____________________
____________________ Instr: ____________________

Explain the purpose of the experiment:

List the first Learning Objective:
OBJECTIVE 1:

After reviewing the results, describe how the objective was validated by this experiment.

List the second Learning Objective:
OBJECTIVE 2:

After reviewing the results, describe how the objective was validated by this experiment.

List the third Learning Objective:
OBJECTIVE 3:

After reviewing the results, describe how the objective was validated by this experiment.

Conclusion:

If required, attach to this form: ☐ Answers to Questions, ☐ Tables, and ☐ Graphs.

CHAPTER 17

EXPERIMENT 17-1

CAPACITIVE REACTANCE

LEARNING OBJECTIVES

At the completion of this experiment, you will be able to:

- Understand how a capacitor reacts in a series *RC* circuit by calculation and measurement.
- Understand the effects of changing frequency upon an *RC* series circuit.
- Plot a graph of frequency versus V_C and V_R for an *RC* series circuit.

SUGGESTED READING

Chapters 16, 17, and 18, *Basic Electronics*, Grob/Schultz, twelfth edition

INTRODUCTION

Capacitors have the ability to store an electric charge, much like a battery. The charge exists between the two plates made of conductive material. These plates are separated by an insulator or dielectric. In a DC circuit, a capacitor will charge up to the potential difference across it and act as an open circuit, blocking any DC current flow. In an AC circuit, the capacitor will charge and discharge in proportion to the frequency of the alternating current, acting like a resistor or, at high frequencies, a short circuit to the alternating current. In fact, one definition of a capacitor is any two conductors separated by an insulator.

A charged capacitor is easily discharged by connecting a conducting path across the dielectric. Placing your fingers across a charged capacitor will discharge the capacitor through your fingers. With large capacitors, this will result in electric shock. *Never* assume that a capacitor is discharged.

Capacitors are manufactured and rated in units called *farads* (F), named after Michael Faraday. When 1 C (coulomb) is stored in the dielectric, with a potential difference of 1 V, the capacitance is equal to 1 F. Typically, capacitors are most often found in picofarad (pF) and microfarad (μF) values. The voltage rating of a capacitor is a maximum voltage that can be placed across the capacitor without damage. For example, a 10-μF capacitor, rated at 10 V, could be easily ruptured (exploded) if 20 V was across the capacitor.

When two capacitors of equal value are placed in series, the value of capacitance is reduced by one-half. This occurs because the dielectric thickness is increased and the plates are farther apart. Similar to resistors in parallel, the formula for capacitors in series is calculated by using the reciprocal method:

$$\frac{1}{C_T} = \frac{1}{C_1} + \frac{1}{C_2} + \cdots$$

When two capacitors of equal value are placed in parallel, the value of capacitance is increased by twice as much. Similar to resistors in series, the formula for capacitors in parallel is calculated by using simple addition:

$$C_T = C_1 + C_2 + \cdots$$

In an AC circuit, a capacitor will allow current to flow. The current does not actually flow through the capacitor; it flows because of the charging and discharging action of the capacitor. Therefore, the capacitive current varies with the frequency of applied AC voltage. As the frequency increases, the capacitor's opposition (in ohms) to current flow decreases. As frequency decreases, the opposition increases. That is, the capacitor reacts to changes in frequency. This is what is meant by capacitive reactance X_C. The formula for the capacitive reactance (in ohms) of any capacitor is

$$X_C = \frac{1}{2\pi fC}$$

where 2π is the sine-wave rotation, f is the frequency in hertz, and C is the value of capacitance in farads. Also, because $^1/(2\pi) = 0.159$,

$$X_C = \frac{0.159}{fC}$$

$$f = \frac{0.159}{CX_C}$$

$$C = \frac{0.159}{fX_C}$$

Similar to an inductor, a capacitor will have a phase difference between the voltage across it and its current. Capacitive current i_C leads v_C by 90°. The term *ICE* is an easy way to remember that *I* (current) leads *C* (capacitive voltage *E*). Figure 17-1.1 illustrates this.

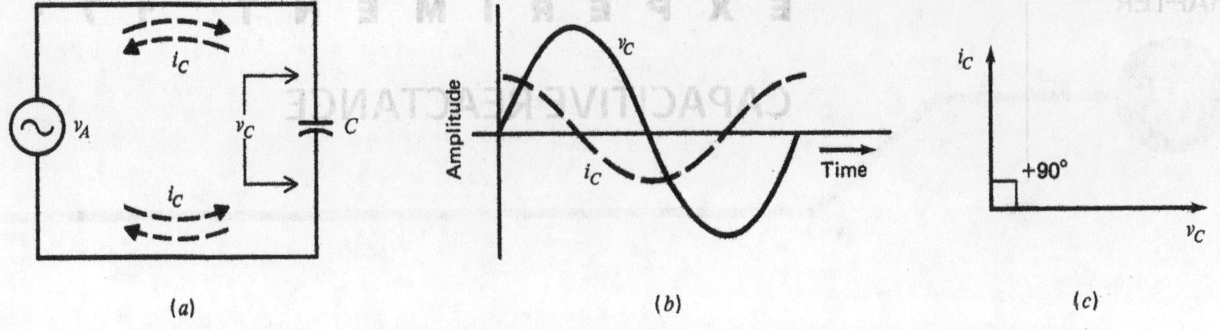

Fig. 17-1.1 *RC* circuit. (*a*) Purely capacitive circuit. (*b*) Waveshapes of i_C lead v_C by 90°. (*c*) Phasor diagram.

In an *RC* series circuit, the value of current is the same in all parts of the circuit. However, each has its own series voltage drop:

$$V_R = I_T \times R \qquad \text{and} \qquad V_C = I_T \times X_C$$

Again, similar to an inductor, the voltages must be added by phasor addition because of the 90° phase shift due to the lagging capacitive voltage. Figure 17-1.2 illustrates these points. In accordance with the triangles in Fig. 17-1.2, the total voltage in an *RC* series circuit is

$$V_T = (V_R^2 + V_C^2)^{1/2} \text{ or } V_T = \sqrt{V_R^2 + V_C^2}$$

and the total impedance (opposition to current flow) is

$$Z_T = (R^2 + X_C^2)^{1/2} \text{ or } Z_T = \sqrt{R^2 + X_C^2}$$

Remember that the circuit phase angle is between the series current and the generator AC voltage as follows:

$$\text{Inverse tan } \frac{X_C}{R} = \text{circuit phase angle}$$

This is because the phase angle is negative due to V_C lagging I_C.

The following procedures will allow you to validate the information given in this introduction.

EQUIPMENT

Signal generator
Oscilloscope

COMPONENTS

(1) 4.7-kΩ resistor
(1) 47-kΩ resistor
(1) 0.0068-μF capacitor
(1) 0.1-μF capacitor
(1) 0.01-μF capacitor

PROCEDURE

1. Connect the circuit of Fig. 17-1.3.

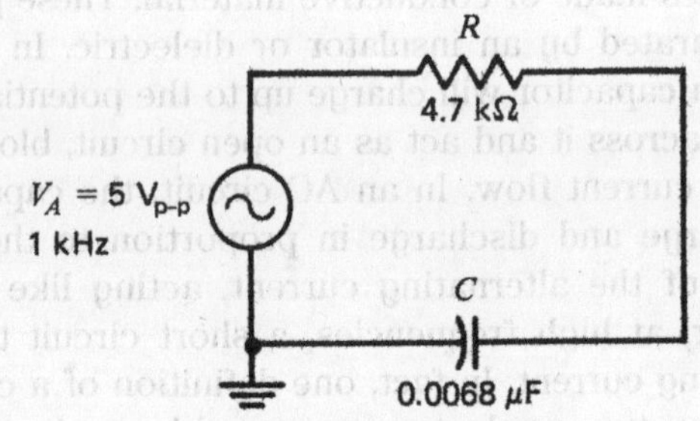

Fig. 17-1.3 *RC* circuit for measurement.

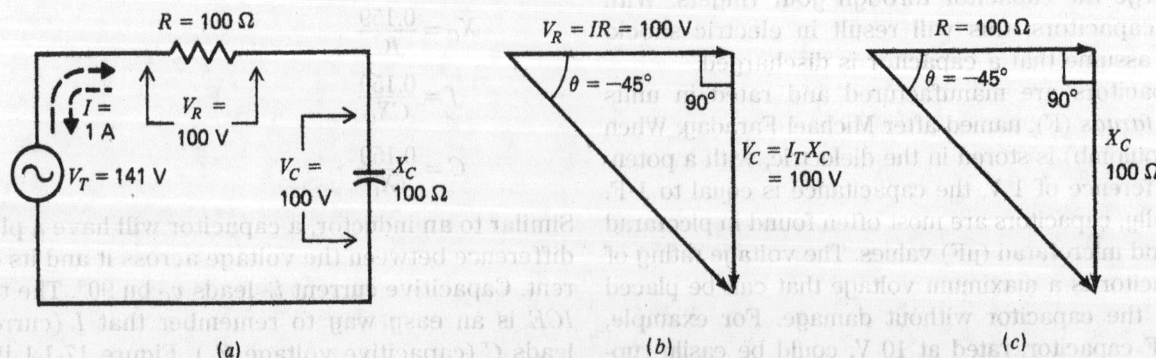

Fig. 17-1.2 (*a*) *RC* series circuit. (*b*) Phasor voltages. (*c*) Impedance triangle.

Note: Remember, the applied voltage (5 $V_{p\text{-}p}$) must be adjusted and maintained only after the entire circuit is connected. The signal generator has its own internal resistance and IR voltage drop that can subtract from the applied voltage. Therefore, constantly monitor the input voltage whenever you change frequency in the following steps.

2. Calculate X_C for this circuit in Table 17-1.1 as

$$X_C = \frac{1}{2\pi fC}$$

3. Measure and record in Table 17-1.1 the peak-to-peak voltage across the capacitor. Be sure the ground points are kept in the same place.

4. Measure and record in Table 17-1.1 the peak-to-peak voltage across the resistor. If necessary, exchange places with the capacitor to keep the ground points in a common place.

5. Calculate and record the circuit current I_T in Table 17-1.1, using the Ohm's law formula. Do this by using the measured voltage for both components.

Note: The values of I_T ((a) and (b)) will be different because R and X_C are determined by different methods. Real-world values can vary by 10 percent and still be valid.

$$\text{(a) } I_T = \frac{V_R}{R}$$

$$\text{(b) } I_T = \frac{V_C}{X_C}$$

6. Calculate and record in Table 17-1.1 the total circuit impedance, using the formula

$$Z_T = (R^2 + X_C^2)^{1/2}$$

7. Refer to the circuit of Fig. 17-1.3. Change the value of the capacitor to 0.1 μF, and repeat steps 1 to 6.

8. Refer to the circuit of Fig. 17-1.3. Change the value of the capacitor to 0.01 μF, and change the value of the resistor to 47 kΩ.

9. Readjust the signal generator to 200 Hz, and increase the frequency in 200-Hz steps from 200 Hz to 1 kHz. Measure and record the voltage across the resistor V_R at each frequency step. Create your own table identical to Table 17-1.1.

10. Increase the frequency from 1 to 15 kHz in 1-kHz steps, and measure and record V_R at each frequency step.

11. Calculate and record X_C at each frequency in steps 9 and 10 above, as $X_C = 1/(2\pi fC)$.

12. Calculate the circuit current I_T as V_R/R for each frequency in step 11 above. Also, calculate V_C as $I_T X_C$ for each frequency.

13. Plot a graph of V_C and V_R versus frequency for the data in steps 10 to 12. Indicate where the V_C and V_R curves intersect at 45°.

OPTIONAL PROCEDURES

Series Capacitance

Connect two capacitors of equal value in series, and repeat steps 9 to 12. Make your own data table and record all the values.

Parallel Capacitance

Connect two capacitors of equal value in parallel, and repeat steps 9 to 12. Make your own data table and record all values.

Note: Use R = 10 kΩ and V_A = 5 $V_{p\text{-}p}$.

NAME DATE

QUESTIONS FOR EXPERIMENT 17-1

Answer the following questions on a separate sheet of paper. Show all work.

1. Using your graphed data, calculate the circuit phase angle at 8 kHz.

2. Using your graphed data, calculate the circuit phase angle at 600 Hz.

3. For a series *RC* circuit, at what frequency would a 10-μF capacitor have $X_C = 100\ \Omega$?

4. For a series *RC* circuit, what value of capacitor would have 31.8 Ω of X_C at 5 Hz?

5. For the circuit in Fig. 17-1.4, the voltage across $C = 25\ V_{p\text{-}p}$. The circuit phase angle is 45°. What is the voltage across *R* and V_A?

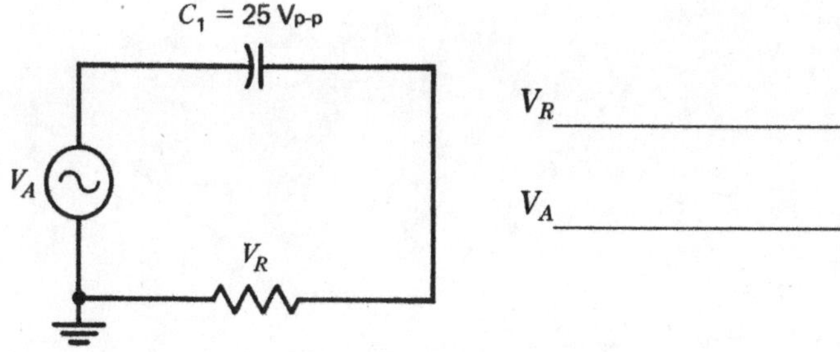

V_R ____________________

V_A ____________________

Fig. 17-1.4 *RC* circuit at 45°.

TABLE FOR EXPERIMENT 17-1

TABLE 17-1.1

Procedure Step	Measurement	Value	Calculations
2	X_C		______ Ω
3	V_C	______ $V_{p\text{-}p}$	
4	V_R	______ $V_{p\text{-}p}$	
5*a*	I_T		______ mA
5*b*	I_T		______ mA
6	Z_T		______ Ω
7	X_C		______ Ω
	V_C	______ $V_{p\text{-}p}$	
	V_R	______ $V_{p\text{-}p}$	
	I_T(a)		______ mA
	I_T(b)		______ mA
	Z_T		______ Ω

The students should prepare their own tables for $R = 47$ kΩ (step 8) and steps 9 to 13.

EXPERIMENT RESULTS REPORT FORM

Experiment No.: ____________________ Name: ____________________

Date: ____________________

Experiment Title: ____________________ Class: ____________________

____________________ Instr: ____________________

Explain the purpose of the experiment:

List the first Learning Objective:
OBJECTIVE 1:

After reviewing the results, describe how the objective was validated by this experiment.

List the second Learning Objective:
OBJECTIVE 2:

After reviewing the results, describe how the objective was validated by this experiment.

List the third Learning Objective:
OBJECTIVE 3:

After reviewing the results, describe how the objective was validated by this experiment.

Conclusion:

If required, attach to this form: ☐ Answers to Questions, ☐ Tables, and ☐ Graphs.

CHAPTER 18

EXPERIMENT 18-1

CAPACITIVE REACTANCE

LEARNING OBJECTIVES

At the completion of this experiment, you will be able to:

- Describe capacitor coupling action.
- Measure the demonstrated effects of capacitive coupling.
- Determine the approximate frequency where a capacitor begins coupling.

SUGGESTED READING

Chapter 18, *Basic Electronics*, Grob/Schultz, twelfth edition

INTRODUCTION

The most common type of coupling in amplifier circuits (Fig. 18-1.1) is *capacitive*. In this type of coupling, the output of one stage is connected to the input of another stage while the signals are sent through a capacitor. As in any coupling circuit that has sensitivity to frequency changes, the main goal of a good capacitive coupling circuit is to pass the signal with little attenuation of desired signals while attenuating the undesired frequencies. An advantage of the capacitive coupling circuit is its ability to block the DC voltage that may be present in the output signal of the first stage. This is important in that, as in transistor amplifiers, the DC operational characteristics of the second stage would be affected if its input signal were not free of possible DC components.

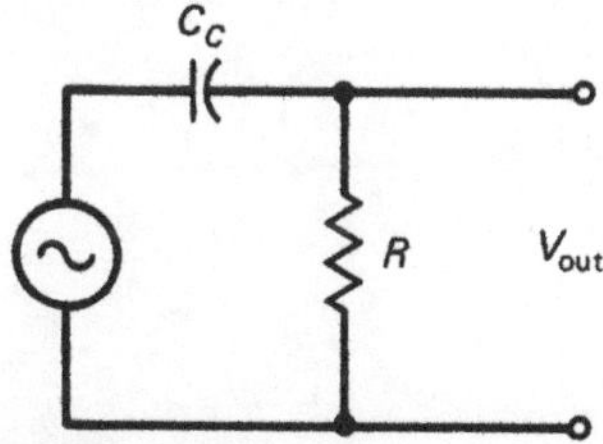

Fig. 18-1.1 AF coupling circuit.

EQUIPMENT

Protoboard
Signal generator
Oscilloscope
Test leads

COMPONENTS

(1) 0.01-μF capacitor
(1) 22-kΩ, 0.25-W resistor

PROCEDURE

1. Connect the circuit of Fig. 18-1.2. The capacitor in this circuit is used in a coupling application. The relatively low reactance of this capacitor allows practically all the generated AC voltage to be dropped across the resistor. Ideally very little of the generated AC voltage is dropped across the coupling capacitor.

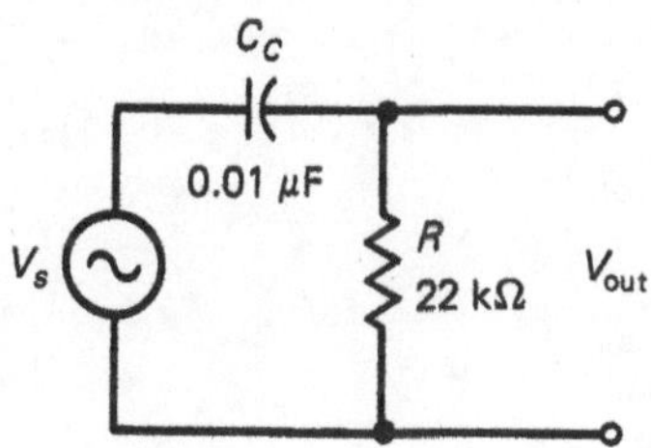

Fig. 18-1.2 AF coupling circuit.

2. Turn on the signal generator. Adjust the generator to 5 $V_{p\text{-}p}$ at 100 Hz. Measure and record in Table 18-1.1 the following AC voltages: source generator, coupling capacitor, and resistor.

3. Complete the measurements, and record the data in Table 18-1.1.

4. The dividing line for C_c to be a coupling capacitor at a specific frequency can be determined when X_c equals one-tenth (or less) of R. At this point the series RC circuit becomes primarily resistive, and all the voltage drop of the signal generator is across the series resistance, with very little voltage dropped across the coupling capacitor. Using the one-tenth rule described above and the information contained in Table 18-1.1, determine the frequency at which the capacitor becomes a *coupling* capacitor. Adjust the signal generator to this frequency, and complete the measurements required in Table 18-1.2.

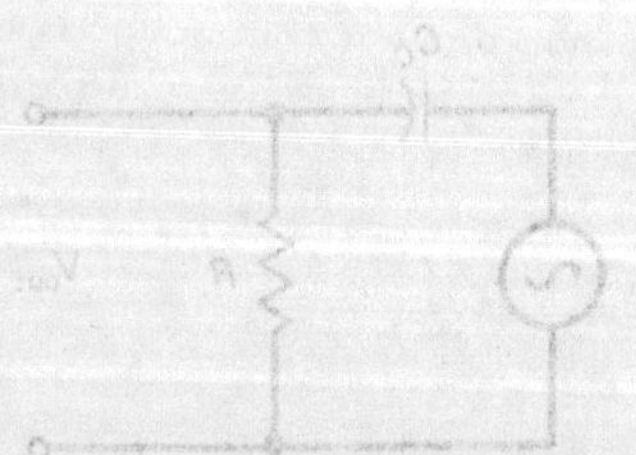

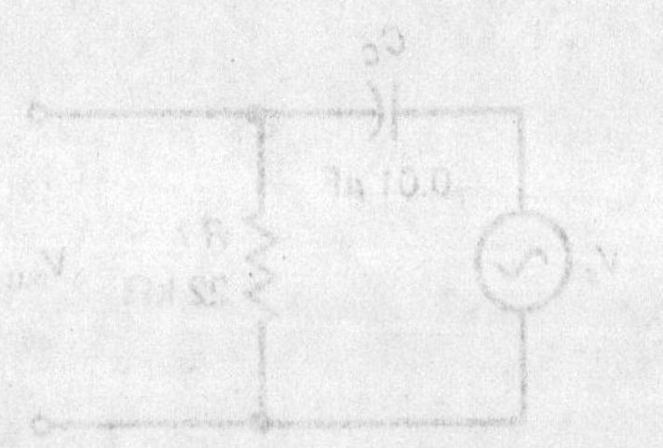

NAME DATE

QUESTIONS FOR EXPERIMENT 18-1

1. **After reading the capacitor coupling section in your book, describe what a coupling capacitor is used for.**

2. **What importance does the one-tenth rule have, and where is it used?**

3. **What effect does a coupling capacitor have on the DC current?**

TABLES FOR EXPERIMENT 18-1

TABLE 18-1.1

Frequency	V_s (p-p)	V_c (p-p)	V_R (p-p)
100 Hz			
200 Hz			
300 Hz			
400 Hz			
500 Hz			
600 Hz			
700 Hz			
800 Hz			
900 Hz			
1.0 kHz			
1.1 kHz			
1.2 kHz			
1.3 kHz			
1.4 kHz			
1.5 kHz			
1.6 kHz			
1.7 kHz			
1.8 kHz			
1.9 kHz			
2.0 kHz			
3.0 kHz			
4.0 kHz			
5.0 kHz			
6.0 kHz			
7.0 kHz			
8.0 kHz			
9.0 kHz			
10 kHz			
15 kHz			
20 kHz			

TABLE 18-1.2

Frequency Where C_c Becomes a Coupling Capacitor	V_s (p-p)	V_c (p-p)	V_R (p-p)
________	________	________	________

EXPERIMENT RESULTS REPORT FORM

Experiment No.: ____________ Name: ____________

Date: ____________

Experiment Title: ____________ Class: ____________

____________ Instr: ____________

Explain the purpose of the experiment:

List the first Learning Objective:

OBJECTIVE 1:

After reviewing the results, describe how the objective was validated by this experiment.

List the second Learning Objective:

OBJECTIVE 2:

After reviewing the results, describe how the objective was validated by this experiment.

List the third Learning Objective:

OBJECTIVE 3:

After reviewing the results, describe how the objective was validated by this experiment.

Conclusion:

If required, attach to this form: ☐ Answers to Questions, ☐ Tables, and ☐ Graphs.

CHAPTER 18

EXPERIMENT 18-2

CAPACITIVE PHASE MEASUREMENTS: USING AN OSCILLOSCOPE

LEARNING OBJECTIVES

At the completion of this experiment, you will be able to:

- Make phase measurements on the oscilloscope, using the dual-trace method.
- Determine the phase angle by measuring voltages and calculating the phase shift.

SUGGESTED READING

Chapters 15 and 18, *Basic Electronics*, Grob/Schultz, twelfth edition

INTRODUCTION

This experiment concentrates on the measurement of a phase angle. Unlike frequency measurements, phase measurements compare two sine-wave signals of similar frequency; the result is a phase angle between two waves.

Consider the drawing (Fig. 18-2.1) where two voltage waveforms are shown 90° out of phase. Figure 18-2.1*a* shows the two sine waves as they might be displayed on a dual-trace oscilloscope. Sine wave *A* starts at 0 V and 0°. Therefore, it is considered the reference signal.

Sine wave *B* actually begins 90° before sine wave *A*. In other words, the peak voltage of sine wave *B* occurs when sine wave *A* is at zero. The phasor diagram of Fig. 18-2.1*b* shows that both signals have about equal magnitude (voltage), indicated by the length of the arrow. The angle is shown with respect to the reference signal *A* on the horizontal axis. Therefore, the phase angle has magnitude and direction; here the direction is +90° compared to 0° of the reference signal. This corresponds to the standard practice of using a counterclockwise rotation as the positive direction of rotation. This also means that signal *B* leads signal *A* by +90°.

Although it is easier to describe phase differences when the signals have the same voltage, consider the drawing in Fig. 18-2.2. Figure 18-2.2*a* shows two sine waves with different magnitudes, but having the same frequency. Figure 18-2.2*b* shows two sine waves 180° out of phase, also with different magnitudes. Notice that both phase measurements can be represented by their corresponding phasor diagrams.

Finally, to measure phase, a phase-shift network will have to be created. This will be a simple *RC* circuit.

EQUIPMENT

Oscilloscope with *xy* capability
DMM
Digital frequency counter (optional)
Signal generator

COMPONENTS

Leads for connecting equipment as required
(1) 1.2-kΩ, 0.25-W resistor
(1) 0.01-μF capacitor

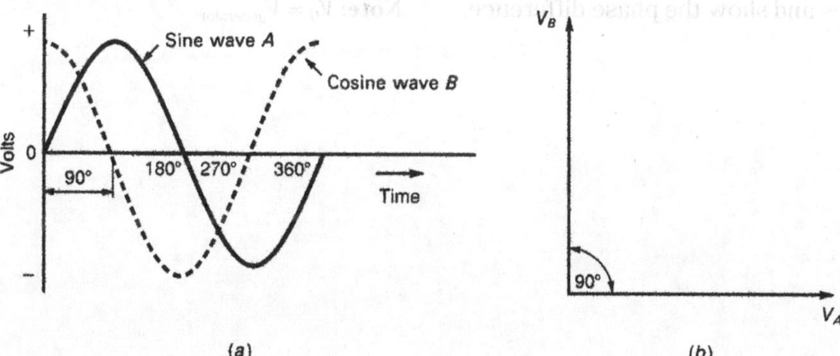

Fig. 18-2.1 Phase angle of two sine waves.

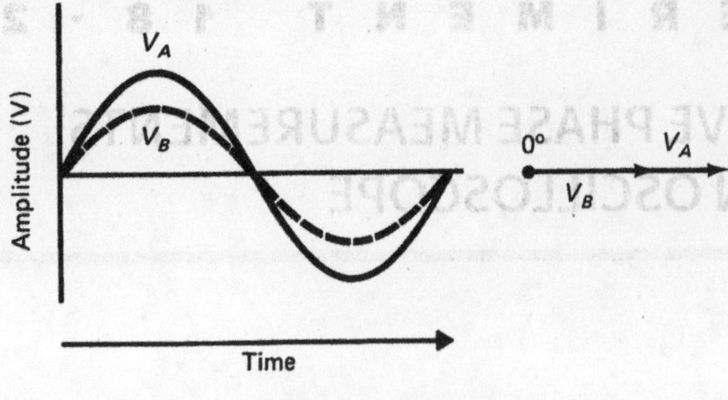

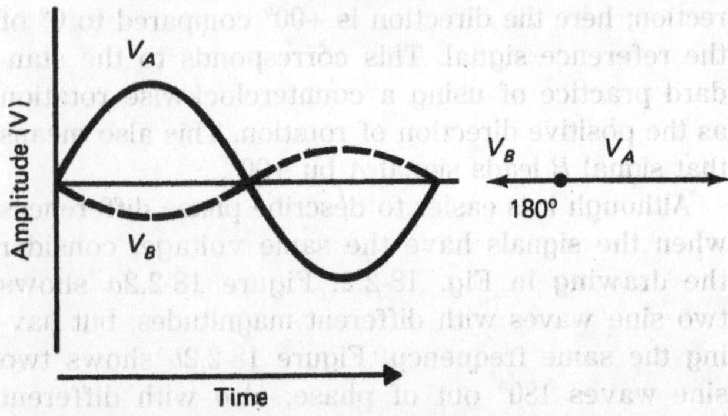

Fig. 18-2.2 Phase angles with different amplitudes.

PROCEDURE

1. Connect the circuit as shown in Fig. 18-2.3, and turn on the power.

2. Measure the p-p voltage across the resistor, and record the results in Table 18-2.1.

3. Adjust the oscilloscope to time-base operation, and set it to display one or two complete cycles of each wave. Set both channels to the same ground at the center graticule. Then measure the phase difference between the two signals: channel 2 = V_G and channel 1 = V_R. Also, you can use the position knobs, and, if applicable, choose the CHOP mode instead of the ALTERNATE or ADD mode.

Draw the traces as best as you can in Table 18-2.1. Label the two traces and show the phase difference, in degrees, where the V_G wave crosses the zero or middle graticule lines. Also, show the degree per division.

4. Change the frequency of the signal generator to 8 kHz, and repeat steps 2 and 3. Use Table 18-2.2 to record your results.

5. Calculate the phase angle using series X_c and R from the impedance triangle: $\tan\theta = -\frac{X_c}{R}$ and record in Table 18-2.1.

Note: Use the calculator inverse tangent button ($\tan^{-1}$).

6. Calculate the phase angle using measured values V_R and V_G: $\cos\theta = \frac{V_R}{V_G}$ and record in Table 18-2.1.

Note: Use the calculator inverse cosine button ($\cos^{-1}$). Also, using measured values will result in a positive phase value but you know it is an *RC* circuit. So, the phase shift is expected between V_G and V_R.

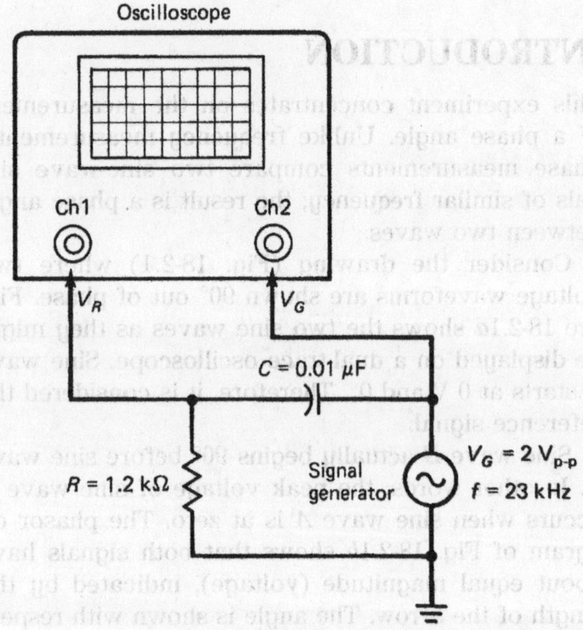

Fig. 18-2.3 Phase measurement of a phase-shift circuit. Note: $V_G = V_{generator}$.

NAME DATE

QUESTIONS FOR EXPERIMENT 18-2

1. Why did changing the frequency of V_G cause a different phase angle?

2. If you didn't have any test equipment, could you calculate the phase angle between V_G and V_R? How?

3. Draw the phasor showing the results for a phase angle where $V_G = 12$ V and $V_R = 8.4$ V.

TABLES FOR EXPERIMENT 18-2

TABLE 18-2.1

Procedure Step	V_G = 23 kHz at 2 $V_{p\text{-}p}$
2	V_R = ______ $V_{p\text{-}p}$
	V_G = ______ $V_{p\text{-}p}$
3	Degrees per DIV = ______
5	θ_z = ______ calculated
6	θ_z = ______ measured

TABLE 18-2.2

Procedure Step	V_G = 8 kHz at 2 $V_{p\text{-}p}$
2	V_R = ______ $V_{p\text{-}p}$
	V_G = ______ $V_{p\text{-}p}$
3	Degrees per DIV = ______
5	θ_z = ______ calculated
6	θ_z = ______ measured

Note: **The negative sign used in the calculated phase angle ($-X_c/R$) in the Grob/Schultz book for series *RC* indicates V_G lags *I*. If you simply calculate the phase angle without the minus sign (S_c/R), then your answer should be similar to the measured calculation using V_R and V_G.**

EXPERIMENT RESULTS REPORT FORM

Experiment No.: ____________________ Name: ____________________

Date: ____________________

Experiment Title: ____________________ Class: ____________________

____________________ Instr: ____________________

Explain the purpose of the experiment:

List the first Learning Objective:

OBJECTIVE 1:

After reviewing the results, describe how the objective was validated by this experiment.

List the second Learning Objective:

OBJECTIVE 2:

After reviewing the results, describe how the objective was validated by this experiment.

Conclusion:

If required, attach to this form: ☐ Answers to Questions, ☐ Tables, and ☐ Graphs.

CHAPTER 19

EXPERIMENT 19-1

INDUCTORS

LEARNING OBJECTIVES

At the completion of this experiment, you will be able to:

- Troubleshoot an inductor to verify that it is functional for continuity.
- Verify the DC and AC response of an inductor.
- Induce a voltage with an inductor and light a neon bulb.

SUGGESTED READING

Chapter 19, *Basic Electronics*, Grob/Schultz, twelfth edition

INTRODUCTION

Inductors are coils of insulated wire. Hence, they are often called *coils*. Coils have a specific number of coated and insulated wire windings around a center core, which is often made of air or iron. Inductors can produce (induce) voltage because of this winding construction, which can oppose or choke off higher-frequency currents. Inductors are rated in units of henrys (H), named after the scientist Joseph Henry. For example, a 1-H coil has the ability to induce 1 V when the current changes at the rate of 1 A/s. Whenever current is flowing in a wire or conductor that is next to another conductor, it is by definition an inductor. Figure 19-1.1 shows some typical inductors.

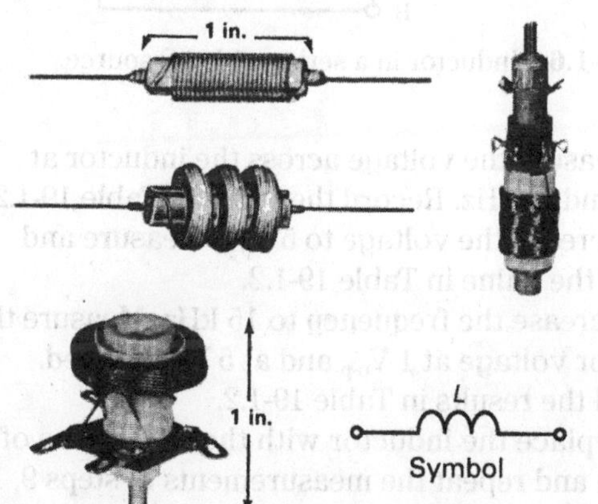

Fig. 19-1.1 Types of inductors and symbol.

The current flowing in the windings of an inductor produces a magnetic field. With so many turns in close proximity, a large polarized magnetic field builds up, which has the ability to create a force field. A common example is a door bell that has a magnet in the center of an inductor. The free-moving magnet is forced out by the field's magnetic opposition and strikes a bell. When the current is stopped, the field collapses and the iron magnet falls back into the center of the inductor.

More windings or turns in a inductor result in a larger magnetic field when current flows. When a DC voltage is applied to an inductor, only the resistance of the wire exists and the current flows easily. In this case, the inductor is a short circuit to DC, with only the resistance of the coil. Because of this, you can verify that a coil is working by using an ohmmeter and checking for continuity. Figure 19-1.2 shows the circuit for verifying an inductor with an ohmmeter.

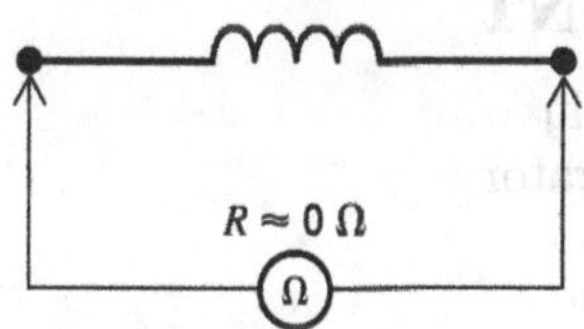

Fig. 19-1.2 Verifying an inductor with an ohmmeter.

When an AC signal is applied to the inductor, there is a reaction to the change in frequency. This reaction is greater when the frequency increases because the inductor opposes the flow of current or *chokes* it off. That is why an inductor is often called a *choke*. Another common use of an inductor, however, is in a power supply, where the buildup of the magnetic field in a primary winding is used to induce a voltage in a secondary winding, often called a *tap*. This is one method of stepping down a higher voltage to a lower voltage; the ratio of the number of turns in the primary to the number of turns in the secondary determines the value of induced voltage. Figure 19-1.3 shows a transformer, used to step down a voltage.

Another form of inductor is very small and is often used in microwave applications. It is called a *toroid*

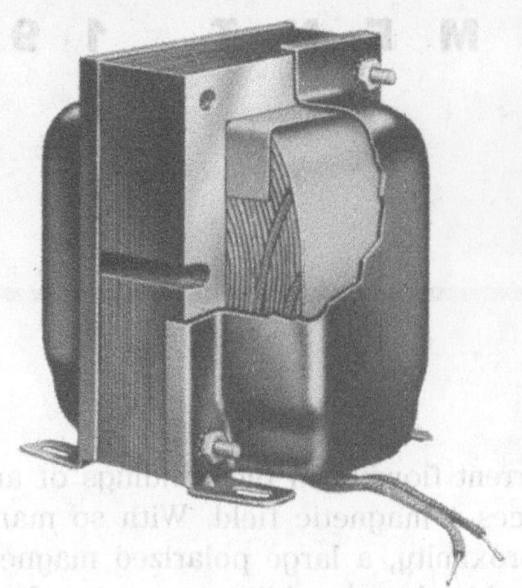

Fig. 19-1.3 Transformer.

and is used to control or filter high frequencies. A toroid, as shown in Fig. 19-1.4, has only a few turns wound around a magnetic ring.

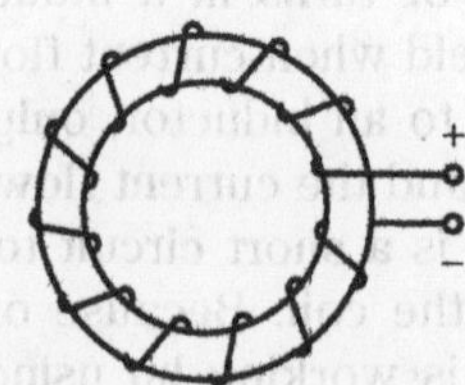

Fig. 19-1.4 Toroid.

EQUIPMENT

DC power supply
AC signal generator
DMM
Oscilloscope

COMPONENTS

Resistors (0.25 W): 100 Ω
Inductors:
(4) coils of varying values from 10 mH to 0.5 H
(1) 0.5-H choke
Transformers: Standard 60-Hz power transformer with two or three taps or a center tap
Neon bulb: Ne2 bulb lighting at approximately 60 V or a similar bulb

PROCEDURE

1. First, obtain a step-down transformer and record its values in Table 19-1.1. Then, verify the resistance of the primary and secondary windings by checking the resistance (continuity). Record the value of DC resistance for each winding in Table 19-1.1. You should record the color of the wires (taps) in the table to identify them.

2. Check the continuity of each inductor, and record the resistance value and the value of the inductor in Table 19-1.2. If any inductor is bad, replace it.

3. Connect the circuit of Fig. 19-1.5, but do not insert the inductor.

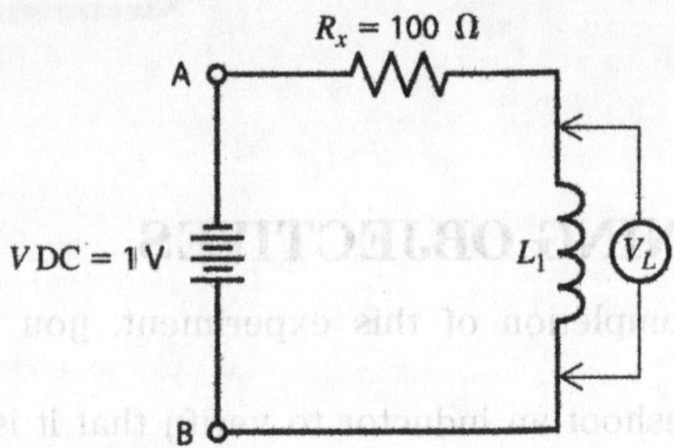

Fig. 19-1.5 Inductor in a series circuit (with current-limiting resistor).

4. Apply 1 V DC across points A–B without the inductor in the circuit. Turn off the power supply and insert the inductor. Turn on the power supply and measure the voltage across the inductor. Record the value in Table 19-1.2.

5. Increase the applied voltage to 5 V. Measure and record the value in Table 19-1.2.

6. Replace the inductor with the next value. Measure the voltage across the inductor at 1 V and at 5 V, and record the results in Table 19-1.2.

7. Repeat step 6 for the remaining inductors.

8. Connect the circuit of Fig. 19-1.6, but do not connect the inductor. Adjust the power supply so that 1 V_{p-p} at 100 Hz appears across A–B.

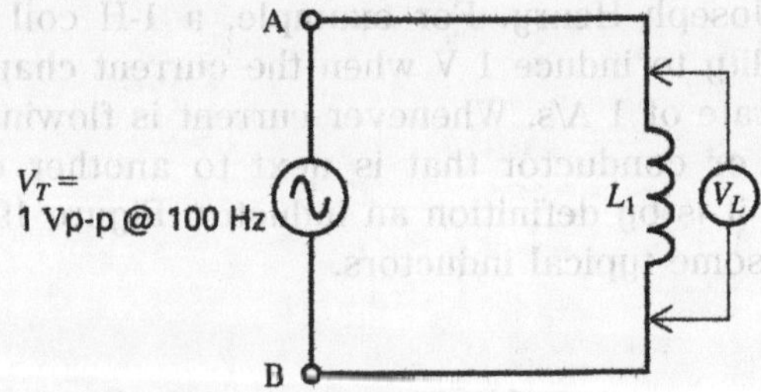

Fig. 19-1.6 Inductor in a series with AC source.

9. Measure the voltage across the inductor at 1 V_{p-p} and 100 Hz. Record the value in Table 19-1.2.

10. Increase the voltage to 5 V_{p-p}. Measure and record the value in Table 19-1.2.

11. Increase the frequency to 15 kHz. Measure the inductor voltage at 1 V_{p-p} and at 5 V_{p-p} applied. Record the results in Table 19-1.2.

12. Replace the inductor with the next value of inductor, and repeat the measurements in steps 9, 10, and 11. Then repeat this procedure for each value of inductor you have.

OPTIONAL

13. Obtain the neon bulb and connect the circuit of Fig. 19-1.7, but do not apply power yet.

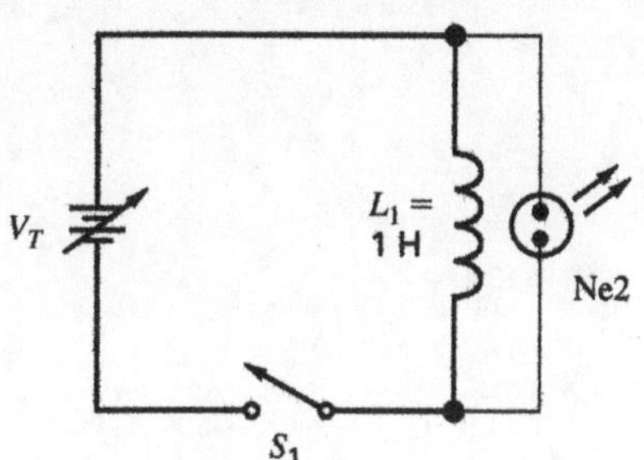

Fig. 19-1.7 Inductor with neon bulb circuit.

14. Apply exactly 2 V DC to the circuit.

15. Toggle (close and open quickly) switch S_1 in an instantaneous manner to create a current pulse. This will induce a voltage in the inductor that should be greater than the source.

16. Did the bulb light or not? Record the results in Table 19-1.3.

17. Increase the voltage in 0.5-V steps, and repeat steps 15 and 16 until the bulb lights. Note the applied voltage that lights the bulb in Table 19-1.3. *Do not exceed the allowable voltage rating (about 10 V is average max) for the neon bulb.*

18. Refer to Table 19-1.1. With the step-down transformer connected to an AC source, verify that the taps are lowering the applied voltage.

Note: Perform this step only with your instructor's approval. Use caution when measuring the outputs of the taps. Do not do this step until you have checked with your instructor about how to safely step down the voltage with a safe input of alternating current.

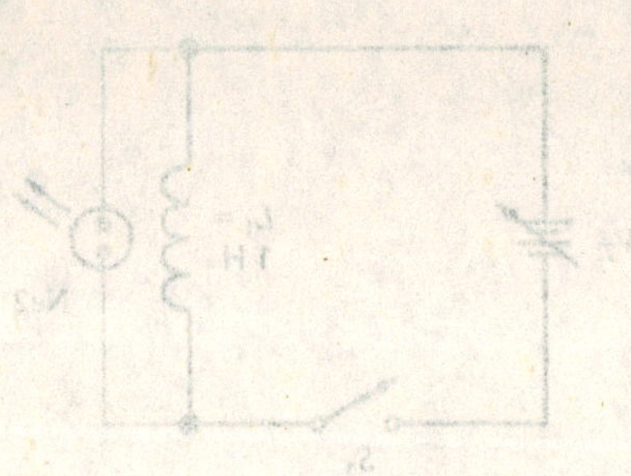

NAME DATE

QUESTIONS FOR EXPERIMENT 19-1

1. How is an inductor constructed?

2. How can you check if an inductor is working?

3. How does a transformer tap work to produce various voltages?

4. Is an inductor sensitive to changes in frequency?

5. If an inductor measured 75 Ω, would it still be good?

6. What was the difference between the inductor voltage at 100 Hz and at 15 kHz?

7. Why is an inductor called a choke?

8. How is voltage induced in a coil?

9. How could you keep the neon bulb lit longer?

10. Is an inductor capable of causing a shock if you touch it?

TABLES FOR EXPERIMENT 19-1

TABLE 19-1.1 Transformer

Inductor	Wire Colors	Resistance
Primary		
Secondary		

TABLE 19-1.2 Inductors

Value (H) and Resistance	$V_T = 1$ V	$V_T = 5$ V	$f = 100$ Hz		$f = 15$ kHz	
	V_L	V_L	$V_{AC} = 1\ V_{p\text{-}p}$	$V_{AC} = 5\ V_{p\text{-}p}$	$V_{AC} = 1\ V_{p\text{-}p}$	$V_{AC} = 5\ V_{p\text{-}p}$

OPTIONAL:

TABLE 19-1.3 Induced Voltage Circuit (Neon Bulb)

Applied DC Voltage												
Volts	2	2.5	3	3.5	4	4.5	5	5.5	6	6.5	7	7.5
Volts	8	8.5	9	9.5	10							

Circle the approximate voltage that lights the bulb.

EXPERIMENT RESULTS REPORT FORM

Experiment No.: ______________________ Name: ______________________

Date: ______________________

Experiment Title: ______________________ Class: ______________________

______________________ Instr: ______________________

Explain the purpose of the experiment:

List the first Learning Objective:

OBJECTIVE 1:

After reviewing the results, describe how the objective was validated by this experiment.

List the second Learning Objective:

OBJECTIVE 2:

After reviewing the results, describe how the objective was validated by this experiment.

List the third Learning Objective:

OBJECTIVE 3:

After reviewing the results, describe how the objective was validated by this experiment.

Conclusion:

If required, attach to this form: ☐ Answers to Questions, ☐ Tables, and ☐ Graphs.

CHAPTER 20

EXPERIMENT 20-1

INDUCTIVE REACTANCE

LEARNING OBJECTIVES

At the completion of this experiment, you will be able to:

- Understand the concept of inductive reactance and validate the formula $X_L = 2\pi fL$.
- Understand phase relationships.
- Plot the frequency response of a series *RL* circuit.

SUGGESTED READING

Chapters 19, 20, and 21, *Basic Electronics*, Grob/Schultz, twelfth edition

INTRODUCTION

To study inductance, cosine waves, inductive reactance, inductive coupling, and many other aspects of an inductor (coil) in an AC circuit would require weeks of laboratory time. Therefore, this experiment will focus on the basic properties of an inductor that are fundamental to a technician.

As its name implies, an inductor produces its own (self-induced) voltage. In fact, any conductor that is close to another conductor (like two wires in close proximity) can be considered an inductor if current is flowing. Thus, as current flows in any inductor, a voltage is induced. This is accomplished because the inductor (coil) produces a magnetic field, and the magnetic field is greater with more turns of wire and/or more changes of current (frequency).

Remember that a coil's inductance uses the symbol *L* (from linkages of magnetic flux) and is measured in henrys (H). An 8-H coil, for example, is a rather large inductor and is used for 60-Hz AC power lines or other low-frequency (audio range) applications. A 1-μH coil, however, is a rather small inductor and is used at higher frequencies. Also, remember that as the frequency increases in an AC circuit, the inductive reactance X_L increases even though the value of the inductor (in henrys) remains the same. In other words, the coil reacts to, or opposes, a change in current. As frequency increases, X_L increases. Therefore, the formula for inductive reactance is

$$X_L = 2\pi fL$$

where *f* is the determining factor. Note that 2π is the constant circular motion from which a sine wave is derived (360°).

As X_L increases, the amount of circuit current decreases because the inductive reactance acts like a variable resistance. That is, for any given frequency, the inductive reactance will have a different resistance in ohms. For this reason, two series inductors of the same value will have twice as much reactance (in ohms), behaving like two series resistors. Similarly, two parallel inductors of the same value will have half as much reactance as one. And, according to Ohm's law, the value of circuit current is

$$I_T = \frac{V_T}{X_L}$$

where X_L is in ohms.

When a voltage is induced in a coil, the current lags the induced voltage by 90°. This lag exists in time, as is best represented by Fig. 20-1.1.

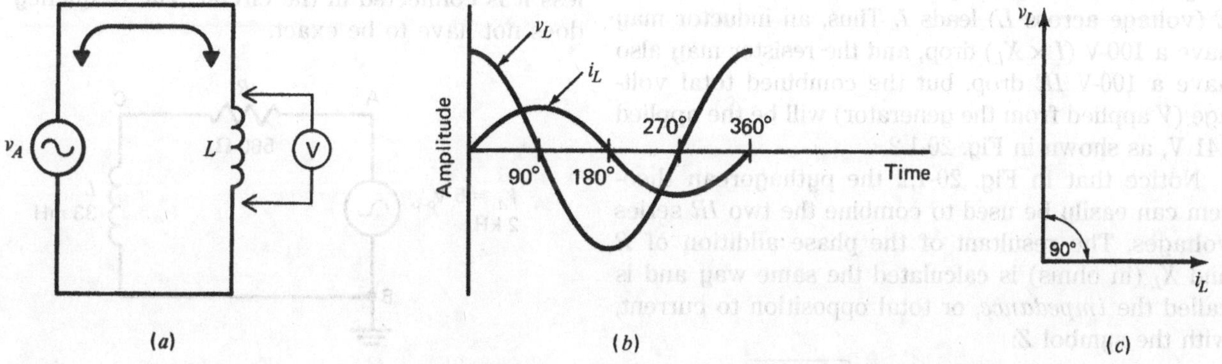

Fig. 20-1.1 Inductive circuit. (*a*) Purely inductive circuit. (*b*) Sine wave i_L lags v_L. (*c*) Phasor diagram.

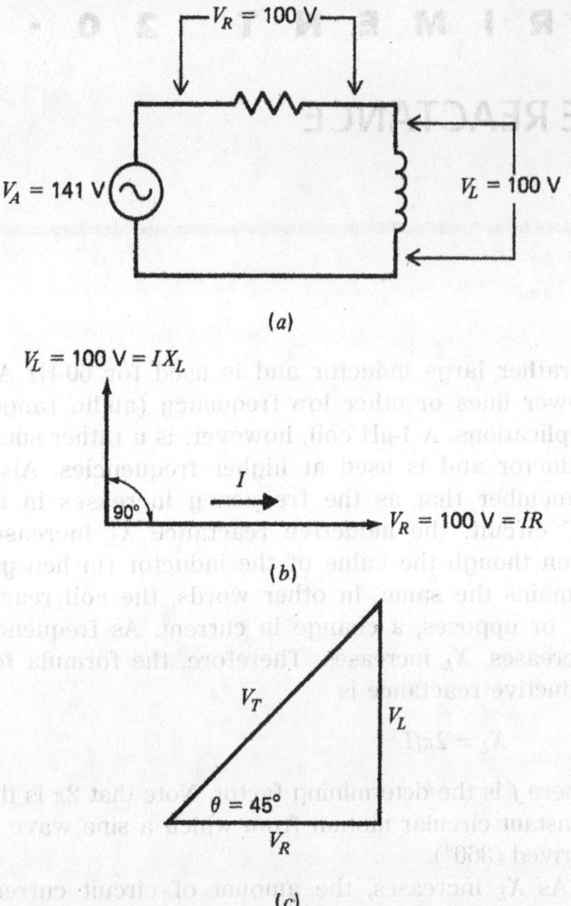

Fig 20-1.2 *RL* series circuit. (*a*) Series *RL* circuit with voltage drops shown. (*b*) Phasor diagram. (*c*) Resultant of two phasors = $V_T = \sqrt{V_R^2 + V_L^2} = 141$ V.

Although the current in the inductor lags by 90°, the value is still the same in all parts of a series circuit. The time lag exists only between the induced voltage and current, and is more theoretical than practical.

When a resistor and an inductor are in series with an AC source, the current is the same in all parts of the circuit, and the inductor and the resistor each have its own voltage drop (*IR*) calculated by Ohm's law. However, the voltage across the inductor leads the voltage across the resistor by 90° due to the lagging current. This is often referred to as *ELI*, where *E* (voltage across *L*) leads *I*. Thus, an inductor may have a 100-V ($I \times X_L$) drop, and the resistor may also have a 100-V *IR* drop, but the combined total voltage (*V* applied from the generator) will be the applied 141 V, as shown in Fig. 20-1.2.

Notice that in Fig. 20-1.2 the pythagorean theorem can easily be used to combine the two *IR* series voltages. The resultant of the phase addition of *R* and X_L (in ohms) is calculated the same way and is called the *impedance*, or total opposition to current, with the symbol *Z*:

$$Z = (R^2 + X_L^2)^{1/2} \text{ or } Z = \sqrt{R^2 + X_L^2}$$

Therefore, if $R = 1$ kΩ and $X_L = 1$ kΩ,

$$Z = (1000^2 + 1000^2)^{1/2} \text{ or } Z = \sqrt{1000^2 + 1000^2} = 1.414 \text{ k}\Omega$$

The phase angle θ (theta) of the complete circuit is calculated as the inverse tangent of X_L/R. In the case of Fig. 20-1.2,

$$\frac{X_L}{R_L} = \frac{1000}{1000} = 1 \text{ and } \tan^{-1} = 45°$$

At 45°, notice that X_L and *R* are equal in resistance and also in *IR* voltages.

The procedures that follow will allow you to validate all the concepts discussed above. However, remember that the effects of inductive reactance are sometimes difficult to comprehend in the same way that magnetism produces somewhat mysterious effects.

EQUIPMENT

Oscilloscope
Signal generator
Protoboard
Leads
DC power supply

COMPONENTS

(1) 560-Ω resistor
(1) 33-mH inductor

PROCEDURE

Note: Show all the calculations on a separate sheet, and label which procedure step they apply to.

1. Using an ohmmeter, measure the DC resistance of the inductor and record the value in Table 20-1.1. Remember, this is the resistance at 0 Hz.
2. Connect the circuit of Fig. 20-1.3.

Note: Adjust the signal generator to 5 $V_{p\text{-}p}$ by placing the oscilloscope across points A and B only after the entire circuit is connected. This is because the signal generator has its own internal resistance. Therefore, V_A can never be adjusted correctly unless it is connected in the circuit. The frequency does not have to be exact.

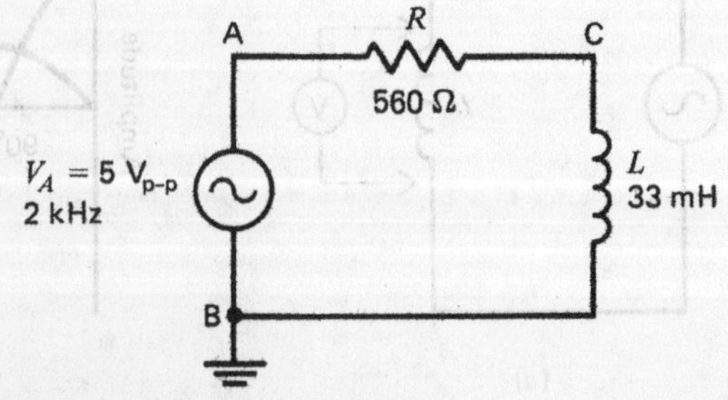

Fig 20-1.3 Series *RL* circuit.

3. Calculate X_L for this circuit as:

$$X_L = 2\pi fL$$

Record the value in Table 20-1.1.

4. Measure and record the peak-to-peak voltage across the inductor. Be sure the oscilloscope probe (+) is at point C.

5. Measure and record in Table 20-1.1 the peak-to-peak voltage across the resistor. Be sure to exchange the resistor's place with the coil so that the oscilloscope ground and the signal generator ground are at the same point.

6. Calculate the circuit current and record it in Table 20-1.1 by using the Ohm's law formula. Do this by using the measured voltages for both components.

Note: The values of I_T will be different because the measured values and the calculations are determined using different methods. Variations up to about 10 percent are common.

$$\text{(a) } I_T = \frac{V_R}{R} \qquad \text{(b) } I_T = \frac{V_L}{X_L}$$

7. Calculate and record in Table 20-1.1 the total circuit impedance, using two formulas:

$$\text{(a) } Z_T = (R^2 + X_L^2)^{1/2} \qquad \text{(b) } Z_T = \frac{V_G}{I_T}$$

8. Calculate and record in Table 20-1.1 the frequency as:

$$f = \frac{X_L}{2\pi L}$$

9. Calculate and record in Table 20-1.1 the inductances as:

$$L = \frac{X_L}{2\pi f}$$

10. Refer to Fig. 20-1.3. Readjust the frequency to 100 Hz. Increase the frequency in 100-Hz steps from 100 Hz to 1 kHz. Measure and record V_L at each frequency step in Table 20-1.2.

11. Beginning at 1 kHz, increase the frequency in 1-kHz steps from 1 kHz to 10 kHz, and measure and record V_L at each step in Table 20-1.2.

12. Calculate X_L for each frequency point in steps 10 and 11 above, and record the values in Table 20-1.2.

13. Plot a graph of frequency versus V_L and X_L for the data of steps 11 and 12 using semilog paper.

OPTIONAL PROCEDURES

Series Inductance

Connect two inductors of equal value in series, and repeat steps 1 to 9. Make your own data table (similar to Table 20-1.1). Or add a column on to Table 20-1.1, and label it as 2 series L.

Parallel Inductance

Connect two inductors of equal value in parallel, and repeat steps 1 to 9. Make your own data table (similar to Table 20-1.1). Or add a column on to Table 20-1.1, and label it as 2 parallel L.

NAME ______________________________ DATE __________

QUESTIONS FOR EXPERIMENT 20-1

Answer true (T) or false (F) to the following:

_____ **1.** As frequency increases, X_L also increases.

_____ **2.** The DC resistance of a coil is always 0 Ω.

_____ **3.** As frequency decreases, X_L also increases.

_____ **4.** Two parallel coils of equal value will have twice as much inductance as one coil.

_____ **5.** Two series coils of equal value will have half as much inductance.

_____ **6.** If, in a series *RL* circuit, $V_R = 2\ V_{p\text{-}p}$ and $V_L = 2\ V_{p\text{-}p}$, the applied voltage would equal $4\ V_{p\text{-}p}$.

_____ **7.** In a series *RL* circuit, the voltage across the resistor always lags the voltage across the inductor by 90°.

_____ **8.** The current in an *RL* series circuit is the same in all parts of the circuit.

_____ **9.** There is a phase relationship in an *RL* circuit because the resistor opposes a change in current.

_____ **10.** The phase angle of an *RL* series circuit, where *R* and X_L are equal, will always be 45°.

TABLES FOR EXPERIMENT 20-1

TABLE 20-1.1

Procedure Step	Circuit Component	Value
1	R_L Measured	________ Ω
3	X_L Calculated	________ Ω
4	V_L Measured	________ $V_{p\text{-}p}$
5	V_R Measured	________ $V_{p\text{-}p}$
6*a*	I_T Calculated	________ mA
6*b*	I_T Calculated	________ mA
7*a*	Z_T Calculated	________ Ω
7*b*	Z_T Calculated	________ Ω
8	*f* Calculated	________ kHz
9	*L* Calculated	________ H

TABLE 20-1.2

F	V_L	X_L
100 Hz		
200 Hz		
300 Hz		
400 Hz		
500 Hz		
600 Hz		
700 Hz		
800 Hz		
900 Hz		
1 kHz		
2 kHz		
3 kHz		
4 kHz		
5 kHz		
6 kHz		
7 kHz		
8 kHz		
9 kHz		
10 kHz		

EXPERIMENT RESULTS REPORT FORM

Experiment No.: ____________________ Name: ____________________

Date: ____________________

Experiment Title: ____________________ Class: ____________________

____________________ Instr: ____________________

Explain the purpose of the experiment:

List the first Learning Objective:
OBJECTIVE 1:

After reviewing the results, describe how the objective was validated by this experiment.

List the second Learning Objective:
OBJECTIVE 2:

After reviewing the results, describe how the objective was validated by this experiment.

List the third Learning Objective:
OBJECTIVE 3:

After reviewing the results, describe how the objective was validated by this experiment.

Conclusion:

If required, attach to this form: ☐ Answers to Questions, ☐ Tables, and ☐ Graphs.

CHAPTER 21

EXPERIMENT 21-1

INDUCTIVE CIRCUITS

LEARNING OBJECTIVES

At the completion of this experiment, you will be able to:

- Measure time and voltage in an inductive circuit.
- Calculate phase shift between signals.
- Determine the Q of an inductor.

SUGGESTED READING

Chapter 21, *Basic Electronics*, Grob/Schultz, twelfth edition

INTRODUCTION

Most inductors are combined with capacitors to create filters. However, inductors can also be used alone as chokes to keep an AC or rf signal out of some part of a circuit, usually the DC bias of an amplifier or a mixer. In this lab, you verify the use of an inductor for feeding DC to one part of a circuit while keeping the rf or AC signal out.

As you previously learned, an ideal inductor is considered a short to DC. Therefore, when an inductor is in a circuit, it will pass DC but will oppose AC. This opposition to AC is a variable impedance, depending upon the AC frequency and the amount of inductance L. Inductors have also some internal resistance that is usually very small compared to the reactance of the inductor. In most cases, as frequency increases, the reactance of the inductor is usually much greater than its resistance.

Because the inductor is a reactive component, the concept of phase angle becomes important, especially with high-frequency or microwave circuits. Therefore, keep in mind the principle that the AC voltage always leads current in the inductor (which is the same as the current always lagging the voltage). In both cases, the difference is always 90 degrees. This 90-degree phase angle is used to describe the constant delay in time between the peak voltage and the peak current in the inductor. This effect occurs because the inductor has many closely spaced turns or windings. However, regardless of the number of turns of the inductor or the frequency, the 90-degree phase difference stays constant between V and I in the inductor itself. For practical purposes current is not measured directly in an inductive circuit but can be calculated from the voltage and the reactance.

Another quality of an inductor is its Q. For any coil, this quality of merit of Q is calculated as X_L/R, where R is the internal DC resistance of the coil. Of course, the reactance X_L is dependent upon frequency. Therefore, the Q is valid only for a given

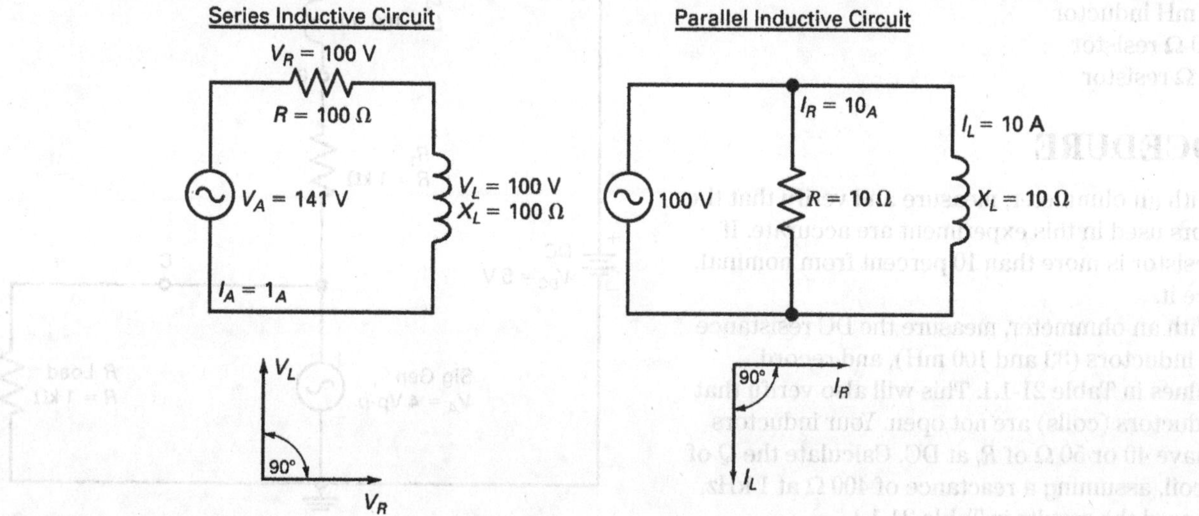

Fig. 21-1.1 Phasor diagram. Left, *RL* series circuit impedance. Right, parallel *RL* circuit response.

frequency. In general, the higher the Q, the greater the quality of the coil. Because Q is a ratio, less R has the most effect.

In review, keep in mind that when R and L are in series, the amount of current is the same in all parts of the circuit but the voltages across L and R are 90 degrees out of phase. This means that at the same point of time their voltages are different: When the resistor voltage is at its peak, the inductor voltage is at zero. To describe this, a phasor diagram is used (Fig. 21-1.1), where the two voltages V_L and V_R are shown on a right triangle. This is also used to describe the series circuit impedance Z of an RL series circuit. When L and R are in parallel, the 90-degree phase angle (shift in time) applies to the branch currents but the voltages are the same. Again, the right triangle and phasor diagram (Fig. 21-1.1) can be used to describe this parallel RL circuit response.

Note: **The ideal phase shift of 90 degrees between R and L is not a practical measurement because the scope is grounded and cannot easily be made to measure across the individual components R and L at the same time. For this reason, measurements in this lab will have phase shifts less than 90 degrees due to the combined RL effects.**

EQUIPMENT

DMM
Oscilloscope: dual-trace
Signal generator
DC supply or Signal Gen
DC offset
Protoboard
Leads

COMPONENTS

(1) 100-mH inductor
(1) 33-mH inductor
(2) 100-Ω resistor
(2) 1-kΩ resistor

PROCEDURE

1. With an ohmmeter, measure and verify that the resistors used in this experiment are accurate. If any resistor is more than 10 percent from nominal, replace it.

2. With an ohmmeter, measure the DC resistance of the inductors (33 and 100 mH), and record the values in Table 21-1.1. This will also verify that the inductors (coils) are not open. Your inductors may have 40 or 50 Ω of R_i at DC. Calculate the Q of each coil, assuming a reactance of 400 Ω at 1 kHz, and record the results in Table 21-1.1.

3. Connect the circuit of Fig. 21-1.2.

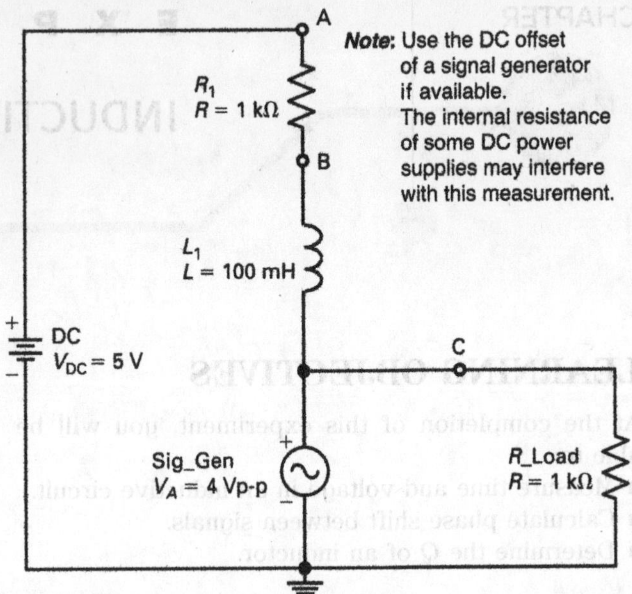

Fig. 21-1.2 Inductive circuit. Step 4.

4. Using a DC voltmeter, measure and record, in Table 21-1.2, the DC voltages to ground across bias resistor R_1 at point A, inductor L_1 at point B, and the 1-kΩ load resistor at point C.

5. Adjust the signal generator to each frequency step shown in Table 21-1.3, and measure the peak voltages (not p-p) to ground across bias resistor R_1 at point A, inductor L_1 at point B, and the load resistor at point C. Record the results in Table 21-1.3.

6. Exchange the position of the resistor R_1 and the inductor L_1 so that the circuit looks like the one shown here in Fig. 21-1.3. Repeat the previous steps for this new configuration, measuring and

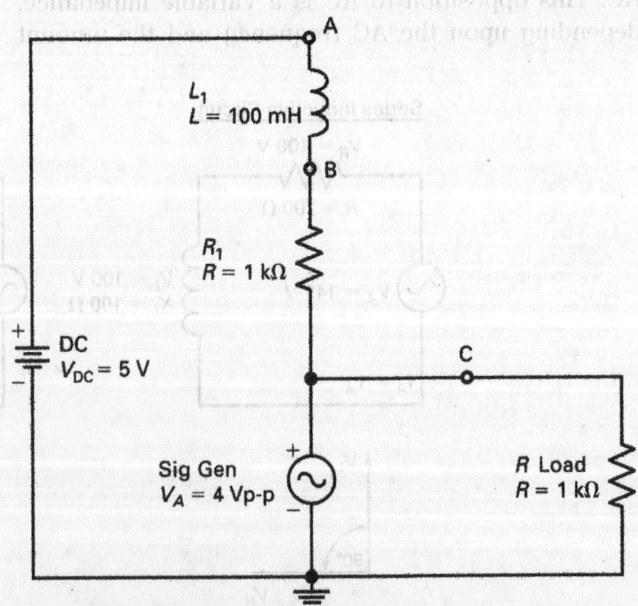

Fig. 21-1.3 Inductive circuit. Step 6.

recording the peak voltage at point C, across the 1-kΩ load resistor as you adjust the signal generator to 4 $V_{p\text{-}p}$ for each step. Refer to Tables 21-1.4 and 21-1.5 for recording results and varying the frequency.

7. Slowly turn down the DC power supply voltage from 5 to 0 volts, and look at the traces on the scope. Note any change that occurs to the AC signal, and be prepared to include this information in your report.

8. Connect the circuit of Fig. 21-1.4 shown here. The signal generator should be set to 200 Hz at 4 volts p-p (peak-to-peak). You will need to connect the oscilloscope (Channel 1) to the circuit to measure the applied signal to get the correct results. Remember that the signal generator has its own impedance that is in parallel with the circuit—so connect it first and then measure 4 $V_{p\text{-}p}$ across the circuit.

Note: The scope time base will be at about 1 ms per div on the x axis for 200 Hz, and you may have to adjust the triggering. The y axis for magnitude (p-p value) can be 0.5 volt/div for channel 1 (4 $V_{p\text{-}p}$) and 0.5 volt/div for channel 2.

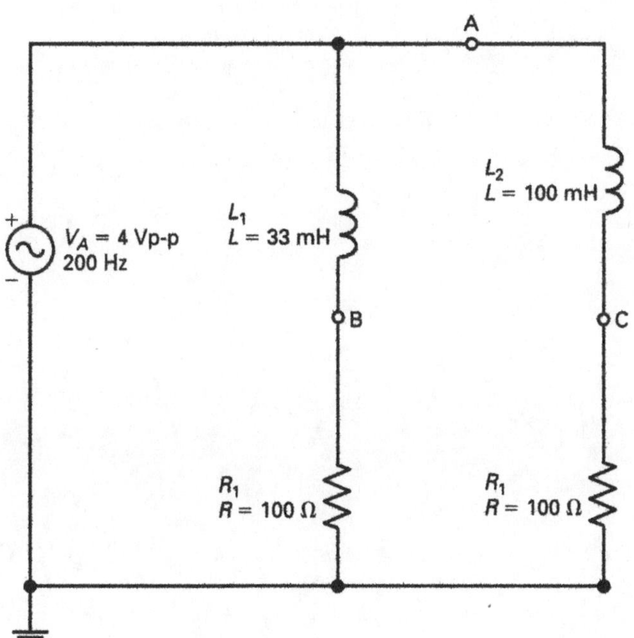

Fig. 21-1.4 Step 8.

9. With the dual-trace oscilloscope displaying about 2 periods each of the p-p signals from ground to point A (channel 1) and from ground to point C (channel 2), draw the two signals as best as you can on the oscilloscope graphs in Table 21-1.6. This will show their phase relationship or time on the x axis. On the graph, label the two signals: V_A is 4 $V_{p\text{-}p}$ and the other is V_C. Also, indicate the difference on the x axis. For example, if the difference is ½ division (between peaks), at 1 ms per division, then the difference in time is about 0.5 ms. Using that value, you can calculate the phase shift by using the example formula shown in Table 21-1.6.

10. Decrease the signal generator frequency to 100 Hz, and lower and notice the results on the oscilloscope between the phase of the two signals—adjusting the scope time base as necessary. Also, be sure to check the 4 $V_{p\text{-}p}$ and adjust it if necessary. Notice what happens and keep this in mind for answering the questions—there is no need to record the results.

11. Increase the signal generator to 1 kHz with 4 Vp-p applied—notice how you adjust the signal generator (up or down in amplitude) and adjust the scope to display about 2 periods of each signal. Draw the two signals on the graph, and indicate which signal is the input and which is point C to ground. Also include the amplitude of V_C and the difference in time between the two signals (x axis time) in Table 21-1.6.

12. Increase the signal generator to 10 kHz, adjusting the applied voltage. Again, in Table 21-1.6, record the traces on the graph with amplitude, time, and phase differences.

13. Return the signal generator to 1 kHz with 4 Vp-p applied. Now, connect the scope channel 2 probe to point B (to ground). Notice how the signal at point B differs from that at point C due (at 1 kHz). Move the scope probe back and forth between points B and C to verify the difference between inductor values (33 mH versus 100 mH). Use this observation in your report.

NAME DATE

QUESTIONS FOR EXPERIMENT 21-1

1. Why is the DC voltage across the load resistor always zero for this experiment?

2. Why is the AC voltage across the load resistor the same as the generator?

3. What would happen if the inductor value (100 mH) increased greatly, for example, to 1 H, in the first two circuits measured?

4. In the last circuit, what was the effect on the phase (time axis) of the two traces as the applied frequency was decreased?

5. In the last circuit, was the applied voltage (4 $V_{p\text{-}p}$) decreased or increased as you increased the frequency for the last steps? Explain why you think this was necessary.

6. In the last step, was the signal across the resistor (point B) greater or lesser for the branch with 33 mH, instead of 100 mH, and why?

TABLES FOR EXPERIMENT 21-1

TABLE 21-1.1 $X_L = 400\ \Omega$

Resistance and *Q* of 33-mH Inductor	Resistance and *Q* of 100-mH Inductor
$R =$ ______	$R =$ ______
$Q =$ ______	$Q =$ ______

Note: Calculate Q as X_K/R.

TABLE 21-1.2

Applied DC Voltage	DC Voltage Point B	DC Voltage Point C
5.0	______	

TABLE 21-1.3

Frequency	A: Peak Volts	B: Peak Volts	C: Peak Volts
100 Hz	______	______	______
300 Hz	______	______	______
500 Hz	______	______	______
700 Hz	______	______	______
1 kHz	______	______	______
3 kHz	______	______	______
5 kHz	______	______	______
7 kHz	______	______	______
9 kHz	______	______	______
10 kHz	______	______	______

TABLE 21-1.4

Applied DC Voltage	DC Voltage Point B	DC Voltage Point C
5.0	______	______

TABLE 21-1.5

Frequency	A: Peak Volts	B: Peak Volts	C: Peak Volts
100 Hz			
300 Hz			
500 Hz			
700 Hz			
1 kHz			
3 kHz			
5 kHz			
7 kHz			
9 kHz			
10 kHz			

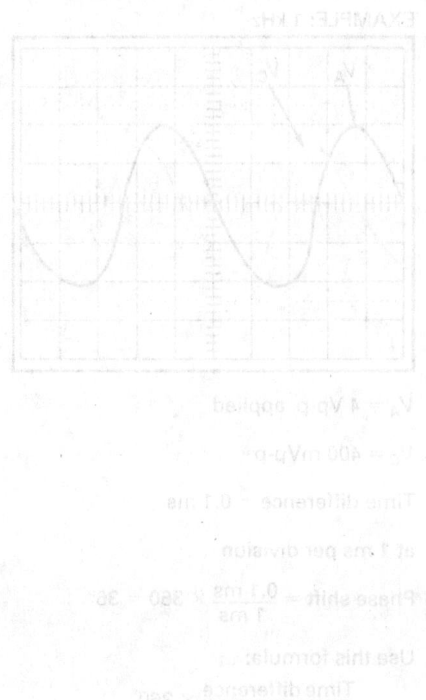

TABLE 21-1.6

STEP 9: 200 Hz

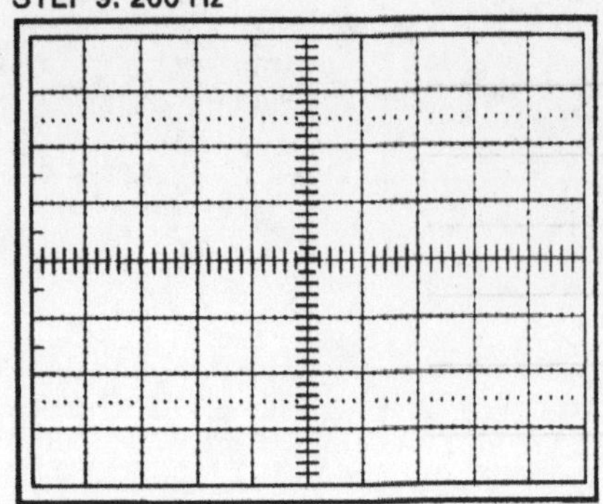

V_A = 4 Vp-p applied

V_C = ________ p-p

Time difference = ________

Phase shift = ________

STEP 11: 1 kHz

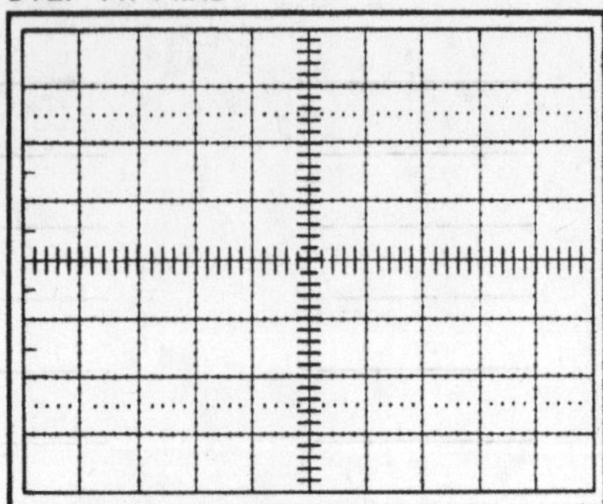

V_A = 4 Vp-p applied

V_C = ________ p-p

Time difference = ________

Phase shift = ________

STEP 12: 10 kHz

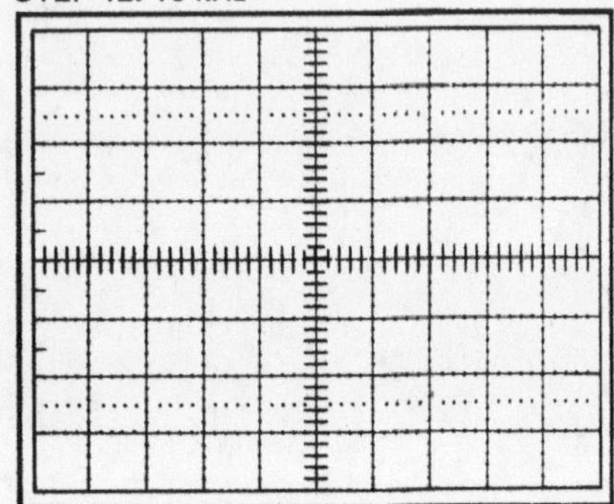

V_A = 4 Vp-p applied

V_C = ________ p-p

Time difference = ________

Phase shift = ________

EXAMPLE: 1 kHz

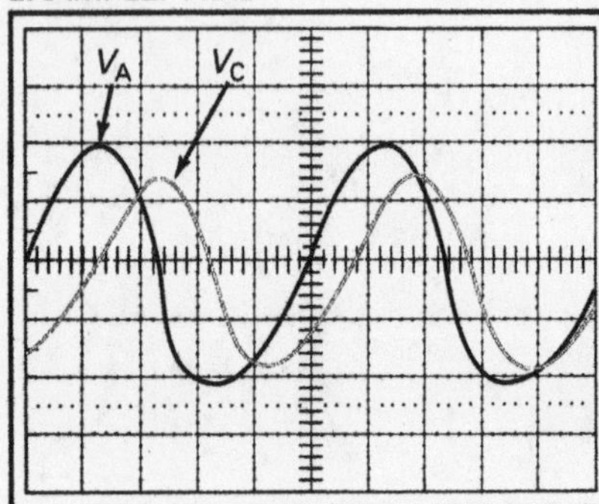

V_A = 4 Vp-p applied

V_C = 400 mVp-p

Time difference = 0.1 ms

at 1 ms per division

Phase shift $= \frac{0.1 \text{ ms}}{1 \text{ ms}} \times 360 = 36°$

Use this formula:

$$\frac{\text{Time difference}}{\text{Time for 1 cycle}} \times 360$$

EXPERIMENT RESULTS REPORT FORM

Experiment No.: ______________________ Name: ______________________

Date: ______________________

Experiment Title: ______________________ Class: ______________________

______________________ Instr: ______________________

Explain the purpose of the experiment:

List the first Learning Objective:

OBJECTIVE 1:

After reviewing the results, describe how the objective was validated by this experiment.

List the second Learning Objective:

OBJECTIVE 2:

After reviewing the results, describe how the objective was validated by this experiment.

List the third Learning Objective:

OBJECTIVE 3:

After reviewing the results, describe how the objective was validated by this experiment.

Conclusion:

If required, attach to this form: ☐ Answers to Questions, ☐ Tables, and ☐ Graphs.

CHAPTER 22

EXPERIMENT 22-1

RC TIME CONSTANT

LEARNING OBJECTIVES

At the completion of this experiment, you will be able to:

- Understand the concept $T = RC$.
- Validate the decaying discharge current in an RC series circuit.
- Plot a graph of V_c and I_c versus time for single, series, and parallel capacitance.

SUGGESTED READING

Chapters 18 and 22, *Basic Electronics*, Grob/Schultz, twelfth edition.

INTRODUCTION

A capacitor in series with a resistance will charge to 63.2 percent of the applied voltage in one time constant. This time constant T (in seconds) is applied to nonsinusoidal waveforms as a transient response, where R is the series resistance in ohms and C is the value of the capacitance in farads, or

$$T = RC$$

For example, the circuit of Fig. 22-1.1 has the following time constant:

$$\begin{aligned} T = RC &= (1{,}000)(4 \times 10^{-6}) \\ &= 4{,}000 \times 10^{-6} \\ &= 4 \times 10^{-3} \\ &= 4 \text{ ms} \end{aligned}$$

Notice that the applied voltage does not affect the value of T. The time constant T is the time for the voltage across C to change by 63.2 percent. If V_A in Fig. 22-1.1 were 100 V, then C would change to 63.2 V in 4 ms. In another 4 ms, the capacitor would change to 63.2 percent of the remaining voltage, or 63.2 percent of 36.8 V. This would repeat until the voltage across C equaled the applied voltage. Also, during this time, the circuit current would steadily decay until V_C reached its maximum voltage, or V_A.

Similarly, if the capacitor were discharging, RC would specify the time it takes C to discharge 63.2 percent of the way down, to the value equal to 36.8 percent of the initial voltage across C at the start of discharge. A capacitor is usually considered charged after approximately five time constants.

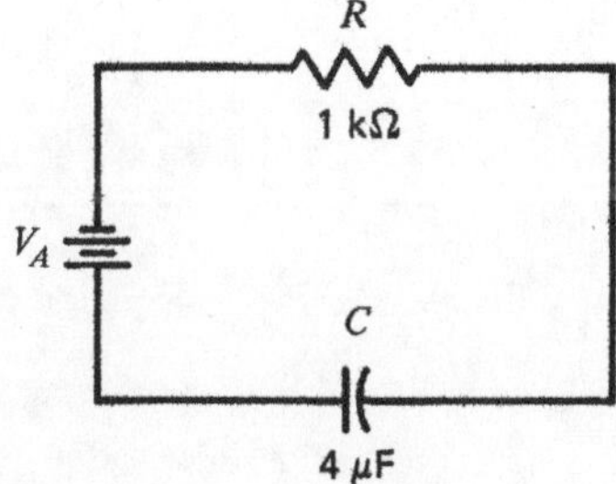

Fig. 22-1.1 *RC* time constant circuit.

EQUIPMENT

DC power supply
DMM with micro-ammeter capability
Leads and a Proto-board
Blank paper or tabular paper and pencil or pen

COMPONENTS

(2) 4 μF or similar capacitors
(1) 1.2 MΩ resistor

PROCEDURE

This experiment should be done by two people: one reading the value on the DMM and one recording it in a table. The readings and recordings occur every two seconds—this is fast. If you are alone, you can do it every 4 seconds and estimate the values in between for every 2 seconds, but it will be difficult and is not recommended.

1. Connect the circuit of Fig. 22-1.2. The switch S_1 can be either a lead or the power supply on/off switch. S_2 is a wire lead.

Caution: Do not turn power on yet.

2. Calculate and record in Table 22-1.1 the amount of circuit current I_T by using Ohm's law. Also, record the value of capacitance C (single capacitor value used).

3. Refer to Fig. 22-1.2, with S_2 closed; apply power (close S_1). Current is now flowing in the circuit. Therefore, measure the total DC current (capacitor is bypassed) and record the value in Table 22-1.2.

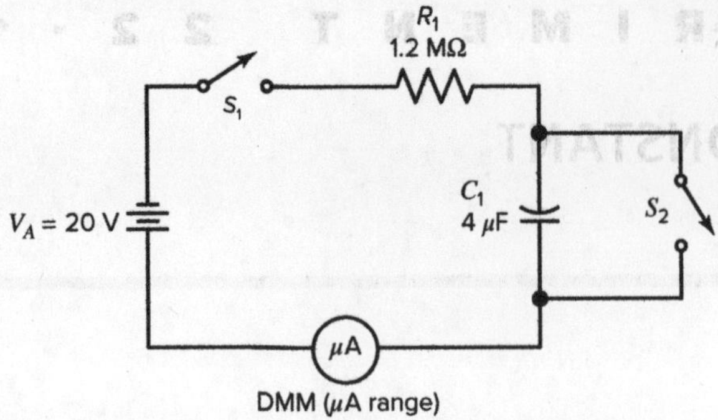

Fig. 22-1.2 *RC* series circuit for *RC* time constant measurement.

4. Discharge the capacitor: Turn off power (same as open S_1) and close S_2 (capacitor is shorted).

5. Now open S_2: the discharged capacitor is now in the circuit with no power applied. Watch your ammeter (DMM) and turn on power (S_1 closed). You should see the value of current decreasing as the capacitor charges in a few seconds it should decrease to zero. The amount of time it takes is the RC time constant.

For the next steps, you will record values of current and then voltage every 2 seconds for 30 seconds. The objective is to observe the current decaying as the capacitor charges and then to observe the voltage increasing—they are inversely related.

6. Now prepare to read and record the current in micro-amps on the meter every 2 seconds as current decays. You will need some blank paper and a partner: one person to call out the value every 2 seconds and another person to record the value. You can use the phrase "ready go" when you apply power at time zero by closing S_1 (keep S_2 open). And then call out the value every 2 seconds as your partner record the value in micro-amps. Be sure your meter is set for micro-amps. When you are ready, discharge the capacitor again and try a practice run.

7. Discharge the capacitor again and begin the measurement: record the value of current every two seconds from time zero to 30 seconds on a separate piece of paper (not the table yet). This is your first run of collecting data.

8. Repeat the procedure (previous step) one or two more times to verify your results—you can average the values—add them and divide by the number of trials. Record the final values of current for every 2 seconds in Table 22-1.1.

8. Connect a voltmeter (DMM) across C_1 and repeat the measurement for voltage across the capacitor every 2 seconds. Be sure to do it two or three times and average the values. Record the final values of increasing voltage across C_1 in Table 22-1.1.

9. Connect two capacitors in series, repeat steps 2 to 8, and record the values in Table 22-1.2.

10. Connect two capacitors in parallel, repeat steps 2 to 8, and record the values in Table 22-1.3.

Now you should have all the data you need.

11. Using the data from the first table (single capacitor), plot a graph of current I_c and voltage V_c. Draw it as best as you can—it will be similar to the graph shown in Fig. 22-1.3, which is a typical plot of voltage and current over time for and *RC* circuit. On your graph, be sure to label the axes values based on your data.

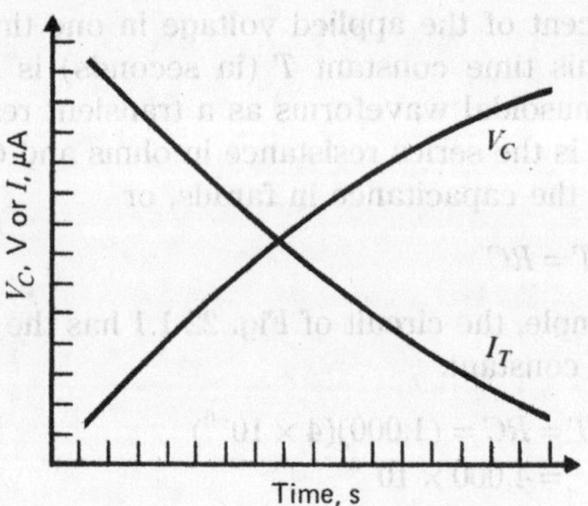

Fig. 22-1.3 Typical plot of time versus amplitude for *RC* time constant. Label both axes as necessary and indicate *RC* values in the title.

NAME ____ DATE ____________

QUESTIONS FOR EXPERIMENT 22-1

Answer true (T) or false (F) to the following:

_____ **1.** The *RC* time constant is expressed in units of farads.

_____ **2.** The formula for the *RC* time constant is $T = 1/RC$.

_____ **3.** The time constant T is the time for the voltage across R to change by 36.8 percent.

_____ **4.** The change of 63.2 percent is constant for all values of R and C as long as a sine wave is the applied voltage.

_____ **5.** For a 1,000-Ω R in series with an 18-μF C, the time constant would be 180 ms.

_____ **6.** The time constant formula is the same for a discharging capacitor.

_____ **7.** After approximately five time constants, a capacitor is considered charged to the applied DC voltage.

_____ **8.** After two time constants in an *RC* circuit where $R = 2$ kΩ and $C = 15$ μF, the capacitor would be charged to about 86 V with an applied voltage of 100 V.

_____ **9.** For question 8, if C were increased to 30 μF and R were increased to 4 kΩ, the capacitor would charge twice as fast.

_____ **10.** An *RC* charge or discharge curve usually has the same shape, regardless of the values of R and C.

TABLES FOR EXPERIMENT 22-1

TABLE 22-1.1 Single Capacitor—Current Decay and Voltage Charge

R_1 = 1.2 MΩ, total current I_t = _________ with 20 volts applied and C = _________ .

Time in Seconds	Current (μA)	Voltage (volts)
0		
2		
4		
6		
8		
10		
12		
14		
16		
18		
20		
22		
24		
26		
28		
30		

TABLE 22-1.2 Two Capacitors in Series—Current Decay and Voltage Charge

R_1 = 1.2 MΩ, total current I_t = _________ with 20 volts applied and C = _________ .

Time in Seconds	Current (μA)	Voltage (volts)
0		
2		
4		
6		
8		
10		
12		
14		
16		
18		
20		
22		
24		
26		
28		
30		

TABLE 22-1.3 Two Capacitors in Parallel—Current Decay and Voltage

Charge $R_1 = 1.2\ \text{M}\Omega$, total current I_t = ________ with 20 volts applied and C = ________.

Time in Seconds	Current (μA)	Voltage (volts)
0		
2		
4		
6		
8		
10		
12		
14		
16		
18		
20		
22		
24		
26		
28		
30		

EXPERIMENT RESULTS REPORT FORM

Experiment No.: ______	Name: ______
	Date: ______
Experiment Title: ______	Class: ______
______	Instr: ______

Explain the purpose of the experiment:

List the first Learning Objective:
OBJECTIVE 1:

After reviewing the results, describe how the objective was validated by this experiment.

List the second Learning Objective:
OBJECTIVE 2:

After reviewing the results, describe how the objective was validated by this experiment.

List the third Learning Objective:
OBJECTIVE 3:

After reviewing the results, describe how the objective was validated by this experiment.

Conclusion:

If required, attach to this form: ☐ Answers to Questions, ☐ Tables, and ☐ Graphs.

CHAPTER 23

EXPERIMENT 23-1

AC CIRCUITS: RLC SERIES

LEARNING OBJECTIVES

At the completion of this experiment, you will be able to:

- Understand series reactance and resistance.
- Determine the net reactance, phase angle, and impedance of an *RLC* circuit.
- Determine the real power of an AC circuit as opposed to apparent power.

SUGGESTED READING

Chapter 23, *Basic Electronics*, Grob/Schultz, twelfth edition

INTRODUCTION

Previous experiments covered *RL* series circuits and *RC* series circuits. In both of these AC circuits, the concept of reactance was investigated. The results showed that the reactive component had its own voltage drop that was out of phase with the resistive component, although the circuit current was the same in all parts of the circuit.

In an *RLC* series circuit, the same concepts apply. However, when X_L and X_C are both in the circuit, the opposite phase angles enable one to cancel the effect of the other. For X_L and X_C in series, the net reactance is the difference between the two series reactances, resulting in less reactance than either one alone. In parallel circuits, the branch currents cancel, resulting in less total current.

Consider the circuit of Fig. 23-1.1. Notice that the circuit current is found by dividing the applied voltage by the total net reactance of the circuit. Here,

$$\frac{120\text{ V}}{20\ \Omega} = 6\text{ A}$$

The net reactance is the difference between $X_L = 60\ \Omega$ and $X_C = 40\ \Omega$. In the same manner, the difference between the two voltages is equal to the applied voltage because the IX_L and IX_C voltages are opposite. If the values were reversed, the net reactance would be 20 Ω X_C. The current would still be 6 A, but it would have a lagging (−90°) instead of a leading (+90°) phase angle.

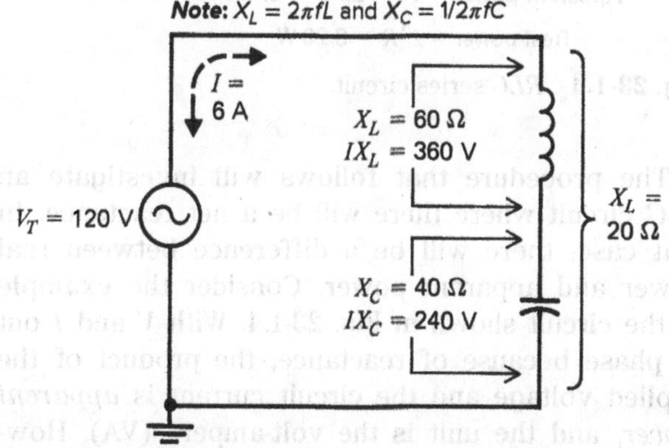

Fig. 23-1.1 *LC* circuit reactances. Net reactance = 60 Ω − 40 Ω = 20 Ω.

When resistance is added to the circuit, the total effect is determined by phasors. The phasor for the circuit of Fig. 23-1.1, if *R* were added, would be as shown in Fig. 23-1.2. Or, if X_L were greater, it would be as shown in Fig. 23-1.3.

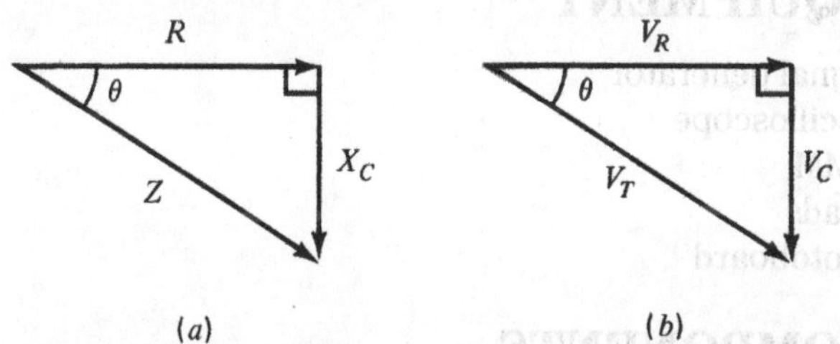

Fig. 23-1.2 Phasor diagrams for net capacitive reactance. (*a*) Impedance. (*b*) Voltage.

When X_L and X_C are equal, the net reactance is 0 Ω, and the result is called *resonance*. Because resonance has a specific application, it will be studied separately in later experiments.

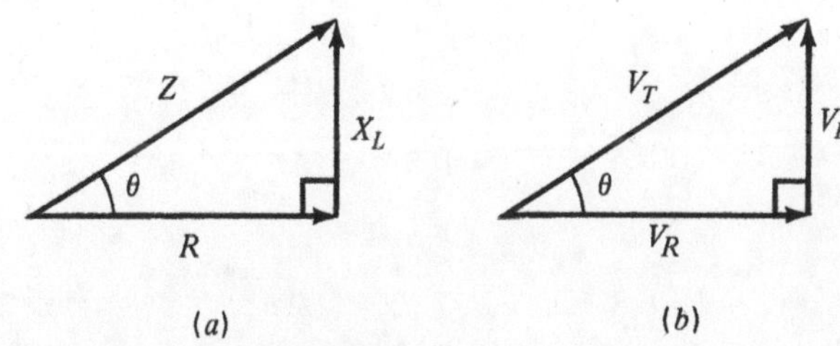

Fig. 23-1.3 Phasor diagrams for net inductive reactance. (*a*) Impedance. (*b*) Voltage.

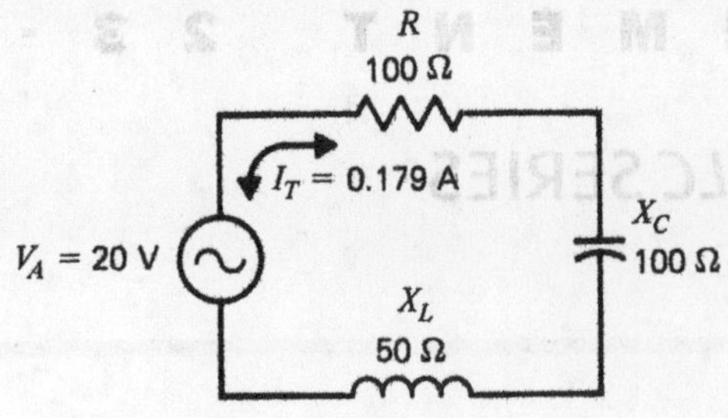

Fig. 23-1.4 *RLC* **series circuit.**

The procedure that follows will investigate an *RLC* circuit where there will be a net reactance. In that case, there will be a difference between real power and apparent power. Consider the example of the circuit shown in Fig. 23-1.4. With *V* and *I* out of phase because of reactance, the product of the applied voltage and the circuit current is *apparent power,* and the unit is the volt-ampere (VA). However, the *real power* is the resistive power, dissipated as heat, and can always be found as I^2R. Finally, the ratio of real power over apparent power is called the *power factor* and is equal to the cosine of the phase angle of the circuit. The power factor is used commercially to bring about efficient distribution of energy in power lines.

EQUIPMENT

Signal generator
Oscilloscope
DMM
Leads
Protoboard

COMPONENTS

(1) 1-kΩ, 0.25-W resistor
(1) 0.01-μF capacitor
(1) 33-mH inductor

PROCEDURE

1. Connect the circuit of Fig. 23-1.5.
2. Use an oscilloscope. Measure and record in Table 23-1.1 the peak-to-peak voltage across the capacitor (V_C), the inductor (V_L), and the resistor (V_R).

Note: Avoid ground loops when measuring voltage with the oscilloscope by moving the resistor and inductor to keep the same ground point.

3. Calculate and record X_L and X_C in Table 23-1.1.
4. Calculate and record the circuit current I_T for V_R/R, V_C/X_C, and V_L/X_L.
5. Calculate and record in Table 23-1.1 the net reactance and the impedance *Z*. Also, draw a phasor diagram for these values and determine the phase angle.
6. Determine the apparent power, real power, and power factor and record in Table 23-1.1.
7. Increase the frequency to 10 kHz and repeat steps 1 to 6. Record in Table 23-1.1.

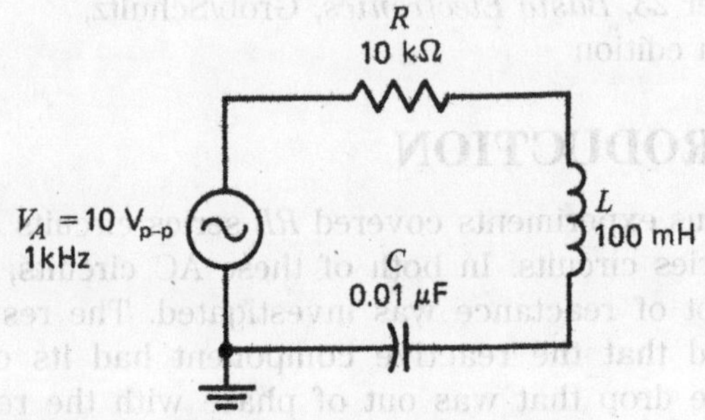

Fig. 23-1.5 *RLC* **series circuit for measuring net reactance.**

OPTIONAL PROCEDURE

Use different values of *R*, *L*, and *C*. Record the values and repeat steps 1 to 7. Also, try different frequencies: 5 kHz, 100 kHz, and 1 MHz. Use a separate sheet to record and discuss the results.

NAME DATE

QUESTIONS FOR EXPERIMENT 23-1

1. In an *RLC* series circuit, why can an inductor have more measured voltage than the applied voltage?

2. If X_L and X_C were zero, what would the phase angle of Fig. 23-1.5 be?

3. Explain the difference between real power and apparent power.

4. If an *RLC* series circuit had four inductors, five resistors, and seven capacitors, all in series, would the value of current be the same in all parts of the circuit?

5. When would the circuit of Fig. 23-1.5 (as shown) have the opposite net reactance?

TABLE FOR EXPERIMENT 23-1

TABLE 23-1.1

Procedure Step	Circuit Component	Steps 2–6: Value at $f = 1$ kHz	Step 7: Value at $f = 10$ kHz
2	V_C measured	________	________
	V_L measured	________	________
	V_R measured	________	________
3	X_L calculated	________	________
	X_C calculated	________	________
4	$I_T = V_R/R$	________	________
	$I_T = V_C/X_C$	________	________
	$I_T = V_L/X_L$	________	________
5	X_O calculated (net reactance)	________	________
	Z calculated		
6	Apparent power	________	________
	Real power	________	________
	Power factor	________	________

5 Draw phasor diagram here:

EXPERIMENT RESULTS REPORT FORM

Experiment No.: ____________________ Name: ____________________
Date: ____________________
Experiment Title: ____________________ Class: ____________________
____________________ Instr: ____________________

Explain the purpose of the experiment:

List the first Learning Objective:
OBJECTIVE 1:

After reviewing the results, describe how the objective was validated by this experiment.

List the second Learning Objective:
OBJECTIVE 2:

After reviewing the results, describe how the objective was validated by this experiment.

List the third Learning Objective:
OBJECTIVE 3:

After reviewing the results, describe how the objective was validated by this experiment.

Conclusion:

If required, attach to this form: ☐ Answers to Questions, ☐ Tables, and ☐ Graphs.

CHAPTER 24

EXPERIMENT 24-1

COMPLEX NUMBERS FOR AC CIRCUITS

LEARNING OBJECTIVES

At the completion of this experiment, you will be able to:

- Use complex numbers to describe circuit impedances.
- Measure and record complex data.
- Calculate the complex impedance of measured circuits.

SUGGESTED READING

Chapter 24, *Basic Electronics*, Grob/Schultz, twelfth edition

INTRODUCTION

Complex numbers are most often used to describe the impedance (Z) of a circuit. Consider an amplifier for example, as shown in Fig. 24-1.1. It may have both resistance and reactance to the input signal that is to be amplified over a specific frequency range. The AC signal is also coming from the output of another stage or circuit, such as a preamp. Therefore, to achieve maximum power transfer, the output impedance of one stage must be matched to the input of the next stage. To do this, you need to determine the amplifier's input impedance (resistance and reactance) so that it can be modified to match the output impedance of the previous stage. Therefore, the use of complex numbers becomes an important way to describe Z and obtain a match.

In general, many rf circuits are designed for off-the-shelf use to have a standard value of impedance:

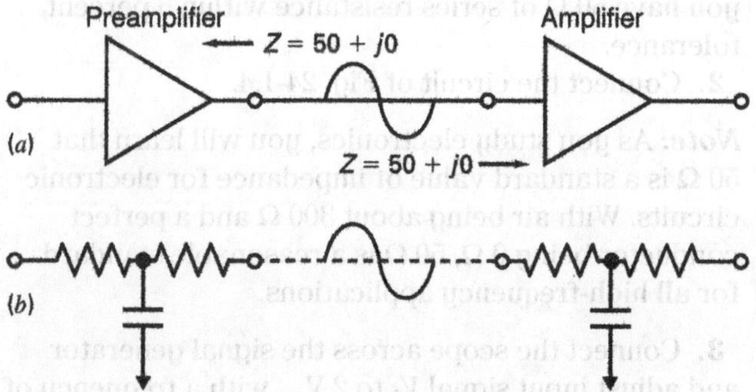

Fig. 24-1.1 Amplifier circuit; (*b*) equivalent circuit.

$50 + j0$. Of course, there is some tolerance. For this reason, you will see test equipment often made with 50 Ω of impedance.

The complex number describes the amount of resistance (real part) and the amount of reactance (imaginary part) due to X_L or X_C in the circuit. The reactance or imaginary part is referred to as the j operator and is often plotted on a polar plot. By describing the impedance Z as a complex number, the test technician can work with the circuit designer to achieve a better match from one output to the input of another stage.

In rf and microwave circuits, the measurement of circuit impedance is usually done with a ***network analyzer***. The network analyzer sends a signal into a circuit and determines the amount of signal that is reflected from the input. Thus, by knowing the input signal and the reflected signal, the impedance can be calculated automatically and plotted on the display. The network analyzer can also measure transmission to determine the amount of signal passing through (gain or loss) a circuit compared to the injected signal. However, this type of expensive test equipment is usually reserved for high-frequency (microwave) study and not often used in the first year of electronics for technicians. But it is possible to measure circuits that represent impedances and then describe them using complex numbers.

As shown in Fig. 24-1.2, the impedance of a circuit can be plotted as a phasor or put on a polar plot.

Using complex numbers to describe impedance has this advantage: The real part from the resistance stays the same regardless of the frequency applied. However, the reactive part of $+j$ or $-j$ Ω is specific to a given frequency. Also, remember that an impedance of $25 - j100$ Ω, for example, is not meaningful unless the frequency is specified. This is because the reactive component (j) is calculated from X_L or X_C.

The circuit of Fig. 24-1.3 shows a mismatch between two circuits. With two different impedances, the AC signal will not transfer with maximum power from one circuit stage to the next, unless the impedances are matched exactly equal (real and imaginary). In addition, there will always be a phase shift between the real and imaginary parts unless there is no reactance at all.

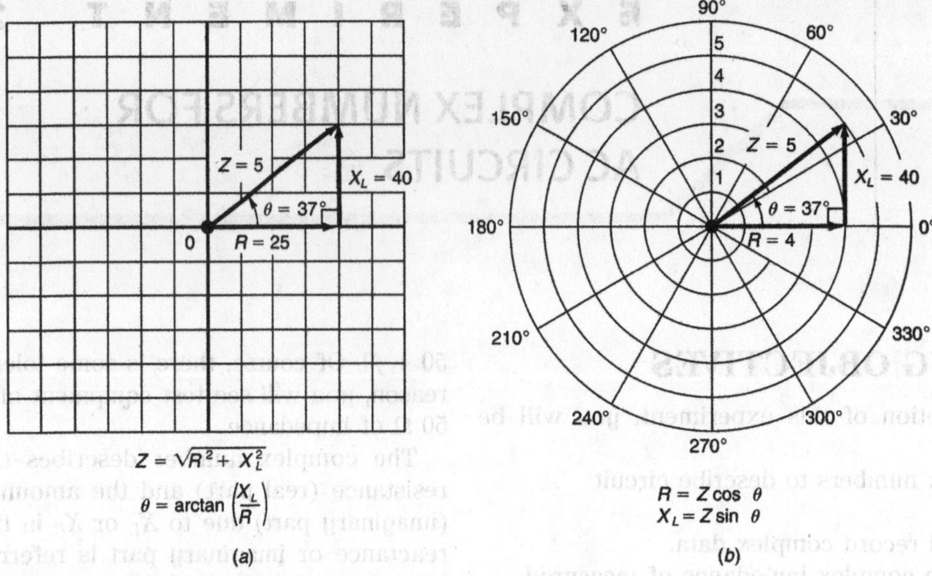

Fig. 24-1.2 Impedance plotted as a phasor (left) and as a polar plot (right).

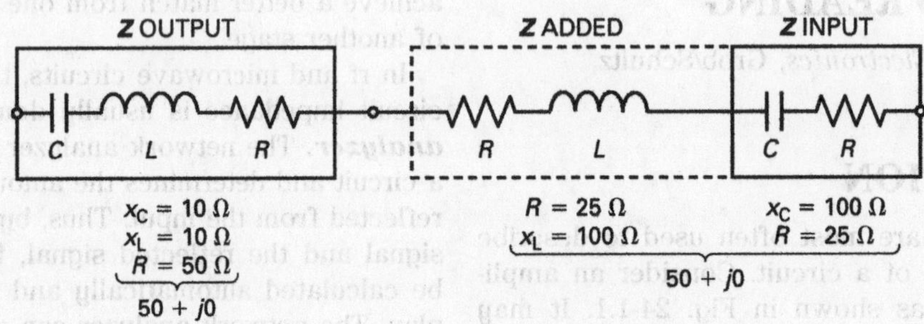

Fig. 24-1.3 A mismatch between two circuits. Solution in the text below.

For the circuit with the mismatch (*Z* input), a resistance of 25 Ω is added to the input; along with an inductor that had 100 Ω of inductive reactance at the frequency of interest to cancel out the 100 Ω of capacitive reactance. Also, keep in mind that when *L* and *C* are in series or parallel, they can also resonate at some frequency so this is another factor that must be considered.

For this experiment, without using a network analyzer, the use of complex numbers to describe circuit impedance will be obtained from measuring voltages and then calculating reactance and writing the complex number to describe the circuits. Afterward, the circuit will be modified to achieve the desired impedance.

EQUIPMENT

DMM
Dual-trace oscilloscope
Breadboard
Signal generator
Leads as needed

COMPONENTS

(1) 50-Ω resistor or (2) 100-Ω resistors connected in parallel
(1) 1-μF capacitor
(1) 33-mH inductor

PROCEDURE

1. With an ohmmeter, measure the 50-Ω resistor or the two 100-Ω resistors in parallel to be sure you have 50 Ω of series resistance within 5 percent tolerance.

2. Connect the circuit of Fig. 24-1.4.

Note: As you study electronics, you will learn that 50 Ω is a standard value of impedance for electronic circuits. With air being about 300 Ω and a perfect conductor being 0 Ω, 50 Ω is a reasonable standard for all high-frequency applications.

3. Connect the scope across the signal generator and adjust input signal V_A to 2 $V_{p\text{-}p}$ with a frequency of 3 kHz the input.

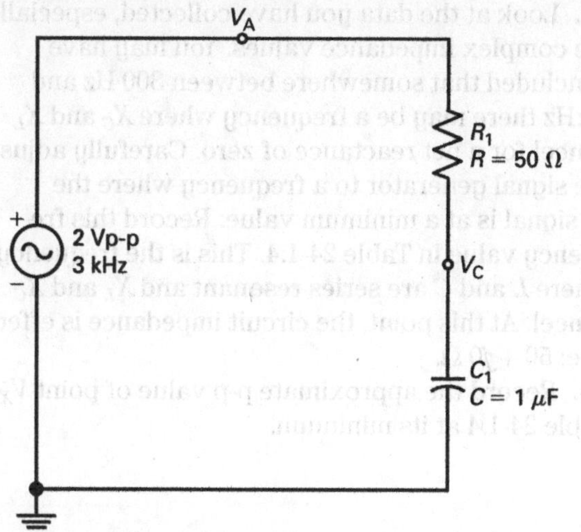

Fig. 24-1.4 Series *RC*. Step 2.

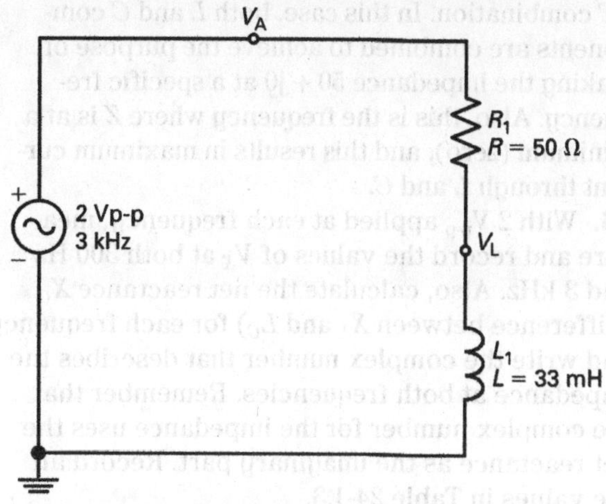

Fig. 24-1.5 Series *RL*. Step 13.

4. Measure the p-p signal voltage across the capacitor at V_C and record the value in Table 24-1.1.

5. Calculate the reactance of the 1-μF capacitor at 3 kHz, and record the value in Table 24-1.1. Use the formula $X_C = 1/(6.28 \times 3\ \text{kHz} \times 1\ \mu\text{F})$.

6. Knowing the R and X_C values, record the complex impedance Z in Table 24-1.1.

7. Look at the two signals V_C and V_A on the scope, and approximate the difference in time between the V_A peak and the V_C peak. For example, if the scope is set to 0.1 ms per division (that is the same as 100 μs), and the peaks are three divisions apart, then the difference in time would be about 300 microseconds. Record the value between peaks in microseconds in Table 24-1.1 as delta time.

8. Notice which peak, V_A or V_C, is leading or lagging, and note this on a separate sheet of paper—even drawing it if you want. You will use this information to answer the questions for this lab.

9. Reduce the frequency of the signal generator to 300 Hz, adjusting the applied voltage from the signal generator to 2 $V_{p\text{-}p}$ input.

10. Measure and record the p-p signal of V_C in Table 24-1.1.

11. Calculate X_C for the signal at 300 Hz and record the complex impedance.

12. Approximate the distance in time between peaks V_A and V_C and record that value also—you may see very little change but try to record it anyway. You will use this information in the report. Also, you may want to move the traces up and down or take the scope out of CAL to line up the peaks for a better view. You may be adjusting the time base also. Always remember to check your scope, signal generator, and circuit connections carefully at each step and reset them as needed. Do not make the mistake of assuming the scope is correctly set—always check your settings before recording a measured value.

13. Connect the *R-L* circuit of Fig. 24-1.5 by replacing the capacitor with a 33-mH inductor.

14. Repeat the steps for this circuit by measuring and recording V_L, calculating X_L and writing the complex impedance, and also approximating and recording the distance between peaks V_L and V_A at both 3 kHz and 300 Hz, according to Table 24-1.2.

Note: At this point in the experiment, you should see that the complex numbers for the impedance of each of the two circuits have some similarities and some differences. Consider these concepts and the information in the introduction for your report and also for the next steps.

15. Connect the circuit of Fig. 24-1.6, where both L and C are now in series with 50 Ω of resistance. Notice that point V_X is across the combined series

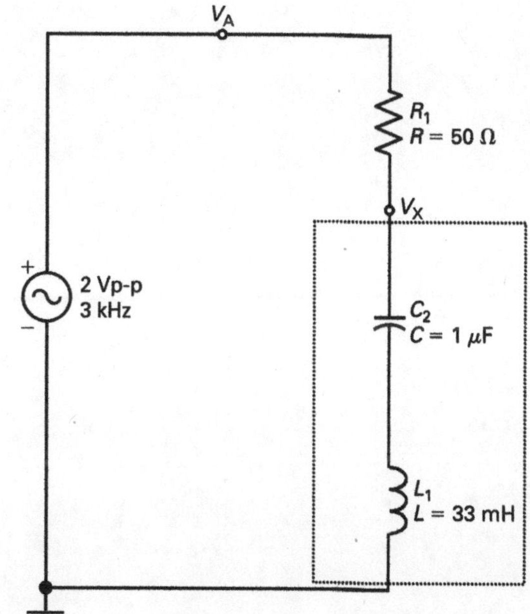

Fig. 24-1.6 Series *RLC* complex *Z* circuit. Step 15.

LC combination. In this case, both L and C components are combined to achieve the purpose of making the impedance 50 + j0 at a specific frequency. Also, this is the frequency where Z is at a minimum (zero), and this results in maximum current through L and C.

16. With 2 $V_{p\text{-}p}$ applied at each frequency, measure and record the values of V_X at both 300 Hz and 3 kHz. Also, calculate the net reactance X_O (difference between X_L and L_C) for each frequency, and write the complex number that describes the impedance at both frequencies. Remember that the complex number for the impedance uses the net reactance as the imaginary part. Record all the values in Table 24-1.3.

17. Look at the data you have collected, especially the complex impedance values. You may have concluded that somewhere between 300 Hz and 3 kHz there may be a frequency where X_C and X_L cancel for a net reactance of zero. Carefully adjust the signal generator to a frequency where the V_X signal is at a minimum value. Record this frequency value in Table 24-1.4. This is the frequency where L and C are series resonant and X_L and X_C cancel. At this point, the circuit impedance is effective: 50 + j0 Ω.

18. Record the approximate p-p value of point V_X in Table 24-1.4 at its minimum.

NAME DATE

QUESTIONS FOR EXPERIMENT 24-1

1. **In the circuit *RC* of Fig. 24-1.4, which signal was leading and which was lagging in time? Explain your answer, and remember that the oscilloscope x axis is in the time domain. Therefore, a signal peak that appears first (starting from the left or earlier time point) is really the leading signal—the one that follows later in time is lagging.**

2. **In the circuit of Fig. 24-1.5, which signal was leading and which was lagging in time?**

3. **Explain the differences between the measured signals at V_A and either V_C or V_L in the *RC* and *RL* circuits.**

4. **In the last circuit (Fig. 24-1.6), why do you think the p-p values of V_X were equal to V_A? Use your data to explain this. *Hint:* Compare the reactance to the 50 Ω of resistance.**

TABLES FOR EXPERIMENT 24-1

TABLE 24-1.1 *RC* Complex Impedance Circuit (Fig. 24-1.4)

Frequency	V_A p-p	V_C p-p	X_C	Delta Time	Complex *Z*
3 kHz	2 $V_{p\text{-}p}$				
300 Hz	2 $V_{p\text{-}p}$				

TABLE 24-1.2 *RL* Complex Impedance Circuit (Fig. 24-1.5)

Frequency	V_A p-p	V_L p-p	X_L	Delta Time	Complex *Z*
3 kHz	2 $V_{p\text{-}p}$				
300 Hz	2 $V_{p\text{-}p}$				

TABLE 24-1.3 *RLC* Complex Impedance Circuit (Fig. 24-1.6)

Frequency	V_A p-p	V_X p-p	X_O	Delta Time	Complex *Z*
3 kHz	2 $V_{p\text{-}p}$				
300 Hz	2 $V_{p\text{-}p}$				

Note: X_O is the net reactance.

TABLE 24-1.4 Minimal *Z* Values (Fig. 24-1.6)

Frequency (within 10 Hz) for $50 + j0$	Approx p-p value of V_X

EXPERIMENT RESULTS REPORT FORM

Experiment No.: ______________________ Name: ______________________

Date: ______________________

Experiment Title: ______________________ Class: ______________________

______________________ Instr: ______________________

Explain the purpose of the experiment:

List the first Learning Objective:

OBJECTIVE 1:

After reviewing the results, describe how the objective was validated by this experiment.

List the second Learning Objective:

OBJECTIVE 2:

After reviewing the results, describe how the objective was validated by this experiment.

List the third Learning Objective:

OBJECTIVE 3:

After reviewing the results, describe how the objective was validated by this experiment.

Conclusion:

If required, attach to this form: ☐ Answers to Questions, ☐ Tables, and ☐ Graphs.

CHAPTER 25

EXPERIMENT 25-1

SERIES RESONANCE

LEARNING OBJECTIVES

At the completion of this experiment, you will be able to:

- Understand the concepts of series resonance.
- Validate the formula for the resonant frequency $f_r = 1/[2\pi(LC)^{1/2}]$.
- Plot a graph of frequency versus circuit current.

SUGGESTED READING

Chapter 25, *Basic Electronics*, Grob/Schultz, twelfth edition

INTRODUCTION

The resonant effect takes place when $X_L = X_C$ for a series inductor and capacitor. The frequency at which the opposite reactances are equal is called the *resonant frequency*, and it can be calculated as:

$$f_r = \frac{1}{2\pi(LC)^{1/2}} \text{ or } f_r = \frac{1}{2\pi\sqrt{LC}}$$

Because of the canceling effect of X_L and X_C (opposite in phase), the resonant effect can occur only at one specific frequency for a given combination of L and C. In general, large values of L and C provide a relatively low resonant frequency. Smaller values of L and C provide a higher value for f_r. Figure 25-1.1 shows (*a*) a series resonant circuit and (*b*) its response curve.

Notice that X_L and X_C are equal at 1,000 kHz. This can be proven by using the formulas for $X_C = 1/(2\pi fC)$ and $X_L = 2\pi fL$. If the frequency of the signal generator (V_T) were to change, the circuit would no longer be resonant. Although the opposite reactances (180° out of phase) would still give a canceling effect, there would still be some net reactance remaining: X_C for a decrease in frequency and X_L for an increase in frequency. In Fig. 25-1.1, the only opposition to current flow is the resistance r_S, which is the resistance of the coil.

The main characteristic of a series resonant circuit is that the circuit current is maximum at the resonant frequency, as shown in Fig. 25-1.1*b*. This is called a *resonant rise* in current. Also, because of the canceling of X_C and X_L, the circuit impedance is minimum at the resonant frequency. And the circuit current is in phase with the generator voltage, meaning that the circuit phase angle is 0° at resonance.

Finally, because the circuit current is maximum, the voltage across either L or C is maximum at resonance. Thus, if the output of Fig. 25-1.1 were taken across either L or C, the result would be maximum voltage.

In addition, the quality of a resonant circuit is determined by the sharpness of the resonant rise in voltage across L or C, called Q, where Q is calculated as X_L/r_S. The greater the ratio of the reactance to the series resistance, the higher the Q and the sharper the resonant effect. Note that the X_L reactance,

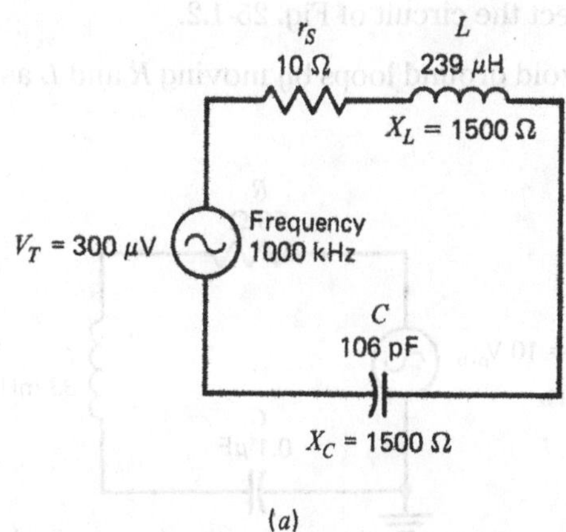

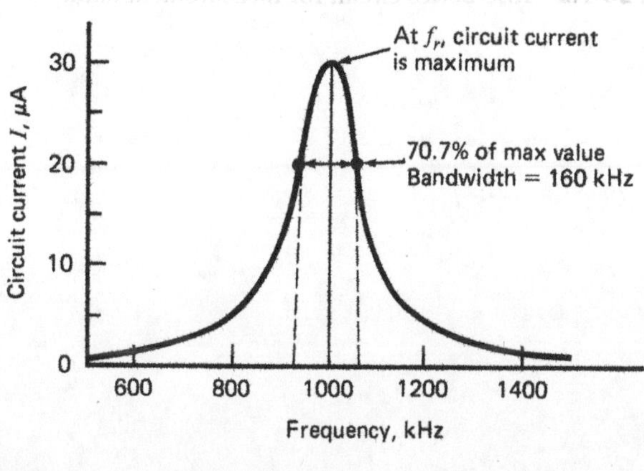

Fig. 25-1.1 Series resonant circuit. (*a*) Schematic diagram. (*b*) Response curve.

rather than X_C, is used to determine Q, due to the DC resistance of the coil. Also, the greater the L/C ratio, the greater the circuit Q. Thus, increasing L and decreasing C can produce a higher Q for any given resonant frequency. Q can also be measured as $Q = V_{in}/V_{out}$, where V_{out} is equal to the voltage across either L or C and V_{in} equals the generator voltage.

Remember, the main purpose of resonant circuits is for tuning, that is, to tune to a desired frequency, as in radio, television, and other forms of communication instrumentation.

EQUIPMENT

Signal generator
Oscilloscope
Protoboard
Leads

COMPONENTS

(1) 56-Ω, 0.25-W resistor
(1) 0.1-μF capacitor
(1) 33-mH inductor

PROCEDURE

1. Connect the circuit of Fig. 25-1.2.

Note: Avoid ground loops by moving R and L as required.

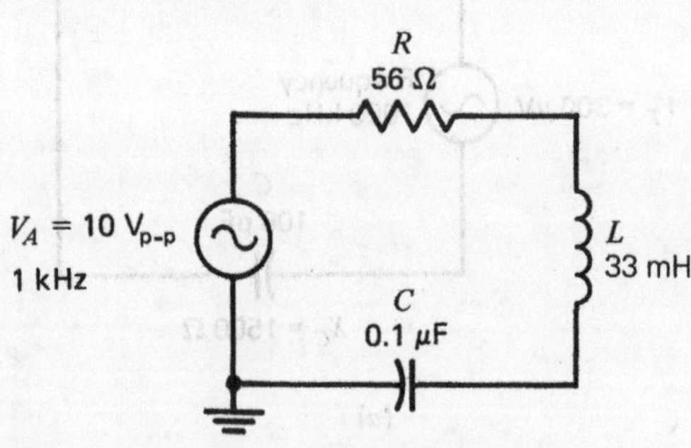

Fig. 25-1.2 *RLC* series circuit for measurement data.

2. Calculate and record the resonant frequency in Table 25-1.1.
3. Measure the voltages across R, L, and C, using an oscilloscope, and record the results in Table 25-1.1.

Note: For steps 4 through 8, create your own data table and graph the results in a response curve.

4. Increase the frequency in 250-Hz steps from 1 to 5 kHz and repeat step 3 at each frequency. Be sure to keep the applied voltage constant at each step. Thus, measure V_A at each step to be sure there is no change.

Note: Knowing the calculated resonant frequency, measure and record V_R, V_L, and V_C at several (3 or 4) extra frequency points on either side of that frequency.

5. Calculate the Q of the circuit. Now determine Q by measurement, and compare both values.
6. Calculate and record the circuit current for each frequency step as $I = V_R/R$.
7. Calculate and record the values of X_L and X_C at the two edges of the bandwidth. This is the same as the two half-power points on either side of the resonant frequency (f_r), where the voltage decreases to 70.7 percent.
8. Calculate and record the circuit impedance Z and the phase angle at f_r and at both half-power points.
9. Plot a graph of frequency versus the voltages V_R, V_L, and V_C. Indicate the bandwidth (half-power points on either side of f_r), where each half-power point is 70.7 percent of I_{max}.

NAME ______________________________ DATE ______________

QUESTIONS FOR EXPERIMENT 25-1

Answer true (T) or false (F) to the following:

______ **1.** The resonant effect occurs when X_L is greater than X_C.

______ **2.** The resonant frequency of a series L_C circuit must always be above 60 Hz.

______ **3.** The circuit impedance Z of a series circuit is maximum at f_r.

______ **4.** X_L and X_C are 90° out of phase at resonance.

______ **5.** The highest value of series circuit current is at resonance.

______ **6.** The greater the circuit Q, the higher the resonant frequency.

______ **7.** $f_r = 1/[2\pi(LC)^{1/2}]$ and $Q = X_L/X_C$.

______ **8.** The resonant frequency of a series RLC circuit with $R = 100\ \Omega$, $L = 8$ H, and $C = 4\ \mu F$ is 28 Hz.

______ **9.** The Q of the circuit values in question 8 is 14.

______ **10.** The resonant frequency of the circuit values of question 8 would not change if $L = 4$ H and $C = 8\ \mu F$.

TABLE FOR EXPERIMENT 25-1

TABLE 25-1.1

Procedure Step	Measurement	Measured Value	Calculation
2	f_r	______ Hz	______ Hz
3	V_R	______ V at 1 kHz	
	V_L	______ V at 1 kHz	
	V_C	______ V at 1 kHz	

GRAPH FOR EXPERIMENT 25-1

EXPERIMENT RESULTS REPORT FORM

Experiment No.: ______________________ Name: ______________________

Date: ______________________

Experiment Title: ______________________ Class: ______________________

______________________ Instr: ______________________

Explain the purpose of the experiment:

List the first Learning Objective:

OBJECTIVE 1:

After reviewing the results, describe how the objective was validated by this experiment.

List the second Learning Objective:

OBJECTIVE 2:

After reviewing the results, describe how the objective was validated by this experiment.

List the third Learning Objective:

OBJECTIVE 3:

After reviewing the results, describe how the objective was validated by this experiment.

Conclusion:

If required, attach to this form: ☐ Answers to Questions, ☐ Tables, and ☐ Graphs.

CHAPTER 25

EXPERIMENT 25-2

PARALLEL RESONANCE

LEARNING OBJECTIVES

At the completion of this experiment, you will be able to:

- Understand the concept of parallel resonance.
- Note the differences between a series and a parallel LC resonant circuit.
- Plot a graph of frequency versus amplitude.

SUGGESTED READING

Chapter 25, *Basic Electronics*, Grob/Schultz, twelfth edition

INTRODUCTION

Similar to a series resonant circuit, the resonant effect takes place in a parallel LC circuit, when $X_L = X_C$. The formulas for f_r and $Q = X_L/R$ are also the same. The cancellation of X_L and X_C, due to the 180° phase difference at the resonant frequency, is similar for both series and parallel circuits. However, the parallel LC circuit has differences due to the source being outside the parallel branches of L and C.

At resonance, the reactive branch currents cancel in the main line and produce minimum current only in the main line. Because the main-line current is minimum, its impedance is maximum at f_r. The parallel branches of L and C (also called a *tank circuit*), however, have maximum current at resonance because they are in separate branches. In a circulating manner, the capacitor discharges into the inductor, which in turn has its field current collapse into the other side of the capacitor. This is called the *flywheel effect* and is possible only because of the ability of L and C to store energy.

Therefore, a parallel resonant circuit, or tank circuit, has both main-line and tank currents. Because the tank current is maximum at resonance, the greatest voltage drop will occur across either L or C at f_r. In the circuit of Fig. 25-2.1*a*, note that the resistances r_{S_1} and r_{S_2} are used to measure either line current or tank current for comparison ($V/R = I$).

Remember that in a parallel resonant circuit, low frequencies will take the path of least resistance, or the L branch. Likewise, high frequencies will take the path of least reactance, the C branch. Therefore, when X_L and X_C are equal, the tank impedance will be maximum. Thus, as a tuning circuit, the maximum voltage can be taken across the tank at the resonant frequency.

A higher circuit Q will result in a sharper response curve or narrower bandwidth. Since the bandwidth is determined by the half-power points (70.7 percent of I_{max}) on either side of the resonant frequency, the bandwidth is equal to f_r/Q. Thus, for f_1 and f_2 in Fig. 25-2.1*b*, the difference between the two frequencies is the bandwidth.

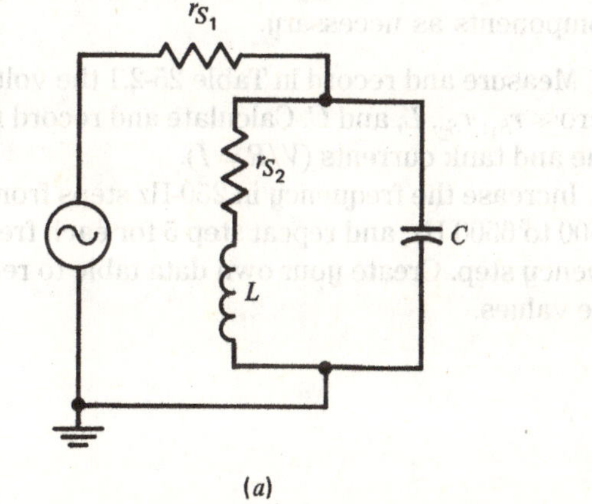

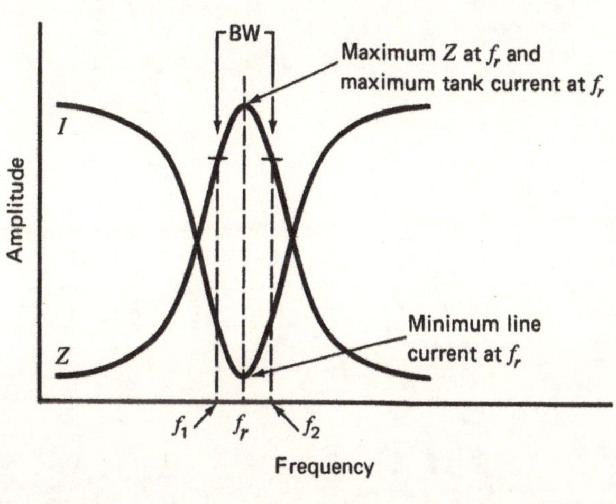

Fig. 25-2.1 Parallel resonant circuit. (*a*) Schematic. (*b*) Graph.

EQUIPMENT

Signal generator
Oscilloscope
Protoboard
Leads

COMPONENTS

(1) 1-MΩ potentiometer (connected as rheostat)
(1) 100-Ω, 0.25-W resistor
(1) 47-kΩ, 0.25-W resistor
(1) 0.01-μF capacitor
(1) 100-mH inductor

PROCEDURE

1. Calculate and record in Table 25-2.1 the resonant frequency of the circuit of Fig. 25-2.2. Show your work.
2. Calculate and record in Table 25-2.1 the Q. Show your work.
3. Calculate and record in Table 25-2.1 the bandwidth (equals f_r/Q).
4. Connect the circuit of Fig. 25-2.2.

Note: **Keep the input voltage constant for all frequencies, and avoid ground loops by moving the components as necessary.**

5. Measure and record in Table 25-2.1 the voltage across r_{S_1}, r_{S_2}, L, and C. Calculate and record main-line and tank currents ($V/R = I$).
6. Increase the frequency in 250-Hz steps from 3500 to 6500 Hz, and repeat step 5 for each frequency step. Create your own data table to record the values.

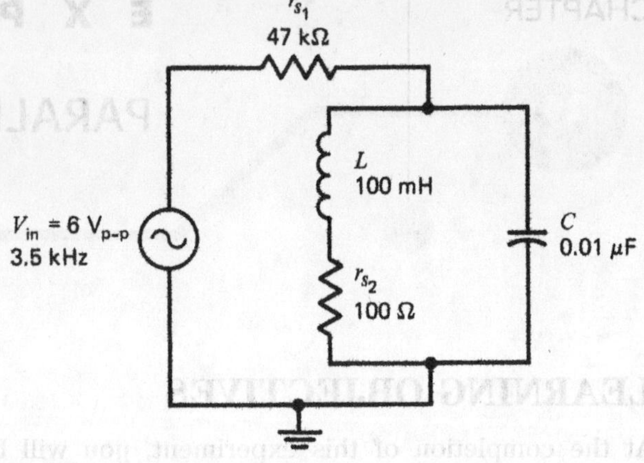

Fig. 25-2.2 Parallel resonant circuit for measurement.

7. Determine Z_T as follows and record the results in Table 25-2.1. Replace r_{S_1} (main-line resistor) with a 1-MΩ rheostat. Adjust the circuit to the resonant frequency, and adjust r_{S_1} so that its voltage equals the voltage across the tank. Record the value.

Note: **When the voltage across the rheostat equals the voltage across the tank, its ohmic value is equal to the tank impedance. This is due to the laws of series circuits, where equal resistances or impedances divide the applied voltage equally.**

8. Plot a graph of frequency versus amplitude for both tank current and main-line current. Indicate the bandwidth and any other significant points on the graph.

NAME DATE

QUESTIONS FOR EXPERIMENT 25-2

Answer true (T) or false (F) to the following:

_____ **1.** The formula for resonance is $1/(2\pi fC)$.

_____ **2.** The higher the Q, the sharper the bandwidth response.

_____ **3.** One difference between a series resonant circuit and a parallel resonant circuit is that the parallel resonant circuit has minimum impedance at f_r, while the series has maximum impedance at f_r.

_____ **4.** Greater values of L and C will result in increased bandwidth.

_____ **5.** If a tank circuit had a resonant frequency of 8 kHz with a 2-kHz bandwidth, the bandwidth could not be decreased without changing the resonant frequency.

_____ **6.** At the resonant frequency, a tank circuit requires minimum input power from the source.

_____ **7.** In a tank circuit, line current is maximum and tank current is minimum at resonance.

_____ **8.** For a tank circuit with $L = 2$ H and $C = 10$ μF, the resonant frequency is approximately 36 Hz.

_____ **9.** For question 8, a resonant frequency of 3600 Hz could be obtained by increasing L to equal 200 H.

TABLE FOR EXPERIMENT 25-2

TABLE 25-2.1

Procedure Step	Measurement	Measured Value	Calculation
1	f_r		________Hz
2	Q		________(X_L = ______Ω)
3	BW		________Hz
5	V_{rS1}	________V_{p-p} at 3.5 kHz	
	V_{rS2}	________V_{p-p} at 3.5 kHz	
	V_L	________V_{p-p} at 3.5 kHz	
	V_C	________V_{p-p} at 3.5 kHz	
	I_{tank}		________A
	I_{line}		________A
7	Z_T	________Ω	

GRAPH FOR EXPERIMENT 25-2

EXPERIMENT RESULTS REPORT FORM

Experiment No.: ____________________ Name: ____________________

Date: ____________________

Experiment Title: ____________________ Class: ____________________

____________________ Instr: ____________________

Explain the purpose of the experiment:

List the first Learning Objective:

OBJECTIVE 1:

After reviewing the results, describe how the objective was validated by this experiment.

List the second Learning Objective:

OBJECTIVE 2:

After reviewing the results, describe how the objective was validated by this experiment.

List the third Learning Objective:

OBJECTIVE 3:

After reviewing the results, describe how the objective was validated by this experiment.

Conclusion:

If required, attach to this form: ☐ Answers to Questions, ☐ Tables, and ☐ Graphs.

EXPERIMENT 26-1

FILTERS

LEARNING OBJECTIVES

At the completion of this experiment, you will be able to:

- Understand how filters separate different frequency components.
- Learn the difference between high-pass, low-pass, bandstop, and bandpass filters.
- Plot a graph of frequency versus amplitude for different filter types.

SUGGESTED READING

Chapter 26, *Basic Electronics*, Grob/Schultz, twelfth edition

INTRODUCTION

Filters are used to separate wanted signals from unwanted signals. For example, a radio can receive many different station broadcasts. But somehow, it must isolate the desired broadcast signal and filter out the frequencies that are not part of the broadcast. Thus, filters are used to allow only those desired frequencies to pass through certain parts of a circuit.

Inductors and capacitors are used, in various configurations, to build filters. Generally, capacitors allow high frequencies to pass, while inductors allow low frequencies to pass through them with very little reactance or opposition to current flow. Also, the manner in which a filter's components are placed topology determines the type of filter it is.

Low-pass filters allow low frequencies to pass through them, while higher frequencies are sent to ground. High-pass filters allow only high frequencies to pass, while blocking low frequencies by sending them to ground.

The following procedures examine four simple filter types.

EQUIPMENT

Signal generator
Oscilloscope
Protoboard
Leads; graph paper

COMPONENTS

(1) 1-kΩ resistor
(1) 56-Ω resistor
(1) 0.01-μF capacitor
(1) 0.1-μF capacitor
(1) 10 μF capacitor
(2) 33-mH inductors

PROCEDURE

1. Connect the circuit of Fig. 26-1.1.

Note: Remember to keep the input voltage constant at each frequency for all the steps in this procedure.

2. Increase the frequency from 1 to 10 kHz in 1-kHz steps. Measure and record in Table 26-1.1 the peak-to-peak output voltage at each step.

3. At 10 kHz, place another 33-mH inductor across points C and D. Measure and record the output voltage.

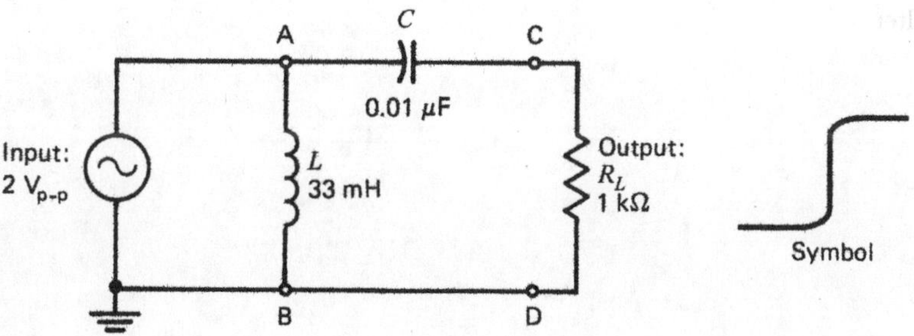

Fig. 26-1.1 High-pass filter.

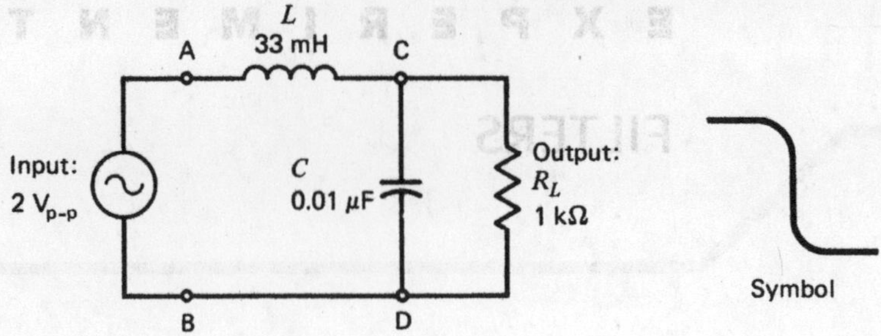

Fig. 26-1.2 Low-pass filter.

4. At 10 kHz, remove one of the inductors and replace the 0.01-μF capacitor with a 0.1-μF capacitor. Measure and record the output voltage.

5. Connect the circuit of Fig. 26-1.2.

6. Decrease the frequency from 10 kHz to 100 Hz in the steps shown in Table 26-1.1. Measure and record the output voltage at each step.

7. Place a second capacitor of 10 μF across points A and B. Measure and record the output voltage for each step shown in Table 26-1.1.

8. Connect the circuit of Fig. 26-1.3.

9. Increase the frequency from 2 to 18 kHz in the steps shown in Table 26-1.1. Measure and record the output voltage at each step.

10. Connect the circuit of Fig. 26-1.4.

11. Increase the frequency from 2 to 18 kHz in the steps shown in Table 26-1.1. Measure and record the output voltage at each frequency step.

12. Plot a graph of frequency versus load voltage for each of the previous four filter circuits.

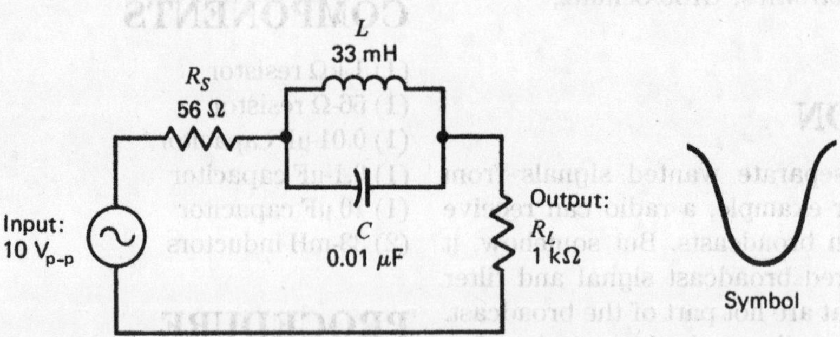

Fig. 26-1.3 Bandstop filter.

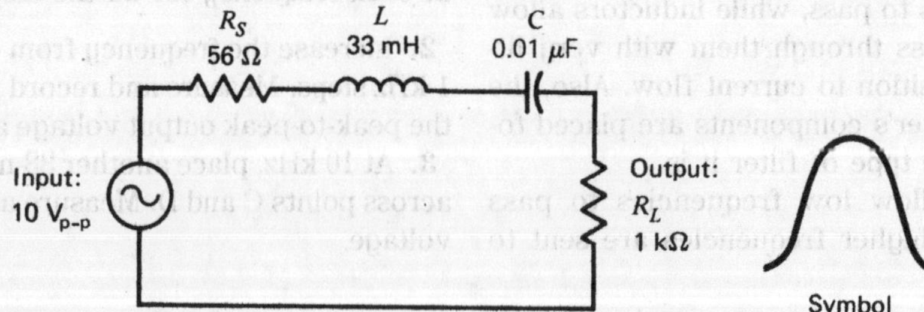

Fig. 26-1.4 Bandpass filter.

NAME ______________________________ DATE ____________

QUESTIONS FOR EXPERIMENT 26-1

1. **Explain what happens to low frequencies in the circuit of Fig. 26-1.1. Why don't they reach the load R_L?**

2. **Explain what happens to high frequencies in the circuit of Fig. 26-1.2. Why don't they reach the load R_L?**

3. **Explain how the resonant effect works as a filter in the circuit of Fig. 26-1.3.**

4. **Explain how the resonant effect works as a filter in the circuit of Fig. 26-1.4.**

5. **Design a filter circuit that (*a*) passes frequencies from approximately 8 to 12 kHz and (*b*) stops or rejects frequencies from 8 to 12 kHz. Show all the values and calculations, where $V_A = 1\ V_{p\text{-}p}$ and $R_L = 8\ \Omega$ (similar to a radio speaker).**

TABLE FOR EXPERIMENT 26-1

TABLE 26-1.1

Procedure Step	Measurement Frequency	Load Voltage, $V_{p\text{-}p}$
2	1 kHz	
	2 kHz	
	3 kHz	
	4 kHz	
	5 kHz	
	6 kHz	
	7 kHz	
	8 kHz	
	9 kHz	
	10 kHz	
3	10 kHz	
4	10 kHz	
6	10 kHz	
	9 kHz	
	8 kHz	
	7 kHz	
	6 kHz	
	5 kHz	
	4 kHz	
	3 kHz	
	2 kHz	
	1 kHz	
	500 Hz	
	200 Hz	
	100 Hz	
	50 Hz	
7	10 kHz	
	5 kHz	
	2 kHz	
	1 kHz	
	500 Hz	
	100 Hz	
	50 Hz	

Procedure Step	Measurement Frequency	Load Voltage, $V_{p\text{-}p}$
9	2.0 kHz	
($C = 0.001\ \mu F$)	3.0 kHz	
$f_c \approx 9$ kHz	4.0 kHz	
	5.0 kHz	
	6.0 kHz	
	6.5 kHz	
	7.0 kHz	
	7.5 kHz	
	8.0 kHz	
	8.5 kHz	
	9.0 kHz	
	9.5 kHz	
	10.0 kHz	
	12.0 kHz	
	14.0 kHz	
	16.0 kHz	
	18.0 kHz	
11	2.0 kHz	
($C = 0.01\ \mu F$	3.0 kHz	
$f_o \approx 9$ kHz)	4.0 kHz	
	5.0 kHz	
	6.0 kHz	
	7.0 kHz	
	8.0 kHz	
	8.5 kHz	
	9.0 kHz	
	9.5 kHz	
	10.0 kHz	
	11.0 kHz	
	12.0 kHz	
	13.0 kHz	
	14.0 kHz	
	15.0 kHz	
	18.0 kHz	

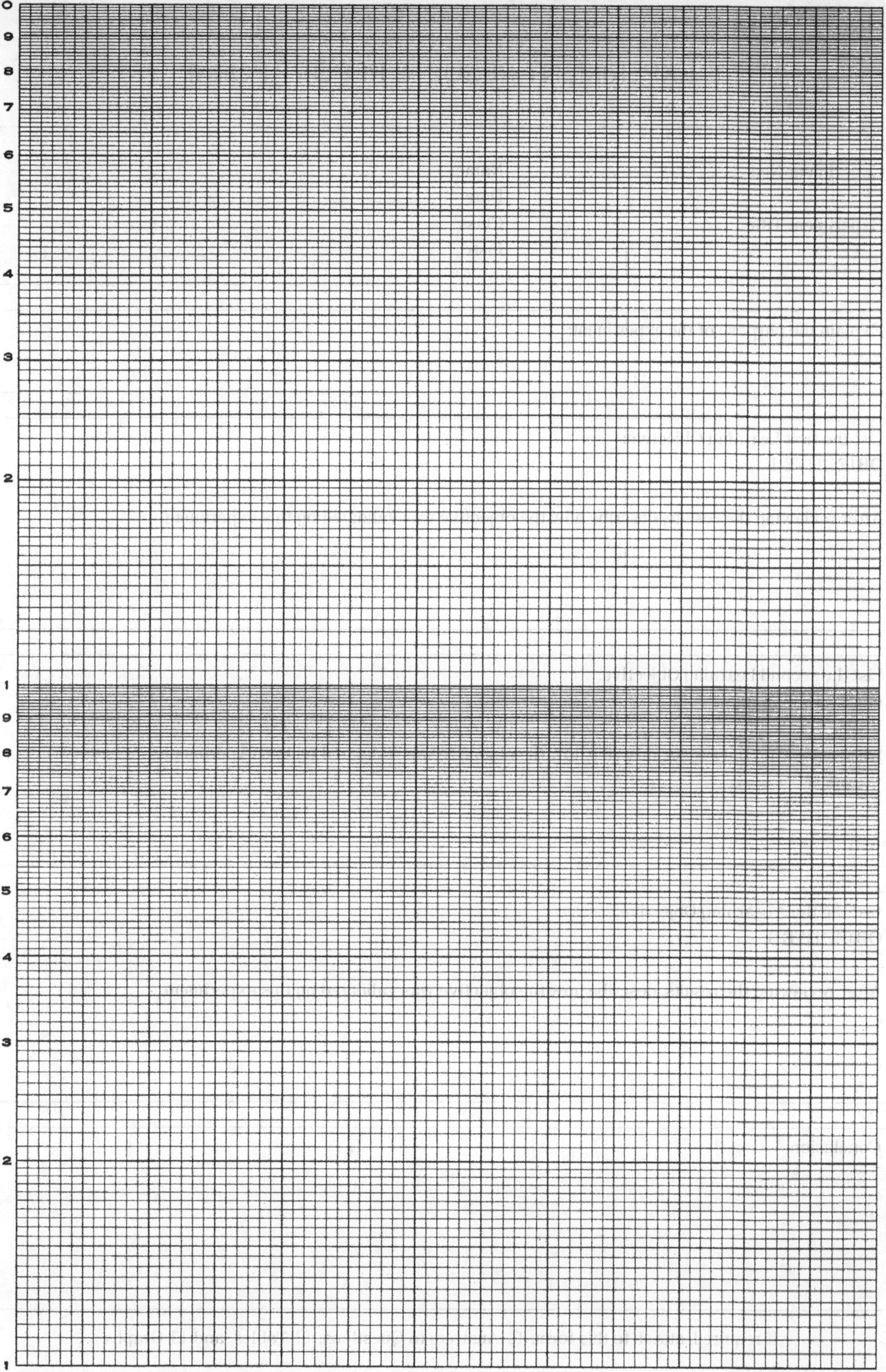
10
9
8
7
6
5
4
3
2
1
9
8
7
6
5
4
3
2
1

EXPERIMENT RESULTS REPORT FORM

Experiment No.: ______________________ Name: ______________________

Date: ______________________

Experiment Title: ______________________ Class: ______________________

______________________ Instr: ______________________

Explain the purpose of the experiment:

List the first Learning Objective:

OBJECTIVE 1:

After reviewing the results, describe how the objective was validated by this experiment.

List the second Learning Objective:

OBJECTIVE 2:

After reviewing the results, describe how the objective was validated by this experiment.

List the third Learning Objective:

OBJECTIVE 3:

After reviewing the results, describe how the objective was validated by this experiment.

Conclusion:

If required, attach to this form: ☐ Answers to Questions, ☐ Tables, and ☐ Graphs.

CHAPTER 26

EXPERIMENT 26-2

FILTER APPLICATIONS

LEARNING OBJECTIVES

At the completion of this experiment, you will be able to:

- Design a resonant filter circuit.
- Measure BW and determine Q for resonant filters.
- Troubleshoot bandstop and bandpass filters.

SUGGESTED READING

Chapters 25 and 26, *Basic Electronics*, Grob/Schultz, twelfth edition

INTRODUCTION

A resonant circuit can be used to tune to a specific resonant frequency. Therefore, it has the effect of filtering out all signals except the resonant frequency. This applies to both series and parallel resonant circuits. Therefore, depending on the circuit configuration and where the resistive load is connected, resonant filters can be bandpass or bandstop in their effects.

Series resonant filters have a maximum current and a minimum impedance at resonance. When connected in series with a load, the series-tuned LC circuit allows the maximum current to flow to the resistive load at resonance. This is a bandpass filter (see Fig. 26-2.1*a*). However, when connected across the load, the tuned LCR circuit becomes a low-impedance shunt path for frequencies at and near resonance, and little or no current flows to the load. This is a bandstop filter (see Fig. 26-2.1*b*).

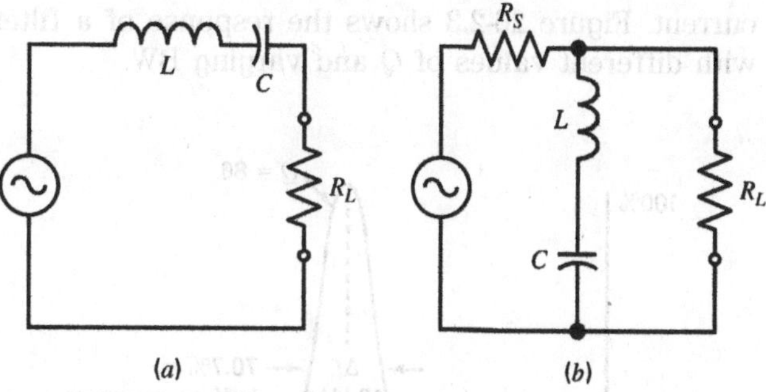

Fig. 26-2.1 Resonant filters (series). (*a*) Bandpass filter. (*b*) Bandstop filter.

Parallel resonant filters (tanks) have a maximum impedance and a minimum current at resonance. When connected in series with a load, the parallel-tuned LC circuit provides maximum impedance in series with the load, stopping the current flow to the resistive load at resonance. This is a bandstop filter (see Fig. 26-2.2*a*). When the load is connected across the parallel resonance, however, the LC circuit becomes an impedance for frequencies at and near resonance and allows maximum current flow to the load. This is a bandpass filter (see Fig. 26-2.2*b*).

Designing a Filter

To design a bandpass or bandstop filter, you need to calculate the resonant frequency that you want to stop or pass. This is done using the resonant formula

$$\text{Resonant frequency } (f_r) = \frac{1}{2\pi\sqrt{LC}}$$

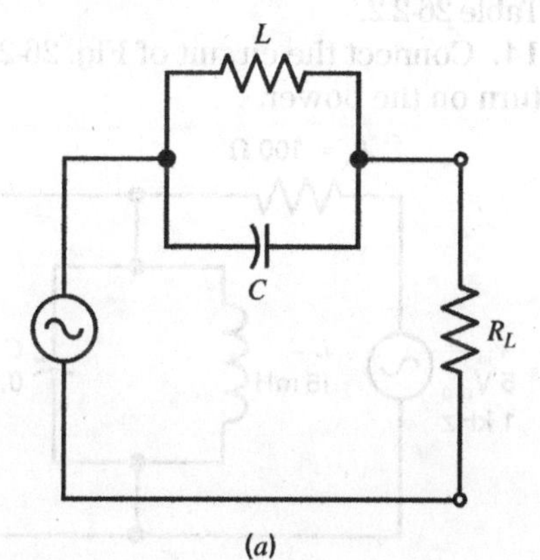

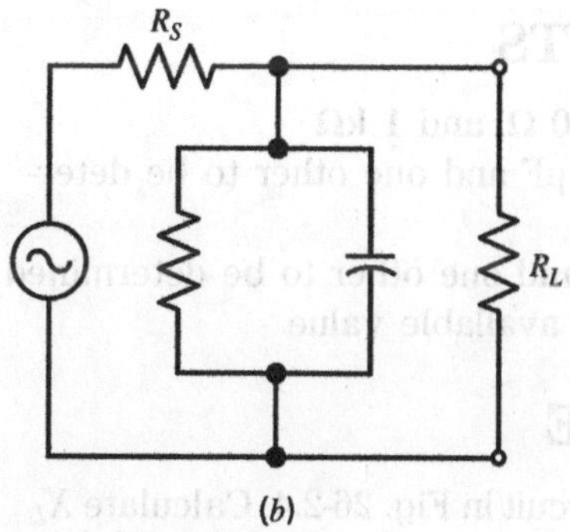

Fig. 26-2.2 Resonant filters (parallel). (*a*) Bandstop filter. (*b*) Bandpass filter.

In addition, the bandwidth (BW) of the resonance can become critical if it is too wide or too narrow. Bandwidth is measured practically as 70.7 percent of the peak of the response. Bandwidth is related to the circuit Q, where Q is a ratio: $Q = X_L/r$ (coil). The idea usually is to have a sharp (narrow) response with reasonable bandwidth: high Q. In a high-Q circuit, the ratio of L to C is also greater, and this means that you can adjust different values of L and C to achieve the same resonance, but with different bandwidths, and different Q rises in voltage or current. Figure 26-2.3 shows the response of a filter with different values of Q and varying BW.

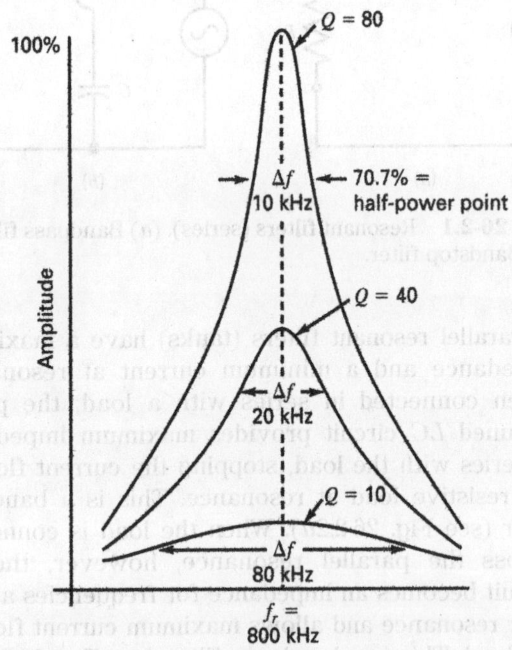

Fig. 26-2.3 Resonant filter response: BW and Q.

EQUIPMENT

Signal generator
DMM
Oscilloscope

COMPONENTS

Resistors: 10 Ω, 100 Ω, and 1 kΩ
Capacitors: 0.0047 µF and one other to be determined by design
Inductors: 15 mH and one other to be determined by design or any available value

PROCEDURE

1. Refer to the circuit in Fig. 26-2.4. Calculate X_L and record the value in Table 26-2.1.

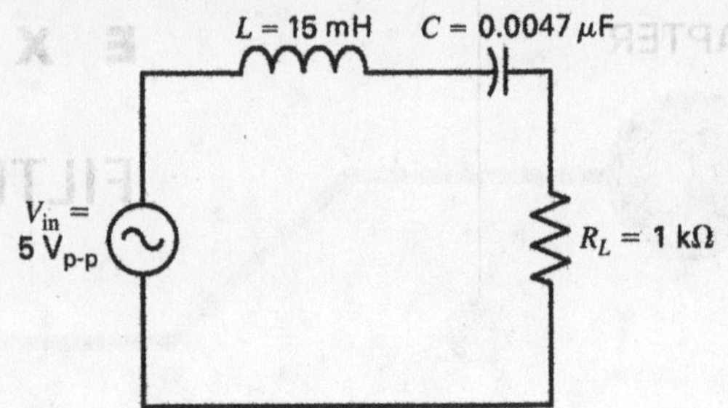

Fig. 26-2.4 Series-tuned LC filter.

2. Measure the resistance of the coil L and record the value in Table 26-2.1.

3. Calculate the circuit Q, the resonant frequency, and determine the bandwidth. Record the values in Table 26-2.1.

4. Connect the circuit of Fig. 26-2.4, but do not turn on power.

5. Apply power and adjust the signal generator frequency so that the maximum voltage appears across the load resistor. Record both the measured resonant frequency and the peak-to-peak voltage across the load resistor in Table 26-2.1.

6. Measure the BW as 70.7 percent of the value in step 5, and record the measurement in Table 26-2.1.

7. Insert a 10-Ω resistor between L and C. Do not make any adjustments to the signal generator voltage. This resistor will represent additional series resistance of the coil, which should change the circuit Q. Add 10 Ω to the measured value, and record this number as R coil measured in Table 26-2.1.

8. Calculate the new values of Q and BW with the added resistance, and record the values in Table 26-2.1.

9. Measure the peak-to-peak voltage across the load resistor, and record the value in Table 26-2.1.

10. Measure the BW as 70.7 percent of the value in step 9, and record it in Table 26-2.1.

11. Refer to the circuit in Fig. 26-2.5. Calculate X_L and record the value in Table 26-2.2.

12. Measure the resistance of the coil L, and record the value in Table 26-2.2.

13. Calculate the circuit Q, the resonant frequency, and the bandwidth, and record the values in Table 26-2.2.

14. Connect the circuit of Fig. 26-2.5, but do not turn on the power.

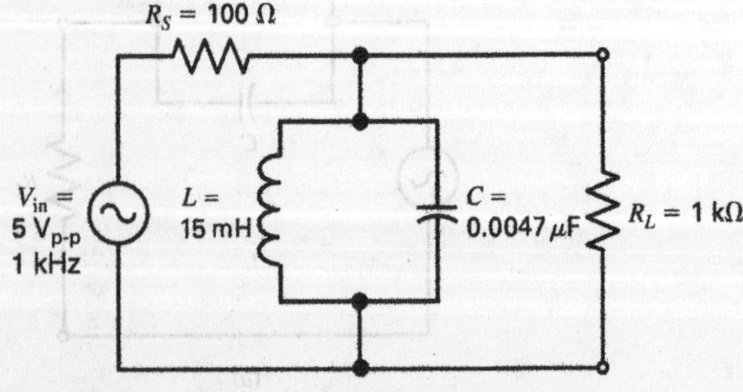

Fig. 26-2.5 Parallel-tuned LC filter.

15. Apply power and adjust the signal generator so that the maximum voltage appears across the load resistor. Record both the measured resonant frequency and the peak-to-peak voltage across the load resistor in Table 26-2.2.

16. Measure the BW as 70.7 percent of the value in step 15, and record the measurement in Table 26-2.2.

17. Insert a 10-Ω resistor in series with the coil L. Do not make any adjustments to the signal generator voltage. This resistor will represent additional series resistance of the coil, which should change the circuit Q. Add 10 Ω to the measured value, and record this number as the measured value in Table 26-2.2.

18. Calculate the new values of Q and BW with the added resistance, and record the values in Table 26-2.2.

19. Measure the peak-to-peak voltage across the load resistor, and record the value in Table 26-2.2.

20. Measure the BW as 70.7 percent of the value in step 19, and record the measurement in Table 26-2.2.

21. *Filter design.* Design a bandstop filter with a resonant frequency of either 5 or 50 kHz, or do both if you have time. The goal is to have a high-Q circuit if possible. On a separate sheet of paper, show all your calculations and draw the circuit. Be sure your lab can supply the values of L and C you need—if not, use different values of L and C. Show all your calculations and then build the circuit. Test your design by measuring 10 frequency points on either side of the resonant frequency, with enough spacing so that you have the data to plot the response on a graph. Graph the results, and indicate BW, resonant frequency, and any other values you want.

Note: $V_{in} = 5\ V_{p\text{-}p}$ or less.

NAME DATE

QUESTIONS FOR EXPERIMENT 26-2

1. What elements control the BW of a resonant filter?

2. What is meant by the Q of a resonant circuit?

3. Why is a high-Q circuit desirable?

4. When would you expect one inductor to have a greater resistance than another?

5. What would happen if the series resistor R_S were not in the circuit of Fig. 26-2.5?

TABLES FOR EXPERIMENT 26-2

TABLE 26-2.1 (Steps 1 through 10)

	Freq. Calc.	Freq. Meas.	V_{RL} Meas.	BW Meas.	R_{coil} Meas.	X_L Calc.	Q Calc.	BW Calc.
With 10-Ω R_{coil} added								

TABLE 26-2.2 (Steps 11 through 20)

	Freq. Calc.	Freq. Meas.	V_{RL} Meas.	BW Meas.	R_{coil} Meas.	X_L Calc.	Q Calc.	BW Calc.
With 10-Ω R_{coil} added								

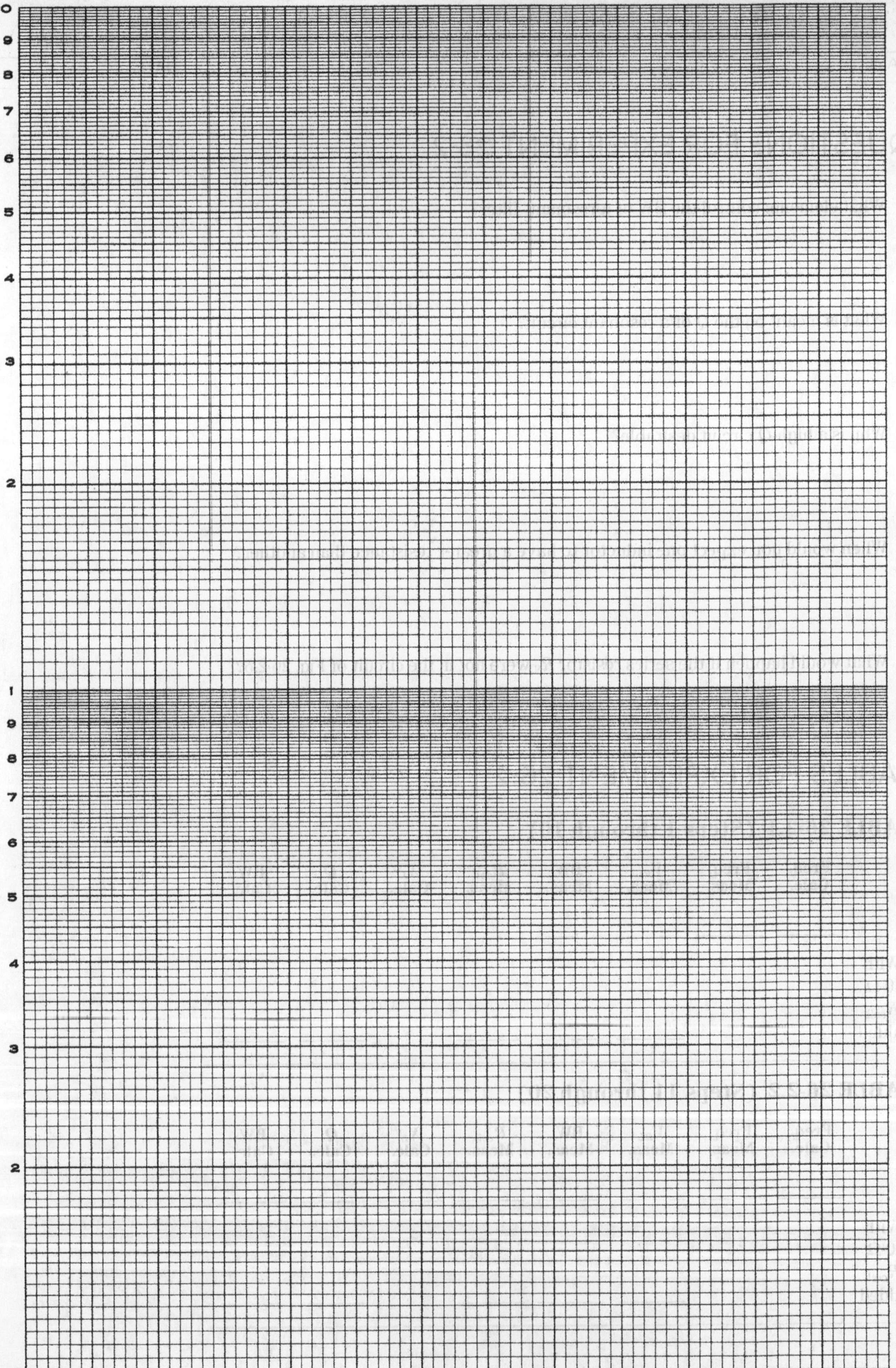
10
9
8
7
6
5
4
3
2
1
9
8
7
6
5
4
3
2
1

EXPERIMENT RESULTS REPORT FORM

Experiment No.: ______________________ Name: ______________________

Date: ______________________

Experiment Title: ______________________ Class: ______________________

______________________ Instr: ______________________

Explain the purpose of the experiment:

List the first Learning Objective:

OBJECTIVE 1:

After reviewing the results, describe how the objective was validated by this experiment.

List the second Learning Objective:

OBJECTIVE 2:

After reviewing the results, describe how the objective was validated by this experiment.

List the third Learning Objective:

OBJECTIVE 3:

After reviewing the results, describe how the objective was validated by this experiment.

Conclusion:

If required, attach to this form: ☐ Answers to Questions, ☐ Tables, and ☐ Graphs.

EXPERIMENT 27-1

THE WYE (Y)-CONNECTED THREE-PHASE AC GENERATOR

LEARNING OBJECTIVES

At the completion of this experiment, you will be able to:

- Explain the mathematical relationship between the line voltage, V_L, and phase voltage, V_ϕ, in a Y-connected three-phase AC generator.
- Measure the load voltage, V_Z, and load current, I_Z, in the Y-Y configuration.
- Measure the neutral current, I_N, in a Y-Y configuration with a balanced load.
- Calculate the load power, P_Z, and total power, P_T, in a balanced, Y-connected load using measured values.

SUGGESTED READING

Chapter 27, *Basic Electronics,* Grob/Schultz, thirteenth edition.

INTRODUCTION

A three-phase AC voltage source has three separate output voltages, with each output voltage being 120° out of phase with the other. There are two types of three-phase AC generators, the wye (Y)-connected generator and the delta (Δ)-connected generator. Which one to use depends on the application. There are also two types of loads used with three-phase AC generators, the Y-connected load and the Δ-connected load. As a result, there are four possible source/load configurations in three-phase AC power systems. The four possible configurations include the Y-Y configuration, the Y-Δ configuration, the Δ-Δ configuration, and the Δ-Y configuration. In this experiment we will concentrate on the Y-Y configuration exclusively.

Caution: *When doing this experiment, it is extremely important that you review the safety procedures and guidelines outlined earlier in the Experiments Manual. You will be working with dangerously high voltages and every safety precaution imaginable must be observed. It is highly recommended that your instructor be present to supervise and/or oversee all circuit connections and measurements. When making voltage measurements, put one hand in your pocket, if possible, or as an alternative, attach the meter leads when the power is off and then restore power to view the measured values. Also, make all current measurements using a clamp-on ammeter. And finally, and perhaps most importantly, follow the proper Lock-out tagout (LOTO) safety procedures when doing this experiment. LOTO ensures that all power is shut off while building the circuit and also while the meters are properly configured for making circuit measurements.*

EQUIPMENT

DMM
Clamp-on ammeter
Enclosed (boxed) wye (Y)-connected 120/208 V three-phase AC voltage source
Dual-Trace Oscilloscope

COMPONENTS

Three 250-W resistive heating elements

Connecting leads or wires

PROCEDURE

Figure 27-1.1 shows a wye (Y)-connected three-phase AC voltage source. As shown, there are multiple output terminals for connecting to phases A, B, and C as well as the neutral point, N. There is also a ground (G) terminal. This three-phase AC voltage source has an external plug, located in the rear, which plugs directly into a standard three-phase AC receptacle.

1. On the Y-connected three-phase AC voltage source, locate the three-phase output terminals A, B, C, and N. Next, using an oscilloscope, measure each

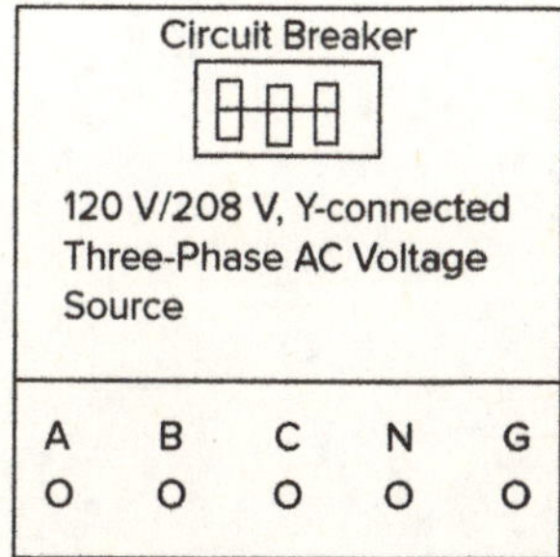

Fig. 27-1.1 120 V/208 V, Y-connected three-phase AC voltage source

phase voltage with respect to the neutral (N) terminal. Draw the measured waveforms in Table 27-1.1. Using the voltage waveform, V_{AN}, as the reference waveform, draw the other two waveforms (V_{BN} and V_{CN}) in the proper phase relation with respect to each other. Contact an instructor for assistance in setting up the triggering function on the oscilloscope if necessary. On your drawing, indicate the peak voltage of each waveform and the degrees of phase difference between them.

2. Using a DMM, measure the phase voltages, V_{AN}, V_{BN}, and V_{CN}. Record your measured values in Table 27-1.2.

3. Next, use your DMM to measure the line voltages, V_{AB}, V_{BC}, and V_{CA}. Record your measured values in Table 27-1.2.

4. Using your measured values, calculate the ratio of the line voltage, V_{AB}, to the phase voltage, V_{AN}. Record your answer in Table 27-1.2.

5. Connect the Y-Y configuration shown in Fig. 27-1.2.

6. Using your DMM, measure and record the load voltages V_{Z_1}, V_{Z_2}, and V_{Z_3}. Record your measured values in Table 27-1.3.

7. Next, measure the load currents, I_{Z_1}, I_{Z_2}, and I_{Z_3}. To make these current measurements, use a clamp-on ammeter. To measure I_{Z_1}, as an example, connect the clamp-on ammeter around the *line wire* connecting terminal **A** on the three-phase generator to terminal **a** on the Y-connected load. Record your measure values in Table 27-1.3.

8. Next, using the clamp-on ammeter, measure the current in the neutral wire. Record your measured value in Table 27-1.3.

9. Remove the neutral wire from the circuit and remeasure the load voltages V_{Z_1}, V_{Z_2}, and V_{Z_3}. Record your measured values in Table 27-1.4.

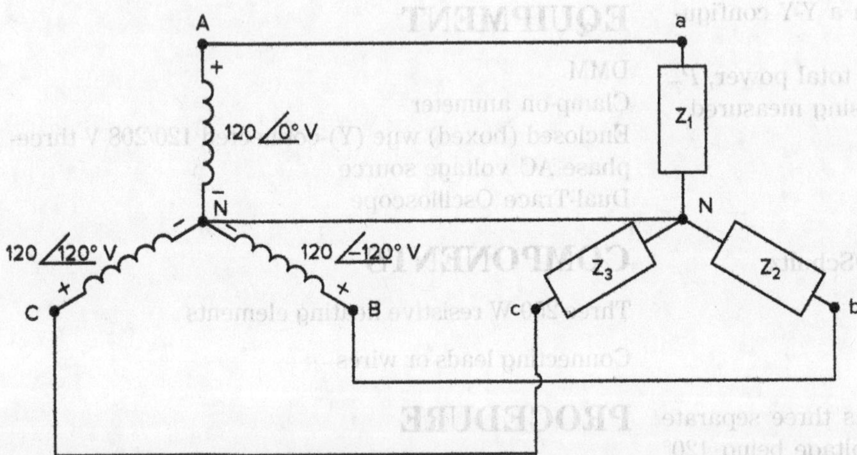

Fig. 27-1.2 Y-Y configuration

NAME DATE

QUESTIONS FOR EXPERIMENT 27-1

1. How are the phase voltage, V_θ, and line voltage, V_L, related in a wye (Y)-connected three-phase AC generator?

2. How are the phase currents, I_θ, and line currents, I_L, related in a Y-Y configuration?

3. In a Y-Y configuration with a balanced load, will removing the neutral wire have any effect on the operation of the circuit?

4. Using the values recorded in Table 27-1.3, calculate the load power, P_Z, and total power, P_T.

   $P_Z =$ ________ $P_T =$ ________

TABLE 27-1.1

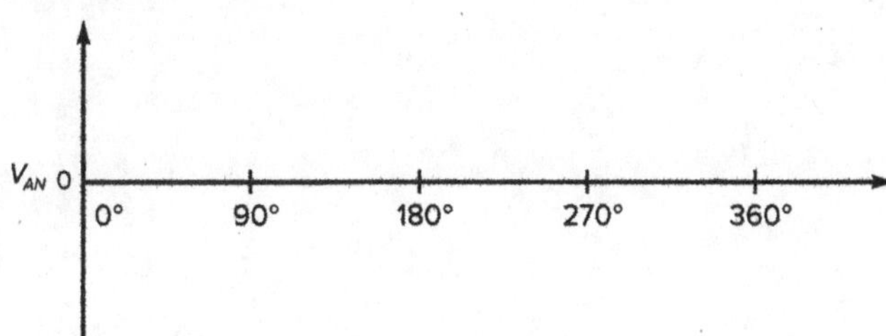

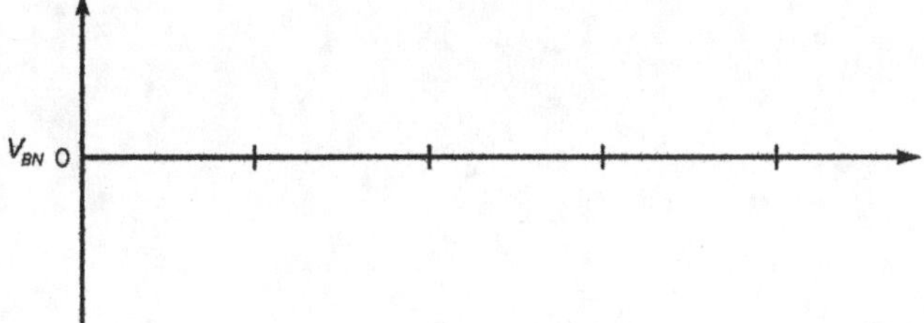

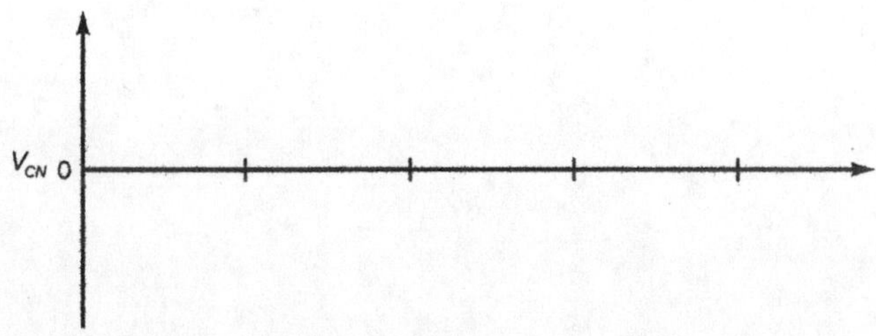

TABLE 27-1.2

V_{AN} = ___________ V_{BN} = ___________ V_{CN} = ___________ (Phase voltages)

V_{AB} = ___________ V_{BC} = ___________ V_{CA} = ___________ (Line voltages)

$\frac{V_{AB}}{V_{AN}}$ = ___________

TABLE 27-1.3

V_{Z_1} = ___________ V_{Z_2} = ___________ V_{Z_3} = ___________ (Load voltages)

I_{Z_1} = ___________ I_{Z_2} = ___________ I_{Z_3} = ___________ (Load currents)

I_N = ___________ (Neutral current)

TABLE 27-1.4

V_{Z_1} = ___________ V_{Z_2} = ___________ V_{Z_3} = ___________ (No neutral wire)

EXPERIMENT RESULTS REPORT FORM

Experiment No.: ______________________ Name: ______________________

Date: ______________________

Experiment Title: ______________________ Class: ______________________

______________________ Instr: ______________________

Explain the purpose of the experiment:

List the first Learning Objective:

OBJECTIVE 1:

After reviewing the results, describe how the objective was validated by this experiment.

List the second Learning Objective:

OBJECTIVE 2:

After reviewing the results, describe how the objective was validated by this experiment.

List the third Learning Objective:

OBJECTIVE 3:

After reviewing the results, describe how the objective was validated by this experiment.

Conclusion:

If required, attach to this form: ☐ Answers to Questions, ☐ Tables, and ☐ Graphs.

CHAPTER 28

EXPERIMENT 28-1

P-N JUNCTION

LEARNING OBJECTIVES

At the completion of this experiment, you will be able to:

- Determine the forward-bias conditions of a diode.
- Determine the reverse-bias conditions of a diode.
- Develop the characteristic curve for a diode.

SUGGESTED READING

Chapter 28, *Basic Electronics*, Grob/Schultz, twelfth edition

INTRODUCTION

This experiment will introduce the characteristics of a silicon diode. The diode will be ohmmeter-tested and then connected to a DC power supply and its operating characteristics graphed.

Basically, a 1N4004 diode will conduct in the forward direction (forward bias) with a low value of internal forward resistance. In the reverse-bias condition, the diode will not conduct and has an almost infinite resistance. However, under extreme conditions of high voltage or current, in either direction, the diode will break down and conduct or else be destroyed.

A manufacturer's specification sheet for this diode has been included in Appendix G. Note the ratings and characteristics for these diodes. These terms should be consistent with the explanations found in most theory textbooks.

EQUIPMENT

Oscilloscope
DMM
DC power supply, 0–10 V
DC power supply, 0–150 V
Protoboard or springboard
Test leads

COMPONENTS

(1) 1N4004

Resistors (all 0.25 W):

(1) 330 Ω	(1) 12 kΩ
(1) 680 Ω	(1) 1.0 MΩ

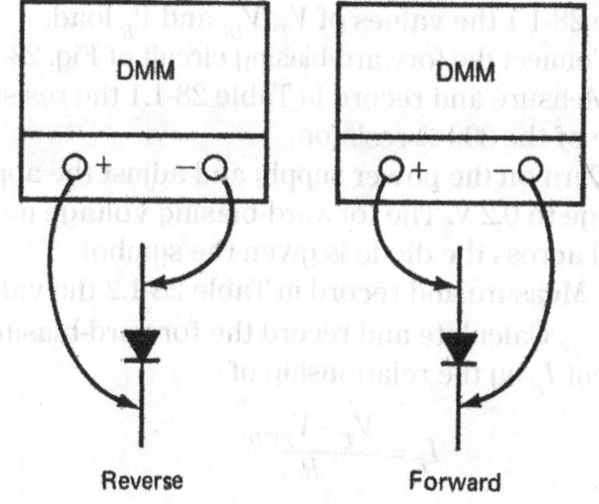

Fig. 28-1.1 Circuit connection for the measurement of forward and reverse resistance.

PROCEDURE

1. With the DMM on the ohm-meter function, record in Table 28-1.1 the forward and reverse resistance of the diode. See Fig. 28-1.1.

2. Connect the circuit of Fig. 28-1.2, with R_{L_1} equal to 12 kΩ. Measure and record in Table 28-1.1 the voltage of the power supply (V_T), the voltage across the diode (V_D), and the voltage across the load resistor (V_{R_1}). Also, measure the voltage across the diode while you slowly increase the power supply from 0 to 4 V, and note that the voltage across the diode should increase to approximately 0.65 V and remain constant, regardless of any further increase in supply voltage.

3. Repeat step 2 for a load resistance of 680 Ω. Be sure to record this information in Table 28-1.1.

4. Reverse the power supply leads so that polarity is reversed for Fig. 28-1.2's circuit. The diode is now reverse-biased. For both values of load

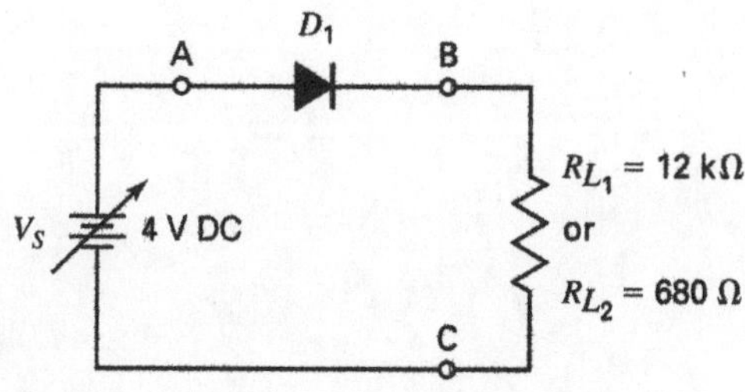

Fig. 28-1.2 Forward-biasing the diode.

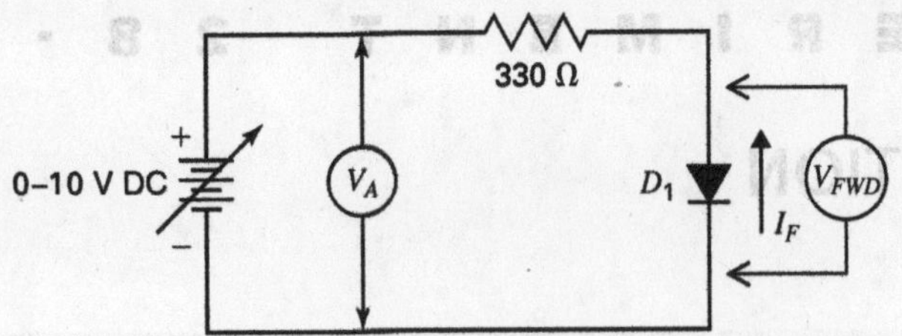

Fig. 28-1.3 Forward-bias condition.

resistance (12 kΩ and 680 Ω), measure and record in Table 28-1.1 the values of V_T, V_D, and V_R load.

5. Connect the forward-biasing circuit of Fig. 28-1.3.

6. Measure and record in Table 28-1.1 the resistance value of the 330-Ω resistor.

7. Turn on the power supply and adjust the applied voltage to 0.2 V. The forward-biasing voltage measured across the diode is given the symbol V_{FWD}. Measure and record in Table 28-1.2 the value of V_{FWD}. Calculate and record the forward-biasing current I_F by the relationship of

$$I_F = \frac{V_T - V_{FWD}}{R}$$

8. Repeat step 7 for the V_A values of:

0.25 V	0.65 V
0.30 V	0.70 V
0.35 V	0.75 V
0.40 V	0.80 V
0.45 V	0.85 V
0.50 V	0.90 V
0.55 V	0.95 V
0.60 V	1.00 V

9. Using the data obtained in steps 7 and 8, plot the forward characteristics of this diode in Fig. 28-1.4 (quadrant 1).

Note: If desired, draw your own plot similar to Fig. 28-1.4 to hand in with your report.

10. Connect the circuit shown in Fig. 28-1.5.

11. Increase the power supply voltage from 0 to 150 V in 50-V steps. This voltage will reverse-bias the diode. The reverse-bias voltage measured across the diode is given the symbol V_{REV}. Measure and record in Table 28-1.3 the corresponding voltage dropped across R_1.

12. Calculate and record the reverse current I_R for each voltage level applied in step 11.

13. Plot I_R versus V_{REV} on Fig. 28-1.4 (quadrant 3).

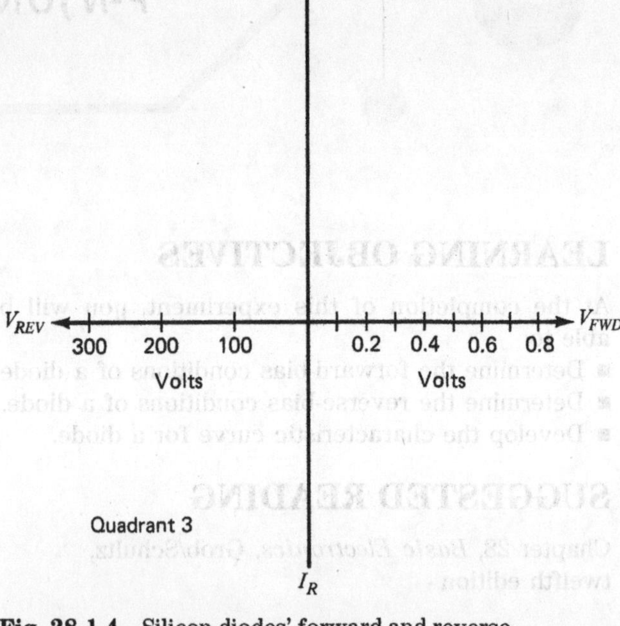

Fig. 28-1.4 Silicon diodes' forward and reverse characteristic curves.

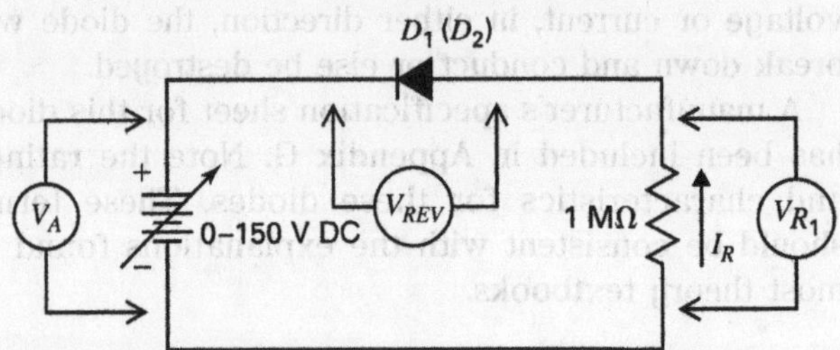

Fig. 28-1.5 Reverse-bias condition.

NAME ______________________ DATE __________

QUESTIONS FOR EXPERIMENT 28-1

1. What general statement could be made for the forward and reverse resistances of a good diode?

2. Under what conditions can the diode be forward-biased?

3. Under what conditions can the diode be reverse-biased?

4. Describe the significance of the curve found in quadrant 1 of Fig. 28-1.4.

5. Describe the significance of the curve found in quadrant 3 of Fig. 28-1.4.

TABLES FOR EXPERIMENT 28-1

TABLE 28-1.1

Procedure Step		R_L	V_T	V_D	$V_{R\ LOAD}$
1	Fwd. bias R = ________ Ω				
	Rev. bias R = ________ Ω				
2	Fwd. bias	12 kΩ	________	________	________
3	Fwd. bias	680 Ω	________	________	________
4	Rev. bias	12 kΩ	________	________	________
		680 Ω	________	________	________
6	330-Ω R (nominal)				
	= ________ Ω (measured)				

TABLE 28-1.2

Procedure Step	V_T	V_{FWD}	Calculated I_F
7	0.20 V	______	______
8	0.25 V		
	0.30 V	______	______
	0.35 V	______	______
	0.40 V	______	______
	0.45 V	______	______
	0.50 V	______	______
	0.55 V	______	______
	0.60 V	______	______
	0.65 V	______	______
	0.70 V	______	______
	0.75 V	______	______
	0.80 V	______	______
	0.85 V	______	______
	0.90 V	______	______
	0.95 V	______	______
	1.00 V	______	______

TABLE 28-1.3

Procedure Step	V_T	V_{REV}	V_{R_1}	Calculated I_R
11 and	50 V	______	______	______
12	100 V	______	______	______
	150 V	______	______	______

EXPERIMENT RESULTS REPORT FORM

Experiment No.: ____________________ Name: ____________________

Date: ____________________

Experiment Title: ____________________ Class: ____________________

____________________ Instr: ____________________

Explain the purpose of the experiment:

List the first Learning Objective:
OBJECTIVE 1:

After reviewing the results, describe how the objective was validated by this experiment.

List the second Learning Objective:
OBJECTIVE 2:

After reviewing the results, describe how the objective was validated by this experiment.

List the third Learning Objective:
OBJECTIVE 3:

After reviewing the results, describe how the objective was validated by this experiment.

Conclusion:

If required, attach to this form: ☐ Answers to Questions, ☐ Tables, and ☐ Graphs.

CHAPTER 28

EXPERIMENT 28-2

TRANSISTOR AS A SWITCH

LEARNING OBJECTIVES

At the completion of this experiment, you will be able to:

- Understand the importance of cutoff and saturation to the operation of a transistor switch.
- Define the purpose of a transistor inverter.
- Identify the function of a transistor switch.

SUGGESTED READING

Chapters 28 and 29, *Basic Electronics*, Grob/Schultz, twelfth edition

INTRODUCTION

The computers of today do not process numbers in the base 10 (i.e., 0, 1, 2, 3, . . . , 9). Computers instead use binary logic of base 2 (0 and 1) to perform their functions. One fundamental circuit is the transistor switch, also known as an *inverter*. Here, a transistor connected in a common-emitter fashion inverts a signal. That is, if a high-input signal is applied, a low-output signal is created. If a low-input signal is applied, then a high-output signal is created. The circuit of Fig. 28-2.1 is an example of a transistor inverter design.

The circuit of Fig. 28-2.1 is also a transistor switch. In a transistor switch circuit, a voltage level applied to the base terminal will control the potential at the collector. In this fashion, the transistor can be used to turn on or off circuitry connected to the collector. This common-emitter circuit is being switched from cutoff to saturation, as shown in the load line of Fig. 28-2.2.

In this experiment, a transistor will be connected to demonstrate this switching ability.

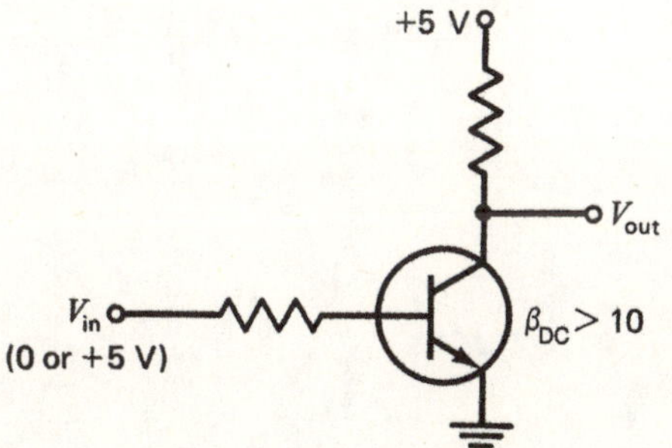

Fig. 28-2.1 Transistor inverter design.

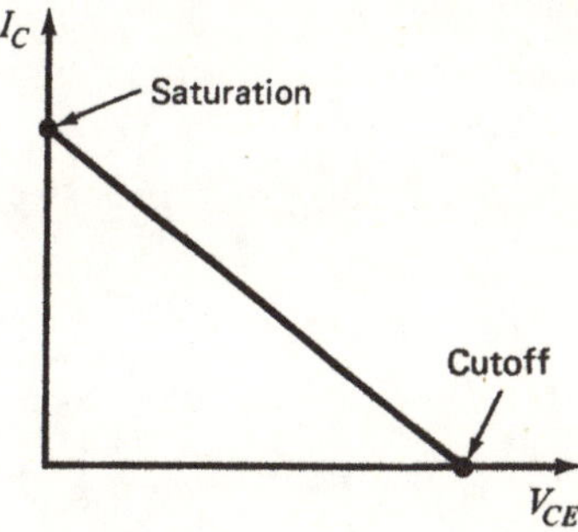

Fig. 28-2.2 Cutoff and saturation plotted on a load line.

EQUIPMENT

DC power supply, 0–10 V
DMM
Protoboard or springboard
Test leads

COMPONENTS

(1) *npn* transistor
(1) *pnp* transistor

Resistors (all 0.25 W):

(1) 1 kΩ (1) 10 kΩ

PROCEDURE

1. Connect the circuit in Fig. 28-2.3. Apply the correct polarity of voltage to V_{CC}.
2. Connect point A to ground. Measure and record in Table 28-2.1 the voltage from point B to ground.
3. Connect point A to +5 V. Measure and record in Table 28-2.1 the voltage from point B to ground.

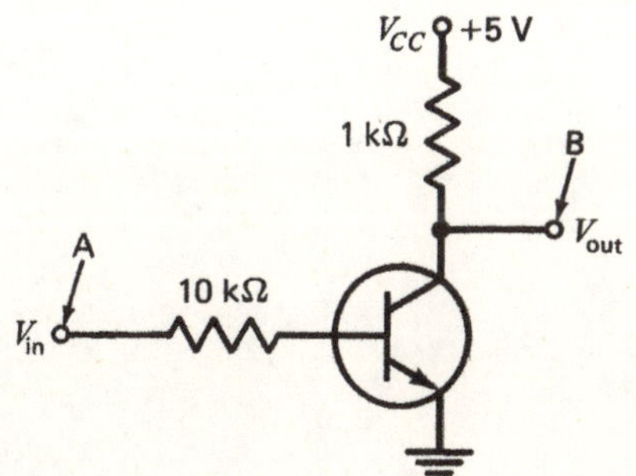

Fig. 28-2.3 Transistor switch.

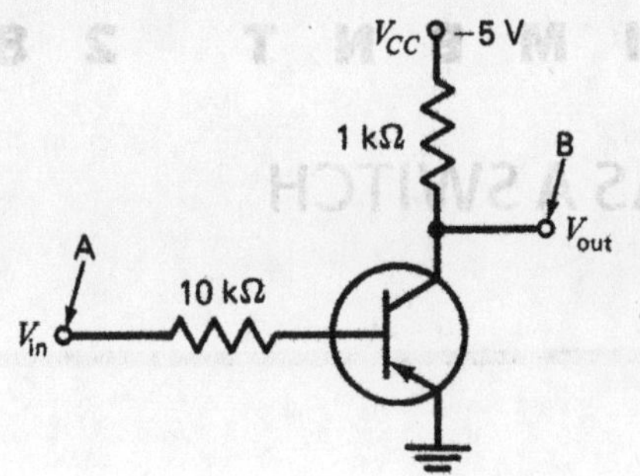

Fig. 28-2.4 Transistor switch.

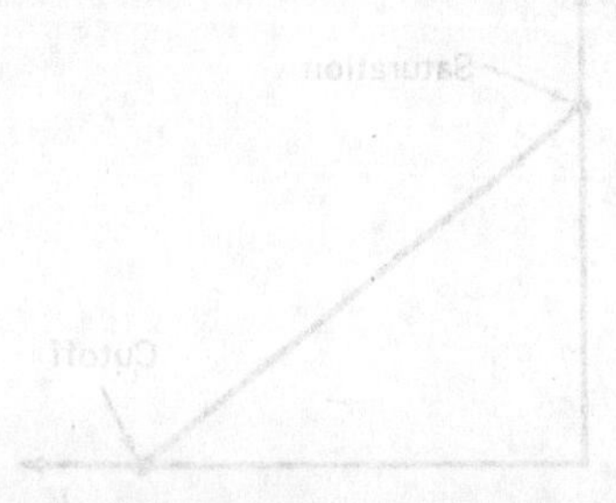

4. Connect the circuit of Fig. 28-2.4. Apply the correct polarity of voltage to V_{CC}.

5. Connect point A to ground. Measure and record in Table 28-2.1 the voltage from point B to ground.

6. Connect point A to −5 V. Measure and record in Table 28-2.1 the voltage from point B to ground.

7. Construct a table of your results that will contrast the two circuits.

NAME DATE

QUESTIONS FOR EXPERIMENT 28-2

1. In the prior circuits, what voltage level would a binary 1 represent? A binary 0? Are the answers the same for both the circuits shown in Figs. 28-2.3 and 28-2.4?

2. What is saturation? How is it demonstrated in this experiment?

3. What is cutoff? How is it demonstrated in this experiment?

4. Are the saturation and cutoff points the same for both the circuits shown in Figs. 28-2.3 and 28-2.4?

5. What are the fundamental differences between the two circuits shown in Figs. 28-2.3 and 28-2.4? Do the differences significantly affect overall outcomes? Explain.

TABLE FOR EXPERIMENT 28-2

TABLE 28-2.1

Procedure Step	Function and Measurement	Value
2	Point B to ground	______ V
3	Point B to ground	______ V
5	Point B to ground	______ V
6	Point B to ground	______ V

EXPERIMENT RESULTS REPORT FORM

Experiment No.: ______________________ **Name:** ______________________

Date: ______________________

Experiment Title: ______________________ **Class:** ______________________

______________________ **Instr:** ______________________

Explain the purpose of the experiment:

List the first Learning Objective:
OBJECTIVE 1:

After reviewing the results, describe how the objective was validated by this experiment.

List the second Learning Objective:
OBJECTIVE 2:

After reviewing the results, describe how the objective was validated by this experiment.

List the third Learning Objective:
OBJECTIVE 3:

After reviewing the results, describe how the objective was validated by this experiment.

Conclusion:

If required, attach to this form: ☐ Answers to Questions, ☐ Tables, and ☐ Graphs.

EXPERIMENT 28-3

DIODE RECTIFIERS

LEARNING OBJECTIVES

At the completion of this experiment, you will be able to:

- Construct a half-wave rectifier circuit.
- Construct a full-wave rectifier circuit.
- Compare a bridge circuit configuration with a center-tap circuit configuration.

SUGGESTED READING

Chapters 28 and 29, *Basic Electronics,* Grob/Schultz, twelfth edition.

INTRODUCTION

Only a battery can supply direct current to a circuit. However, AC voltage can be rectified (turned into DC). For practical purposes, rectified AC voltage will always contain some amount of the AC component, regardless of circuit design.

At this point, the diode has been forward-biased and reverse-biased under DC bias conditions. Therefore, the characteristics under ideal circumstances have been established. The following procedure will use AC voltages in rectifier circuits. The diode characteristics are similiar, while the results are different.

EQUIPMENT

Oscilloscope
1:1 probe
DMM
120:12.6-V center-tapped transformer
120:6.3-V filament transformer
Protoboard or springboard
Test leads

COMPONENTS

(4) 1N4004 silicon diodes

Resistors (all 0.25 W):

(2) 560 Ω (1) 12 kΩ

PROCEDURE

1. Connect the circuit of Fig. 28-3.1.

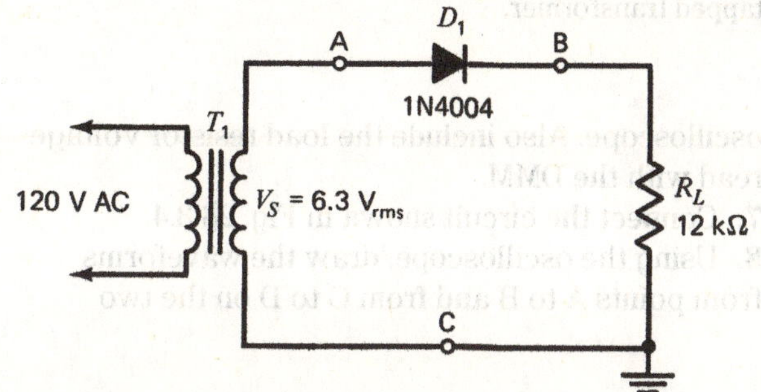

Fig. 28-3.1 Half-wave rectifier circuit.

2. Using the oscilloscope, draw the waveforms from points A to C and from B to C on the two graticule displays in Fig. 28-3.5 (at the end of this experiment). On these graticule displays, be sure to include the peak-to-peak voltages read on the oscilloscope. Also include the load resistor voltage read with the DMM.
3. Connect the circuit shown in Fig. 28-3.2.

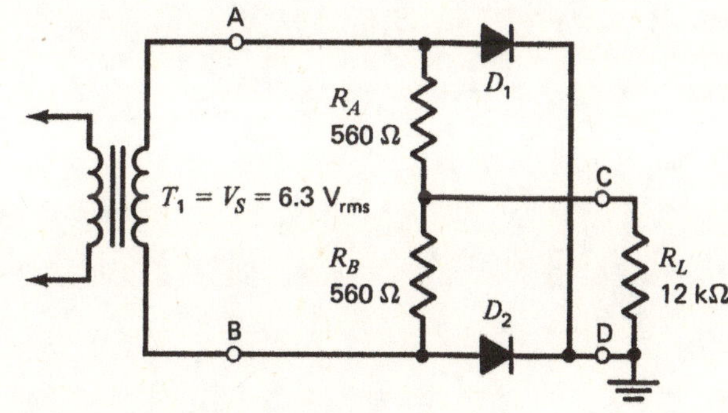

Fig. 28-3.2 Full-wave rectifier circuit.

4. Using the oscilloscope, draw the waveforms from points A to B and from C to D on the two graticule displays in Fig. 28-3.6 (at the end of this experiment). On these graticule displays, be sure to include the peak-to-peak voltages read on the oscilloscope. Also include the load resistor voltage read with the DMM.
5. Connect the circuit shown in Fig. 28-3.3.
6. Using the oscilloscope, draw the waveforms from points A to B and from C to D on the two graticule displays in Fig. 28-3.7 (at the end of this experiment). On these graticule displays, be sure to include the peak-to-peak voltages read on the

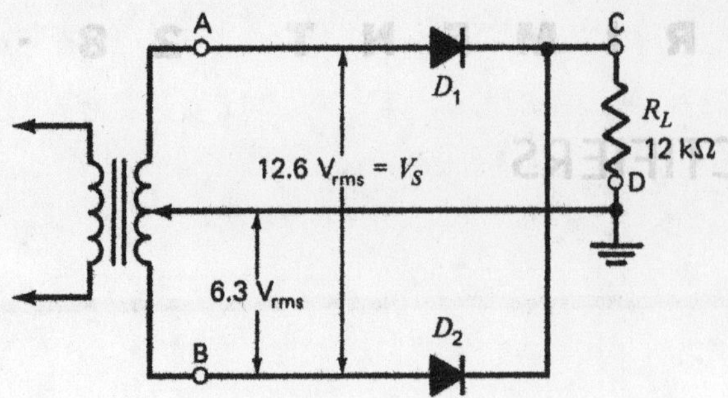

Fig. 28-3.3 Full-wave rectifier circuit using a center-tapped transformer.

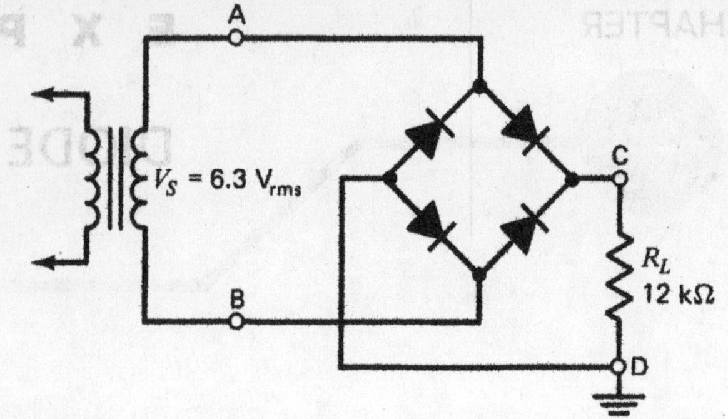

Fig. 28-3.4 Full-wave diode connected in a four-diode bridge arrangement.

oscilloscope. Also include the load resistor voltage read with the DMM.

7. Connect the circuit shown in Fig. 28-3.4.

8. Using the oscilloscope, draw the waveforms from points A to B and from C to D on the two graticule displays in Fig. 28-3.8 (at the end of this experiment). On these graticule displays, be sure to include the peak-to-peak voltages read on the oscilloscope. Also include the load resistor voltage read with the DMM.

NAME DATE

QUESTIONS FOR EXPERIMENT 28-3

1. How does a full-wave bridge rectifier circuit compare with a center-tap circuit?

2. Concerning each of the circuits studied, how do the graticule displays compare?

3. How does the oscilloscope input coupling affect the displayed waveform?

4. Make a general statement concerning the resistance of a *p-n* junction diode.

5. Which of the studied circuits would be best as a DC power supply?

GRAPHS FOR EXPERIMENT 28-3

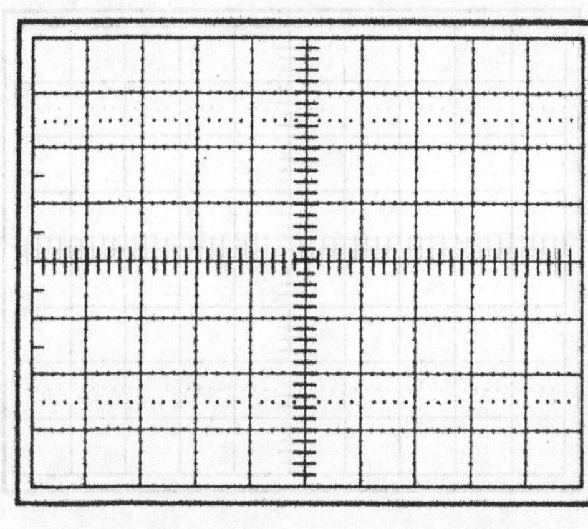

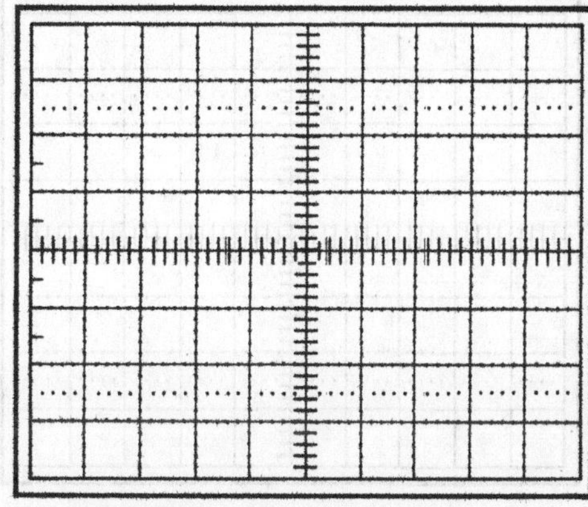

$V_S = V_{in} =$ ________ $V_{R_L} = V_{out} =$ ________

$V_{R_L} =$ ________ DC value

Fig. 28-3.5 Graticule for step 2.

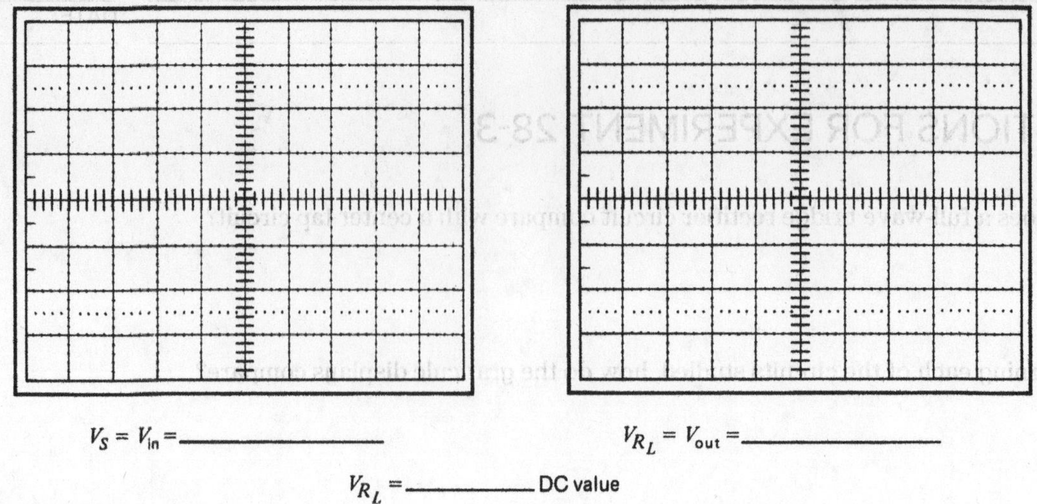

$V_S = V_{in}$ = ________________ $V_{R_L} = V_{out}$ = ________________

V_{R_L} = __________ DC value

Fig. 28-3.6 Graticule for step 4.

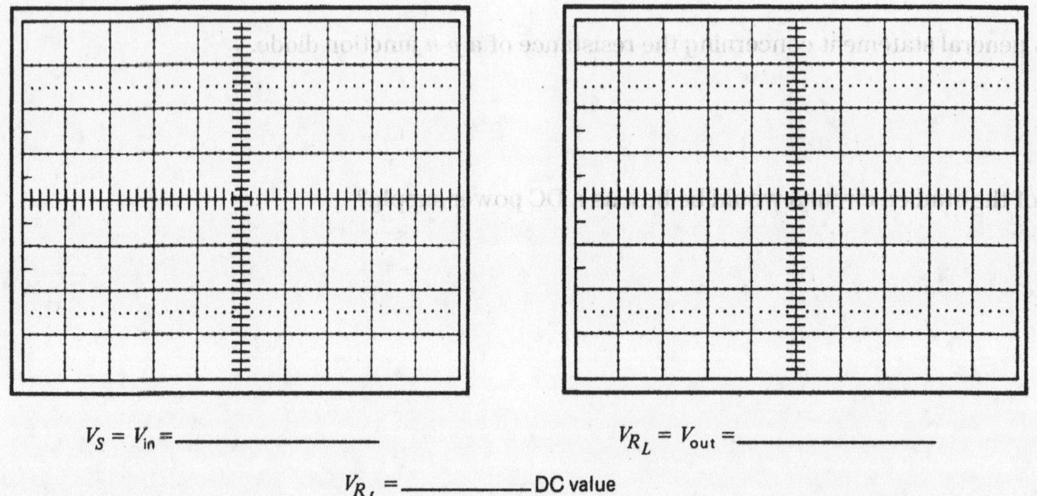

$V_S = V_{in}$ = ________________ $V_{R_L} = V_{out}$ = ________________

V_{R_L} = __________ DC value

Fig. 28-3.7 Graticule for step 6.

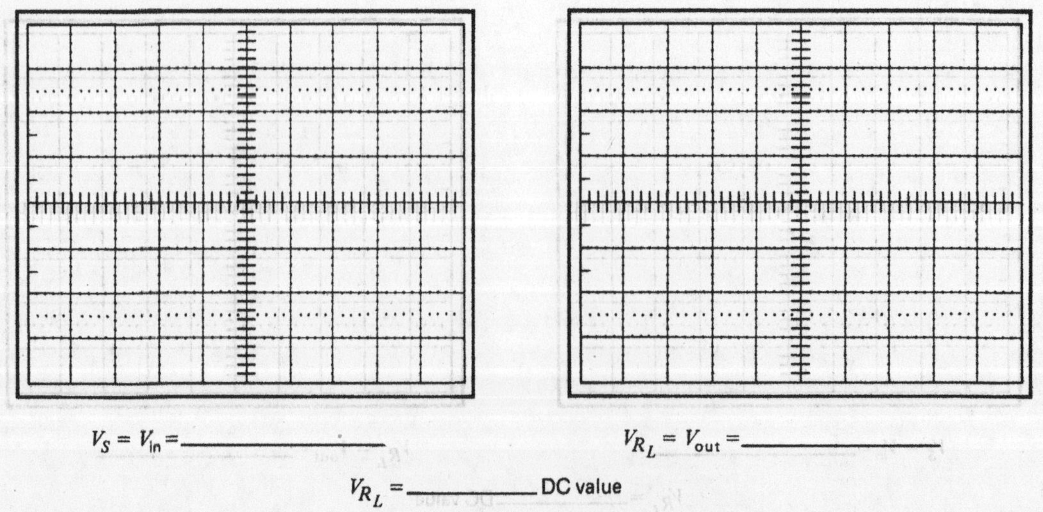

$V_S = V_{in}$ = ________________ $V_{R_L} = V_{out}$ = ________________

V_{R_L} = __________ DC value

Fig. 28-3.8 Graticule for step 8.

EXPERIMENT RESULTS REPORT FORM

Experiment No.: ______________________ Name: ______________________

Date: ______________________

Experiment Title: ______________________ Class: ______________________

______________________ Instr: ______________________

Explain the purpose of the experiment:

List the first Learning Objective:

OBJECTIVE 1:

After reviewing the results, describe how the objective was validated by this experiment.

List the second Learning Objective:

OBJECTIVE 2:

After reviewing the results, describe how the objective was validated by this experiment.

List the third Learning Objective:

OBJECTIVE 3:

After reviewing the results, describe how the objective was validated by this experiment.

Conclusion:

If required, attach to this form: ☐ Answers to Questions, ☐ Tables, and ☐ Graphs.

CHAPTER 28

EXPERIMENT 28-4

RECTIFICATION AND FILTERS

LEARNING OBJECTIVES

At the completion of this experiment, you will be able to:

- Compare the average value, also called the *DC value*, of the half-wave and the full-wave voltages across the load resistor.
- Observe the effects of filtering on a rectified AC voltage.
- Compare the difference between capacitive versus inductive filtering circuits.

SUGGESTED READING

Chapter 28, *Basic Electronics*, Grob/Schultz, twelfth edition

INTRODUCTION

Rectified AC voltage can be filtered to resemble the unidirectional quality of battery DC voltage, although it will always contain some small amount of the AC component, regardless of filtering. This AC voltage component is called *ripple*, and it is a measure of the quality of rectification and filtering.

This experiment investigates the use of reactive components as filters for smoothing the rectified AC voltage. Capacitive reactance, inductive reactance, *RC* time constant, and circuit configuration all affect the smoothing of the ripple.

EQUIPMENT

Oscilloscope
DMM
120:12.6-V, center-tapped transformer
120:6.3-V, filament transformer
Test leads

COMPONENTS

(4) 1N4004 diodes or equivalent
(1) 1-H inductor (choke) or larger up to 8 H
(1) 47-μF capacitor
(1) 10-μF capacitor
(1) 680-Ω, 0.25-W resistor
(1) 2.2-kΩ, 0.25-W resistor

PROCEDURE

1. Connect the circuit shown in Fig. 28-4.1 without the capacitor.

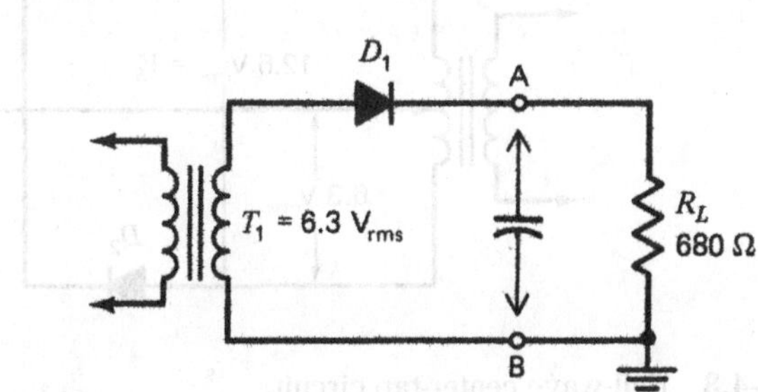

Fig. 28-4.1 Half-wave rectifier.

2. Using an oscilloscope, measure and record in Table 28-4.1 the AC output voltage across R_L.

3. Using a DMM, measure and record the DC value of the voltage across R_L.

4. Connect a 47-μF filtering capacitor across points A and B in the Fig. 28-4.1 circuit. Repeat steps 2 and 3 for measuring the output voltage.

5. Determine and record in Table 28-4.1 the percent of ripple for the filtered circuits only, using the following formula.

$$\%\text{ripple} = \frac{\text{rms value of ripple voltage across } R_L}{\text{DC component across } R_L} \times 100$$

6. Connect the full-wave rectification bridge circuit of Fig. 28-4.2.

7. Repeat steps 2 to 5 for the full-wave rectifier bridge of Fig. 28-4.2.

8. Connect the full-wave, center-tap circuit shown in Fig. 28-4.3 with the load resistor only (omit the "arrowed" components).

Note: Procedures 9 to 14 refer to Fig. 28-4.3 as adjusted for each specific filter configuration. Do not remove a filter component unless you are instructed to do so. The inductor is also called a "choke."

9. Using the oscilloscope, measure and record the AC output voltage across R_L.

10. Measure the DC value of the voltage across R_L, and, only if filtered, compute the percent of ripple.

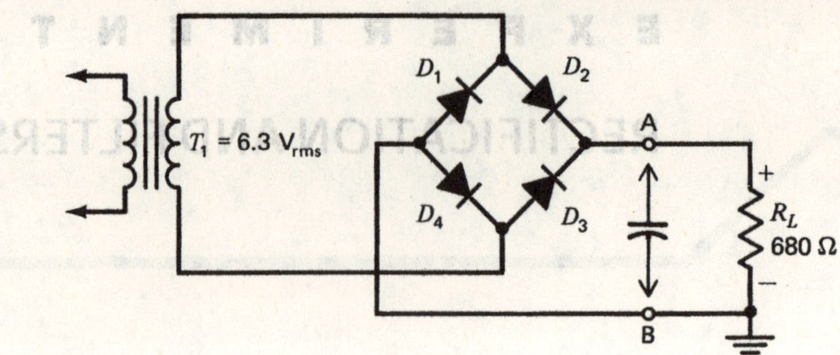

Fig. 28-4.2 Full-wave bridge circuit.

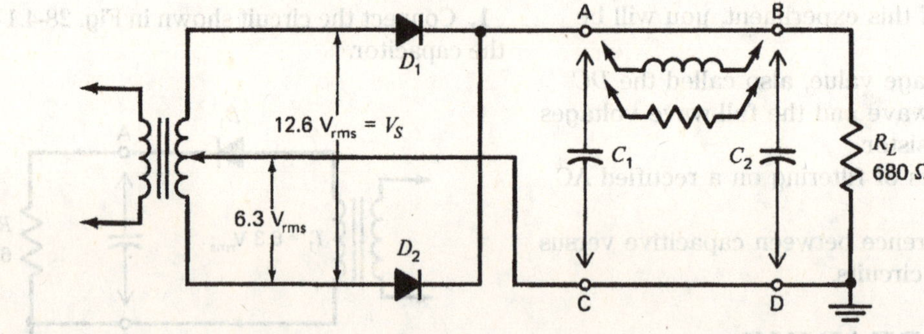

Fig. 28-4.3 Full-wave center-tap circuit.

11. Connect a 47-µF capacitor (C_1) across points A and C, and repeat steps 9 and 10.

12. Remove the jumper between points A and B, and in its place (in series with the load), connect the choke. Repeat steps 9 and 10.

13. Connect a 10-µF capacitor (C_2) across R_L (points B and D), and repeat steps 9 and 10.

14. Replace the choke with a 2.2-kΩ resistor, leaving the capacitors in place, and repeat steps 9 and 10.

NAME DATE

QUESTIONS FOR EXPERIMENT 28-4

1. How does a full-wave bridge rectifier compare with a center-tap circuit in this experiment?

2. Concerning each of the circuits studied in this experiment, how do the displayed oscilloscope waveforms compare with one another?

3. How does the oscilloscope input coupling affect the displayed waveform?

4. Does the resistance of the *p-n* junction diodes affect the operation of the rectified circuits studied?

5. With respect to this experiment, what would be a definition of filtration?

TABLE FOR EXPERIMENT 28-4

TABLE 28-4.1

Procedure Step	Measurement	Value	Ripple Calculation
2	V_{R_L}(oscill.)	______ $V_{p\text{-}p}$	
3	V_{R_L}(DMM)	______ V	
4	V_{R_L}(oscill.)	______ $V_{p\text{-}p}$	
	V_{R_L}(DMM)	______ V	
5	% ripple		______ %
7	V_{R_L}(oscill.)	______ $V_{p\text{-}p}$	
	V_{R_L}(DMM)	______ V	
	V_{R_L}(oscill.)	______ $V_{p\text{-}p}$	
	V_{R_L}(DMM)	______ V	
	% ripple		______ %
9	V_{R_L}(AC)	______ $V_{p\text{-}p}$	
10	V_{R_L}(DC)	______ V	
11	V_{R_L}(AC)	______ $V_{p\text{-}p}$	
	V_{R_L}(DC)	______ V	______ %
12	V_{R_L}(AC)	______ $V_{p\text{-}p}$	
	V_{R_L}(DC)	______ V	______ %
13	V_{R_L}(AC)	______ $V_{p\text{-}p}$	
	V_{R_L}(DC)	______ V	______ %
14	V_{R_L}(AC)	______ $V_{p\text{-}p}$	
	V_{R_L}(DC)	______ V	______ %

EXPERIMENT RESULTS REPORT FORM

Experiment No.: ____________________ Name: ____________________

Date: ____________________

Experiment Title: ____________________ Class: ____________________

____________________ Instr: ____________________

Explain the purpose of the experiment:

List the first Learning Objective:
OBJECTIVE 1:

After reviewing the results, describe how the objective was validated by this experiment.

List the second Learning Objective:
OBJECTIVE 2:

After reviewing the results, describe how the objective was validated by this experiment.

List the third Learning Objective:
OBJECTIVE 3:

After reviewing the results, describe how the objective was validated by this experiment.

Conclusion:

If required, attach to this form: ☐ Answers to Questions, ☐ Tables, and ☐ Graphs.

CHAPTER 28

EXPERIMENT 28-5

TROUBLESHOOTING POWER SUPPLIES

LEARNING OBJECTIVES

At the completion of this experiment, you will be able to:

- Review the design of a basic power supply.
- Evaluate power supply design.
- Evaluate other power supply designs.

SUGGESTED READING

Chapters 11, 26, 28, and 29, *Basic Electronics*, Grob/Schultz, twelfth edition

INTRODUCTION

Most modern electronic devices require a DC voltage source or DC power supply, like a battery, to operate correctly. The power supply modifies the existing power in some way. There are AC power supplies as well as DC supplies.

Almost all power supplies that today's electronic technicians troubleshoot and repair have some form of regulation. This means that their outputs are designed to provide a constant level of voltage, regardless of how much current is needed to operate the equipment. Remember that connecting a load of any kind to a power supply is like connecting a resistor in series with a battery. Therefore, current will flow, and a voltage drop will be present across the load (resistor) and the battery (its internal resistance). A regulated power supply is designed to eliminate the effects of the supply's loading (voltage drop), so that the load will always have a constant fixed voltage across it, no matter how much current flows in the circuit. Although this explanation is extremely simple, it provides a foundation to get started.

In summary, regulation in a power supply means the ability to maintain the output voltage level under various load conditions. However, before the supply voltage can be regulated, an *unregulated* voltage must be provided. To do this, the DC power supply will have to use a rectifier and filter system that helps convert the alternating current, like the 110-V, 60-Hz line voltage, to a relatively smooth direct current, as a battery provides, for the product (load) to operate.

There are many different types of power supply circuits. However, most supplies have some basic components that perform the fundamental tasks required of all power supplies. Therefore, let's review these tasks and the components that perform them.

Transformers

Most power supplies use a transformer to transform or change the line voltage from one level to another. Usually, this is a reduction in voltage from the large line voltage (110 V) to a smaller voltage used for the power supply. This is usually referred to as a *step-down transformer* because it steps down or decreases the line voltage. Also remember that a transformer consists of two separate coils or inductors (a primary and a secondary) that are actually lengths of wire wrapped around a core. Besides reducing the voltage, the transformer isolates the secondary voltage from the line voltage.

In troubleshooting, it is common practice to check the continuity of the transformer's windings with an ohmmeter, to be sure that the wires have not shorted or opened. Always do this with the power off, and be sure to unplug the equipment.

Fuses

Most power supplies have line fuses connected in series with the transformer. Typically, fuses protect the product's circuitry from any great changes in AC line voltage or in the amount of current drawn by the load. Although it may seem obvious to check a fuse whenever a product fails to operate, several common problems are often overlooked by many technicians. For example, fuse ratings must be checked and adhered to for proper operation. If a fuse is supposed to be 1 A and a 0.5-A fuse is used, the product may work for a while and suddenly appear to break. Also, if a slow-blow fuse is specified but not used, the product may continually blow the fuse, no matter what rating is used. *Slow-blow fuses cannot or should not be replaced with regular fuses, and regular fuses should not be replaced with slow-blow fuses.* This simple rule cannot be overemphasized. Always check the fuse when you are troubleshooting.

Rectification

The term *rectification* refers to the concept of changing AC voltage to DC voltage. Therefore, it is common to have a rectifier, typically made of diodes that conduct in only one direction, at the output of the secondary. A wide variety of rectifiers are used in DC power supplies. The type of rectification system employed depends on the design and power demands of the circuit used.

Ripple

Remember that the purpose of any filtering system is to remove the AC component of the rectified AC line voltage. The ideal DC output would be like a battery: pure direct current with no trace of alternating current. However, because no filter can remove 100 percent of the AC component, there is always some amount of what is called *ripple* on the direct current. This ripple on the DC level is usually represented by a percentage:

$$\text{Ripple factor, \%} = \frac{V\,\text{ripple}}{V\,\text{DC, out}} \times 100$$

For troubleshooting purposes, the ripple factor of any filtered output should never exceed 1 percent. In fact, 0.2 percent or less is a good value.

Safety

There is nothing more important for a technician than safety. Regardless of how much knowledge or experience a person has, one small mistake or accident can lead to injury or even death from electric shock.

Electricians are trained to measure high voltages with one hand. If you do not use one hand, you are putting your body (especially your heart and lungs) in the middle—between your two hands—when you measure high voltages. So be careful when you are dealing with line voltages. And never stay near water when high voltages are present.

If someone receives electric shock, always remove the person from the source as quickly as possible. You can do this by shutting off the product's line switch or by simply pulling the plug. Then proper medical attention should be administered as soon as possible. In the case of heart failure or severe shock, CPR or artificial respiration should be administered immediately.

However, the best way to avoid injury or worse is to follow these simple rules:

1. Never work near water.
2. Use a separate isolation transformer with a 1:1 voltage ratio. Plug the transformer into the AC line; then plug the product and/or equipment into the transformer. Even a fused power distribution bar serves as protection.
3. Never work on equipment while you have metal jewelry on your fingers, wrists, neck, or arms.
4. Never walk away from equipment or products that are exposed. Always turn off the power, and put some cover over the opening to prevent accidental discharge of high-voltage capacitors.
5. Never use equipment that is underrated for the work you are doing. This is especially true for high-voltage probes used to test CRT anodes.
6. Always refer to the service manual or call the manufacturer for advice if no manual is available.
7. Always replace components with recommended parts. Do not use components that are out of tolerance or underrated. And always discharge capacitors with an insulated-handle screwdriver before you try to remove them.
8. Never work on equipment that you are not qualified to troubleshoot or repair. Even if you do not injure yourself, you could injure or ruin the equipment.

Basic Power Supply Troubleshooting

Although experience and factory-type training courses are the best way to learn how to test and repair equipment, two troubleshooting summaries are presented below. One is a general set of principles, many of which are reinforced in the experiments in this manual.

Power supplies are often the cause of most consumer product failures because their circuits use high currents and large voltages and dissipate a lot of heat, which can cause component damage.

1. Check Fuses. Obviously, this is the easiest repair you can make. But don't be too quick to judge a bad fuse. Remember that slow-blow fuses cannot be replaced by regular quick-blow types. Always check the amperage rating.
2. Check the Transformer Windings. This is not often a problem with newer equipment. However, older products typically have had more time running under higher temperatures. Therefore, a greater chance exists that the coating around windings has melted or worn away, causing a short, or there is a chance that heat has caused an open. Disconnect the primary and any secondaries from the circuit, and use an ohmmeter to check for continuity. Don't forget that you should always check the input and output voltages to verify their levels with the schematic.
3. Check the Rectifier. A defective diode can cause lowered voltage or no voltage. Use the voltmeter to check AC input and DC output voltages.
4. Check Filter Capacitors. Excessive audible noise, called *hum,* and low output voltage can be caused by a bad filter capacitor. Check the capacitors out of the circuit for open resistance. And check their input and output voltages with an oscilloscope, to verify their filtering (smoothing) ability.

5. Check the Regulator Circuitry. This is often the most difficult part of troubleshooting a regulated power supply. If the rectifier and filter elements are good, then the regulator portion is suspected. You will have to check all components, one at a time, to verify their nominal values or their operation. There is no standard way to check this circuitry except to use a variable high-wattage load resistance and to simulate operation with the real load disconnected. Voltages around the transistor and any other components can then be measured. Use an oscilloscope or voltmeter and any schematic provided. Remember that the theory you just read in the previous section provides a basis for analyzing the measurement results. For example, if you decrease the load and draw more current, you should see some kind of change in the bias around the pass regulator. If there is no change, you can be sure that the regulator circuit is faulty. At that point, you may have to replace components, one at a time, if you cannot determine the faulty one. Many experienced technicians use this method.

6. Check the Load. Connect sections of the load circuit, one at a time, if possible. Remember that a short in the load can cause the power supply to fail. Also, if the load is faulty, the power supply operation may appear to be faulty but is not.

Typical Power Supply Problem

No Voltage: This problem could be caused by a blown fuse or a failure in the filter circuitry.

Solutions: Measure the primary voltage V_{in}. Visually inspect the supply for heat damage—look for darkened components. Replace the fuses. Use the ohmmeter to check for continuity or opens where applicable. If the supply has a total resistance that is low, it is suggested that the load be removed. If the total resistance is high, the problem is then due to the load. After this, reconnect the supply and power on again. If the fuse blows, the supply may have a short. Earlier, we discussed how to check all the discrete components for typical resistance and voltage readings. We then discussed how to connect each part of the circuit one at a time to determine where the short was. Finally, more possibilities were discussed, including the use of a Variac (vary input voltage), before coming to a conclusion.

The circuit you will develop is a low-voltage DC power supply. As Fig. 28-5.1 shows, it is a full-wave bridge rectifier with a pi filter.

The output of this power supply is approximately 12 V. The load is indicated as purely resistive and is designated as R_1. Resistor R_1 will draw a maximum current of 150 mA; R_b is a bleeder resistor and is always connected.

Resistor R_1 is designed to draw 150 mA at 12 V. From Ohm's law R_1 is calculated at 80 Ω. Given the value of R_1, V_{out}, and the load current, the power delivered to the load P_1 can be calculated at 1.8 W. Note that P_1 is not equivalent to peak power P_{peak}. And P_{peak} can be determined if V_{peak} and I_{peak} are known. So V_{peak} equals $V_{rms}/0.707 = 17.8$ V. And I_{peak} equals V_{peak}/R_1, which is 0.222 A. Therefore P_{peak} is equivalent to the product of V_{peak} and I_{peak} and is 3.95 W. Normally a power supply is derated in terms of power by a factor of 2, so an 8-W power rating is advised.

In determining a value of R_b, I_b must be calculated. Bleeder currents are normally one-tenth (0.10) of the load current. In this case I_b is 0.015 A. Now R_b can be determined from Ohm's law as 12 V/0.015 A, or 800 Ω. Power dissipated by R_b can be calculated again as V_{peak}/R_b and is 0.0178 A. Also, P_{peak} is determined as the product of V_{peak} and I_{peak} and is 0.316 W.

In the schematic, the filter capacitors have been specified at 50 μF. Their DC working voltage rating must be at least 12 V. A standard value of 15 DC working voltage makes a good choice.

The next component to be selected is the filter inductor. The inductance value has been specified as 1 H. The current rating must be at least 165 mA and should be derated to at least 300 mA. Note that using an inductor with a higher than required current rating reduced DC losses due to the larger size of the wire used in the windings.

The diodes used in the full-wave rectifier have two main ratings: maximum current and peak inverse voltage. The normal average current for this circuit approaches 200 mA. However, when it is first energized, a large surge exists because the uncharged state of the capacitor provides little opposition to current. Allowing for a margin of safety, diodes with a 1-A rating should be sufficient.

The last component to be selected is the transformer, and it has a specified secondary voltage of

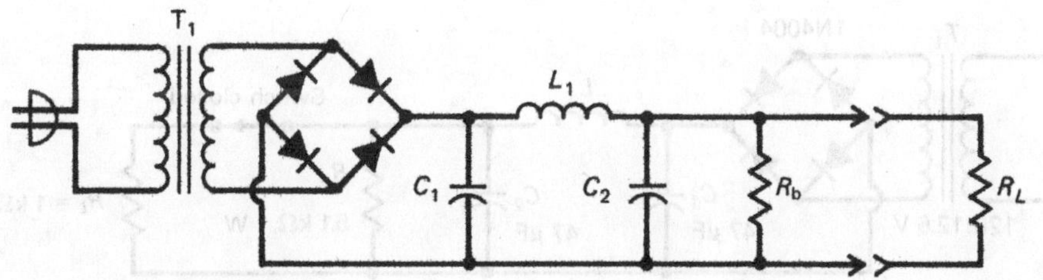

Fig. 28-5.1 Full-wave bridge rectifier circuit.

12.6 V. We have already estimated an average power supply current of approximately 200 mA, and most transformers with a specified voltage far exceed that capacity. Therefore, a 12.6-V transformer rated at 0.5, 1.0, or 1.5 A will be satisfactory.

EQUIPMENT

Oscilloscope
DMM
Protoboard
Test leads
Filament transformer (120 V:12.6 V)

COMPONENTS

(4) 1N4004 diodes or equivalent
(2) 47-μF capacitors
(1) 1 H inductor (choke) or largest available
(1) 1000-Ω, 1-W resistor
(1) 5.1-kΩ, 1-W resistor
(1) 100-kΩ, 1-W resistor
(1) SPST switch

PROCEDURE

1. Build the circuit shown in Fig. 28-5.2. Do not energize. Double-check all component connections.

2. When you are satisfied that the circuit is connected correctly, energize it.

3. Measure the unloaded DC and AC peak-to-peak ripple voltages. Record these measurements in Table 28-5.1 in the "Results" section of this experiment.

4. Disconnect the power to the supply.

5. Connect the load resistor to the output terminals of the circuit, as shown in Fig. 28-5.3.

6. Measure and record the fully loaded DC and AC peak-to-peak ripple voltages in Table 28-5.1.

7. Disconnect the power to the supply.

8. From the information measured, calculate the percentage of regulation by the formula

$$\%\ \text{regulation} = \frac{V_{\text{no load}} - V_{\text{full load}}}{V_{\text{full load}}} \times 100$$

Record the percentage of regulation in Table 28-5.1. However, note that a low percentage value would indicate an output voltage that does not vary much from full-load to no-load condition. A low percentage is preferable and indicates a more constant output voltage.

9. With the power disconnected to the supply, remove R_L from the circuit. Connect a voltmeter across R_b. Energize and deenergize the supply circuit. Record how long it takes for C_2 to discharge.

The *RC* time constant for the R_b and C_2 components should be totally discharged in 5 time constants. The length of time required can be calculated by

$$\text{Time} = 5T \quad \text{where} \quad T = RC$$
$$= 5R_bC_2$$
$$= 1.2\ \text{s}$$

10. With the circuit deenergized, change R_b to 100 kΩ. Energize and deenergize the circuit, and note the discharge time. Calculate the discharge time for this circuit.

Note: If there were no bleeder resistor, how long would circuit discharge take? In some circuits, a discharge time of days is possible. For this reason, never operate a power supply without a bleeder resistor.

11. When the capacitor has fully discharged (power off), replace the 100-kΩ resistor with the original 1-kΩ resistor.

12. With the circuit deenergized and R_L disconnected, connect an oscilloscope across the power supply output

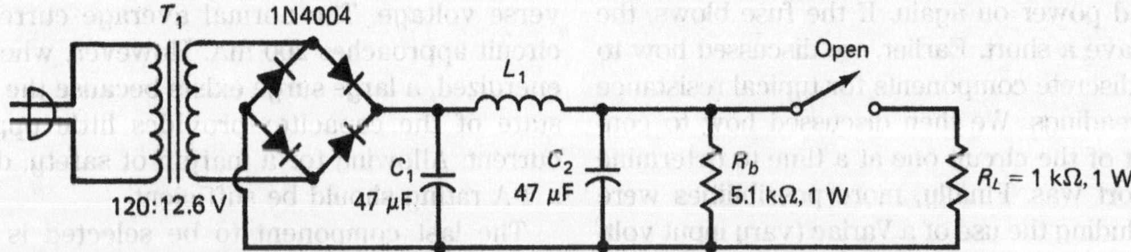

Fig. 28-5.2 Full-wave test circuit.

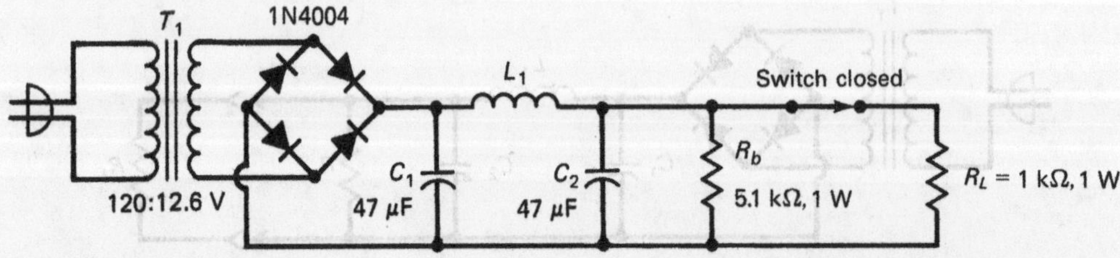

Fig. 28-5.3 Full-wave circuit loaded.

terminals. Energize the circuit. With the oscilloscope correctly adjusted, measure the peak-to-peak ripple voltage across R_b. Record this value and the calculated value of V_{rms} in Table 28-5.1.

The ripple voltage and ripple percentage provide an indication of the quality or smoothness of a DC voltage. Ripple must be measured under both load and no-load conditions. Ripple is measured with an oscilloscope. The results will be in peak-to-peak values; after the measurement is taken, convert to rms values.

13. Deenergize the circuit.

14. With the circuit deenergized, connect R_L across R_b. Energize and record the peak-to-peak ripple voltage in Table 28-5.1. Calculate the V_{rms} equivalent and record it in Table 28-5.1. Deenergize the circuit.

The full-load ripple should be larger and have a different waveshape than no-load ripple. The ripple waveshape should change from a sawtooth to a sine wave. Under no load, the capacitor remains charged longer, and the inductor provides little effect. When the load is increased, the capacitor discharges faster and the inductor begins filtering more.

15. Determine and record in Table 28-5.1 the ripple percentage from

$$\text{Ripple, \%} = \frac{V_{rms}}{V_{DC}} \times 100$$

A low ripple percentage indicates a high-quality filter. When the load is increased, the ripple voltage increases, and the DC voltage level decreases.

Power Supply Defect

The most common defect in a power supply is an open fuse or circuit breaker. The next most probable problem is an open diode.

Shorted, open, and leaking capacitors are common. The symptoms produced by these defects are usually recognizable to a skilled technician.

16. With the power supply deenergized, remove one diode, as shown in Fig. 28-5.4. Making sure that the circuit is under full load, energize the circuit. Measure and record the DC and ripple levels; note the ripple waveshape.

17. Compare the output characteristics of the open-diode circuit in step 16 with the original circuit characteristics.

A power supply overload or a shorted filter capacitor will result in excessive diode current. This excessive diode current can cause a diode to open. If this happens in a full-wave bridge-type circuit, the circuit will operate as a half-wave circuit. This defect will result in obvious symptoms. The output voltage will drop, the ripple voltage will change, and the ripple frequency should decrease from 120 to 60 Hz.

18. With the circuit energized, measure the DC voltage across the open diode. As soon as you are finished examining the circuit and taking measurements, deenergize and replace the diode.

19. With the circuit deenergized and the diode replaced, remove capacitor C_1, as shown in Fig. 28-5.5. This simulates an open capacitor condition. The capacitor is supposed to charge to the peak value of the input signal. Therefore an open capacitor should cause an output voltage drop. Removal of the capacitor will also cause the ripple to increase.

20. With the load R_L connected, energize; then measure, and record the DC output voltage, ripple voltage, and ripple frequency in Table 28-5.1.

21. Deenergize the circuit and reconnect capacitor C_1.

22. With the circuit deenergized, remove C_2, as shown in Fig. 28-5.6.

23. Energize the circuit and take measurements of V_{DC} and V_{ripple}. Capacitor C_2 is important in removing ripple. Its effect on the DC voltage level may not be

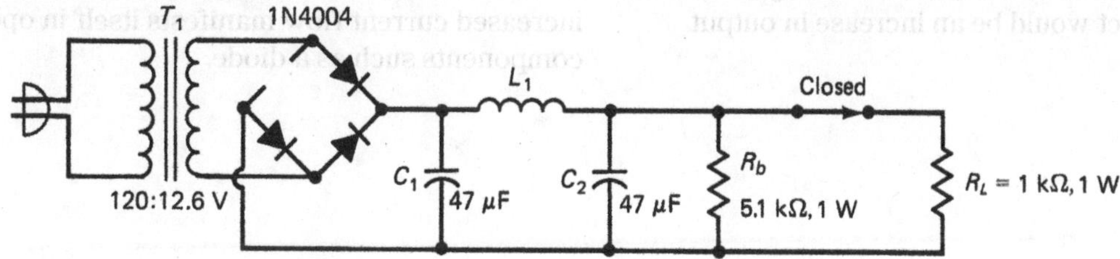

Fig. 28-5.4 Full-wave open diode test circuit.

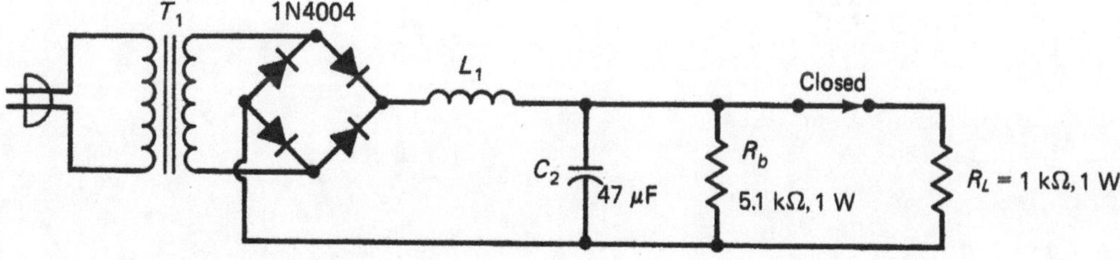

Fig. 28-5.5 Full-wave open input capacitor test circuit.

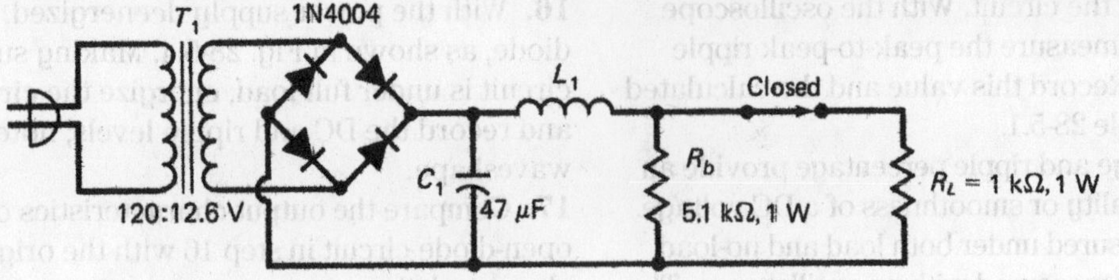

Fig. 28-5.6 Full-wave open output capacitor test circuit.

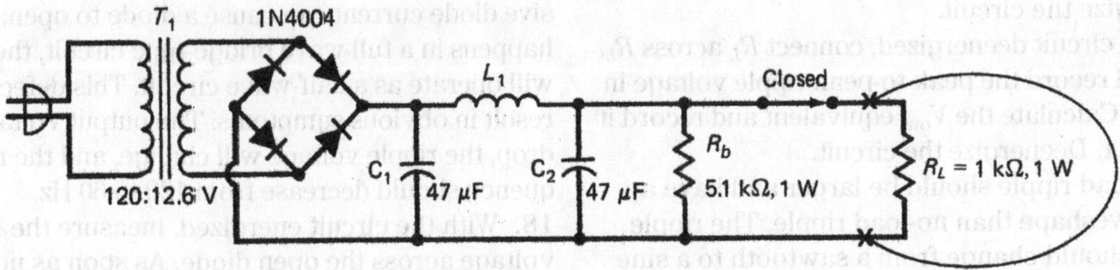

Fig. 28-5.7 Full-wave shorted output circuit (***do not construct!***).

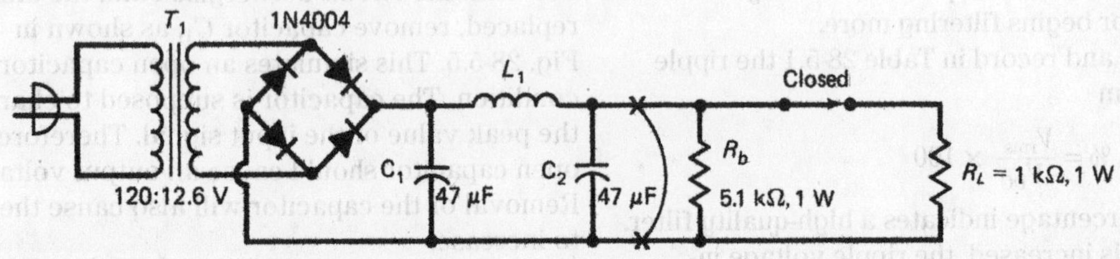

Fig. 28-5.8 Full-wave shorted capacitor circuit (***do not construct!***).

as noticeable compared to the effects of the input capacitor C_1. Both capacitors have an effect on the ripple and DC average, but when compared, the input capacitor C_1 has the greatest effect overall.

24. Deenergize the circuit and disconnect. The effect on short circuits was not discussed in this experiment. It was not presented because electric shorts can cause damage. Consider the shorted circuit in Fig. 28-5.7. The overall effect would be an increase in output current.

The sudden increase in current would continue until the maximum ratings of the components were reached. Circuit opens would occur, most likely a diode.

Another circuit condition (see Fig. 28-5.8) is a shorted or leaky capacitor. A leaky capacitor occurs when the dielectric strength deteriorates and allows more current to flow through the capacitor. Again increased current flow manifests itself in opened components such as a diode.

NAME DATE

QUESTIONS FOR EXPERIMENT 28-5

1. What are the circuit characteristics of a power supply with an open output capacitor?

2. What are the circuit characteristics of a power supply with an open input capacitor?

3. What are the circuit characteristics of a power supply with an open rectifier diode?

4. What does the percentage of full-load ripple indicate?

5. Why is a bleeder resistor important in a power supply?

6. What safety factors must be observed when you are working with a power supply?

TABLE FOR EXPERIMENT 28-5

TABLE 28-5.1

Procedure Step	Value	Measured	Calculated
3	V_{DC}		
3	V_{ripple}		
6	V_{DC}		
6	V_{ripple}		
8	% Regulation		
9	T for C_2		
10	T for C_2		
12	V_{ripple}		
12	V_{rms}		
14	V_{ripple}		
14	V_{rms}		
15	% Ripple		
16	V_{DC}		
	V_{ripple}		
18	V_{DC}		
20	V_{DC}		
	V_{ripple}		
	f_{ripple}		
23	V_{DC}		
	V_{ripple}		

EXPERIMENT RESULTS REPORT FORM

Experiment No.: ____________________ Name: ____________________

Date: ____________________

Experiment Title: ____________________ Class: ____________________

____________________ Instr: ____________________

Explain the purpose of the experiment:

List the first Learning Objective:

OBJECTIVE 1:

After reviewing the results, describe how the objective was validated by this experiment.

List the second Learning Objective:

OBJECTIVE 2:

After reviewing the results, describe how the objective was validated by this experiment.

List the third Learning Objective:

OBJECTIVE 3:

After reviewing the results, describe how the objective was validated by this experiment.

Conclusion:

If required, attach to this form: ☐ Answers to Questions, ☐ Tables, and ☐ Graphs.

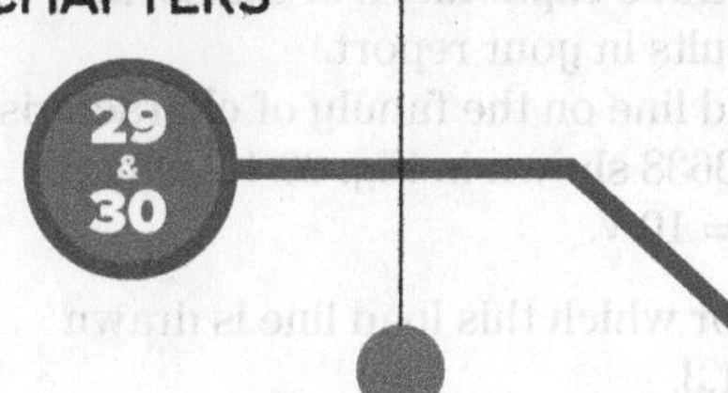

EXPERIMENT 29-1

COMMON-EMITTER AMPLIFIER

LEARNING OBJECTIVES

At the completion of this experiment, you will be able to:

- Study the phase relationship of the input signal versus the output signal of a common-emitter amplifier.
- Investigate the purpose of biasing in a common-emitter amplifier.
- Plot a load line on the characteristic curves of a transistor.

SUGGESTED READING

Chapters 29 and 30 *Basic Electronics*, Grob/Schultz, twelfth edition

INTRODUCTION

This experiment uses a bipolar junction transistor (BJT) to amplify a small value of audio-range signal. The transistor polarity will be verified (*pnp* or *npn*), and the bias supply will be adjusted to allow a maximum value of undistorted output voltage for the sine-wave input. The concept of gain (A_V) will be examined by varying the value of the collector resistor. The fundamental aspects of the common-emitter configuration can be determined from an analysis of the data obtained in the experiment.

Bias is defined as electrical force applied to a semiconductor transistor for the purpose of establishing a reference level for the operation of the device. In this experiment, a common-emitter *pnp* transistor amplifier is biased for operation under different levels of reference. The reference levels are established with respect to the load line and the *Q* point of operation. Consistent with the principles discussed in theory textbooks, amplification of a sine wave will have varying output results depending upon the biasing of the transistor. The results obtained in this experiment should provide sufficient data for analysis of the concept of transistor bias.

EQUIPMENT

Signal generator
Oscilloscope
Low-voltage power supply
Microammeter
Test leads
DMM

COMPONENTS

(1) 100-kΩ potentiometer
(2) 10 μF at 25-V capacitors
(1) 2N3638 transistor or equivalent

Resistors (all 0.25 W):

(1) 470 Ω	(1) 10 kΩ
(1) 1 kΩ	(1) 4.7 kΩ

PROCEDURE

1. Using a DMM test the transistor to determine the emitter-to-base and base-to-collector junctions. This is a test of the diode action of the transistor. Forward bias is positive potential to the *p* material and negative potential to the *n* material, indicated by a low-resistance reading with the ohmmeter. Emitter-to-collector junctions will have a high resistance reading.

2. Connect the circuit shown in Fig. 29-1.1.

3. With the power supply (V_{CC}) adjusted at −10 V, obtain −5 V at the collector by adjusting the potentiometer (R_B). This means that one-half the voltage (V_{CC}) of the supply is developed both across the collector to ground and across the collector resistor itself.

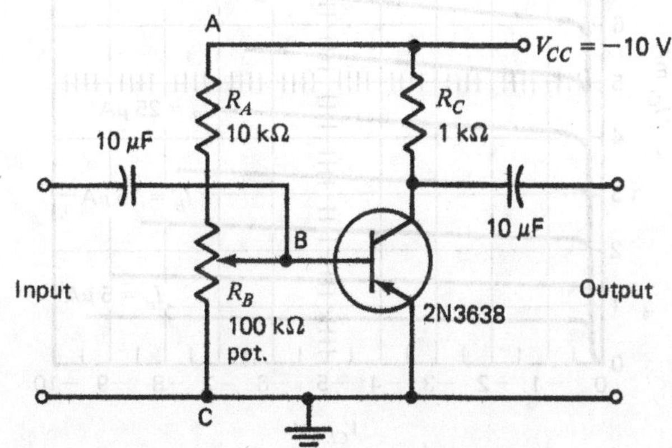

Fig. 29-1.1 *PNP* common-emitter amplifier.

Note: The wiper of the potentiometer divides the potentiometer so that part of its resistance is added to the value of R_A. The remaining resistance is equal to R_B. This is called *voltage divider bias.*

4. Apply a 30-$mV_{p\text{-}p}$ input signal to the amplifier.

Note: It may be necessary to attenuate the output of the signal generator. This can be done by creating a voltage divider network across the terminals of the signal generator. For example, a 10-kΩ resistor in series with a 1-kΩ resistor will divide the voltage into 11 equal parts, 1 part across 1 kΩ and 10 parts across 10 kΩ. The frequency of the input signal should be set at 1 kHz.

5. Measure and record in Table 29-1.1 the output voltage from the amplifier.
6. Measure and record in Table 29-1.1 the DC voltage from the base to the emitter.
7. Calculate the gain of the amplifier (A_V):

$$A_V = \frac{V_{\text{out}}}{V_{\text{in}}}$$

Record the results in Table 29-1.1.

8. Disconnect the power supply from the circuit (V_{CC} and ground). Carefully remove the potentiometer wiper lead (point B) from the circuit, and measure the resistance of R_A (points A and B) and R_B (points B and C). Record the results in Table 29-1.1.
9. Change the value of R_C to 470 Ω and repeat steps 2 to 8. Record the results in Table 29-1.2.
10. Change the value of R_C to 4.7 kΩ and repeat steps 2 to 8. Record the results in Table 29-1.3.
11. Determine the phase relationship of the input and output signals by obtaining a second oscilloscope lead and using the dual-trace capabilities of the oscilloscope. Note the results in your report.
12. Draw a DC load line on the family of characteristic curves for a 2N3638 shown in Fig. 29-1.2, where R_C = 1 kΩ and V_{CC} = 10 V.

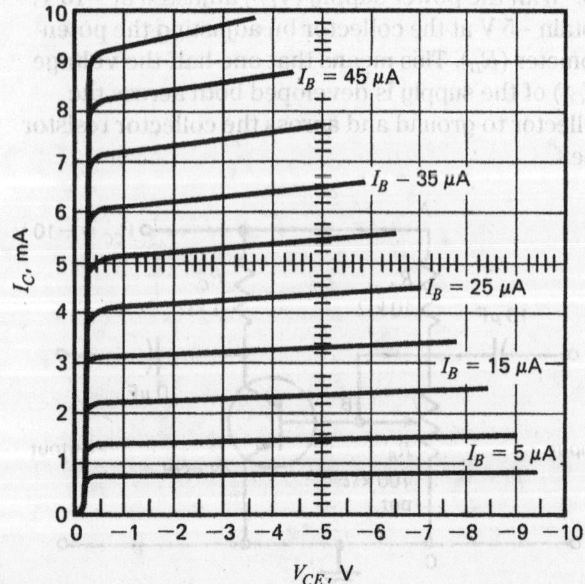

Fig 29-1.2 2N3638 family of curves (approximate).

Note: The circuit for which this load line is drawn appears in Fig. 29-1.3.

13. Connect the circuit of Fig. 29-1.3.

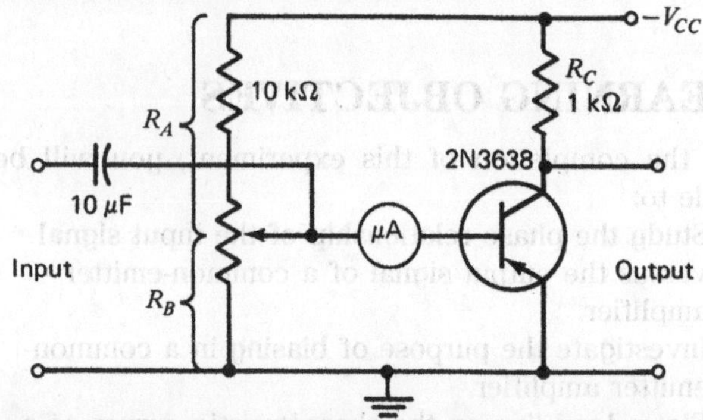

Fig. 29-1.3 Common-emitter amplifier.

Note: Voltages are negative for the ***pnp*** transistor, and for convenience, the following abbreviations will be used:

Q = quiescent, the point or conditions under which the currents and voltages in a transistor exist when no signal is applied = DC bias conditions
I_B = base current
I_C = collector current
V_{CC} = power supply voltage
V_{CE} = collector-to-emitter voltage
V_C = collector-to-ground voltage
V_{BE} = base-to-emitter voltage

DC Characteristics

14. Check all the connections and polarity. Apply –10 V, which is equal to V_{CC}.
15. Adjust the bias potentiometer to obtain –5.0 V, which is equal to V_{CEQ} = ½ V_{CC}.
16. Measure and record I_{BQ} and I_{CQ} on the characteristic curve sheet shown in Fig. 29-1.4.

Note: I_{CQ} cannot be measured directly. Use Ohm's law to determine the current through R_C.

17. Measure and record V_{BEQ} in Table 29-1.4.
18. Disconnect V_{CC}, and carefully disconnect the microammeter from the circuit in order to measure the values of R_A and R_B. Measure and record these values in Table 29-1.4.
19. Reconnect the circuit, and repeat steps 16 to 18, with the bias potentiometer set at V_{CEQ} = –8.5 V and –1.5 V. (Record results in Tables 29-1.5 and 29-1.6, respectively.)

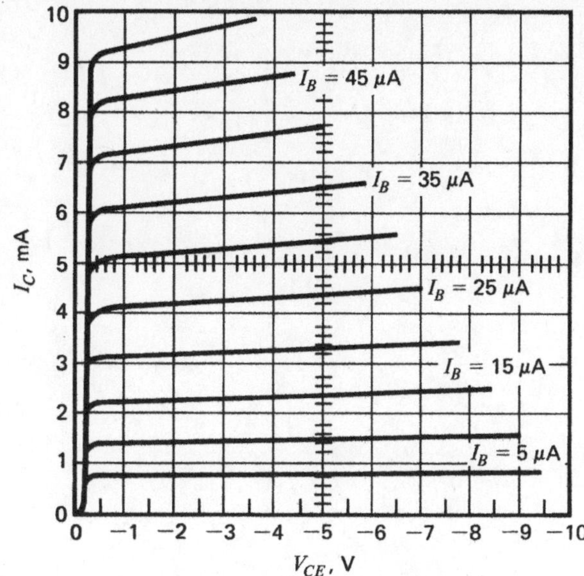

Fig. 29-1.4 2N3638 family of curves (approximate).

AC Characteristics

20. Disconnect or short-circuit the microammeter. Reconnect the power supply and adjust the bias potentiometer to obtain $V_{CEQ} = 5.0$ V.

21. Apply an input signal at 1 kHz and adjust the signal amplitude until the output voltage is equal to

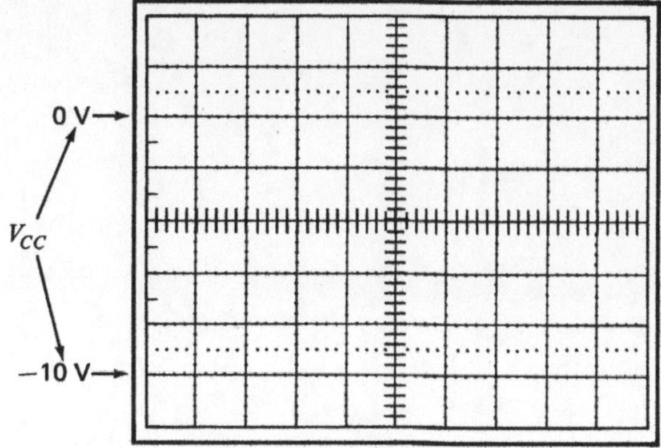

Fig. 29-1.5 Oscilloscope graticule.

6.0 $V_{p\text{-}p}$. Measure and record the input signal (V_{in}), and use this same value for the following procedures (Table 29-1.7).

22. Record the output waveform of the amplifier with the oscilloscope set at DC coupling. When drawing the output waveforms, indicate the parameters of the transistor bias as shown on the graticule in Fig. 29-1.5.

23. Repeat steps 21 and 22 with the bias potentiometer adjusted to obtain $V_{CEQ} = 8.5$ V and 1.5 V. Be sure to use the previous value of input signal. The output signal will change. Use Table 29-1.7 and Fig. 29-1.6 for these data.

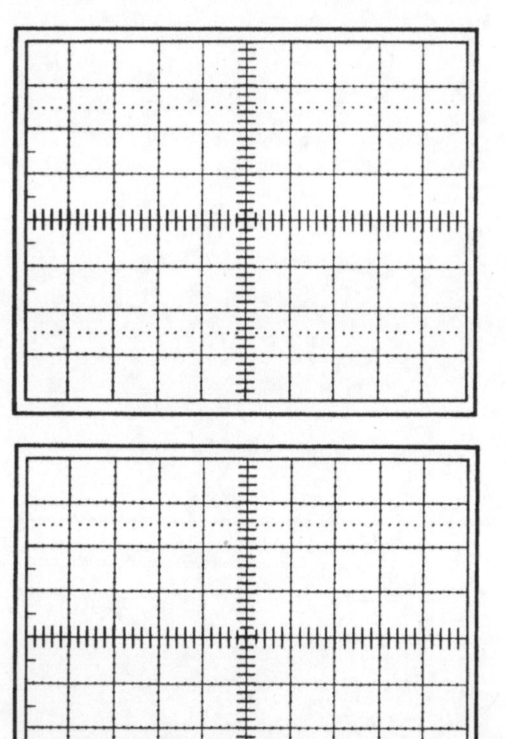

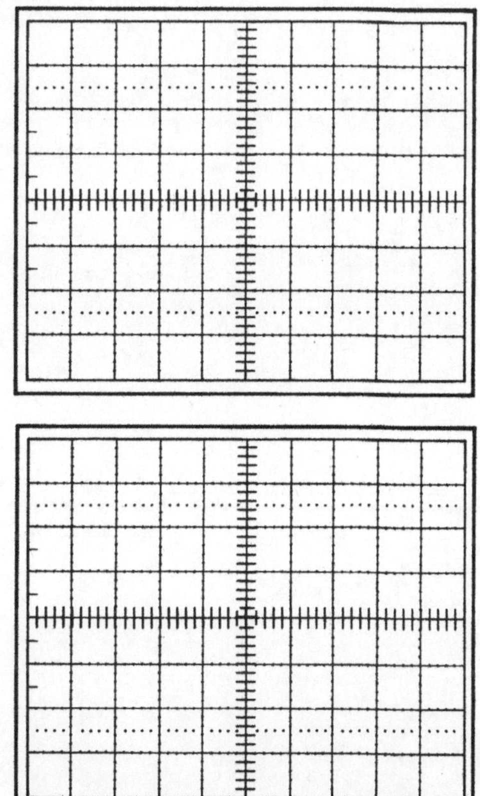

Fig. 29-1.6 Additional oscilloscope graticules.

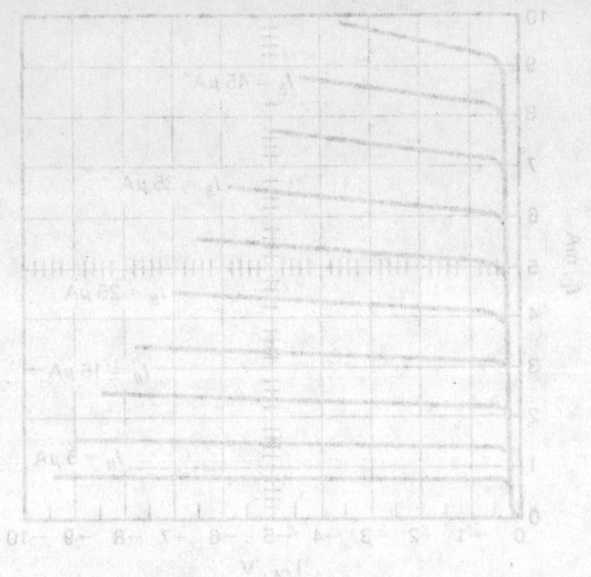

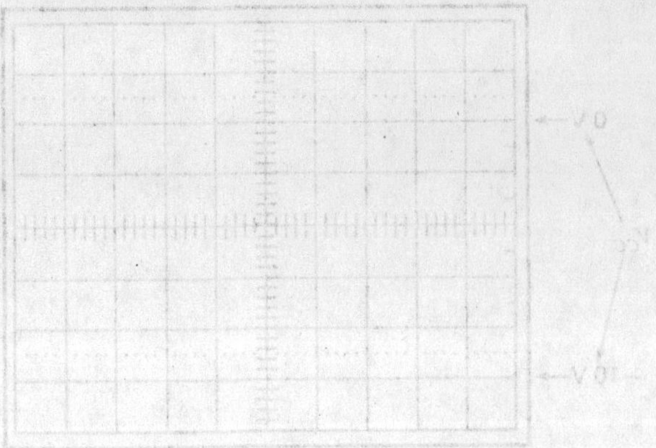

NAME DATE

QUESTIONS FOR EXPERIMENT 29-1

1. What is the purpose of biasing the transistor?

2. What were the values of biasing resistors used?

3. What is the difference between using *npn* and *pnp* transistors with respect to bias supply voltage?

4. What is the phase relationship of the input voltage versus the output voltage?

5. Explain the differences between DC and AC characteristics.

TABLES FOR EXPERIMENT 29-1

TABLE 29-1.1 $V_{in} = 30\ mV_{p-p}$

Procedure Step	Measurement	Value
5	V_{out}	______ V_{p-p}
6	V_{BE}	______ V
7	A_V	______
8	R_A	______ Ω
	R_B	______ Ω

TABLE 29-1.2 $R_C = 470\ \Omega$

Procedure Step	Measurement	Value
9	V_{out}	______ V_{p-p}
	V_{BE}	______ V
	A_V	______
	R_A	______ Ω
	R_B	______ Ω

TABLE 29-1.3 $R_C = 4.7\ k\Omega$

Procedure Step	Measurement	Value
10	V_{out}	______ V_{p-p}
	V_{BE}	______ V
	A_V	______
	R_A	______ Ω
	R_B	______ Ω

TABLE 29-1.4 $V_{CEQ} = -5\ V$

Procedure Step	Measurement	Value
17	V_{BEQ}	______ V
18	R_A	______ Ω
	R_B	______ Ω

TABLE 29-1.5 $V_{CEQ} = -8.5\ V$

Procedure Step	Measurement	Value
19	V_{BEQ}	______ V
	R_A	______ Ω
	R_B	______ Ω

TABLE 29-1.6 $V_{CEQ} = -1.5\ V$

Procedure Step	Measurement	Value
19	V_{BEQ}	______ V
	R_A	______ Ω
	R_B	______ Ω

TABLE 29-1.7

Procedure Step	Measurement	Value
21	V_{in}	______ V_{p-p}
23	$V_{CEQ} = 8.5\ V$	V_{out} = ______ V_{p-p}
	$V_{CEQ} = 1.5\ V$	V_{out} = ______ V_{p-p}

OSCILLOSCOPE GRATICULES FOR EXPERIMENT 29-1

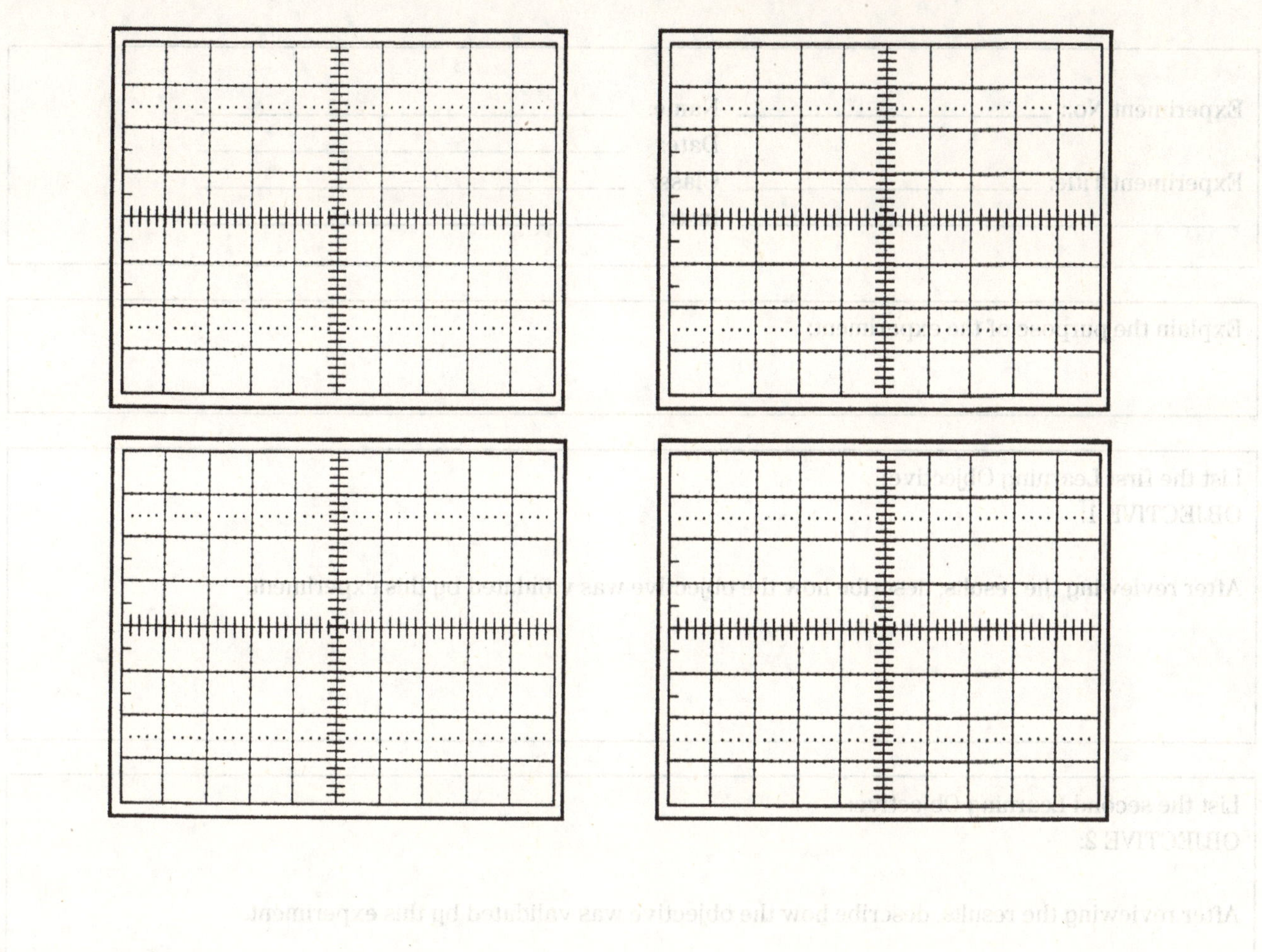

EXPERIMENT RESULTS REPORT FORM

Experiment No.: ______________________ Name: ______________________________

Date: ______________________________

Experiment Title: ______________________ Class: ______________________________

______________________________ Instr: ______________________________

Explain the purpose of the experiment:

List the first Learning Objective:

OBJECTIVE 1:

After reviewing the results, describe how the objective was validated by this experiment.

List the second Learning Objective:

OBJECTIVE 2:

After reviewing the results, describe how the objective was validated by this experiment.

List the third Learning Objective:

OBJECTIVE 3:

After reviewing the results, describe how the objective was validated by this experiment.

Conclusion:

If required, attach to this form: ☐ Answers to Questions, ☐ Tables, and ☐ Graphs.

CHAPTER 31

EXPERIMENT 31-1

FET AMPLIFIER

LEARNING OBJECTIVES

At the completion of this experiment, you will be able to:

- Identify the polarity of a FET.
- Study the circuit components and power supply polarity necessary to permit a FET to function as an amplifier.
- Evaluate the performance of a FET amplifier.

SUGGESTED READING

Chapter 31, *Basic Electronics*, Grob/Schultz, twelfth edition

INTRODUCTION

A junction field-effect transistor (usually referred to as a FET) is a semiconductor device that is controlled like a vacuum tube. In its simplest form, the FET is a layer or channel of *n*-type material that acts like a resistor between its two end terminals called the *source* and the *drain*. On both sides of the channel, *p*-type material forms a gate through which electrons flow (drain current) from source to drain. With current flowing through the *n* channel, the FET is considered to be a *normally on* device, which is also referred to as in an *enhancement mode*. However, when a control voltage is applied to the gate which is a reverse-bias gate-to-source condition, the channel is depleted and the current through the channel decreases. Once the channel has been depleted enough, the FET is said to be *pinched off;* that is, no further drain current will flow. It is important to note, however, that current can still flow while the channel is pinched off. The point is that no further amount of current will flow. See the family of curves in Fig. 31-1.1. It is also important to note that a FET amplifier (common source) will have a *Q* point of operation well into the pinch-off region, allowing for a symmetrical (positive to negative) output voltage swing.

EQUIPMENT

Signal generator
Oscilloscope
DC power supply, 0–15 V
DMM
Test leads

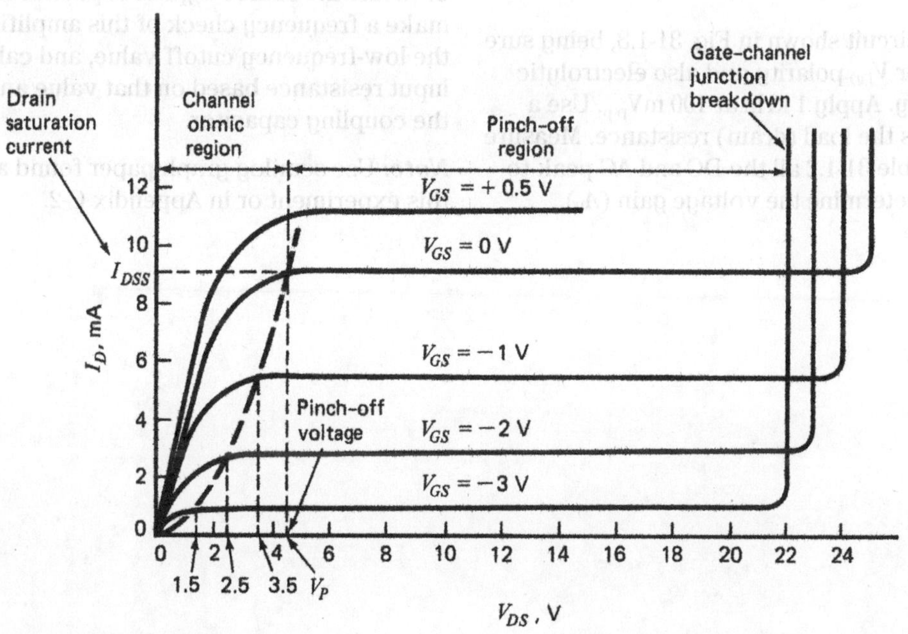

Fig. 31-1.1 *N*-channel FET drain curves.

COMPONENTS

(1) 2N3823 FET, or equivalent

Resistors (all 0.25 W):

(1) 1 kΩ	(1) 5.6 kΩ
(1) 2.2 kΩ	(1) 8.2 kΩ
(1) 4.7 kΩ	(1) 470 kΩ

(2) 0.01-μF capacitors
(1) 100-μF capacitor

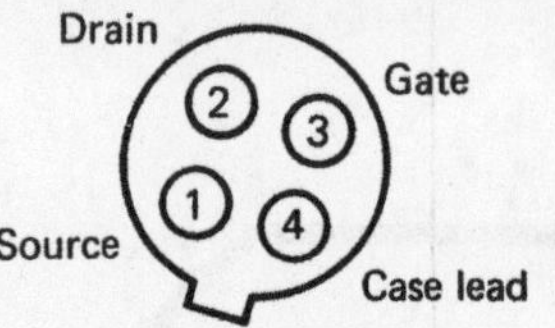

PROCEDURE

1. **Use an ohmmeter on the R × 1000 range, and measure *S – G*, *G – D*, and *S – D* in both directions of lead polarity. Record results in Table 31-1.1. A diode junction should be indicated when measuring *S – G* and *G – D*. The polarity of the FET should be indicated by the forward-bias indication of the ohmmeter. A *p* channel will require a positive-ground, negative-drain potential. An *n*-channel FET will require a negative-ground, positive-drain potential. See Fig. 31-1.2 for the correct lead orientation for *n*- and *p*-channel FETs.**

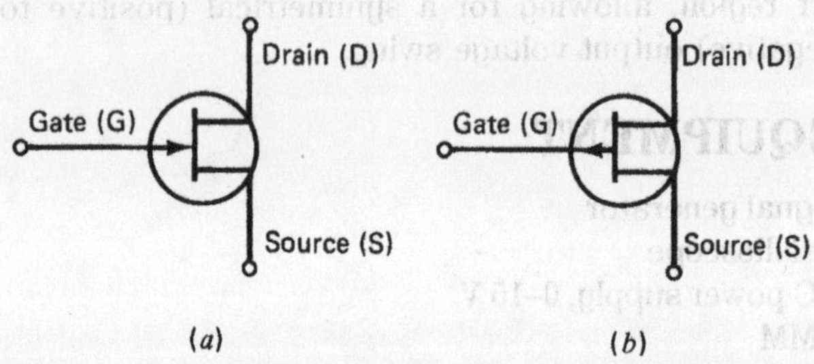

Fig. 31-1.2 (*a*) *N*-channel FET. (*b*) *P*-channel FET.

2. **Connect the circuit shown in Fig. 31-1.3, being sure to observe proper V_{DD} polarity and also electrolytic capacitor polarity. Apply 1 kHz at 100 $mV_{p\text{-}p}$. Use a 2.2-kΩ resistor as the load (drain) resistance. Measure and record in Table 31-1.2 all the DC and AC peak-to-peak voltages. Determine the voltage gain (A_V).**

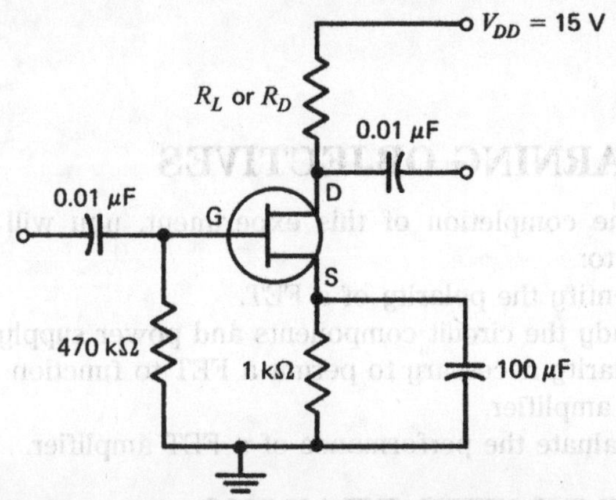

Fig. 31-1.3 *N*-channel FET amplifier using a 2N3823 FET.

3. **Replace the load resistor with 4.7-kΩ, 5.6-kΩ, and 8.2-kΩ resistances, and record all the DC and AC peak-to-peak values around the circuit. Determine A_V. Record in Table 31-1.2.**

4. **Using the 8.2-kΩ resistor as a drain load, increase the power supply voltage to 30 V DC, and note any effects (in Table 31-1.3) upon the amplifier.**

5. **Remove the source bypass capacitor, and note the effect (in Table 31-1.3) upon gain and DC parameters.**

Note: Use V_{DD} = 15 V and R_L = 5.6 kΩ.

6. **While the source bypass capacitor is removed, make a frequency check of this amplifier, determine the low-frequency cutoff value, and calculate the input resistance based on that value and the value of the coupling capacitor.**

Note: Use semilog graph paper found at the end of this experiment or in Appendix C-2.

NAME ____________________ DATE ____________

QUESTIONS FOR EXPERIMENT 31-1

1. What was the polarity of the FET used in this experiment?

2. What is the function of the bypass capacitor? What happens when it is removed from the amplifier circuit?

3. What factors influence the input resistance of the amplifier?

4. What is the overall effect that is created by increasing the value of load resistance?

5. What effects to the amplifier are created by increasing the value of the power supply voltage?

TABLES FOR EXPERIMENT 31-1

TABLE 31-1.1

Procedure Step	Measurement	R, Ω
1	S to G	______
	G to D	______
	S to D	______
Leads reversed	S to G	______
	G to D	______
	S to D	______

TABLE 31-1.2

Procedure Step	R_D, kΩ	V_G(DC)	V_D(DC)	V_S(DC)	V_G(AC)	V_D(AC)	V_S(AC)	A_V
2	2.2	______	______	______	______	______	______	______
3	4.7	______	______	______	______	______	______	______
	5.6	______	______	______	______	______	______	______
	8.2							

TABLE 31-1.3

Procedure Step	V_{DD}	Noted Effects
4	30 V DC	______
		______
		______
5	15 V DC	Capacitor removed: ______
		______
		______

SEMILOG PAPER FOR EXPERIMENT 31-1

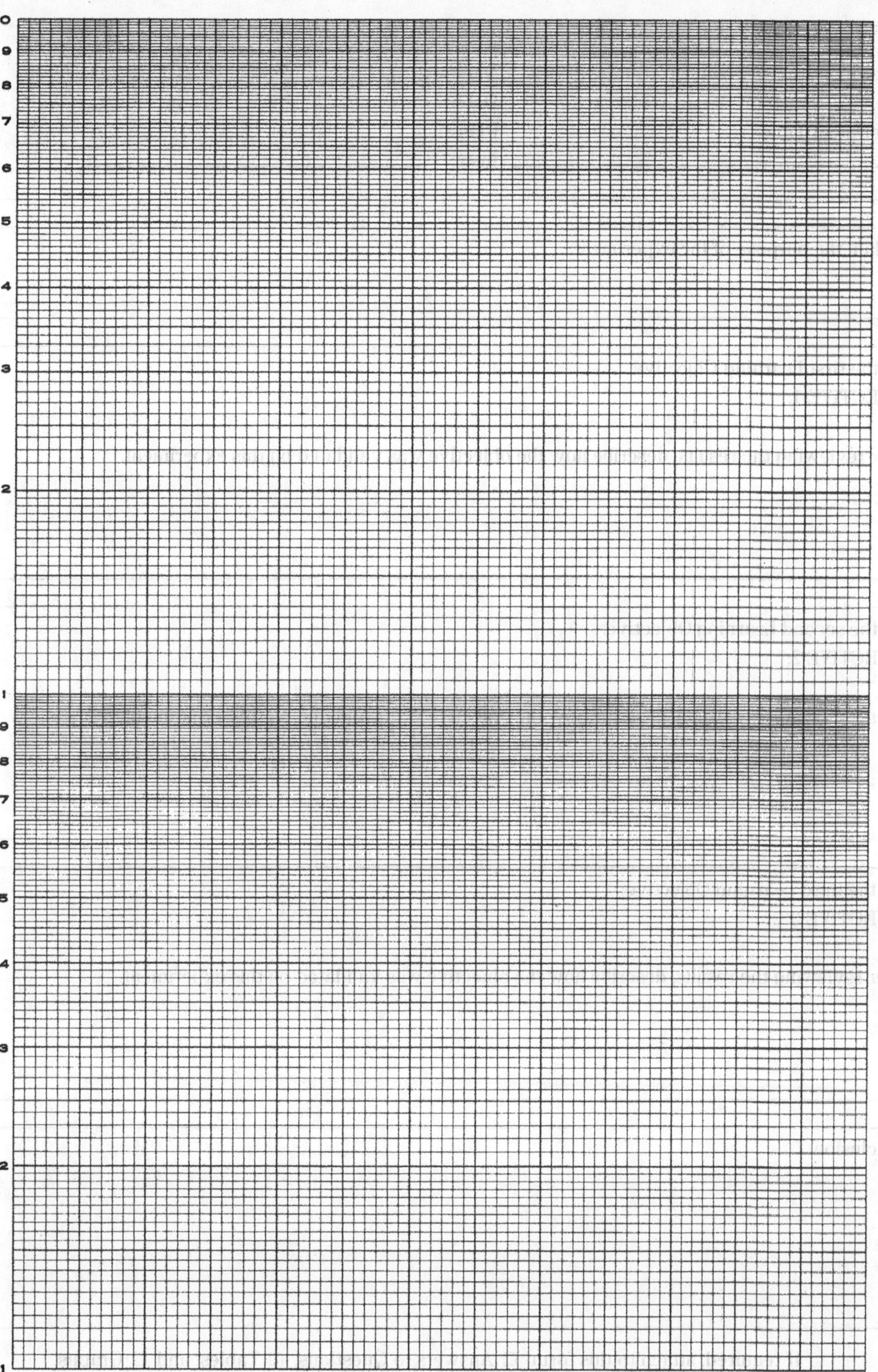

EXPERIMENT RESULTS REPORT FORM

Experiment No.: ______________ Name: ______________

Date: ______________

Experiment Title: ______________ Class: ______________

______________ Instr: ______________

Explain the purpose of the experiment:

List the first Learning Objective:

OBJECTIVE 1:

After reviewing the results, describe how the objective was validated by this experiment.

List the second Learning Objective:

OBJECTIVE 2:

After reviewing the results, describe how the objective was validated by this experiment.

List the third Learning Objective:

OBJECTIVE 3:

After reviewing the results, describe how the objective was validated by this experiment.

Conclusion:

If required, attach to this form: ☐ Answers to Questions, ☐ Tables, and ☐ Graphs.

CHAPTER 32

EXPERIMENT 32-1

TWO-STAGE TRANSISTOR AMPLIFIER

LEARNING OBJECTIVES

At the completion of this experiment, you will be able to:

- Build and test a two-stage amplifier.
- Design a two-stage amplifier for voltage gain.
- Determine the difference between one- and two-stage frequency bandwidth responses.

SUGGESTED READING

Chapter 32, *Basic Electronics*, Grob/Schultz, twelfth edition

INTRODUCTION

Transistors are usually connected in various configurations or stages to perform different tasks. By connecting two transistor amplifiers in the common-emitter configuration, more gain can be obtained from one supply voltage. To accomplish this, the output signal from the first stage is input or coupled into the next stage. This type of amplifier is called a *cascaded amplifier* because it uses more than one transistor.

All the concepts that applied to the single-stage transistor amplifier are applicable in this experiment. The transistor must be properly biased in each stage, and both transistors will have the same polarity: *npn* or *pnp*.

In the circuit shown in Fig. 32-1.1, note the bias resistors on Q_1, the first-stage transistor. Here R_a and R_b provide voltage divider bias, R_c is the collector resistor, and R_e is the emitter resistor.

The bias resistors on Q_2 act exactly the same as those used for Q_1. In fact, they can be the exact same values where R_a and R_b provide voltage divider bias, R_c is the collector resistor, and R_e is the emitter resistor.

Also notice that capacitors C_1 and C_3 block the DC current from the input and the output. Capacitor C_2 is a coupling capacitor that keeps the DC currents isolated between stages and passes or couples the AC output signal from stage 1 into stage 2.

Designing a Two-Stage Amplifier

In a previous experiment, you may recall that a single-stage amplifier's voltage gain was equal to V_{out}/V_{in}. In this experiment, the voltage gain is controlled by the relationship of R_c/R_e. For example, if R_c = 10 kΩ and R_e = 1 kΩ, then the voltage gain is equal to 10, assuming the transistor is operating in the middle of the load line and the common-emitter configuration is used.

Although an amplifier can be designed (through mathematical calculations) to produce a specific

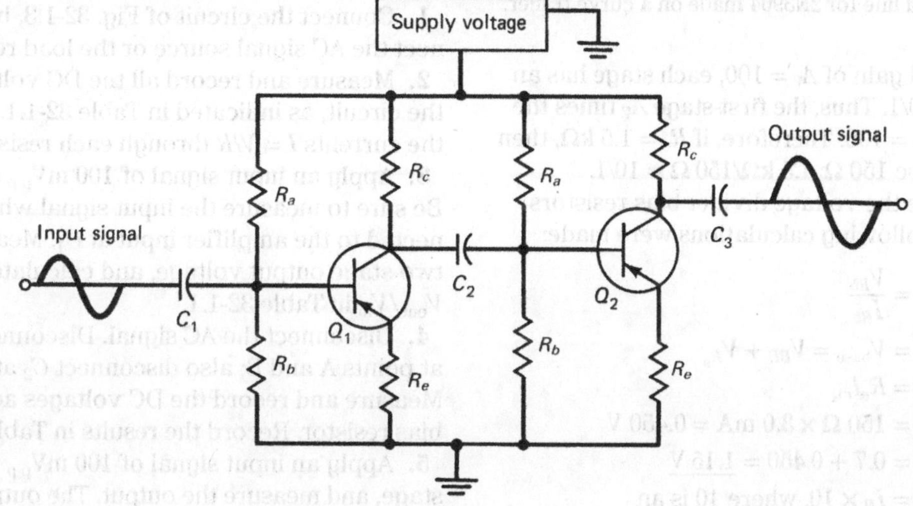

Fig. 32-1.1 Two-stage (cascaded) transistor amplifier.

result, for example, a specific voltage gain, the bias resistor values are often adjusted during testing of the prototype to achieve the desired gain. In this experiment, the test amplifier will produce a voltage gain of about $A_V = 70$ after the bias resistors are adjusted. However, the original design was calculated for each stage to have a voltage gain of 10. Thus, stage 1 gain × stage 2 gain equals 100, as described in the five steps below:

1. A load line was drawn for a single stage where $V_{cc} = 12$ V, $V_{ce} = 6$ V, I_c (max) ≈ 6 mA, where R_c is assumed to be 1.5 kΩ. This load line was drawn on a typical family of curves for a 2N3904, made on a curve tracer.

2. The Q point (middle of the load line) shows $I_c = 3.0$ mA and $I_B = 12.5$ μA in Fig. 32-1.2.

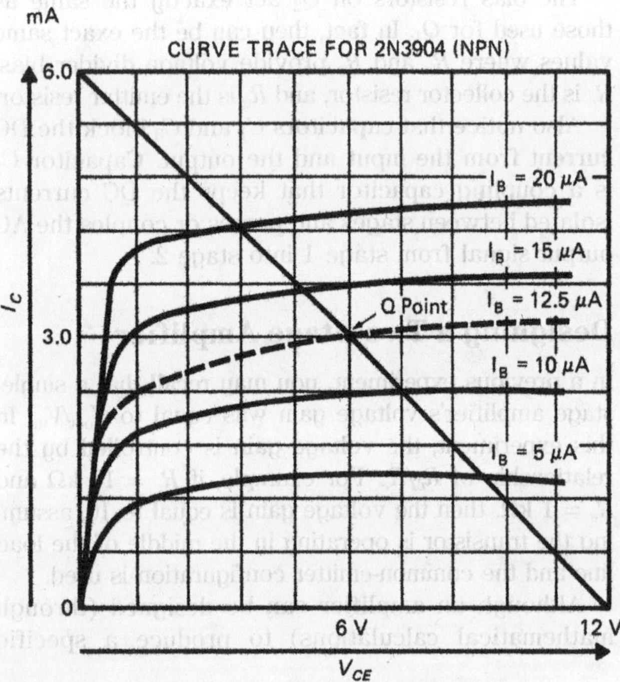

Fig. 32-1.2 Load line for 2N3904 made on a curve tracer.

3. To get a total gain of $A_V = 100$, each stage has an R_c/R_e ratio of 10/1. Thus, the first-stage A_V times the second-stage $A_V = 100$. Therefore, if $R_c = 1.5$ kΩ, then R_e is chosen to be 150 Ω: 1.5 kΩ/150 Ω ≈ 10/1.

4. To determine the voltage divider bias resistors R_a and R_b, the following calculations were made:

$$R_b = \frac{V_{Rb}}{I_{Rb}}$$

$$V_{R_b} = V_{\text{base}} = V_{BE} + V_{R_e}$$

where $V_{R_e} = R_e I_{cQ}$

$$V_{R_e} = 150\ \Omega \times 3.0\ \text{mA} = 0.450\ \text{V}$$

$$V_{R_b} = 0.7 + 0.450 = \underline{1.15\ \text{V}}$$

$I_{R_b} = I_B \times 10$, where 10 is an approximation

$$12.5\ \mu\text{A} \times 10 = \underline{0.125\ \text{mA}}$$

$$R_b = \frac{1.15\ \text{V}}{0.125\ \text{mA}} = \underline{9.2\ \text{k}\Omega}$$

$$R_a = \frac{V_{Ra}}{I_{Ra}}$$

$$V_{R_a} = V_{cc} - V_{R_b} = 12 - 1.15 = \underline{10.85\ \text{V}}$$

$$I_{R_a} = I_{R_b} + I_b = 0.125\ \text{mA} + 12.5\ \mu\text{A} = \underline{0.1375\ \text{mA}}$$

$$R_a = \frac{10.43\ \text{V}}{0.1375\ \text{mA}} = \underline{76\ \text{k}\Omega}$$

5. Finally, the capacitors are chosen so that their reactance will not create any appreciable voltage drop at the amplifier's lowest operating frequency. A typical approximation is made by using the formula $C = 10/(2\pi f_1 Z_{in})$, where f_1 is the lowest frequency and Z_{in} is the estimated input impedance of the amplifier. For example,

$$C = \frac{10}{6.28(100\ \text{Hz})(1.5\ \text{k}\Omega)} = \underline{10\ \mu\text{F}}$$

EQUIPMENT

Oscilloscope (dual trace)
Signal generator
Variable DC power supply
DMM

COMPONENTS

(2) Transistors—2N3904 *npn* silicon
Resistors, all 0.25 W:
(2) 82 kΩ (2) 1.5 kΩ (1) 1 kΩ (2) 150 Ω
(3) 10 kΩ (1) 100 Ω
(3) 10 μF at 25-V DC capacitors

PROCEDURE

1. Connect the circuit of Fig. 32-1.3, but don't connect the AC signal source or the load resistance yet.

2. Measure and record all the DC voltages around the circuit, as indicated in Table 32-1.1. Also calculate the currents $I = V/R$ through each resistor.

3. Apply an input signal of 100 $mV_{p\text{-}p}$ at 5 kHz. Be sure to measure the input signal when it is connected to the amplifier input at C_1. Measure the two-stage output voltage, and calculate the gain as V_{out}/V_{in} in Table 32-1.1.

4. Disconnect the AC signal. Disconnect the stages at points A and B; also disconnect C_2 at point D. Measure and record the DC voltages across each bias resistor. Record the results in Table 32-1.1.

5. Apply an input signal of 100 $mV_{p\text{-}p}$ to each stage, and measure the output. The output of stage 1 is at point D. The input to stage 2 is at point C.

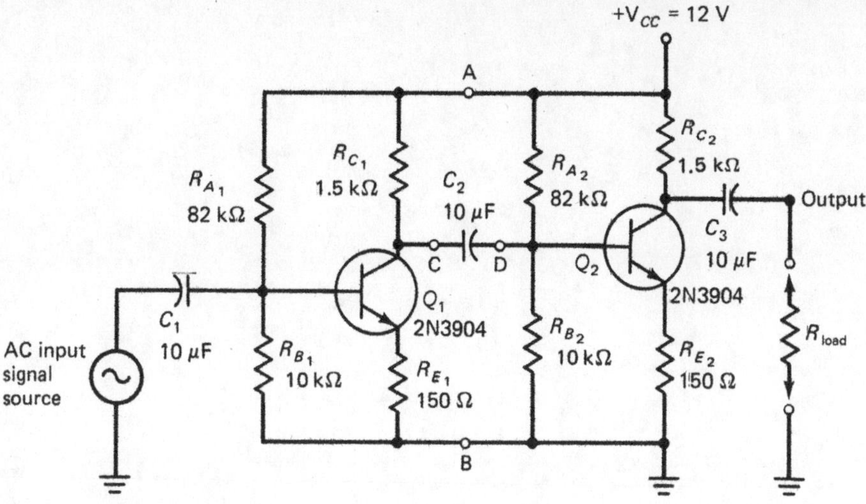

Fig. 32-1.3 Two-stage voltage amplifier.

Record the output voltages, and calculate the gain for each stage in Table 32-1.1.

6. Reconnect both stages, and perform a frequency response test. To do this, apply a 5-kHz signal at 100 mV and decrease the frequency until the 3-dB (half-power) point is reached. The output of stage 2 will show about 70 percent of its previous output voltage. Then increase the frequency until the signal output is 70 percent of its value at 5 kHz. For example, if the 5-kHz output is 10 V, it will roll off to 7 V at the low (f_1) and high (f_2) ends of the amplifier's bandwidth. Calculate the bandwidth (BW) as $f_2 - f_1$. Record all the values according to Table 32-1.1.

7. Connect the load resistor $R_L = 1\ k\Omega$, and measure the output voltage at 10 kHz with $V_{in} = 100$ mV. Record the results in Table 32-1.1.

8. Replace R_L with a 100-Ω resistor and repeat step 7.

9. Replace R_L with a 10-kΩ resistor and repeat step 7.

OPTIONAL

10. Adjust the bias resistors to obtain a gain of 100 as calculated in the introductory material. This step will require you to use a *trial-and-error* process. Try replacing R_{C_2} as a start. On a separate sheet of paper, describe what you did to obtain and verify a gain of 100.

11. Perform a frequency response (bandwidth) test on the single-stage amplifier. Compare the results with the two-stage bandwidth response in your report.

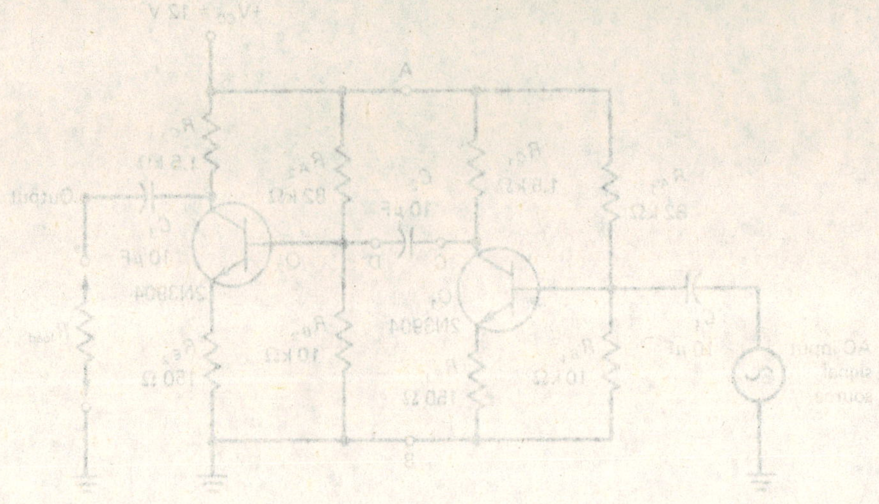

NAME DATE

QUESTIONS FOR EXPERIMENT 32-1

1. With reference to amplifiers, define the term *gain.*

2. What is the general purpose or reason for the creation of the cascading amplifier circuit?

3. When determining the calculated voltage gain of an amplifier, which two resistance values can be used to determine its value? How is this different from a single-stage amplifier? Explain your answer.

4. Describe the importance of the Q point. Explain your answer.

5. Describe the process used to determine the value of the capacitors used for the cascaded amplifier.

TABLES FOR EXPERIMENT 32-1

TABLE 32-1.1

	Resistor	Nominal Value	Measured Voltage	Calculated Current	Voltages with Stages Disconnected (Step 4)
Step 2	R_{a_1}	82 kΩ	______	______	______
	R_{b_1}	10 kΩ	______	______	______
	R_{c_1}	1.5 kΩ	______	______	______
	R_{e_1}	150 Ω	______	______	______
	R_{a_2}	82 kΩ	______	______	______
	R_{b_2}	10 kΩ	______	______	______
	R_{c_2}	1.5 kΩ	______	______	______
	R_{e_2}	150 Ω	______	______	______

Two Stages

Step 3 $V_{in} = 100$ mV, $V_{out} =$ ______, $A_V =$ ______

Note: V_{out} = stage 2 output

Single Stages

Step 5 Stage 1: $V_{in} = 100$ mV, $V_{out} =$ ______ V, $A_V =$ ______

Stage 2: $V_{in} = 100$ mV, $V_{out} =$ ______ V, $A_V =$ ______

Bandwidth

Step 6 At $V_{in} = 100$ mV, $f_1 =$ ______ Hz at ______ V_{out}

BW = ______ $f_2 =$ ______ Hz at ______ V_{out}

Step 7 $R_L = 1$ kΩ, $V_{out} =$ ______ V, $A_V =$ ______

Step 8 $R_L = 100$ Ω, $V_{out} =$ ______ V, $A_V =$ ______

Step 9 $R_L = 10$ kΩ, $V_{out} =$ ______ V, $A_V =$ ______

EXPERIMENT RESULTS REPORT FORM

Experiment No.: ____________________ Name: ____________________

Date: ____________________

Experiment Title: ____________________ Class: ____________________

____________________ Instr: ____________________

Explain the purpose of the experiment:

List the first Learning Objective:

OBJECTIVE 1:

After reviewing the results, describe how the objective was validated by this experiment.

List the second Learning Objective:

OBJECTIVE 2:

After reviewing the results, describe how the objective was validated by this experiment.

List the third Learning Objective:

OBJECTIVE 3:

After reviewing the results, describe how the objective was validated by this experiment.

Conclusion:

If required, attach to this form: ☐ Answers to Questions, ☐ Tables, and ☐ Graphs.

CHAPTER 33

EXPERIMENT 33-1

LIGHT-SENSITIVE DIODES

LEARNING OBJECTIVES

At the completion of this experiment, you will be able to:

- Understand how light energy can be converted to electric current.
- Use a photodiode.
- Understand the construction of a photodiode.

SUGGESTED READING

Chapter 33, *Basic Electronics*, Grob/Schultz, twelfth edition

INTRODUCTION

Although many devices are sensitive to light, the most common types are light-sensitive diodes. These diodes, or photo-sensitive cells, have been used for years to convert or transduce energy from one form to another. For example, a photodiode converts light to electric current, and a light-emitting diode (LED) converts electric current to light.

Changes in the communications industry have created a demand for light-sensitive diodes to be used with fiber-optic cable. This cable, made of a special type of glass, is replacing copper wire in many applications throughout the world.

Instead of varying (modulating) the frequency and amplitude of an AC signal, the intensity of light is modulated and sent through this fiber-optic cable. Thus, diodes are used to convert the light to electric current so that audio, visual, and data transmission can be used in electronic systems.

Although the light in fiber-optic cable is provided by a laser, light-emitting diodes are similar because they convert, or use, electric energy to create light. Light-sensitive diodes, or photodiodes, are almost always found in the receiving end of fiber cable.

Figure 33-1.1 shows the typical construction and use of a light-sensitive diode. This diode, also called a solar cell, converts light to electric current. Notice that the symbol is like a battery or cell, except that the Greek letter λ ("lambda") is shown next to the cell to indicate light or light wave. Also notice that the *p*-type material is very thin so that the light energy can penetrate and excite the *p-n* junction, causing electrons to flow, as the following procedure demonstrates.

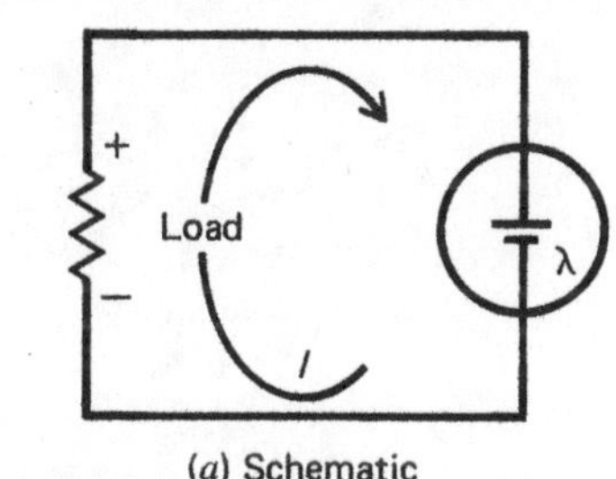

(*a*) Schematic

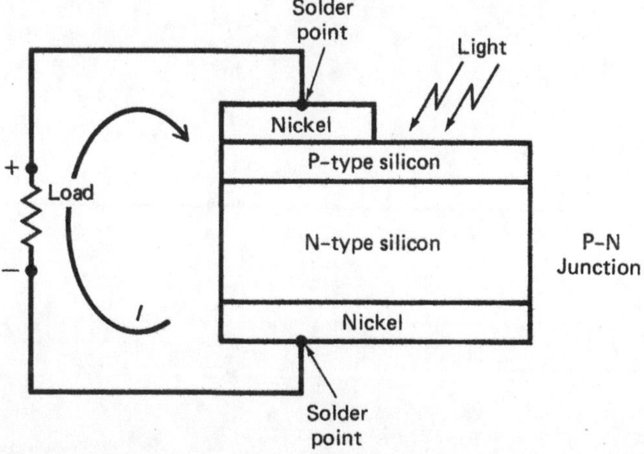

(*b*) Cross section

Fig. 33-1.1 Solar cell or light-sensitive diode schematic and cross section.

EQUIPMENT

DMM
Lamp (light source)
Leads or solder and soldering iron

COMPONENTS

(2) Solar cells (IR#S1M) or any similar *pn* silicon photodiodes
(1) 330-Ω, 1.0-W resistor
(4) Bulbs, one each, 25, 60, 100, and 150 W

PROCEDURE

1. Connect the circuit shown in Fig. 33-1.2, using the 150-W bulb.

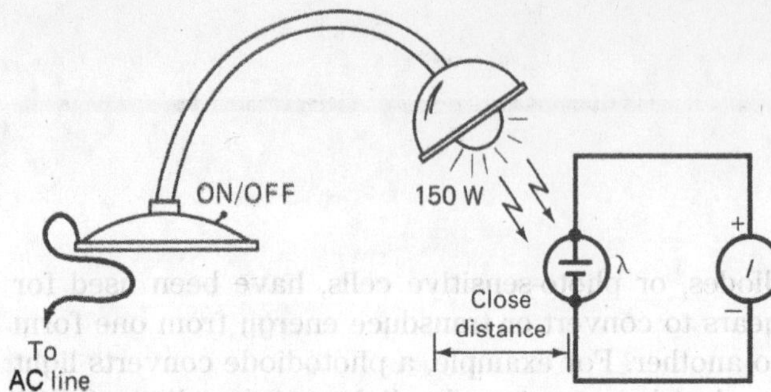

Fig. 33-1.2 Light-sensitive circuit.

Note: Any variable light source can be used instead of four bulbs, as long as it supplies enough light (variable) to produce adequate results.

2. Position the lamp (ON) as close as possible to the diode or cell so that the maximum current flows without overheating the circuit. Record the current in Table 33-1.1.

3. Remove the ammeter, and put a 330-Ω resistor in place of the ammeter. Measure and record in Table 33-1.1 the voltage across the resistor. Keep the lamp in the same position.

4. Turn off the lamp, and replace the 150-W bulb with a 100-W bulb; repeat steps 2 and 3.

5. Turn off the lamp, and replace the 100-W bulb with a 60-W bulb. Repeat steps 2 and 3.

6. Turn off the lamp, and replace the 60-W bulb with a 25-W bulb. Repeat steps 2 and 3.

7. Connect a second diode or cell in series (as close as possible) with the first cell. Repeat steps 2 to 6.

8. Connect a second diode or cell in parallel (as close as possible) to the first cell. Repeat steps 2 to 6.

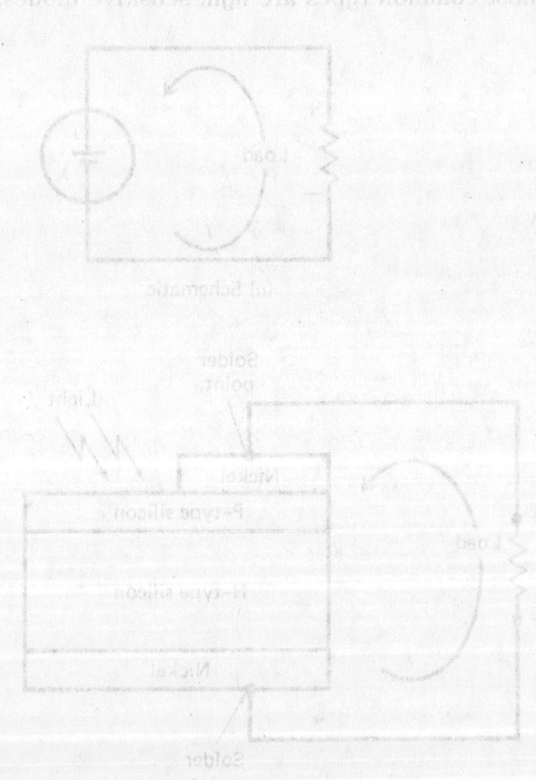

NAME DATE

QUESTIONS FOR EXPERIMENT 33-1

1. **Describe how these types of photocells could be used in three different applications.**

2. **Describe the difference between a zener diode and a regular diode.**

TABLE FOR EXPERIMENT 33-1

TABLE 33-1.1

Step(s)	Lightbulb or Wattage	Current I	Volts across 330 Ω
2 and 3			
4			
5			
6			
Two Devices in Series			
7			
Two Devices in Parallel			
8			

EXPERIMENT RESULTS REPORT FORM

Experiment No.: ______________________ Name: ______________________

Date: ______________________

Experiment Title: ______________________ Class: ______________________

______________________ Instr: ______________________

Explain the purpose of the experiment:

List the first Learning Objective:

OBJECTIVE 1:

After reviewing the results, describe how the objective was validated by this experiment.

List the second Learning Objective:

OBJECTIVE 2:

After reviewing the results, describe how the objective was validated by this experiment.

List the third Learning Objective:

OBJECTIVE 3:

After reviewing the results, describe how the objective was validated by this experiment.

Conclusion:

If required, attach to this form: ☐ Answers to Questions, ☐ Tables, and ☐ Graphs.

CHAPTER

EXPERIMENT 33-2

ZENER DIODES FOR REGULATION AND PROTECTION

LEARNING OBJECTIVES

At the completion of this experiment, you will be able to:

- Understand how zener diodes operate.
- Use a zener diode to protect circuits from excessive current or voltage.
- Use a zener diode as a simple voltage regulator.

SUGGESTED READING

Chapter 33, *Basic Electronics*, Grob/Schultz, twelfth edition

INTRODUCTION

Zener diodes are similar to, but also different from, ordinary silicon *p-n* junction diodes. Zener diodes are similar because both these types of diodes are made from *p* and *n* material. However, they differ because in the zener diode, the *p* and *n* material is "doped" so that it can be used in the reverse direction. Recall that a typical *p-n* junction diode (silicon) allows current to flow in the forward direction when about 0.7 V is across it, causing forward bias.

Generally, a *p-n* junction diode is found in a rectifier circuit where it passes only one direction of current flow from an AC signal. Current is blocked when the *p-n* junction diode is reverse-biased, unless the reverse voltage reaches the breakdown level, which is usually around 5 V. However, the zener diode is designed to perform a different task. It is used in its reverse direction so that current will flow through it when a particular voltage reverse-biases the zener diode. Like the *p-n* junction diode, the zener diode allows current to flow when about 0.7 V is across it in the forward direction. The characteristics of an 8-V zener diode are shown in Fig. 33-2.1.

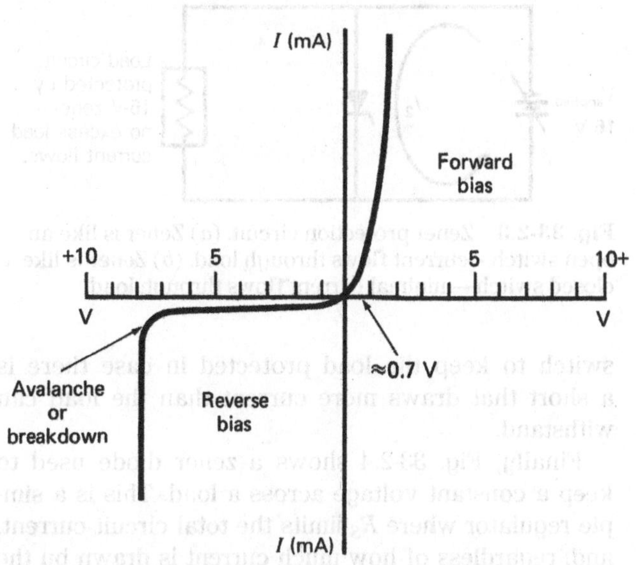

Fig. 33-2.1 The 8-V zener diode characteristics.

Notice that when 8 V is across the zener diode, in the reverse direction, an "avalanche" or breakdown of the *p-n* junction occurs, and the zener is then like a closed switch. Before 8 V is across the zener, only a small leakage current flows. Figure 33-2.2 shows the symbol for a zener diode where current will flow when 8 V is across the zener.

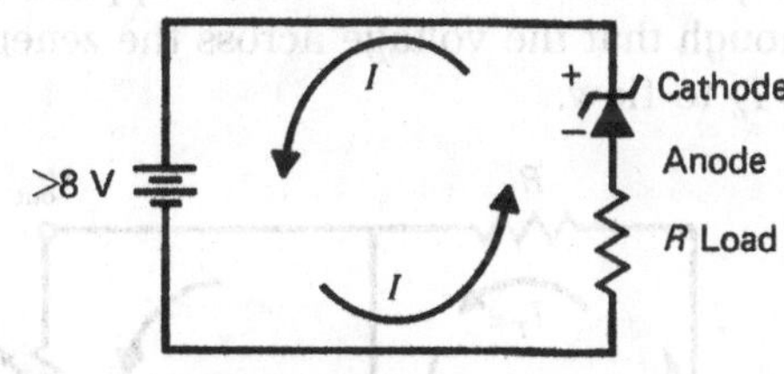

Fig. 33-2.2 Zener diode used as a switch.

The zener diode cathode (−) is on the positive side of the battery, and its anode (+) is on the negative side. When 8 V or more is across the diode, current will flow. Of course, the load has some resistance, which is why more than 8 V is required here.

However, zener diodes are not usually used in this configuration. They are usually put in parallel with other components requiring voltage regulation or protection.

In Fig. 33-2.3, the zener was selected to protect the load from too much current. The load circuit will be destroyed if the voltage across it is 15.5 V or more, resulting in too much current: $I_L = V_Z/R_L$. While V applied is less than 15 V, the zener is like an open switch and the current I_L is all flowing through the load. (See Fig. 33-2.3*a*.) However, when V applied reaches 15 V, the zener acts as a closed switch (0 Ω) so that very little current can flow through the load (see Fig. 33-2.3*b*). In this way, the zener protects the load circuit. In other words, the zener acts as a protection

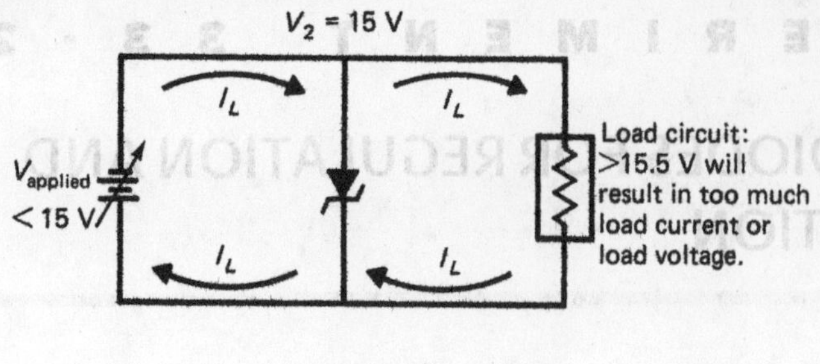

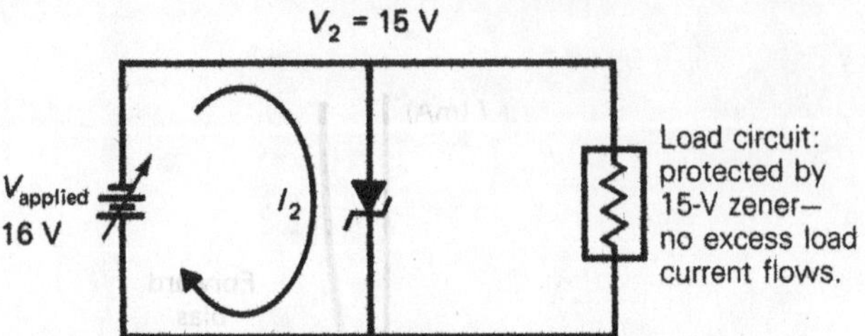

Fig. 33-2.3 Zener protection circuit. (*a*) Zener is like an open switch—current flows through load. (*b*) Zener is like a closed switch—minimal current flows through load.

switch to keep the load protected in case there is a short that draws more current than the load can withstand.

Finally, Fig. 33-2.4 shows a zener diode used to keep a constant voltage across a load. This is a simple regulator where R_S limits the total circuit current, and, regardless of how much current is drawn by the load, the zener diode keeps the load voltage constant. However, this circuit assumes that V applied remains high enough that the voltage across the zener allows current I_Z to flow.

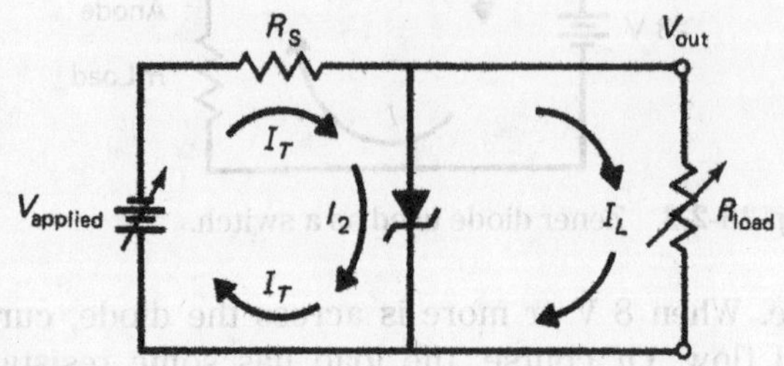

Fig. 33-2.4 Simple zener diode voltage regulator.

Although the zener diode protection circuit and the zener diode regulator circuit may appear similar, remember that the difference is simple. In Fig. 33-2.3, the protection zener is *not* biased ON during normal operation—it conducts only when too much voltage is seen across it or the load. In Fig. 33-2.4, the zener is purposely biased ON during normal operation so that it keeps a constant voltage across the load.

The following procedure will validate the basic operation of a zener diode. However, keep in mind that not all aspects of a zener diode are covered here. A zener diode, like any semiconductor device, has numerous properties and characteristics that require extensive testing to verify. This experiment will introduce you to the most widely used application of a zener diode.

EQUIPMENT

Variable DC power supply (0–30 V)
DMM
Ammeter (or additional DMM if available)

COMPONENTS

(1) 1-kΩ resistor (0.25 W)
(1) 330-Ω resistor (1 W)
(1) 4.7-kΩ resistor (0.25 W)
(1) 100-Ω resistor (0.25 W)
(1) Zener diode, 5 V (1 W)

PROCEDURE

1. Connect the circuit shown in Fig. 33-2.5.

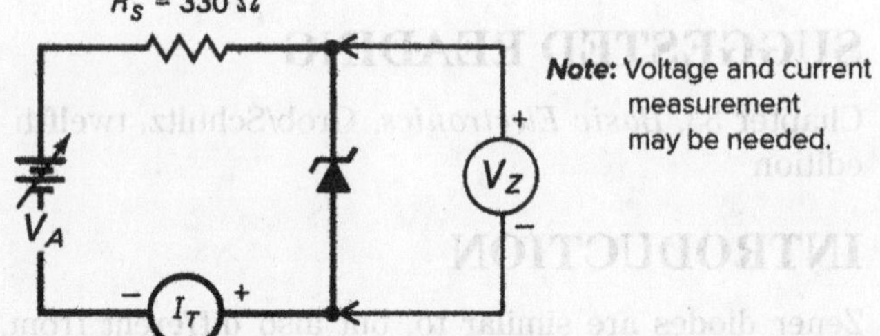

Fig. 33-2.5 Zener diode circuit.

2. Turn on the power supply, and slowly increase the applied voltage until 5 V appears across the zener diode. Measure and record the applied voltage at this point in Table 33-2.1.

3. Measure and record the current I_T in Table 33-2.1.

4. Measure the voltage across R_S, and record the results in Table 33-2.1. Calculate the current through R_S, and record the results next to I_T (measured) in Table 33-2.1.

5. Increase the applied voltage by 2 V, and repeat steps 3 and 4. Use the voltmeter to verify that V applied is 2 V greater. Be sure to measure and record V_Z and V_A.

6. Increase the applied voltage by 2 V more (4 V greater than step 2), and repeat steps 3 and 4. Be sure to measure and record V_Z.

7. Connect a 4.7-kΩ resistor across the zener diode, keeping the rest of the circuit the same, as shown in Fig. 33-2.5.

8. Repeat steps 2 through 4 for the circuit with the 4.7-kΩ load connected. Turn off the power when you have finished.

9. Repeat steps 2 through 4 with a 100-Ω load resistor. Turn off the power when you have finished.

10. Remove any load resistors, and connect the zener in the opposite polarity. Vary the power supply voltage between 0 and 8 V, and measure the voltage across the zener at 1-V intervals. Also, measure I_T and record the results in Table 33-2.2.

NAME ______________________________ DATE __________

QUESTIONS FOR EXPERIMENT 33-2

1. How does a zener diode act when it is connected in the forward-bias manner?

2. How does a zener diode operate when it is connected as a voltage regulator?

3. How does a zener diode operate when it is connected as a protection device?

4. Explain the difference between a zener diode and a *p-n* junction diode.

5. What was the purpose of R_S in the circuit in Fig. 33-2.5?

TABLES FOR EXPERIMENT 33-2

TABLE 33-2.1

Step(s)	V_Z Measured, V	V Applied (V_a) Measured, V	I_T Measured, mA	I_T Calculated, mA	V_{RS} Measured, V
2–4	5	______	______	______	______
5					
$V_a + 2$ V	______	______	______	______	______
6					
$V_a + 4$ V	______	______	______	______	______
8					
$R_L = 4.7$ kΩ	5	______	______	______	______
9					
$R_L = 100$ Ω	5	______	______	______	______

TABLE 33-2.2 Step 10

V Applied (V_a), V	V_Z, V	I_T, mA
0		
1		
2		
3		
4		
5		
6		
7		
8		

EXPERIMENT RESULTS REPORT FORM

Experiment No.: ______________________ Name: ______________________

Date: ______________________

Experiment Title: ______________________ Class: ______________________

______________________ Instr: ______________________

Explain the purpose of the experiment:

List the first Learning Objective:

OBJECTIVE 1:

After reviewing the results, describe how the objective was validated by this experiment.

List the second Learning Objective:

OBJECTIVE 2:

After reviewing the results, describe how the objective was validated by this experiment.

List the third Learning Objective:

OBJECTIVE 3:

After reviewing the results, describe how the objective was validated by this experiment.

Conclusion:

If required, attach to this form: ☐ Answers to Questions, ☐ Tables, and ☐ Graphs.

CHAPTER 33

EXPERIMENT 33-3

SILICON-CONTROLLED RECTIFIER

LEARNING OBJECTIVES

At the completion of this experiment, you will be able to:

- Understand how silicon-controlled rectifier (SCR) devices operate.
- Use an Ohmmeter to test an SCR.
- Use an SCR in both DC and AC circuits.

SUGGESTED READING

Chapter 33, *Basic Electronics*, Grob/Schultz, twelfth edition

Important Operational Note: The SCR is a high-current device and will not usually work with low-current lab equipment. Therefore, it is suggested that you use the MultiSim program for this experiment. Or, consult with your instructor about using a real SCR, if applicable.

INTRODUCTION

An SCR or Silicon Controlled Rectifier (trade name by General Electric) is a four-layer (*p-n-p-n*) current-controlling device, also known as a type of thyristor because it has four layers. You can think of it as two transistors: one *npn* and one *pnp* connected together at the internal *p-n* junctions, as shown in Fig. 33-3.1. Notice the three terminals: anode, gate, and cathode (K is for the cathode, not to be confused with C for collector).

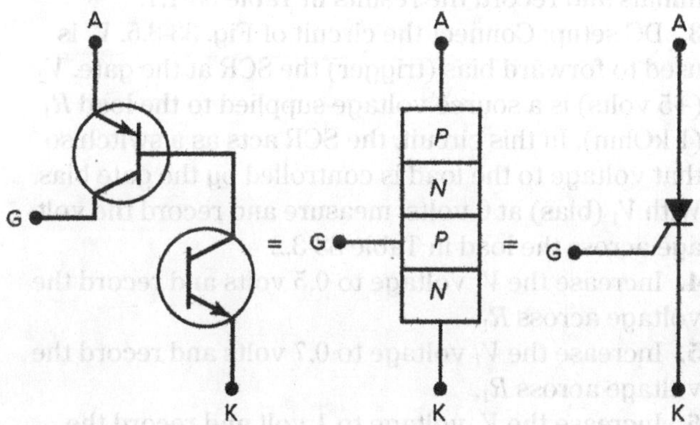

Fig. 33-3.1 SCR circuit and *pnpn* junction.

The gate is connected to the *p* section between two N material sections. Also, an SCR is considered a unidirectional device, which means current flows in one direction when the SCR is triggered or turned on. To trigger an SCR, a small voltage at the gate is applied, and this will allow a large current flow through the device: anode to cathode.

Because the SCR is constructed to handle high power or large current, it is often used in power supplies, motors, heating and air-conditioning systems, power protection circuits, and other high-current flow applications. In fact, some SCRs are much larger in size than the diodes or transistors you have used in these exercises and have built-in heat sinks (metal) to dissipate heat easily, as the large amount of current is flowing as shown in Fig. 33-3.2.

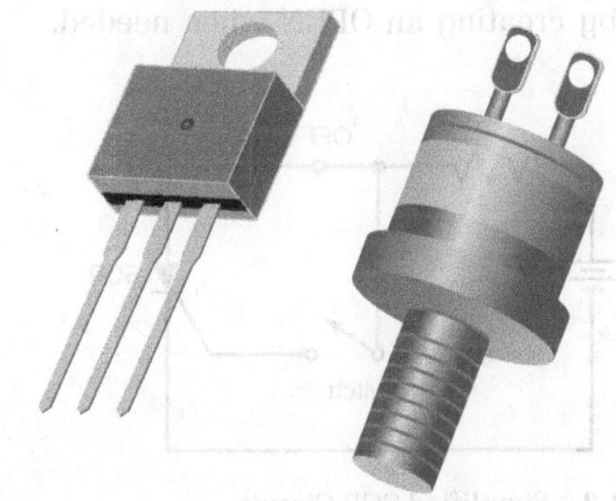

Fig. 33-3.2 Generic SCRs (low power and high power).

Since the SCR is like two transistors (not two diodes), the triggering voltage at the gate drives the *npn* transistor to conduct current. When this happens, the base voltage of the *pnp* section is negative, and this causes the *pnp* to conduct current. So, the *pnp* supplies the base current to the *npn*, and this keeps the SCR conducting current even if the original trigger voltage is removed.

Once an SCR has been triggered at the gate, it is often called *latched* and can remain on even if the triggering gate voltage is removed. Or, it will also remain on until the current flowing through the anode-cathode falls below a level known as its holding current. This means that SCRs will conduct with or without gate voltage if there is enough current flowing, and this is called *breakover* voltage. Another way to explain this is that the upper transistor is still conducting in that situation. Therefore, SCRs are chosen for circuits so that they have a greater *breakover* voltage than ever expected so that they cannot be turned on except at the gate.

Many SCRs are very sensitive when in operation. They may even appear to conduct in both directions at the gate-cathode junction when tested. Never assume an SCR is faulty if you test it with a DMM. Also, some SCRs have a resistance built into the gate-cathode junction and may not even appear to conduct when the typical 0.7 volt is applied to the gate. This resistance is used to protect it from turning on when not desired.

In practical application, no current should flow through an SCR, except for a very small leakage current similar to other *pn* devices. The SCR is only triggered to turn on if the voltage at the gate breaks down the barrier, like most diode junctions, or if the cathode voltage exceeds its specified threshold level.

Figure 33-3.3 is a simplified circuit in which the SCR is connected in series with a DC supply and a resistor. Here, current can flow negative at the cathode. Notice the two switches (ON and OFF). The OFF switch is actually on here so it connects the anode to the DC supply. It is used to turn OFF the flow of current by creating an OPEN when needed.

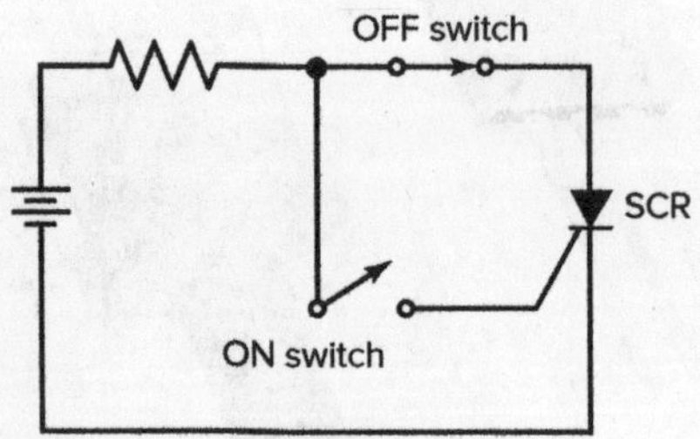

Fig. 33-3.3 Simplified SCR Circuit

However, the ON switch is normally an OPEN. It is pressed ON to connect the gate to the anode to trigger the SCR on (latched) so that current flows from the cathode to the anode. At that point, no further triggering is required. This means even if the ON switch is released (creating an open), the SCR will still be conducting (latched). To turn it off, the OFF button should be released to create an open circuit. Although this circuit demonstrates how an SCR works, typical school lab equipment will not usually provide enough current from the supply to latch the SCR or trigger it on. For that reason, the procedure that follows is best done with the MultiSim program.

EQUIPMENT

Recommended: MultiSim simulation software

Or if using an appropriate SCR:

DMM ohmmeter (capable of SCR testing)
Voltmeter (DMM)
Oscilloscope (dual trace)
Variable high-current DC power supply (two outputs)
Signal generator (capable of supplying enough current)

COMPONENTS

(1) 1 kΩ resistor
(1) SCR: low-current only

PROCEDURE

1. Test the SCR with an Ohmmeter (continuity). As shown here, test the continuity of the gate-to-cathode junction. Remember that the gate is *p* material and the cathode is *n* material. So, connect the cathode to the negative or common input and the gate to the positive or Ohm function input. With the meter in the Ohm or resistance mode, record the results in Table 33-3.1.

Note: Use multisim or a DMM

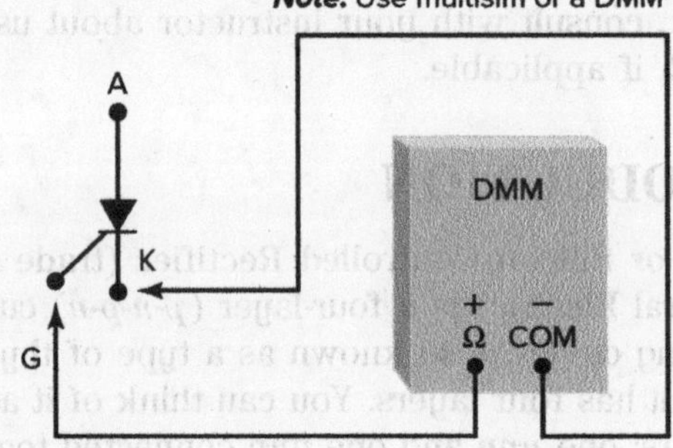

Fig. 33-3.4 Ohmmeter SRC test.

2. Repeat the resistance or continuity test for the gate-to-anode terminals and the anode-to-cathode terminals and record the results in Table 33-1.1.

3. DC setup: Connect the circuit of Fig. 33-3.5. V_1 is used to forward bias (trigger) the SCR at the gate. V_2 (+5 volts) is a source voltage supplied to the load R_1 (1 kOhm). In this circuit, the SCR acts as a switch so that voltage to the load is controlled by the gate bias. With V_1 (bias) at 0 volts, measure and record the voltage across the load in Table 33-3.2

4. Increase the V_1 voltage to 0.5 volts and record the voltage across R_1.

5. Increase the V_1 voltage to 0.7 volts and record the voltage across R_1.

6. Increase the V_1 voltage to 1 volt and record the voltage across R_1.

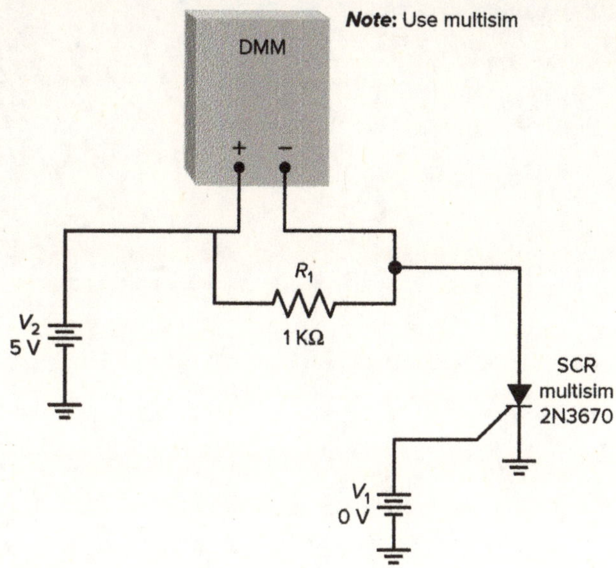

Fig. 33-3.5 DC test for SRC.

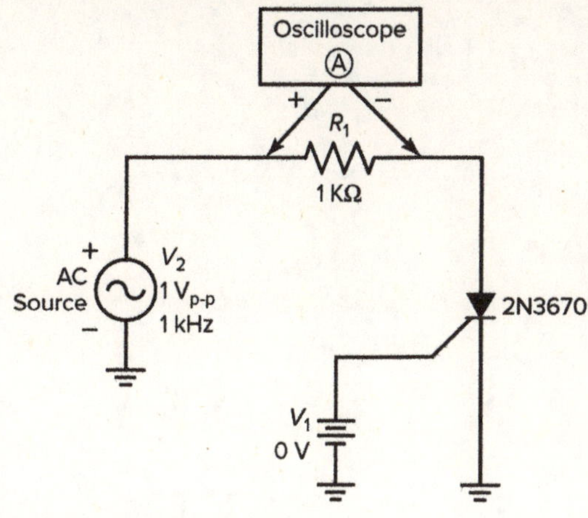

Fig. 33-3.6 AC test for SRC.

7. Increase the V_1 voltage to 2 volts and record the voltage across R.

8. Connect the circuit of Fig. 33-3.6. The oscilloscope leads are across the load R_1. Set the time base to 500 usec (microseconds) per division and the amplitude scale to 200 mV per division. Set the V_2 signal generator to 1 V-peak (2 volts p-p). Set the V_1 gate trigger to 0 volts. Adjust the oscilloscope time and amplitude as you prefer. Record the peak voltage across R_1 in Table 33-1.1.

9. Increase the V_1 voltage to 0.5 volt and record the voltage across R_1.

10. Increase the V_1 voltage to 0.7 volt and record the voltage across R_1.

11. Increase the V_1 voltage to 1 volt and record the voltage across R_1.

12. Increase the V_1 voltage to 2 volt and record the voltage across R.

13. Draw 2 waveforms for step 12 in Table 33-1.1 and label the amplitude and the time.

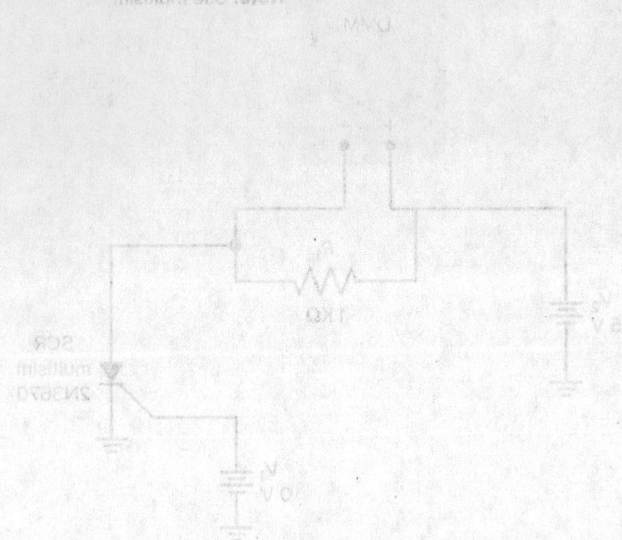

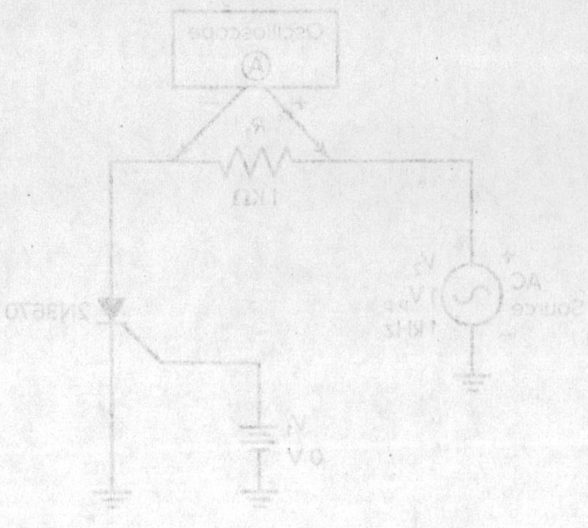

NAME ______________________________ DATE ______________

QUESTIONS FOR EXPERIMENT 33-3

1. What is the difference between an SCR and a transistor?

2. Why are SCRs used to control motors?

3. How can an SCR be turned on?

TABLE FOR EXPERIMENT 33-3

Ohmmeter test (continuity)

Gate to cathode resistance Ω: ____________

Gate to anode resistance Ω: ____________

Anode to cathode resistance Ω: ____________

DC test: voltages across load R_1

$V_1 = 0$ V and load voltage $V_{R_1} =$ ____________V

$V_1 = 0.5$ V and load voltage $V_{R_1} =$ ____________V

$V_1 = 0.7$ V BS load voltage $V_{R_1} =$ ____________V

$V_1 = 1$ V and load voltage $V_{R_1} =$ ____________V

$V_1 = 2$ V and load voltage $V_{R_1} =$ ____________V

AC test: voltages across load R_1 with 1 V peak at 1 KHz

$V_1 = 0$ V and load voltage = ____________Vp

$V_1 = 0.5$ V and load voltage = ____________Vp

$V_1 = 0.7$ V and load voltage = ____________Vp

$V_1 = 1$ V and load voltage = ____________Vp

$V_1 = 2$ V and load voltage = ____________Vp

Waveform for step 12:

EXPERIMENT RESULTS REPORT FORM

Experiment No.: ______________________ Name: ______________________

Date: ______________________

Experiment Title: ______________________ Class: ______________________

______________________ Instr: ______________________

Explain the purpose of the experiment:

List the first Learning Objective:

OBJECTIVE 1:

After reviewing the results, describe how the objective was validated by this experiment.

List the second Learning Objective:

OBJECTIVE 2:

After reviewing the results, describe how the objective was validated by this experiment.

List the third Learning Objective:

OBJECTIVE 3:

After reviewing the results, describe how the objective was validated by this experiment.

Conclusion:

If required, attach to this form: ☐ Answers to Questions, ☐ Tables, and ☐ Graphs.

CHAPTER 34

EXPERIMENT 34-1

OPERATIONAL AMPLIFIERS

LEARNING OBJECTIVES

At the completion of this experiment, you will be able to:

- Demonstrate the operation of an inverting amplifier.
- Demonstrate the operation of a noninverting amplifier.
- Demonstrate the operation of a buffer amplifier.

SUGGESTED READING

Chapter 34, *Basic Electronics*, Grob/Schultz, twelfth edition

INTRODUCTION

The first effort of this experiment will be directed toward the definition of terms used in conjunction with the operational amplifier. The operational amplifier, sometimes shortened to *op amp*, is a highly sophisticated linear integrated circuit (IC) direct-current amplifier, demonstrating high-gain, high-input impedance, and low-output impedance. Originally, the term referred to high-gain, high-performance vacuum tube direct-current amplifiers that were designed to perform mathematical operations with predetermined voltage levels. Operational amplifiers were the basic building blocks of analog computers because they perform the mathematical operations of amplification, addition, subtraction, integration, and differentiation. The operational amplifier used today can still be used to perform these mathematical operations; however, more useful circuit designs have been created. In combination with nonlinear elements such as diodes, they may be used as limiters, level detectors, and nonlinear function generators. By designing operational amplifier circuits that include other active components such as transistors, it is even possible to multiply and divide analog voltages by taking the logarithms and antilogarithms of input voltages.

The modern-day device tends to operate at lower voltages and does not have any of the common problems associated with vacuum tubes. Today's operational amplifier is in an IC format and still resembles the high-gain, direct-current amplifier that uses external feedback for controlled responses. When working with operational amplifiers, the user will find that they lend themselves easily and adapt well to a variety of industrial applications. They can be designed to function as filters, oscillators, pulse modulators, peak detectors, signal-function generators, small-signal rectifiers, instrumentation amplifiers, and a seemingly endless variety of specialized circuit applications. It has been determined that the operational amplifier is the most commonly used IC found in industry today. Figure 34-1.1 shows an IC comparison chart for commonly used linear devices.

When the schematic diagram of an op-amp IC is displayed, it will take the form of a simple triangle. Some European and Japanese manufacturers will use a modified triangle schematic symbol that looks very similar to the symbol used for a digital gate circuit. We will not be using the modified symbol for this text; instead, we will be using a simple triangle symbol, and it is used in all schematics presented for your use in this experiment. The symbol that will be used is represented in Fig. 34-1.2.

To correctly address input signal information to the operational amplifier, two input terminals are made available. They are traditionally drawn on the left-hand side of the schematic diagram (see Fig. 34-1.3). The input terminals are connected internally to a differential amplifier located inside the IC casing. These terminals are referred to as the *inverting* and *noninverting* inputs and carry the symbols of − and +, respectively. In addition to the input terminals, the device must make use of an output terminal. The operational amplifier will use only one terminal for this function, and it is drawn at the apex of the triangle, located on the right-hand side of the schematic diagram. Again direct your attention to Fig. 34-1.3, and notice the placement of the output terminal.

Under normal operating conditions, if an AC signal is applied to the inverting terminal (−) with reference to ground, a 180° inversion, out-of-phase, signal would be seen at the output terminal. This inversion may be difficult to see with the use of a conventional single-trace scope, but when a dual-trace oscilloscope is used to compare the input versus the output signal, the 180° out-of-phase signal is easy to see. An AC signal applied to the noninverting terminal (+) with reference to ground would cause no inversion or phase shift in the signal being measured at the output

Device	Operating Temp. Range °C Min	Max	A_{VOL} Min K	R_i Min Ω	P_D mW	$\lvert I_{io}\rvert$ Max nA	I_b Max nA	CMV_i Min V	†Typ. Slew Rate SR V/μSec	V_o Min V	$\lvert V_{io}\rvert$ Max mV	Offset Adjust	Internal Compen-sation	Output Protec-tion	Input Protec-tion	JEDEC Package Type
709A	−55	+125	25	350 K	108	50	200	±8	0.3	±12	1	no	no	no	no	TO-91, 99, 116
709B	−55	+125	25	150 K	165	200	500	±8	0.3	±12	5	no	no	no	no	TO-91, 99, 116
709C	0	+70	15	50 K	200	500	1500	±8	0.3	±12	10	no	no	no	no	TO-91, 99, 116
739C (Dual)	0	+70	6.5	37 K	420	1000	2000	±10	1.0	+12, −14	6	no	no	yes	yes	TO-116
741B	−55	+125	50	300 K	85	200	500	±12	0.5	±12	5	yes	yes	yes	yes	TO-91, 99, 116
741C	0	+70	20	150 K	85	200	500	±12	0.5	±12	6	yes	yes	yes	yes	TO-91, 99, 116
747B (Dual)	−55	+125	50	300 K	85	200	500	±12	0.5	±12	5	yes	yes	yes	yes	TO-101, 116
747C (Dual)	0	+70	20	150 K	85	200	500	±12	0.5	±12	6	yes	yes	yes	yes	TO-101, 116
748B	−55	+125	50	300 K	85	200	500	±12	0.5	±12	5	yes	no	yes	yes	TO-99
748C	0	+70	20	150 K	85	200	500	±12	0.5	±12	6	yes	no	yes	yes	TO-99
749B (Dual)	−55	+125	25	100 K	220	400	750	±11	1.5	+12, −14.5	3	no	no	yes	yes	TO-116
749C (Dual)	0	+70	15	70 K	330	500	1000	±11	1.5	+12, −14.5	6	no	no	yes	yes	TO-116
800B	−55	+125	10	250 K	180	100	1000	±4	—	±6	50	no	no	no	no	TO-101
800D	−55	+125	10	100 K	180	200	2000	±4	—	±12	—	no	no	no	no	TO-101
801B	−55	+125	10	250 K	180	100	1000	±4	—	±12	50	no	no	no	no	TO-100
801D	−55	+125	10	100 K	180	200	2000	±4	—	±12	—	no	no	no	no	TO-100
805B	−55	+125	30	500 K	225	50	500	±8	2.5	±12	5	no	no	no	yes	TO-91, 99
805C	0	+100	10	100	225	100	1000	±8	2.5	±12	10	no	no	no	yes	TO-91, 99
806B	−55	+125	30	500	225	50	500	±8	2.5	±9	5	no	no	no	yes	TO-91, 99
806C	0	+100	10	100	225	100	1000	±8	2.5	±9	10	no	no	no	yes	TO-91, 99
807B	−55	+125	30	500	225	50	500	±8	2.5	±12	2.5	no	no	no	yes	TO-91, 99
808A	−55	+125	25	1 M	225	15	50	±8	2.5	±12	5	no	no	no	yes	TO-91, 99
808B	−55	+125	25	1 M	225	30	50	±8	2.5	±12	10	no	no	no	yes	TO-91, 99
809B	−55	+125	10	100	150	100	500	±10	—	±10	10	no	no	yes	yes	TO-99, 116
809C	0	+100	10	50	150	350	1000	±10	—	±10	10	no	no	yes	yes	TO-99, 116
810B (Dual)	−55	+125	10	100	150	100	500	±10	—	±10	10	no	no	yes	yes	TO-116
810C (Dual)	0	+100	10	50	150	350	1000	±10	—	±10	10	no	no	yes	yes	TO-116
715B	−55	+125°C	15	IM (typ)	210	250	750	±15	18	±10	±5	yes	yes	yes	yes	TO-100
715C	0	+70°C	10	IM (typ)	300	1500	250	±15	18	±10	±7.5	yes	no	yes	yes	TO-100
846B	−55	+125	100	25 M	90	5	30	±12.5	2.0	±12	±3	yes	no	yes	yes	TO-99
846C	0	+70	50	15 M	75	15	50	±12	2.0	±12	±5	yes	no	yes	yes	TO-99
LM101A	−55	+125	50	1.5 M	120	10	25	±12	0.5	±12	±2	yes	no	yes	yes	TO-99
LM101B	−55	+125	50	300 K	120	200	500	±12	0.5	±12	±5	yes	no	yes	yes	TO-99
LM201A	−25	+85	50	1.5 M	120	10	75	±12	0.5	±12	±2	yes	no	yes	yes	TO-99
LM201C	−25	+85	20	300 K	90	200	500	±12	0.5	±12	±7.5	yes	no	yes	yes	TO-99
LM301A	0	+70	25	500 K	120	50	250	±12	0.5	±12	±7.5	yes	no	no	yes	TO-99
LM307D	0	+70	25	500 K	120	50	250	±12	0.5	±12	±7.5	yes	no	no	yes	TO-99
811B	−55	+125	10	100	150	100	500	±10	—	±10	10	no	no	yes	yes	TO-99, 116
811C	0	+100	10	50	—	350	1000	±10	—	±10	10	no	no	yes	yes	TO-99, 116
813C	0	+70	6	—	120	2000	5000	±5	—	—	4	no	no	yes	yes	TO-99, 116
819B	−55	+125	5	50 K	25	100	500	±4	—	±4	10	no	no	yes	yes	TO-99
841B	−55	+125	50	300	85	200	500	±12	0.5	±12	5	no	no	yes	yes	TO-99
841C	0	+100	20	150 K	85	200	500	±12	0.5	±12	6	no	no	yes	yes	TO-99
844B	−55	+125	100	25 M	75	5	30	±12.5	2.0	±12	±3	yes	yes	yes	yes	TO-99
844C	0	+70	50	15 M	90	15	50	±12	2.0	±12	±5	yes	yes	yes	yes	TO-99
LM107B	−55	+125	50	1.5 M	120	10	75	±12	0.5	±12	±2	yes	yes	yes	yes	TO-99
LM207C	−25	+85	50	1.5 M	120	10	75	±12	0.5	±12	±2	yes	yes	yes	yes	TO-99
MC1437C (Dual)	0	+70	15	50 K	200	500	1500	±8	0.3	±12	10	no	no	no	no	TO-116
MC1439C	0	+70	15	100	200	100	1000	±11	4.2	±10	7.5	no	no	yes	yes	TO-99, 116
MC1458C (Dual)	0	+70	20	150 K	85	200	500	±12	0.5	±12	6	no	yes	yes	yes	TO-99, 116
MC1537B (Dual)	−55	+125	25	150 K	165	200	500	±8	0.3	±12	5	no	no	no	no	TO-91, 99, 116
MC1539B	−55	+125	50	150	150	60	500	±11	4.2	±10	3	no	no	yes	yes	TO-99, 116
MC1558B (Dual)	−55	+125	50	300 K	85	200	500	±12	0.5	±12	5	yes	yes	yes	yes	TO-101, 116

†Unity Gain.

Fig. 34-1.1 IC comparison chart for operational amplifiers.

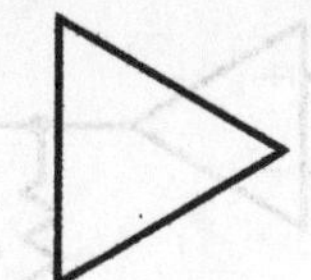

Fig. 34-1.2 Schematic symbol for an op amp.

terminal. A comparison of AC input signals of sine, triangle, and square waves applied to the inverting and noninverting terminals of a typical operational amplifier is shown in Fig. 34-1.4.

Of course, the operational amplifier will need to have supply voltages and other components added in order to operate efficiently.

Circuit Configurations

Circuits using operational amplifiers commonly display properties radically different from those of the individual devices themselves. For example, you will find that the circuit's closed-loop gain, A_{CL}, is only a fraction of the device's internal open-loop gain, A_{OL}. In addition, the circuit input impedance is often much different from the operational amplifier's internal input impedance (although for some circuits you will find it to be of the same magnitude). Output impedance of the operational amplifier circuit is usually less than that of the op-amp device, but the bandwidth is usually greater. It is, therefore, advisable to make the first analysis of this circuit by assuming an ideal op-amp situation and then modifying the analysis for the imperfections existing in the real world.

An ideal operational amplifier would display the following five characteristics:

1. Infinite open-loop gain, A_{OL} = infinite
2. Infinite input impedance, Z_{in} = infinite
3. Zero output impedance, $Z_{out} = 0$
4. Zero offset voltage, $V_{OS} = 0$
5. Zero bias current, $I_b = 0$

Although these approximations are by no means conclusive, they are the values that influence most other characteristics. The approximations will simplify

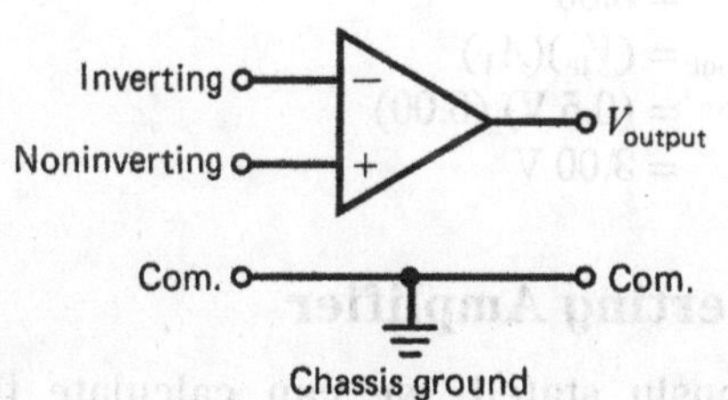

Fig. 34-1.3 The inverting and noninverting terminals for an op amp.

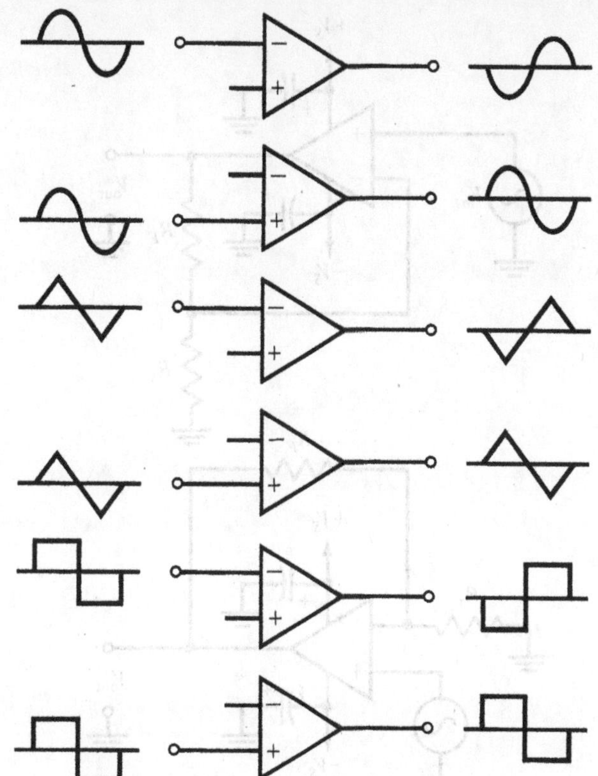

Fig. 34-1.4 Op-amp input-output signal comparison.

the analysis of operational circuitry. For example, the assumption of an infinite input impedance allows us to ignore the loss of any signal current into the amplifier's input terminal. The lack of bias current enables us to neglect the effect of this variable. These assumptions of the ideal amplifier can now be applied to the following circuits.

The Noninverting Amplifier

The noninverting operational amplifier circuit will usually be drawn as the schematic configuration shown in Fig. 34-1.5. Notice that both schematics are basically the same and are just redrawn for easier interpretation.

The voltage appearing at the inverting input terminal is at the same potential as V_{in}. Therefore, the voltage across R_1 will be the same as the potential of V_{in}, or

$$V_{R_1} = V_{in}$$

Knowing the above relationship, we can determine that the amount of current flowing through R_1 will be equal to the current flowing through R_F, the feedback resistor; it would then follow that

$$I_{R_1} = \frac{V_{in}}{R_1}$$

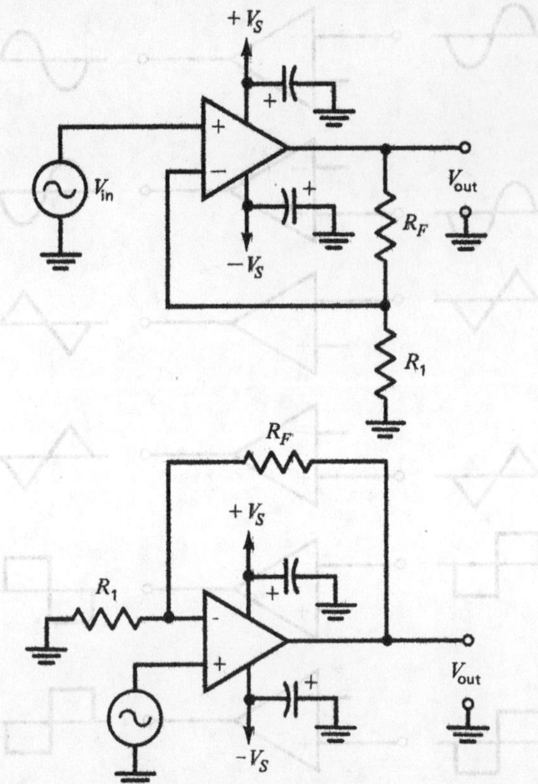

Fig. 34-1.5 Op-amp noninverting schematic diagrams.

Voltage across R_F would be determined by the current flowing through itself, and this can be related back to the input voltage:

$$V_{R_F} = (I)\,(R_F)$$
$$= \frac{(V_{in})(R_F)}{R_1}$$
$$= V_{in}\,\frac{R_F}{R_1}$$

Since the left end of resistor R_F is at the input potential, the voltage across R_F is

$$V_{R_F} = V_{in}\,\frac{R_F}{R_1}$$

This signal appears to be noninverting at the left end of the resistor, and the output voltage V_{out} can be calculated as

$$V_{out} = V_{in} + V_S\,\frac{R_F}{R_1}$$
$$= V_{in}\left\{1 + \frac{R_F}{R_1}\right\}$$

We now have two dependable methods for determining the voltage gain of a noninverting operational amplifier.

The first method makes use of the formula that states that the voltage gain of an amplifier is equal to the ratio of the output voltage V_{out} to the input voltage V_{in}, or

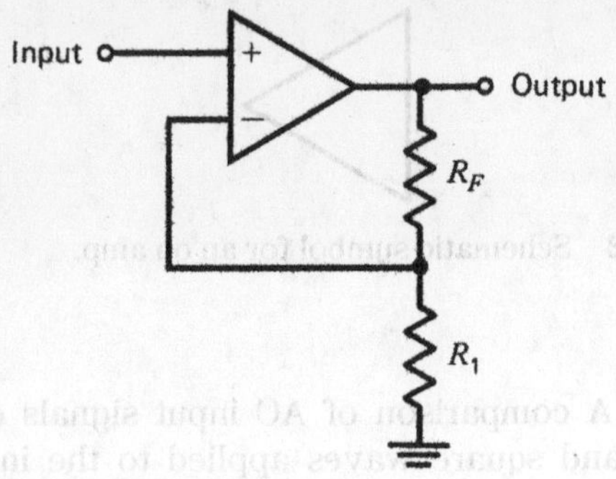

Fig. 34-1.6 Noninverting op amp.

$$A_V = \frac{V_{out}}{V_{in}}$$

Combining this equation with one previous one, we can now realize that the following equation holds:

$$A_V = 1 + \frac{R_F}{R_1}$$

This demonstrates a unique aspect of the noninverting operational amplifier, which is that the voltage gain of this particular amplifier will always be greater than 1, or greater than that of a unity-gain amplifier. Therefore,

$$A_V > 1 \quad \text{(for a noninverting amplifier)}$$

Sample Problem 1. In a noninverting amplifier, where $R_1 = 10\text{ k}\Omega$ and $R_F = 50\text{ k}\Omega$, what will the output signal level be with an input signal (V_{in}) of 0.5 V? Draw the schematic diagram and show all work.

Given: A noninverting operational amplifier of

$$R_F = 50\text{ k}\Omega$$
$$R_1 = 10\text{ k}\Omega$$
$$V_{in} = 0.5\text{ V} = 500\text{ mV}$$

Find: Schematic diagram and V_{out}.

Solution: See Fig. 34-1.6.

$$A_V = 1 + \frac{R_F}{R_1}$$
$$= 1 + \frac{50\text{ k}\Omega}{10\text{ k}\Omega}$$
$$= 6.00$$
$$V_{out} = (V_{in})(A_V)$$
$$= (0.5\text{ V})\,(6.00)$$
$$= 3.00\text{ V}$$

The Inverting Amplifier

As previously stated, we can calculate the voltage gain of any amplifier by noting the relationship existing between the input and output signal

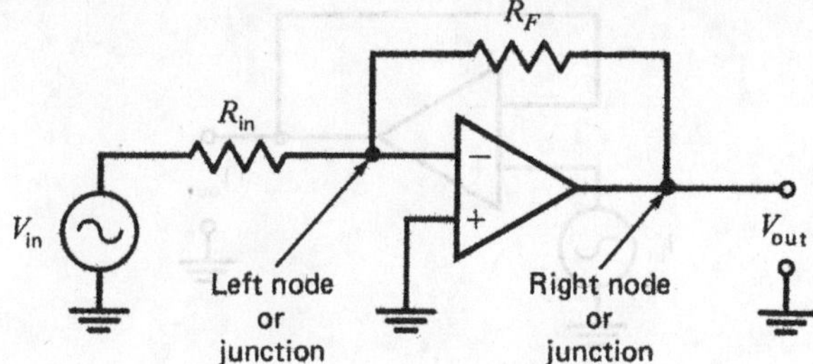

Fig. 34-1.7 The inverting op-amp circuit.

magnitudes. The following equation is valid for any amplifier:

$$A_V = \frac{V_{out}}{V_{in}}$$

Operational amplifiers usually use feedback in the design circuitry in order for them to perform linearly. Negative feedback enables the designer to easily select and control the voltage gain. The inverting operational amplifier schematic diagram is shown in Fig. 34-1.7. Note in Fig. 34-1.7 that the power supply connections, $V+$ and $V-$, are implied but not shown.

Voltage at the inverting input terminal is at the same potential as the noninverting input terminal, or zero. This point is also said to be at a virtual ground reference level. The current through R_{in} will be equal in magnitude to the power supply voltage level divided by the resistance of R_{in}. This relationship is represented by the formula

$$I_{R_{in}} = \frac{V_{in}}{R_{in}}$$

The current of $I_{R_{in}}$ also flows through R_F because the current will not enter the operational amplifier; therefore, the voltage potential across R_F is related to the current through R_F. This is also related to the parameters revolving around the input resistor R_{in}. Therefore,

$$V_{R_F} = I_{R_{in}} \times R_F$$

$$= \frac{V_{in}}{R_{in}} R_F$$

The left side, or node, of R_F is at a zero reference point, and the right node of R_F must be equal to the output voltage V_{out}. But in reality the voltage across R_F is related to

$$V_{in} \frac{R_F}{R_{in}} = V_{out}$$

with the right node negative with respect to the input voltage V_{in}, represented by the formula

$$\frac{V_{out}}{V_{in}} = -\frac{R_F}{R_{in}}$$

where

$$Z_{in} = R_{in}$$

Through rearranging, we have

$$V_{out} = -\frac{R_F}{R_{in}} V_{in}$$

and solving for the voltage gain,

$$A_V = \frac{V_{out}}{V_{in}} = \frac{R_F}{R_{in}}$$

In summary, it can now be realized that the voltage gain can be equal to, less than, or greater than 1:

$$A_V = 1$$
$$A_V < 1$$
$$A_V > 1$$

Typically, it will be found that R_{in} is at least 1 kΩ and that for the operational amplifiers the device's input impedance is equal to the input resistance of R_{in}. Therefore,

$$Z_{in} = R_{in}$$

Sample Problem 2. Notice in the circuit of Fig. 34-1.8 that a 1.00-mV peak AC signal is applied to the circuit and that an oscilloscope is connected to monitor the output signal. What would the expected magnitude and phase of the output signal be?

Given:

$$R_F = 100\ \text{k}\Omega$$
$$R_{in} = 1\ \text{k}\Omega$$
$$V_{in} = 1.00\ \text{mV}$$

Find: Phase of V_{out}.

Solution:

$$A_V = -\frac{R_F}{R_{in}}$$
$$= \frac{100\ \text{k}\Omega}{1\ \text{k}\Omega}$$
$$= -100$$
$$V_{out} = (A_V)(V_{in})$$
$$= (-100)(1.00\ \text{mV})$$
$$= -0.1\ \text{V}$$

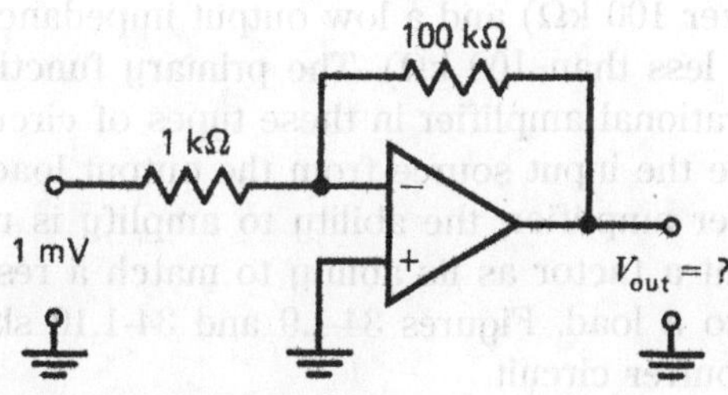

Fig. 34-1.8 Inverting op amp.

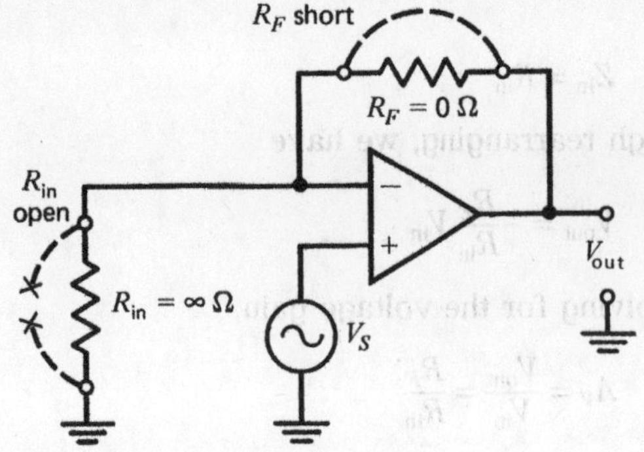

Fig. 34-1.9 The noninverting amplifier used as a voltage follower (buffer) circuit.

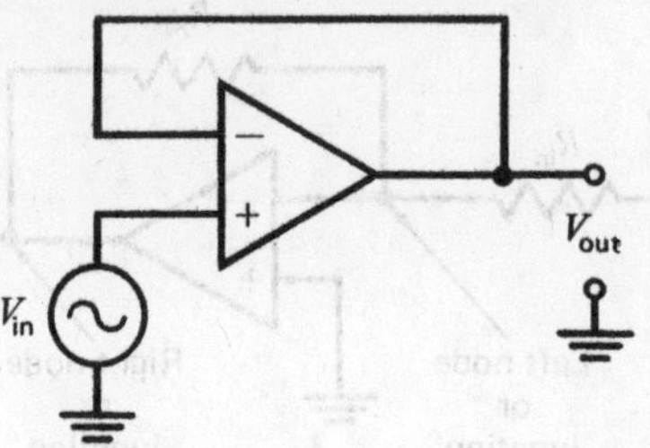

Fig. 34-1.10 Voltage follower stage with the elimination of $R_F + R_{in}$.

The output signal displays a phase shift of 180° out of phase from its input voltage V_{in} reference potential, hence the name *inverting amplifier.*

Sample Problem 3. In Sample Problem 2, what loading effect does this circuit present to the AC signal generator feeding it?

Given:

$R_F = 100\ \text{k}\Omega$
$R_{in} = 1\ \text{k}\Omega$
$V_{in} = 1.00\ \text{mV peak}$
$A_V = -100$
$V_{out} = -0.1\ \text{V}$

Find: Z_{in}.

Solution:

$$Z_{in} = R_{in}$$
$$= 1\ \text{k}\Omega$$

The Buffer Amplifier

Buffering is the process by which the output load does not affect the operating conditions of the previous stage. The circuit employed to provide buffering is known as a buffer stage of a voltage follower (output voltage follows the input) stage. Operational amplifiers are commonly used for this function when a certain degree of isolation is needed between audio amplifier stages. The buffer amplifier will always yield a voltage gain close to 1. This type of op-amp circuit results in a very high input impedance Z_{in} (typically over 100 kΩ) and a low output impedance Z_{out} (usually less than 100 kΩ). The primary function of the operational amplifier in these types of circuits is to isolate the input source from the output load. For the buffer amplifier, the ability to amplify is not as important a factor as its ability to match a resistive source to a load. Figures 34-1.9 and 34-1.10 show a typical buffer circuit.

EQUIPMENT

Signal generator
Oscilloscope
DC power supply ±15 V
Springboard or protoboard
Test leads

COMPONENTS

(1) 741 operational amplifier

Resistors (all 0.25 W):

(2) 10 kΩ	(1) 47 kΩ
(1) 27 kΩ	(1) 82 kΩ
(1) 39 kΩ	

PROCEDURE

1. Connect the circuit shown in Fig. 34-1.11. Apply power, and observe and record the input and output traces on the oscilloscope.

2. Adjust the amplitude of the signal generator to 1.00 $V_{p\text{-}p}$ at 1.00 kHz. Notice the difference between the input and output waveshapes of the amplifier.

3. Verify that the gain of the amplifier is equal to 1.

4. Connect the circuit shown in Fig. 34-1.12.

5. Apply power, and adjust the amplitude of the signal generator to 1.00 $V_{p\text{-}p}$ at 1.00 kHz. Compare the input to the output signal. Record the input and output traces on graticule paper.

6. Determine the voltage gain of the amplifier such that

$$A_V = 1 + \frac{R_F}{R_1}$$

Record this information.

7. Maintain the input amplitude from the signal generator at 1.00 $V_{p\text{-}p}$, and change the value of R_1 to 27, 39, 47, and 82 kΩ. Determine the results of the output signal level and the voltage gain for each change of RF in Fig. 34-1.12, and record the results in Table 34-1.1.

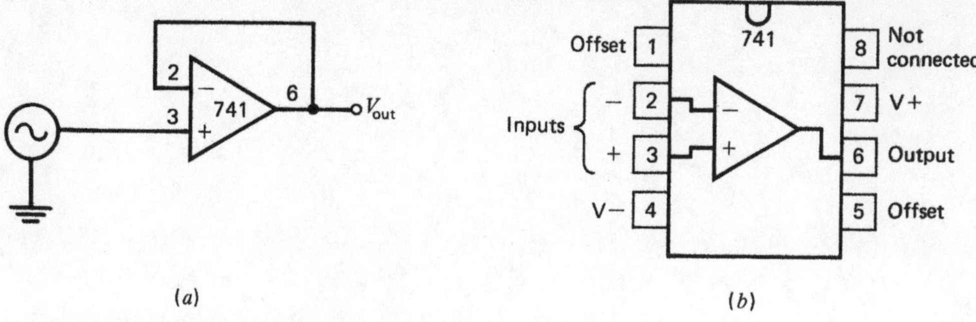

Fig. 34-1.11 (*a*) Voltage follower and (*b*) pinouts for a 741C.

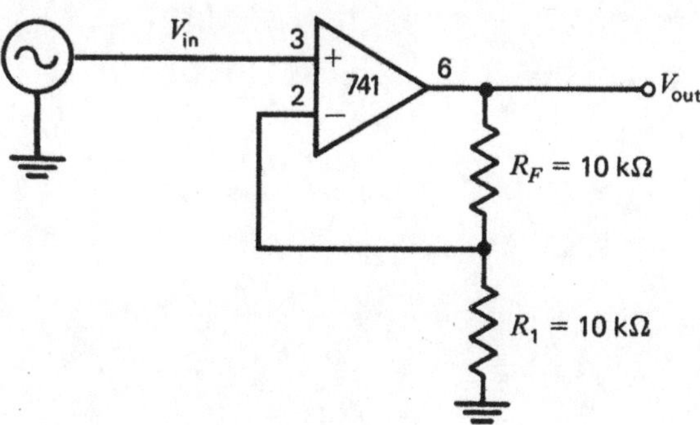

Fig. 34-1.12 Noninverting amplifier.

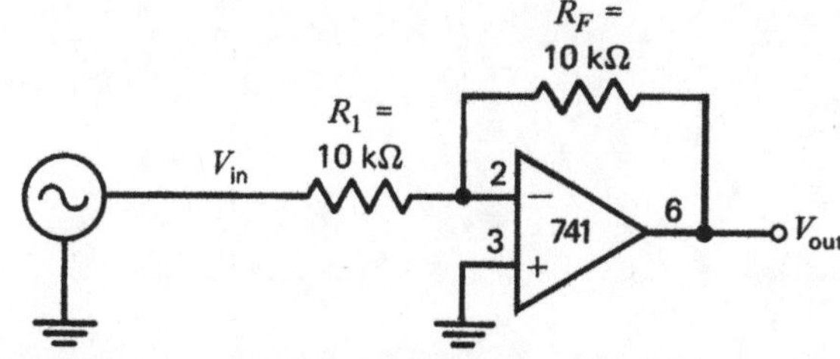

Fig. 34-1.13 Inverting amplifier.

8. Connect the circuit shown in Fig. 34-1.13. Apply power to the circuit, and adjust the amplitude of the signal generator to 1.00 $V_{p\text{-}p}$ at a frequency of 1.00 kHz.

9. Compare the input to output signal of the amplifier. Record the input and output traces on graticule paper.

10. Determine the voltage gain of the amplifier.

11. Repeat step 7 (for Fig. 34-1.13), and record this information in Table 34-1.2.

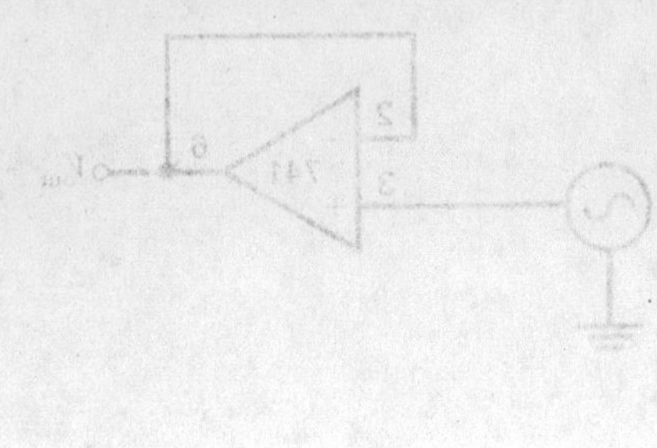

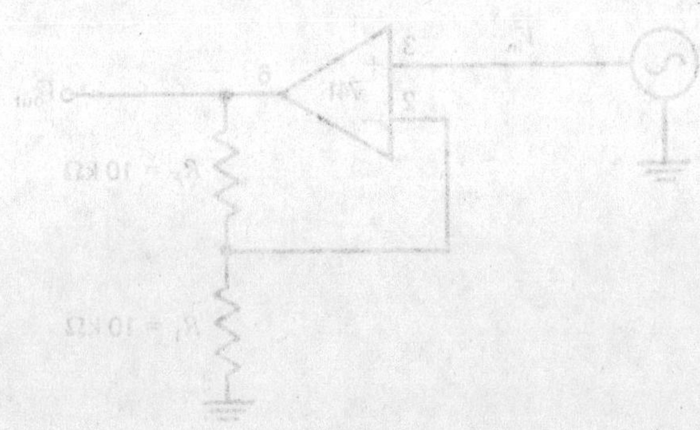

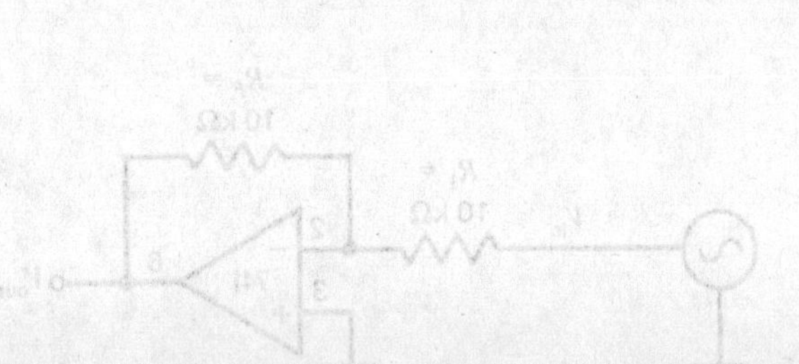

NAME DATE

QUESTIONS FOR EXPERIMENT 34-1

1. **For a voltage follower circuit, what is the difference between the input and output signals for phase and gain?**

2. **For the noninverting amplifier, how does the measured voltage gain compare with the formula for determining gain? What is the percentage of error between these two values?**

3. **For the inverting amplifier, do the experimental results agree with the gain equation?**

4. **What would be the purpose for a buffer amplifier?**

5. **What voltage gain can be expected from a noninverting amplifier?**

TABLES FOR EXPERIMENT 34-1

TABLE 34-1.1 $V_{in} = 1\ V_{p\text{-}p}$

R_1, kΩ	V_{out}	A_V
27		
39		
47		
82		

TABLE 34-1.2 $V_{in} = 1\ V_{p\text{-}p}$

R_1, kΩ	V_{out}	A_V
27		
39		
47		
82		

EXPERIMENT RESULTS REPORT FORM

Experiment No.: ______________________ Name: ______________________

Date: ______________________

Experiment Title: ______________________ Class: ______________________

______________________ Instr: ______________________

Explain the purpose of the experiment:

List the first Learning Objective:

OBJECTIVE 1:

After reviewing the results, describe how the objective was validated by this experiment.

List the second Learning Objective:

OBJECTIVE 2:

After reviewing the results, describe how the objective was validated by this experiment.

List the third Learning Objective:

OBJECTIVE 3:

After reviewing the results, describe how the objective was validated by this experiment.

Conclusion:

If required, attach to this form: ☐ Answers to Questions, ☐ Tables, and ☐ Graphs.

SUPPLEMENTAL EXPERIMENT A-1

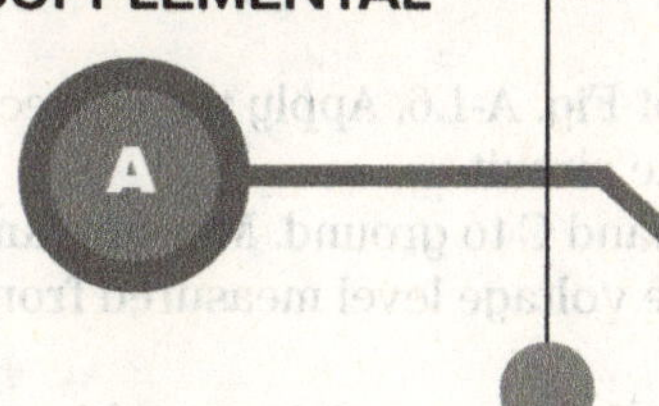

DIGITAL OR AND AND GATES

LEARNING OBJECTIVES

At the completion of this experiment, you will be able to:

- Build a diode OR gate.
- Build a diode AND gate.
- Construct a truth table for a gate circuit.

SUGGESTED READING

Basic Electronics, Grob/Schultz, twelfth edition

INTRODUCTION

The computers of today do not process numbers in the base 10 (that is, 0, 1, 2, 3, . . . , 9). Computers instead use binary logic of base 2 (0 and 1) to perform their functions. Two fundamental circuits are the OR and AND gates.

The OR gate has two or more input signals but creates only one output signal. If any input signal is high (binary 1), the output signal is high. OR gates can be discrete, that is, made from several components such as diodes and resistors. OR gates can also come packaged as integrated circuits or ICs. The OR gate circuit constructed in this experiment is discrete. Figure A-1.1 shows a discrete OR gate and its truth table.

Truth tables are data tables that quickly show the overall circuit action of the gate under investigation. For the circuit of Fig. A-1.1, a binary 1 is equivalent to +5 V, and a binary 0 is equivalent to ground potential. Studying the truth table of Fig. A-1.1 reveals that for a two-input OR gate, the circuit is off only when points A and B are both at ground potential. Another way of explaining the results of the truth table is to say that the OR gate is a "mostly on" device. In the case of the truth table of Fig. A-1.1, the two-input OR gate is on 75 percent of the time. The OR gate has a symbol, and it is shown in Fig. A-1.2.

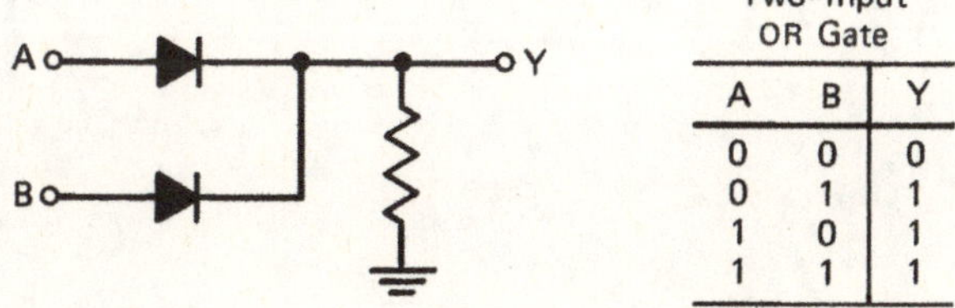

Two-Input OR Gate

A	B	Y
0	0	0
0	1	1
1	0	1
1	1	1

Fig. A-1.1 Two-input OR gate and truth table.

Fig. A-1.2 OR gate symbol.

The AND gate has two or more input signals but creates only one output signal. All the inputs must be high to get a high output signal. AND gates can also be of either discrete or IC design. Figure A-1.3 shows a discrete AND gate and its truth table.

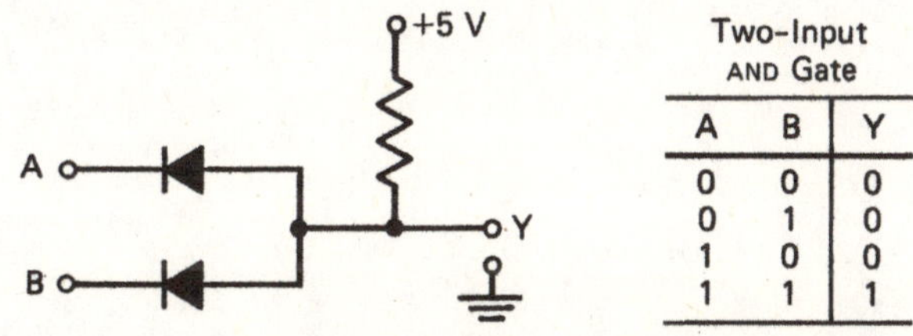

Two-Input AND Gate

A	B	Y
0	0	0
0	1	0
1	0	0
1	1	1

Fig. A-1.3 Two-input AND gate and truth table. Note that 0 = low = 0 V and 1 = high = 5 V.

The AND gate circuit is on only when points A and B are both at a +5 V level, or at a binary 1. Another way of describing the results found in the truth table is to say that the AND gate is a "mostly off" device. In the case of the truth table of Fig. A-1.3, the two-input AND gate is off 75 percent of the time. The AND gate has a symbol, and it is shown in Fig. A-1.4.

EQUIPMENT

DC power supply, 0–10 V
DMM
Protoboard or springboard
Test leads

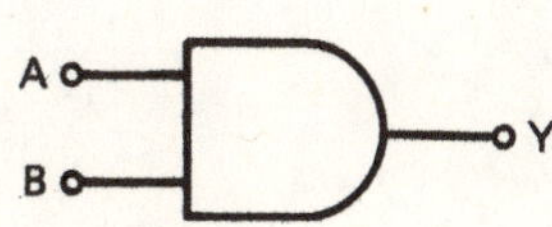

Fig. A-1.4 AND gate symbol.

COMPONENTS

(3) 1N4004 silicon diodes
(1) 1-kΩ, 0.25-W resistor

PROCEDURE

1. Connect the circuit in Fig. A-1.5. Apply the correct polarity of voltage to the circuit.

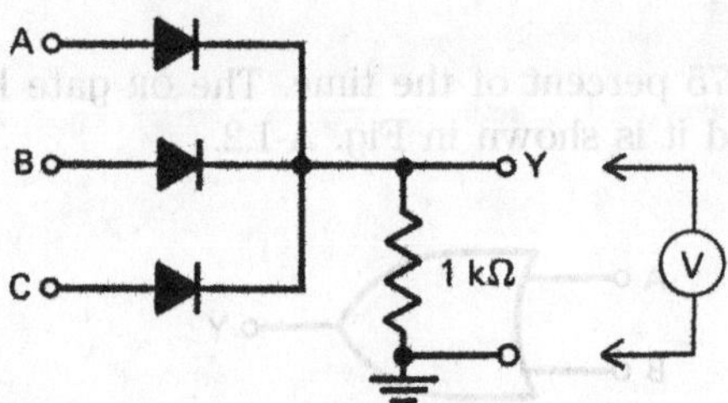

Fig. A-1.5 Three-input OR gate.

2. Connect points A, B, and C to ground. Measure and record in Table A-1.1 the voltage level measured from point Y to ground.

Note: Remember that a 1 = +5 V and a 0 = ground potential.

3. Complete Table A-1.1 for the remaining combinations of A, B, and C.

4. Connect the circuit of Fig. A-1.6. Apply the correct polarity of voltage to the circuit.

5. Connect points A, B, and C to ground. Measure and record in Table A-1.2 the voltage level measured from point Y to ground.

6. Complete Table A-1.2 for the remaining combinations of A, B, and C.

Note: Remember that a 1 = +5 V and a 0 = ground potential.

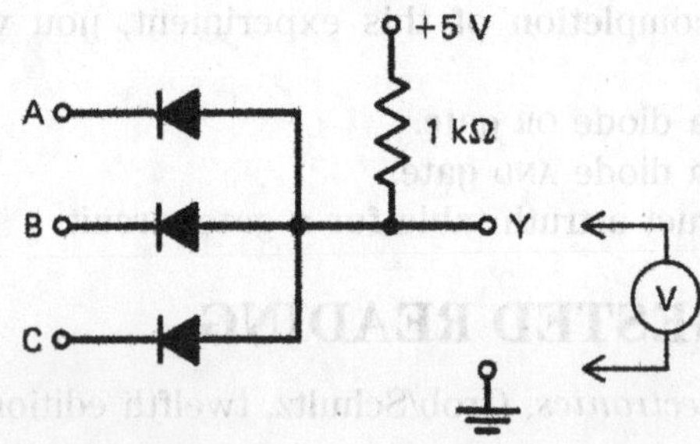

Fig. A-1.6 Three-input AND gate.

NAME ____________________ DATE ____________

QUESTIONS FOR EXPERIMENT A-1 AND TABLES

1. In the prior circuits, what voltage level would a binary 1 represent? A binary 0? Are the answers the same for both circuits of Figs. A-1.5 and A-1.6?

2. What is an OR gate? For what percentage of time is the three-input circuit on?

3. What is an AND gate? For what percentage of time is the three-input circuit off?

4. What function do the diodes perform in the OR and AND gates?

5. What are the fundamental differences between the two circuits shown in Figs. A-1.5 and A-1.6? Do the differences significantly affect overall outcomes? Explain.

TABLES FOR EXPERIMENT A-1

TABLE A-1.1

A	B	C	Y
0	0	0	______
0	0	1	______
0	1	0	______
0	1	1	______
1	0	0	______
1	0	1	______
1	1	0	______
1	1	1	______

TABLE A-1.2

A	B	C	Y
0	0	0	______
0	0	1	______
0	1	0	______
0	1	1	______
1	0	0	______
1	0	1	______
1	1	0	______
1	1	1	______

EXPERIMENT RESULTS REPORT FORM

Experiment No.: ______________________ Name: ______________________

Date: ______________________

Experiment Title: ______________________ Class: ______________________

______________________ Instr: ______________________

Explain the purpose of the experiment:

List the first Learning Objective:
OBJECTIVE 1:

After reviewing the results, describe how the objective was validated by this experiment.

List the second Learning Objective:
OBJECTIVE 2:

After reviewing the results, describe how the objective was validated by this experiment.

List the third Learning Objective:
OBJECTIVE 3:

After reviewing the results, describe how the objective was validated by this experiment.

Conclusion:

If required, attach to this form: ☐ Answers to Questions, ☐ Tables, and ☐ Graphs.

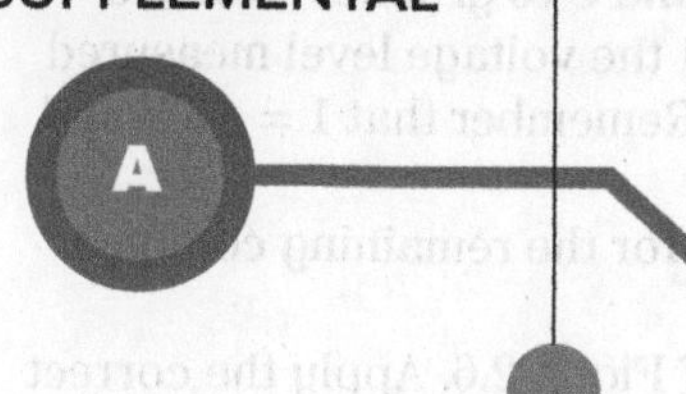

SUPPLEMENTAL **E X P E R I M E N T A - 2**

DIGITAL NOR AND NAND GATES

LEARNING OBJECTIVES

At the completion of this experiment, you will be able to:

- Build a diode NOR gate.
- Build a diode NAND gate.
- Construct a truth table for gate circuits.

SUGGESTED READING

Basic Electronics, Grob/Schultz, twelfth edition

INTRODUCTION

Two fundamental circuits are the NOR and NAND gates. The NOR gate has two or more input signals but creates only one output signal. If any input signal is high (binary 1), the output signal is low. NOR gates can be discrete, that is, made from several components such as diodes and resistors. NOR gates can also come packaged as integrated circuits (ICs). The NOR gate circuit constructed in this experiment is discrete. Figure A-2.1 shows a discrete NOR gate and its truth table.

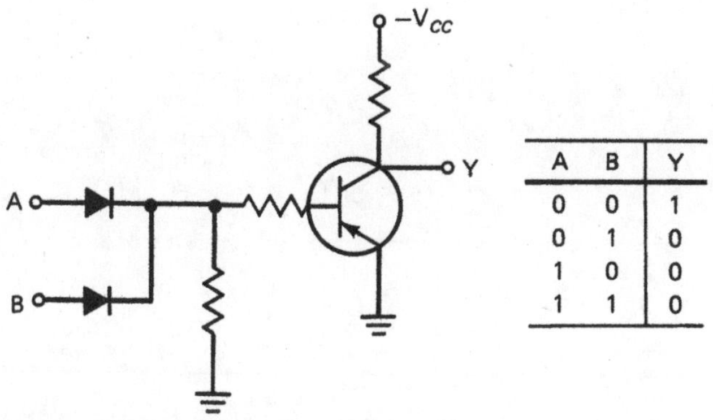

A	B	Y
0	0	1
0	1	0
1	0	0
1	1	0

Fig. A-2.1 Two-input NOR gate and truth table.

Truth tables are data tables that quickly show the overall circuit action of the gate under investigation. For the circuit of Fig. A-2.1, a binary 1 is equivalent to +5 V, and a binary 0 is equivalent to ground potential. Studying the truth table of Fig. A-2.1 reveals that for a two-input NOR gate, the circuit is on only when points A and B are both at ground potential. Another way of explaining the results of the truth table is to say that the NOR gate is a "mostly off" device. In the case of the truth table of Fig. A-2.1, the two-input NOR gate is off 75 percent of the time. The NOR gate has a schematic symbol, and it is shown in Fig. A-2.2.

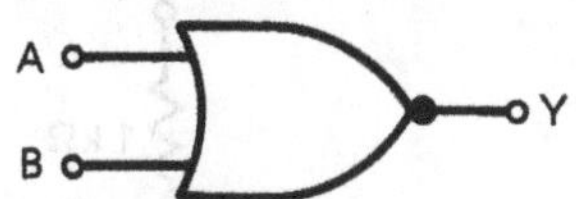

Fig. A-2.2 NOR gate symbol.

The NAND gate has two or more input signals but creates only one output signal. NAND gates can also be either of discrete or IC design. Figure A-2.3 shows a

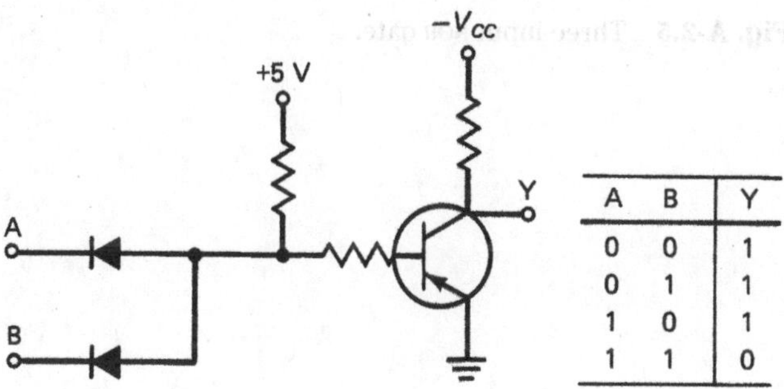

A	B	Y
0	0	1
0	1	1
1	0	1
1	1	0

Fig. A-2.3 Two-input NAND gate and truth table.

discrete NAND gate and its truth table. The NAND gate circuit is off only when points A and B are both at a +5-V level, or at a binary 1. Another way of describing the results found in the truth table is to say that the NAND gate is a "mostly on" device. In the truth table of Fig. A-2.3, the two-input NAND gate is on 75 percent of the time. The NAND gate has a schematic symbol, and it is shown in Fig. A-2.4.

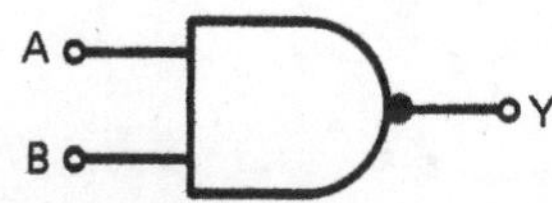

Fig. A-2.4 NAND gate symbol.

EQUIPMENT

Power supply 0–10 V DC
DMM

Protoboard/springboard
Test leads

COMPONENTS

(3) 1N4004 silicon diodes
(1) 2N3904 transistor
(2) 1-kΩ, 0.25-W resistors
(1) 10-kΩ, 0.25-W resistor

PROCEDURE

1. Connect the circuit in Fig. A-2.5. Apply the correct polarity of voltage to the circuit.

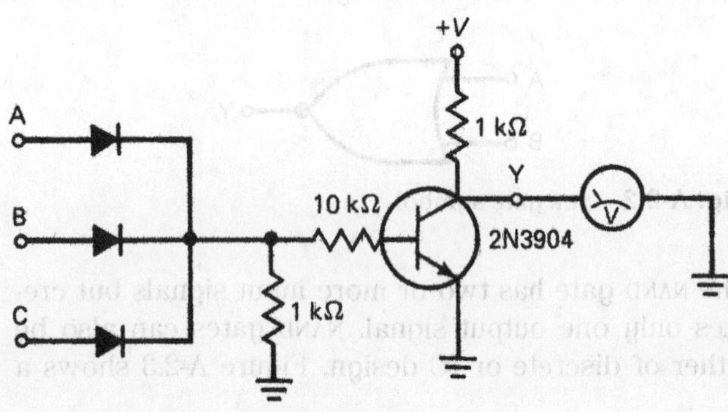

Fig. A-2.5 Three-input NOR gate.

2. Connect points A, B, and C to ground. Measure and record in Table A-2.1 the voltage level measured from point Y to ground. Remember that 1 = +5 V and 0 = ground potential.
3. Complete Table A-2.1 for the remaining combinations of A, B, and C.
4. Connect the circuit of Fig. A-2.6. Apply the correct polarity of voltage to the circuit.

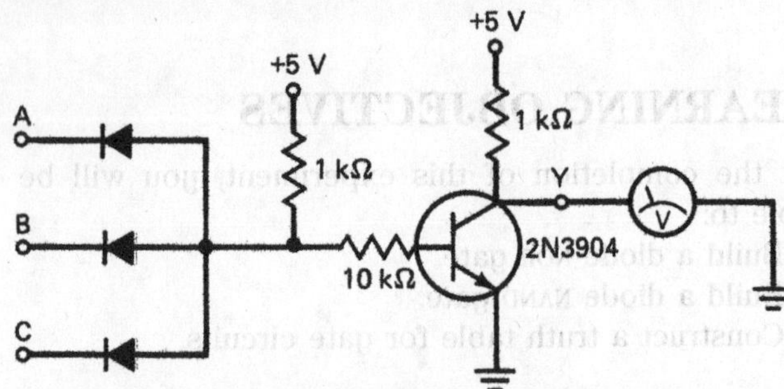

Fig. A-2.6 Three-input NAND gate.

5. Connect points A, B, and C to ground. Measure and record in Table A-2.2 the voltage level measured from point Y to ground. Again remember that 1 = +5 V and 0 = ground potential.
6. Complete Table A-2.2 for the remaining combinations of A, B, and C.

NAME DATE

QUESTIONS FOR EXPERIMENT A-2

1. In the circuits in Figs. A-2.5 and A-2.6, what voltage level would a binary 1 represent? Binary 0? Are the answers the same for the circuits in both these figures?

2. What is a NOR gate? What percentage of the time is the three-input circuit off?

3. What is a NAND gate? What percentage of the time is the three-input circuit on?

4. What function do the diodes perform in the NOR and NAND gates?

5. What are the fundamental differences between the two circuits shown in Figs. A-2.3 and A-2.4? Do the differences significantly affect overall outcomes? Explain.

TABLES FOR EXPERIMENT A-2

TABLE A-2.1

A	B	C	Y
0	0	0	
0	0	1	
0	1	0	
0	1	1	
1	0	0	
1	0	1	
1	1	0	
1	1	1	

TABLE A-2.2

A	B	C	Y
0	0	0	
0	0	1	
0	1	0	
0	1	1	
1	0	0	
1	0	1	
1	1	0	
1	1	1	

EXPERIMENT RESULTS REPORT FORM

Experiment No.: ______________________ Name: ______________________

Date: ______________________

Experiment Title: ______________________ Class: ______________________

______________________ Instr: ______________________

Explain the purpose of the experiment:

List the first Learning Objective:

OBJECTIVE 1:

After reviewing the results, describe how the objective was validated by this experiment.

List the second Learning Objective:

OBJECTIVE 2:

After reviewing the results, describe how the objective was validated by this experiment.

List the third Learning Objective:

OBJECTIVE 3:

After reviewing the results, describe how the objective was validated by this experiment.

Conclusion:

If required, attach to this form: ☐ Answers to Questions, ☐ Tables, and ☐ Graphs.

SUPPLEMENTAL

A

EXPERIMENT A-3

MULTIVIBRATOR FLIP-FLOP

LEARNING OBJECTIVES

At the completion of this experiment, you will be able to:

- Construct a basic multivibrator circuit.
- Test for proper operation in an astable multivibrator circuit.
- Determine which components determine the frequency of operation of a multivibrator.

SUGGESTED READING

Basic Electronics, Grob/Schultz, twelfth edition

INTRODUCTION

Multivibrators are also known as *latches*, *registers*, and *flip-flops*.

A *flip-flop* is a circuit that receives an input signal that is either 5 V or at ground potential. The circuit then produces and remembers a predictable output signal which is also either 5 V or at ground potential. The circuit can maintain its memory even after the input signal has been removed from the device.

The simplest astable multivibrator is formed by connecting two inverting gates back to back through two capacitors, as shown in Fig. A-3.1. This connection produces a circuit which switches from low to high repeatedly. The time between switching, or the frequency of operation, is determined by the value of the capacitors.

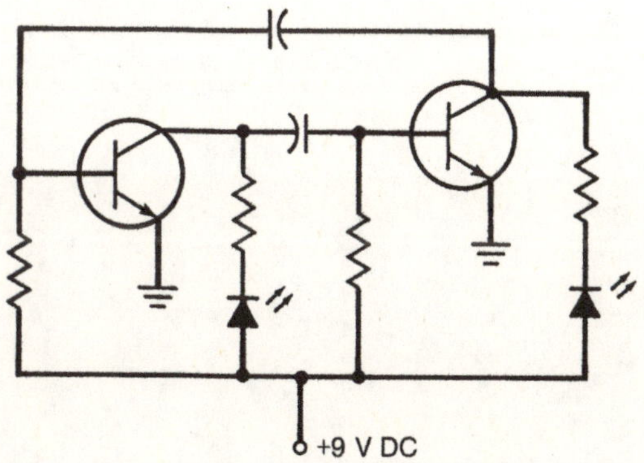

Fig. A-3.1 Basic astable multivibrator circuit.

EQUIPMENT

DC power supply, 0–10 V
DMM
Protoboards
Test leads

COMPONENTS

(2) 2N3904
(2) LEDs
(2) 220-Ω, 0.25-W resistors
(2) 82-kΩ, 0.25-W resistors
(2) 25-μF, 10-V DC electrolytic capacitors
(2) 47-μF, 10-V DC electrolytic capacitors

PROCEDURE

1. Construct the astable multivibrator circuit of Fig. A-3.2 using the two 25-μF, 10-V DC electrolytic capacitors.

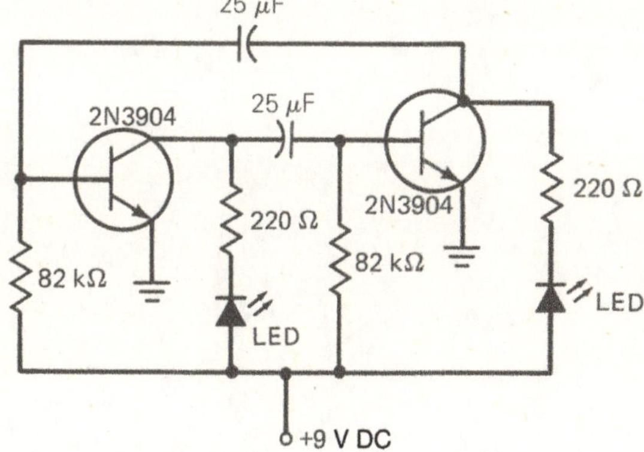

Fig. A-3.2 Astable multivibrator.

2. Turn on the power supply. Determine the frequency of the circuit.
3. Turn off the power supply.
4. Change the value of the capacitors to 47 μF. Turn on the power supply. Determine the output frequency, and note the difference in circuit action. Describe the circuit action of this step in the "Results" section of this experiment. How do the results of steps 2 and 4 differ?

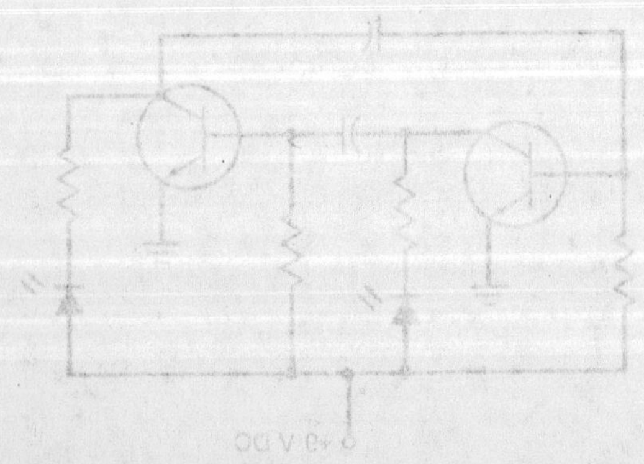

NAME DATE

QUESTIONS FOR EXPERIMENT A-3

1. What components determine the frequency of operation?

2. What does a multivibrator do?

EXPERIMENT RESULTS REPORT FORM

Experiment No.: ______________________ Name: ______________________

Date: ______________________

Experiment Title: ______________________ Class: ______________________

______________________ Instr: ______________________

Explain the purpose of the experiment:

List the first Learning Objective:
OBJECTIVE 1:

After reviewing the results, describe how the objective was validated by this experiment.

List the second Learning Objective:
OBJECTIVE 2:

After reviewing the results, describe how the objective was validated by this experiment.

List the third Learning Objective:
OBJECTIVE 3:

After reviewing the results, describe how the objective was validated by this experiment.

Conclusion:

If required, attach to this form: ☐ Answers to Questions, ☐ Tables, and ☐ Graphs.

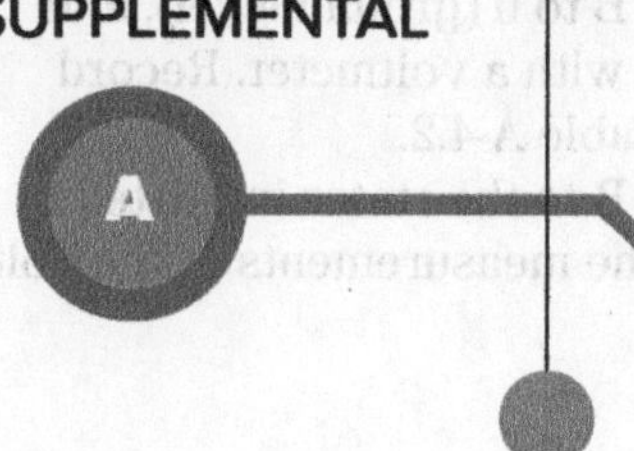

SUPPLEMENTAL **E X P E R I M E N T A - 4**

INTEGRATED LOGIC CIRCUITS

LEARNING OBJECTIVES

At the completion of this experiment, you will be able to:

- Verify the proper operation of an integrated circuit (IC) OR gate.
- Verify the proper operation of an IC AND gate.
- Describe the fundamental differences between AND gates and OR gates.

SUGGESTED READING

Basic Electronics, Grob/Schultz, twelfth edition

INTRODUCTION

The computers of today do not process numbers in base 10 (that is, 0, 1, 2, 3, . . . , 9). Computers instead use binary logic, base 2 (0 and 1), to perform their functions. Two fundamental circuits are the IC OR gate and the IC AND gate.

Integrated circuits contain a complete circuit, consisting of active and passive components connected in a unique circuit configuration contained in a space no larger than a transistor. The active components are transistors and diodes; the passive components are the resistors and capacitors. The components of an IC are not discrete parts wired together to form a complete circuit. Instead the IC is a complete circuit formed by a photographic process, on a small slab of silicon.

The purpose of this experiment is to introduce integrated logic circuits. The IC OR gate has two or more input signals but creates only one output signal. If any input signal is high, the output signal is high. The OR gate circuit constructed in this experiment has an IC orientation. Figure A-4.1 shows an IC OR gate and its truth table.

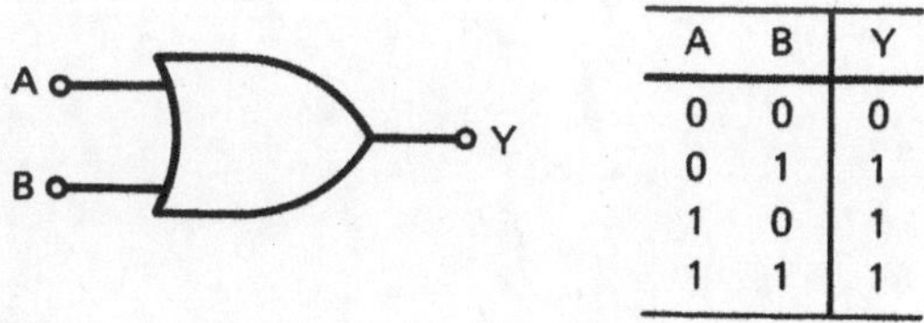

A	B	Y
0	0	0
0	1	1
1	0	1
1	1	1

Fig. A-4.1 IC OR gate and truth table.

The IC AND gate has two or more input signals but creates only one output signal. All the inputs must be high to get a high output signal. Figure A-4.2 shows an IC AND gate and its truth table.

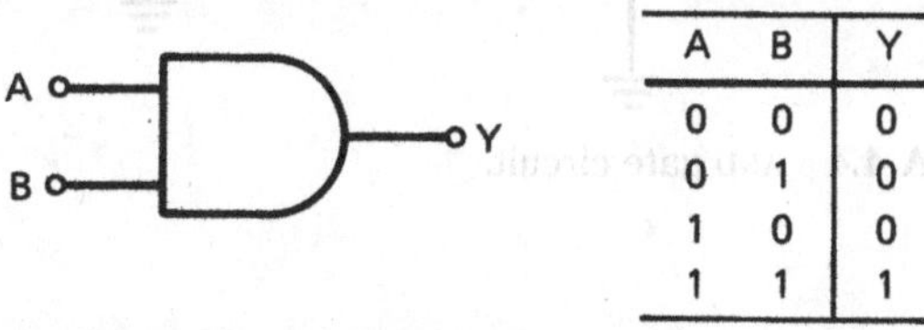

A	B	Y
0	0	0
0	1	0
1	0	0
1	1	1

Fig. A-4.2 IC AND gate and truth table.

It is important to check the manufacturer's specification sheets for proper pin identification before use. The 7400 series ICs are available in the 14- or 16-pin format. The 7400 family of logic ICs uses pins 7 and 8 for the ground connection and pins 14 and 16 for +5 V.

EQUIPMENT

DC power supply, 0 to 10 V
DMM
Protoboard
Test leads

COMPONENTS

(1) 7432 IC quad OR gate
(1) 7408 IC quad AND gate

PROCEDURE

1. Construct the circuit of Fig. A-4.3. Be sure to connect the ground and +5 V to the proper pins (see your instructor).

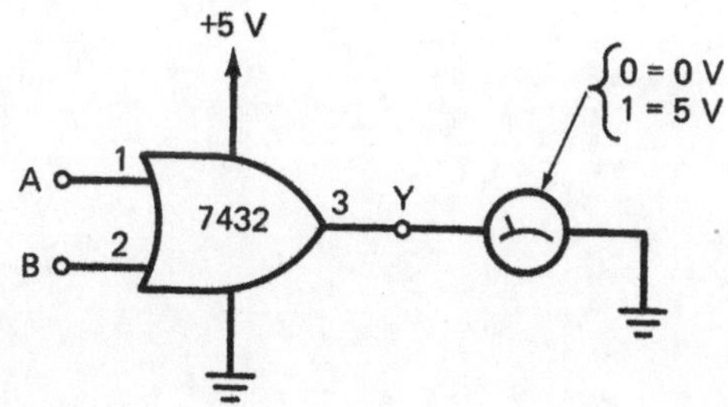

Fig. A-4.3 OR gate circuit.

2. Connect inputs A and B to 0 (ground level), and measure the output level with the VTVM. Record these measurements in Table A-4.1.

3. Connect inputs A and B to the states indicated in Table A-4.1, and record the measurements of the VTVM in Table A-4.1.

4. Construct the circuit of Fig. A-4.4. Be sure to connect the ground and +5 V to the proper pins (see your instructor).

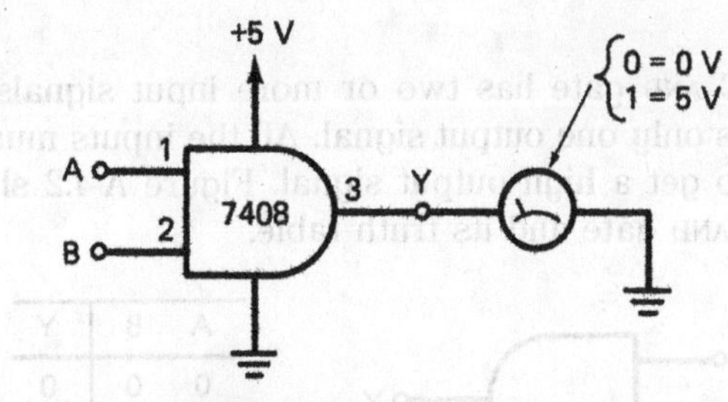

Fig. A-4.4 AND gate circuit.

5. Connect inputs A and B to 0 (ground level), and measure the output level with a voltmeter. Record these measurements in Table A-4.2.

6. Connect inputs A and B to the states indicated in Table A-4.2, and record the measurements of the voltmeter in Table A-4.2.

NAME ______________________________ DATE ______________

QUESTIONS FOR EXPERIMENT A-4

1. In the circuits shown in Figs. A-4.3 and A-4.4, what voltage level would binary 1 represent? Binary 0? Are the answers the same for both circuits?

2. What is an OR gate?

3. In a three-input OR gate circuit, what percentage of the time is the circuit on?

4. What is an AND gate?

5. In a three-input AND gate circuit, what percentage of the time is the circuit off?

6. What are the fundamental differences between the two circuits constructed in this experiment?

TABLES FOR EXPERIMENT A-4

TABLE A-4.1

A	B	Y
0	0	
0	1	
1	0	
1	1	

TABLE A-4.2

A	B	Y
0	0	
0	1	
1	0	
1	1	

EXPERIMENT RESULTS REPORT FORM

Experiment No.: ______________________ Name: ______________________

Date: ______________________

Experiment Title: ______________________ Class: ______________________

______________________ Instr: ______________________

Explain the purpose of the experiment:

List the first Learning Objective:

OBJECTIVE 1:

After reviewing the results, describe how the objective was validated by this experiment.

List the second Learning Objective:

OBJECTIVE 2:

After reviewing the results, describe how the objective was validated by this experiment.

List the third Learning Objective:

OBJECTIVE 3:

After reviewing the results, describe how the objective was validated by this experiment.

Conclusion:

If required, attach to this form: ☐ Answers to Questions, ☐ Tables, and ☐ Graphs.

SUPPLEMENTAL

B

EXPERIMENT B-1

VACUUM TUBE AMPLIFIER

LEARNING OBJECTIVES

At the completion of this experiment, you will be able to:

- Study the operation of a tube as an amplifier.
- Measure the voltage gain of a vacuum tube amplifier.
- Study a family of characteristic curves for a vacuum tube.

SUGGESTED READING

Basic Electronics, Grob/Schultz, twelfth edition

INTRODUCTION

Vacuum tubes are able to operate efficiently over a wide range of frequencies. Their operation is based upon their ability to control, almost instantaneously, the flow of millions of electrons. The main difference between vacuum tubes and transistors is that vacuum tubes are controlled by voltage, whereas transistors, typically, are controlled by current.

Vacuum tubes can be used as (1) voltage amplifiers, (2) power amplifiers, (3) detectors, (4) oscillators, (5) frequency detectors, (6) regulators, (7) modulators, and (8) rectifiers. As a rectifier, a vacuum tube, like a silicon diode, will allow current to flow in only one direction. The vacuum tube operates as follows.

Within the evacuated (void of air) tube, there are electric elements called a *cathode* (an electron emitter, given the sign −) and a *plate* (the electron acceptor, given the sign +). If a positive voltage is applied to the plate, electrons emitted from the cathode are attracted to the plate. In this case, the cathode must be heated in order to release the electrons. One method of heating the cathode so that it will release electrons is through the use of a filament supply, typically 6.3 V_{rms}. With the cathode heated, and a positive voltage applied to the plate (positive electrode), electrons will flow from the cathode (−) to the plate (+), creating the same results as if a *p-n* junction diode was forward-biased.

When a third electrode, called a *control grid,* is added inside a vacuum tube, the tube is then called a *triode.* It is located between the plate and the cathode, as shown in Fig. B-1.1. The control grid actually controls the number of electrons allowed to flow from the cathode to the plate. Again, voltage is the controlling factor. If the control grid is made more negative than the cathode, some of the electrons going toward the plate will be repelled by the grid, thus reducing the plate current. If the plate is made as positive as possible and the maximum number of electrons given off by the cathode is attracted to the plate, the triode is said to be *saturated.* That is, plate current has reached a saturation point, and any increase in the plate voltage will not increase plate current.

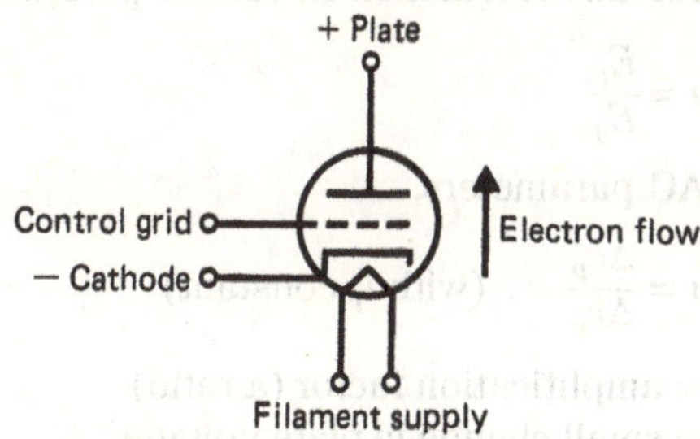

Fig. B-1.1 Triode.

Conversely, if the control grid is biased negative so as to prohibit the flow of electrons from cathode to plate, the triode is said to be a *cutoff* triode. In this case, the flow of current is stopped or cut off.

The operation of a triode tube can be plotted on a load line similar to that of a transistor, with the plate voltage on the x axis, plate current on the y axis, and differing values of grid voltage as a family of curves.

Plate Resistance

The plate resistance is the measure of opposition to the flow of current through the tube. The formula is basically Ohm's law. For DC parameters,

$$R_P = \frac{E_p}{I_p}$$

where E_p = DC plate voltage
I_p = DC plate current

For AC parameters,

$$r_p = \frac{\Delta e_p}{\Delta i_p} \quad \text{(with } e_g \text{ constant)}$$

where r_p = AC plate resistance
Δe_p = small change in plate voltage
Δi_p = small change in plate current

Amplification Factor

The amplification factor is a measure of how much more control the grid voltage exerts over plate current compared to the plate voltage. The symbol used to denote amplification is the Greek letter mu (μ). The formula for this calculation is, for DC parameters,

$$\mu = \frac{E_p}{E_g}$$

and for AC parameters,

$$\mu = \frac{\Delta e_p}{\Delta e_g} \quad \text{(with } i_p \text{ constant)}$$

where μ = amplification factor (a ratio)
Δe_p = small change in plate voltage
Δe_g = small change in grid voltage

Transconductance

The transconductance is a factor that relates the amount of plate current change that is caused by a grid voltage change. The formula is basically one of conductance, where $G = I/E$. For DC parameters,

$$G_m = \frac{I_p}{E_g}$$

For AC parameters,

$$g_m = \frac{\Delta i_p}{\Delta e_g} \quad \text{(with plate voltage held constant)}$$

where g_m = transconductance, in siemens
Δi_g = small change in plate current
Δe_g = small change in grid voltage

EQUIPMENT

High-voltage DC power supply (0–300 V)
Low-voltage DC power supply (0–10 V)
Filament power supply
Signal generator
Test leads
Tube socket for the 6SN7 or 6J5
Test leads
DMM (high voltage input)

COMPONENTS

(1) Tube 6J5 or 6SN7
(2) 0.01-μF capacitor (disc)
(1) 22-kΩ, 2-W resistor
(1) 100-kΩ, 0.25-W resistor

PROCEDURE

1. Connect the circuit shown in Fig. B-1.2. Be careful with the high-voltage power supply.

Note: B+ is the standard symbol for the plate supply voltage. Here, B+ = 300 V DC. The 22-kΩ resistor is referred to as R_p, plate resistor. The 100-kΩ resistor is referred to as R_g, grid resistor. Directly below the grid resistor is the variable power supply, V_{gg}. This is a separate power source that controls the grid voltage, which in turn controls the flow of electrons from the heated cathode (−) to the plate, also referred to as the *anode* (+). Also, note that the filament is not drawn on the schematic of Fig. B-1.2 (it is assumed to be connected).

2. With the circuit of Fig. B-1.2 connected, adjust the grid voltage to −2 V. Also, adjust the signal generator to 1 kHz at 2 Vp-p output.

3. Measure and record in Table B-1.1 the signal voltage from grid to cathode.

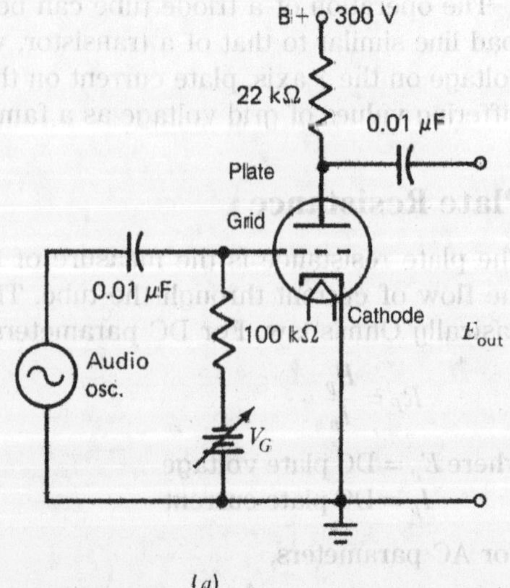

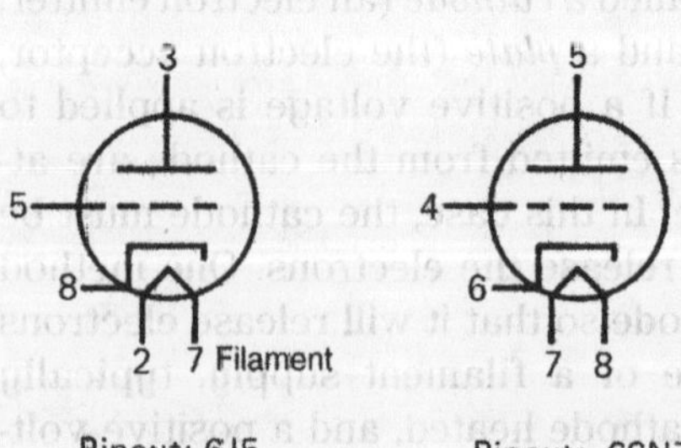

Fig. B-1.2 (*a*) Triode amplifier. (*b*) Pinouts for a 6J5 and a 6SN7 tube.

4. Measure and record in Table B-1.1 the input and output signals for grid voltages of −2 V, −4 V, and −6 V. Calculate and record the voltage gain (A_V) for each one.

5. With the grid voltage set at −4 V, decrease the frequency of oscillation to 20 Hz, and record the peak-to-peak output voltage in Table B-1.1. Also, record the peak-to-peak output voltages with the frequency at 15 and 50 kHz.

6. Calculate the voltage gain (A_V) of the amplifier by using the following formula, where $\mu = 20$ for the 6J5 or 6SN7 vacuum tube. Also, $r_p = 7.0\ \text{k}\Omega$ and is the AC plate resistance. Record this in Table B-1.1.

$$A_V = \frac{\mu \times R_p}{R_p + r_p}$$

7. Compare the calculated voltage gain with the measured voltage gain. Calculate and record in Table B-1.1 the percentage of error between the measured and calculated voltage gains.

8. Replace the 22-kΩ plate resistor with an 8.2-kΩ resistor. For the corresponding grid voltages of −2 V, −4 V, −6 V, and −8 V, measure and record in Table B-1.1 the corresponding plate voltages.

9. With the 8.2-kΩ resistor as R_p, repeat steps 2 to 8. Record results in a separate data table of your own, similar to Table B-1.1.

NAME DATE

QUESTIONS FOR EXPERIMENT B-1

1. Define the purpose and function of each of the elements found in a triode.

2. In this experiment, which step had the largest effect on gain?

3. For this tube experiment, define the condition causing saturation.

4. For this tube experiment, define the condition causing cutoff.

5. How do the calculated values of voltage gain compare to the actual measured values? How are differences accounted for?

TABLE FOR EXPERIMENT B-1

TABLE B-1.1

Procedure Step	Measurement	Value V	Calculated A_V	Measured A_V
3	V_{GK}			
4	V_{in} at $V_G = -2$ V			
	V_{out} at $V_G = -2$ V			
	V_{in} at $V_G = -4$ V			
	V_{out} at $V_G = -4$ V			
	V_{in} at $V_G = -6$ V			
	V_{out} at $V_G = -6$ V			
5	V_{out} at 20 Hz			
	V_{out} at 15 kHz			
	V_{out} at 50 kHz			
6				
7	Percent of error = ____________ %			
8	V_P with $V_G = -2$ V			
	V_P with $V_G = -4$ V			
	V_P with $V_G = -6$ V			
	V_P with $V_G = -8$ V			

EXPERIMENT RESULTS REPORT FORM

Experiment No.: ______________________ Name: ______________________

Date: ______________________

Experiment Title: ______________________ Class: ______________________

______________________ Instr: ______________________

Explain the purpose of the experiment:

List the first Learning Objective:

OBJECTIVE 1:

After reviewing the results, describe how the objective was validated by this experiment.

List the second Learning Objective:

OBJECTIVE 2:

After reviewing the results, describe how the objective was validated by this experiment.

List the third Learning Objective:

OBJECTIVE 3:

After reviewing the results, describe how the objective was validated by this experiment.

Conclusion:

If required, attach to this form: ☐ Answers to Questions, ☐ Tables, and ☐ Graphs.

Appendix A Applicable Color Codes

TABLE A-1 Color Code for Carbon Composition Resistors

COLOR	DIGIT 1st	DIGIT 2nd	MULTIPLIER	TOLERANCE
Black	0	0	1	—
Brown	1	1	10	—
Red	2	2	100	—
Orange	3	3	1,000	—
Yellow	4	4	10,000	—
Green	5	5	100,000	—
Blue	6	6	1,000,000	—
Violet	7	7	10,000,000	—
Gray	8	8	100,000,000	—
White	9	9	1,000,000,000	—
Gold	—	—	0.1	± 5%
Silver	—	—	0.01	±10%
No band	—	—	—	±20%

TABLE A-2 Color Code for Carbon-Film Resistors

COLOR	DIGITS 1st	DIGITS 2nd	DIGITS 3rd	MULTIPLIER	TOLERANCE
Black	0	0	0	1	
Brown	1	1	1	10	±1%
Red	2	2	2	100	±2%
Orange	3	3	3	1,000	
Yellow	4	4	4	10,000	
Green	5	5	5	100,000	±5%
Blue	6	6	6	1,000,000	±0.25%
Violet	7	7	7	10,000,000	±0.10%
Gray	8	8	8		±0.05%
White	9	9	9		±5%
Gold	—	—	—	0.1	±10%
Silver	—	—	—	0.01	

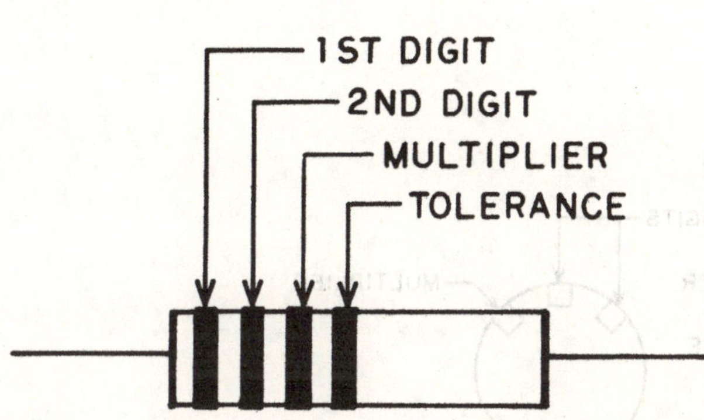

FIG. A-1 Color coding on a carbon composition resistor.

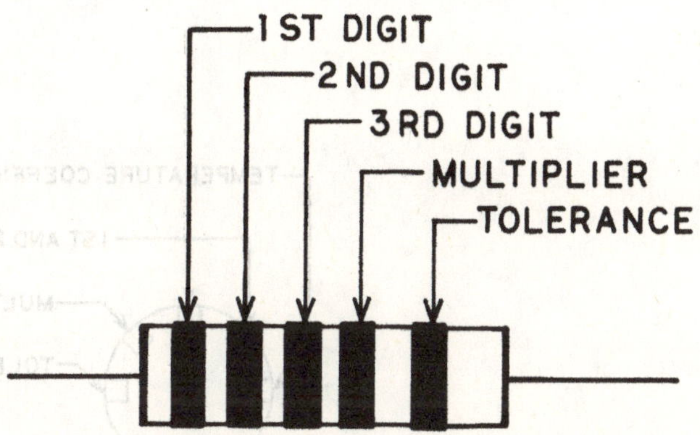

FIG. A-2 Color coding on a carbon-film resistor.

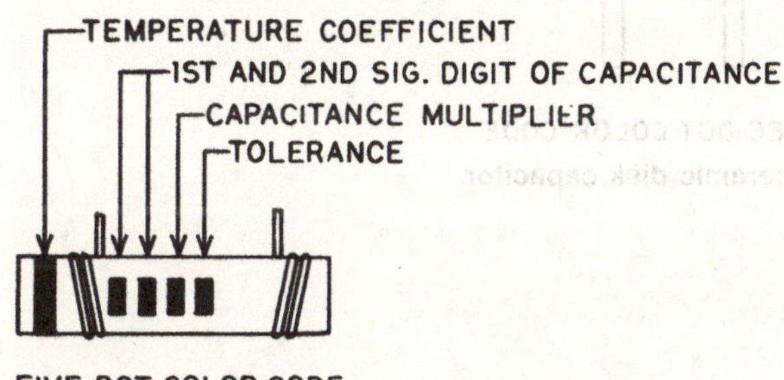

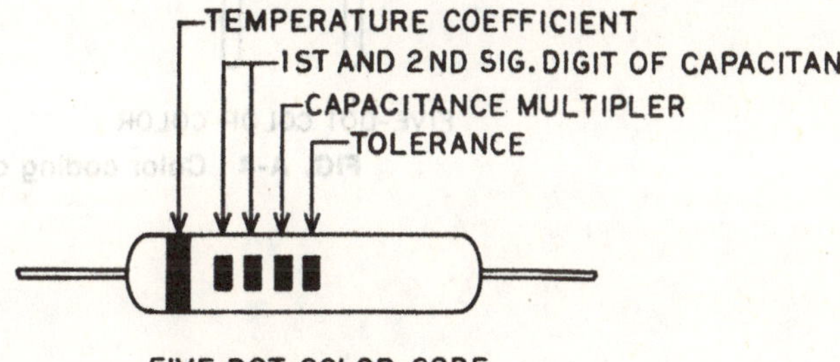

TEMPERATURE COEFFICIENT
TEMPERATURE COEFFICIENT MULTIPLIER
1ST AND 2ND SIG. DIGIT OF CAPACITANCE
CAPACITANCE MULTIPLIER
TOLERANCE

SIX-DOT COLOR CODE
RADIAL LEAD

FIG. A-3 Color coding on tubular ceramic capacitors.

Reproduced from Baer/Ottaway, *Electrical and Electronic Drawing,* fifth edition, McGraw-Hill, 1986.

TABLE A-3 Color Codes for Tubular and Disk Ceramic Capacitors

BAND OR DOT COLOR	CAPACITANCE IN PICOFARADS: SIGNIFICANT DIGITS 1st	2nd	CAPACITANCE MULTIPLIER	TOLERANCE ≤10 pF	> 10 pF	TEMPERATURE COEFFICIENT ppm°C (5-DOT SYSTEM)	TEMPERATURE COEFFICIENT 6-DOT SYSTEM SIG. FIG.	TEMPERATURE COEFFICIENT MULTIPLIER
Black	0	0	1	±2.0 pF	±20%	0	0.0	−1
Brown	1	1	10	±0.1 pF	±1%	−33		−10
Red	2	2	100		±2%	−75	1.0	−100
Orange	3	3	1000		±3%	−150	1.5	−1000
Yellow	4	4				−230	2.0	−10000
Green	5	5		±0.5 pF	±5%	−330	3.3	+1
Blue	6	6				−470	4.7	+10
Violet	7	7				−750	7.5	+100
Gray	8	8	0.01	±0.25 pF		+150 to −1500		+1000
White	9	9	0.1	±1.0 pF	±10%	+100 to −75		+10000
Silver	—	—						
Gold	—	—						

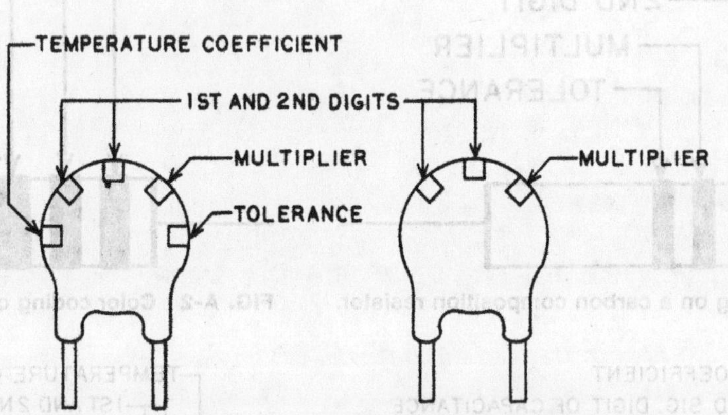

FIG. A-4 Color coding on a ceramic disk capacitor.

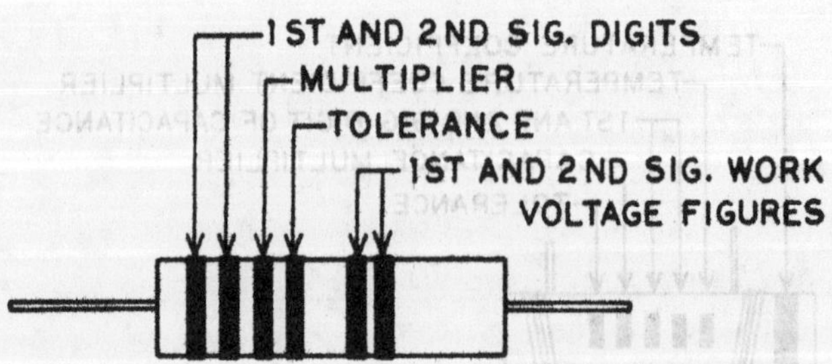

FIG. A-5 Color coding on a paper tubular capacitor.

TABLE A-4 Color Code for Molded Paper Tubular Capacitors

	(CAPACITANCE IN PICOFARADS)				
	SIGNIFICANT DIGITS				
COLOR	1st	2nd	MULTIPLIER	TOLERANCE	VOLTAGE
Black	0	0	1	±20%	—
Brown	1	1	10		100
Red	2	2	100		200
Orange	3	3	1000	±30%	300
Yellow	4	4	10000		400
Green	5	5			500
Blue	6	6			600
Violet	7	7			700
Gray	8	8			800
White	9	9			900
Gold	—	—			1000
Silver	—	—		±10%	—

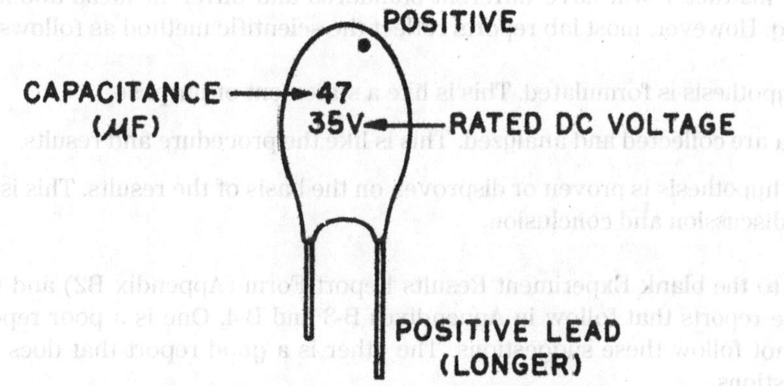

FIG. A-6 Marking of a tantalum capacitor.

Appendix B Lab Report Preparation

Appendix B-1: How to Write Lab Reports

1. Write short sentences whenever possible.
2. Use a dictionary and spell all the words correctly.
3. Write neatly in blue or black ink.
4. Avoid personal pronouns: *I*, *you*, *we*, etc.
5. Never discuss results unless they are part of your data.
6. Do not copy or paraphrase textbook theory.
7. Label all data tables and graphs with titles, numbers, proper units, and column headings.
8. Never write in the margins of your paper.
9. Leave enough space between sentences for the instructor to make corrections.
10. Technical accuracy and completeness are the most important part of a lab report. Unless the report is well organized and easy to read, it is of little value.

Every instructor will have different standards and different ideas about report writing. However, most lab reports reflect the scientific method as follows:

- A hypothesis is formulated. This is like a statement of purpose.
- Data are collected and analyzed. This is like the procedure and results.
- The hypothesis is proven or disproven on the basis of the results. This is like the discussion and conclusion.

Refer to the blank Experiment Results Report Form (Appendix B2) and the two sample reports that follow in Appendixes B-3 and B-4. One is a poor report that does not follow these suggestions. The other is a good report that does use the suggestions.

EXPERIMENT RESULTS REPORT FORM

Experiment No.: ______________________ Name: ______________________

Date: ______________________

Experiment Title: ______________________ Class: ______________________

______________________ Instr: ______________________

Explain the purpose of the experiment:

List the first Learning Objective:

OBJECTIVE 1:

After reviewing the results, describe how the objective was validated by this experiment.

List the second Learning Objective:

OBJECTIVE 2:

After reviewing the results, describe how the objective was validated by this experiment.

List the third Learning Objective:

OBJECTIVE 3:

After reviewing the results, describe how the objective was validated by this experiment.

Conclusion:

If required, attach to this form: ☐ Answers to Questions, ☐ Tables, and ☐ Graphs.

Appendix B-3 An Example of a Poorly Written Report

EXPERIMENT RESULTS REPORT FORM

Experiment No.: *3*

Name: *J. Doe*

Date:

Experiment Title: *Ohm's Law*

Class: *Elec.*

Instr: *Jones*

Explain the purpose of the experiment:

To do the Ohm's Law experiment

List the first Learning Objective:

OBJECTIVE 1:

After reviewing the results, describe how the objective was validated by this experiment.

The more voltage I had the more current I had when I measured.

List the second Learning Objective:

OBJECTIVE 2:

After reviewing the results, describe how the objective was validated by this experiment.

When we changed the resistors I saw more current but it was hard to measure because the needle was bent a little.

List the third Learning Objective:

OBJECTIVE 3:

After reviewing the results, describe how the objective was validated by this experiment.

$I = \frac{V}{R}$ *is Ohm's law.*

Conclusion:

Ohm's law works good.

If required, attach to this form: ☐ Answers to Questions, ☐ Tables, and ☐ Graphs.

Appendix C-1 Blank Graph Paper

Appendix B-4 An Example of a Better Written Student Report

EXPERIMENT RESULTS REPORT FORM

Experiment No.: 3
Name: John Doe
Date:
Experiment Title: Ohm's Law
Class: Elec. 101
Instr: Mrs. R. Jones

Explain the purpose of the experiment:

To validate the ohm's law expressions of $V = I \cdot R$, $I = \frac{V}{R}$ and $R = \frac{V}{I}$

List the first Learning Objective:

OBJECTIVE 1: *Validate $I = \frac{V}{R}$*

After reviewing the results, describe how the objective was validated by this experiment.

With resistance "R" held constant, current varied in direct proportion to any changes in applied voltage.

List the second Learning Objective:

OBJECTIVE 2: *Validate $V = I \cdot R$*

After reviewing the results, describe how the objective was validated by this experiment.

With the voltage held constant, the current was inversely proportional to any changes in circuit resistance.

List the third Learning Objective:

OBJECTIVE 3: $R = \frac{V}{I}$

After reviewing the results, describe how the objective was validated by this experiment.

Resistance increased as current decreased, when the voltage was held constant.

Conclusion:

Ohm's law is valid based upon the results of this experiment. Current is directly proportional to voltage and inversely proportional to resistance.

If required, attach to this form: ☐ Answers to Questions, ☐ Tables, and ☐ Graphs.

Appendix C-1 Blank Graph Paper

Appendix C-1 Blank Graph Paper

Appendix C-1 Blank Graph Paper

Appendix C-1 Blank Graph Paper

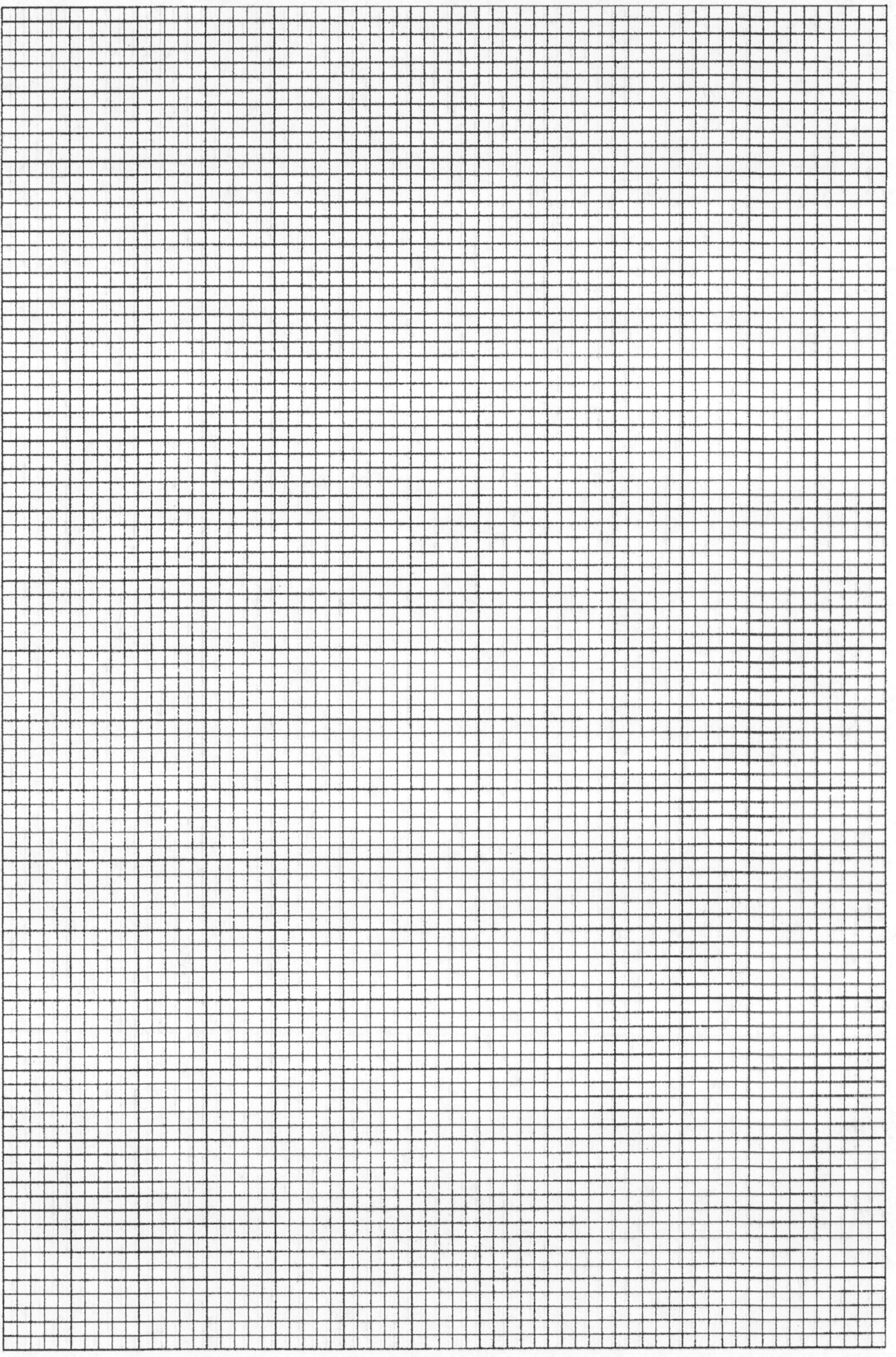

Appendix C-1 Blank Graph Paper

Appendix C-2 Blank Log Graph Paper

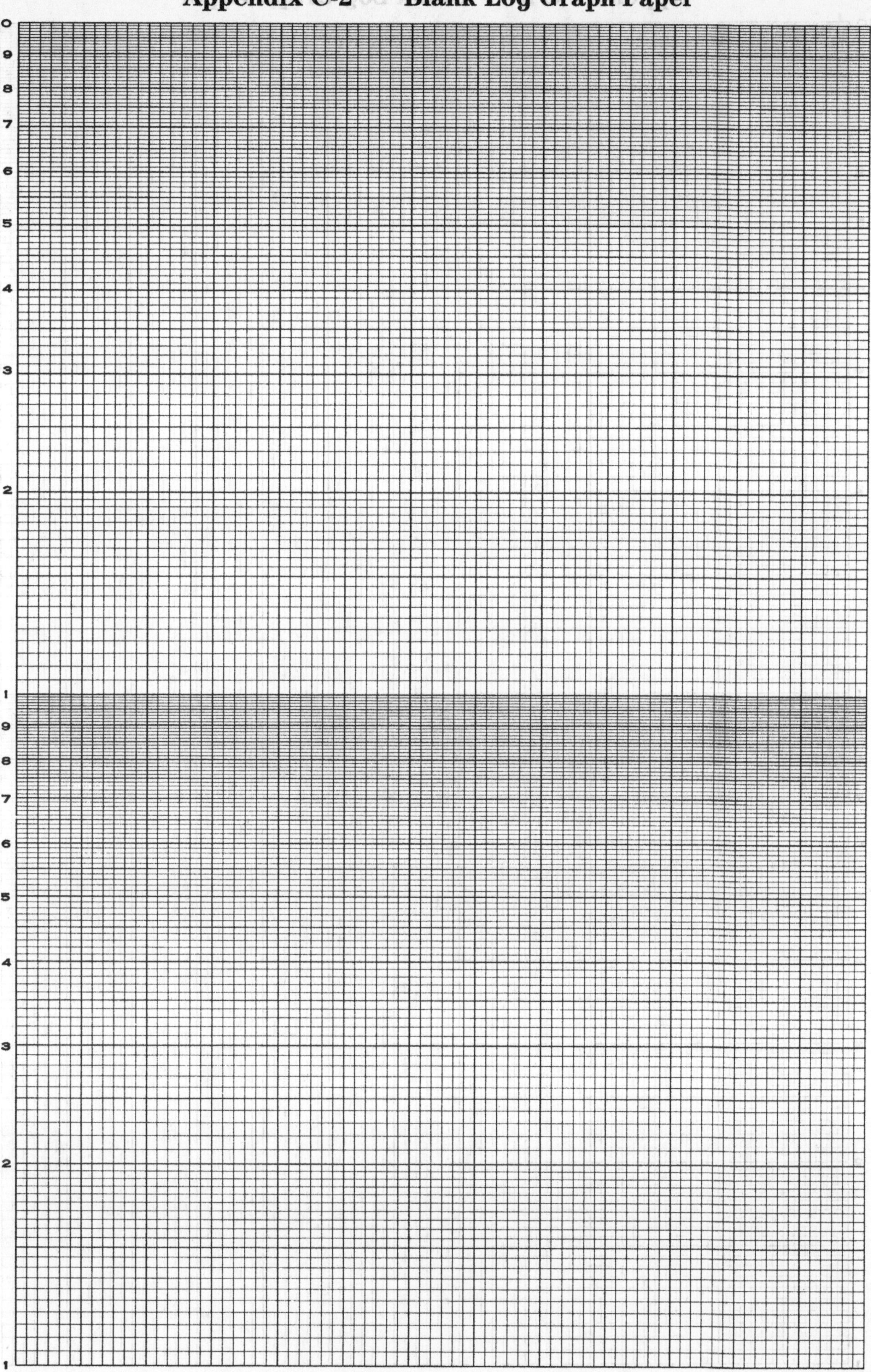

Appendix C-2 Blank Log Graph Paper

10
9
8
7
6
5
4
3
2
1
9
8
7
6
5
4
3
2
1

Appendix C-2 Blank Log Graph Paper

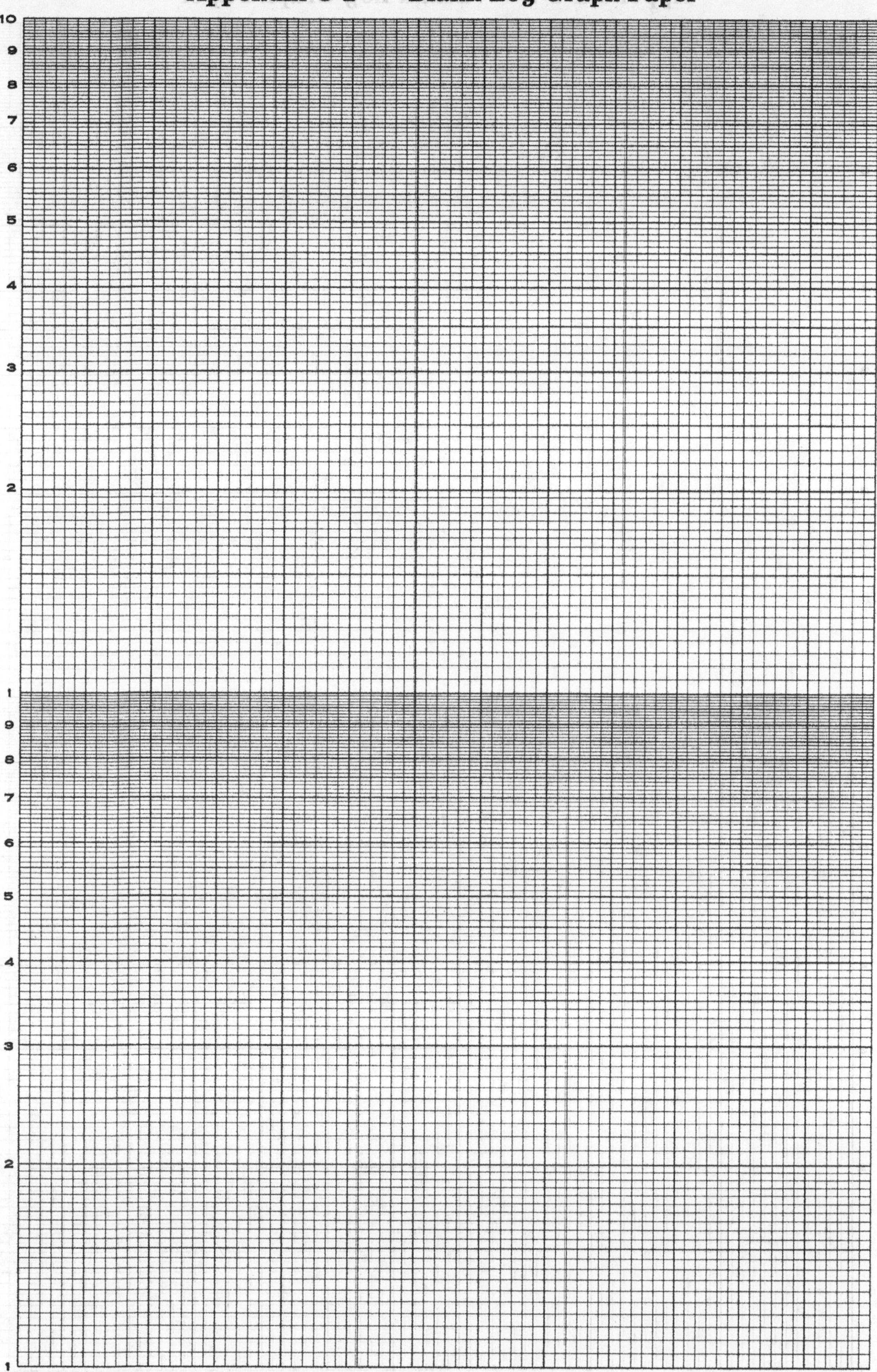

Appendix C-2 Blank Log Graph Paper

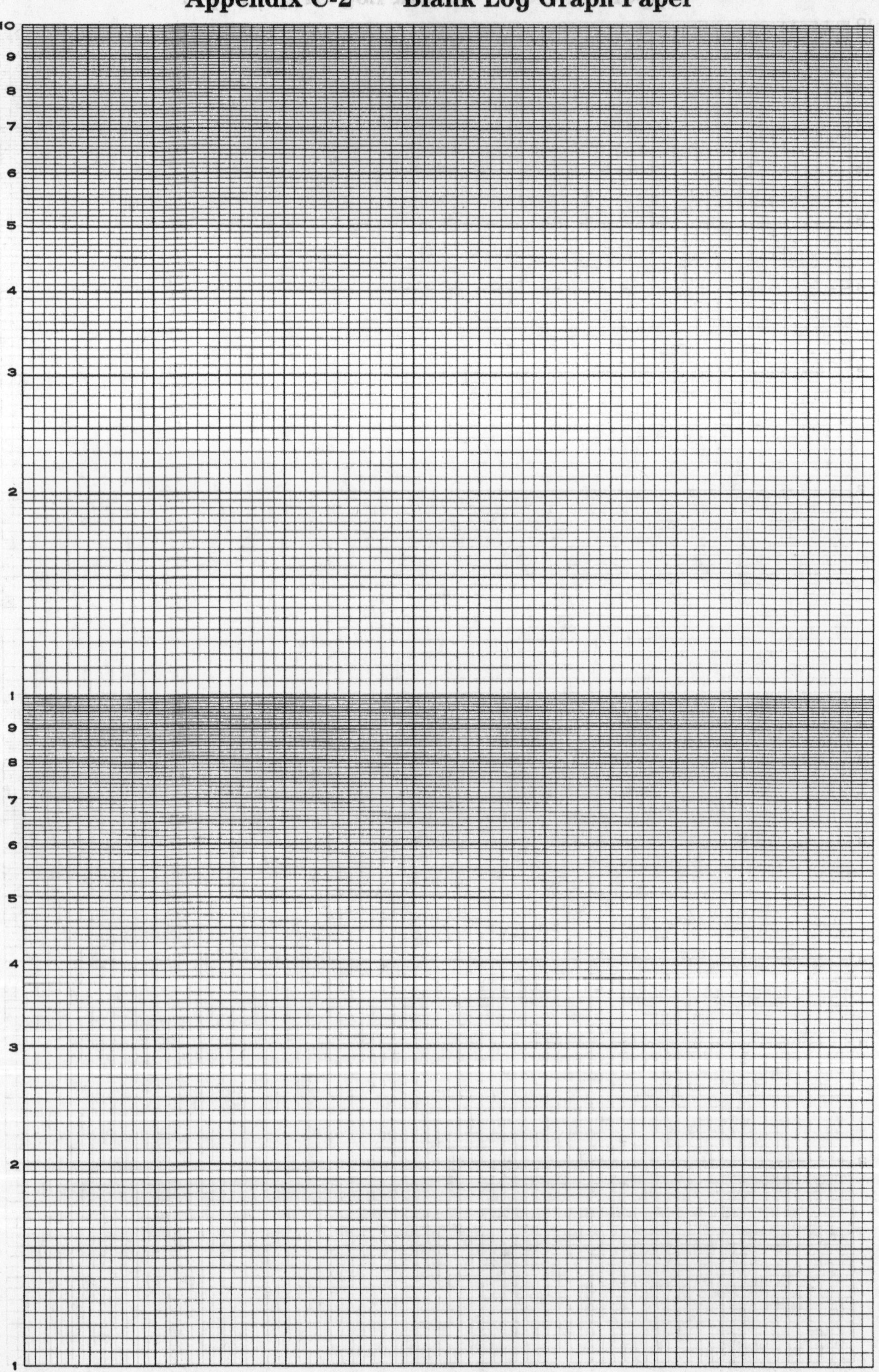

Appendix C-2 Blank Log Graph Paper

10
9
8
7
6
5
4
3
2
1
9
8
7
6
5
4
3
2
1

Appendix D How to Make Graphs

Graphs usually show the relationships of two or more variables. The relationship is actually the curve or line that results. Here are some things to remember when making graphs.

1. Be neat and complete.
2. Never connect points. Always show the characteristic of the curve.
3. There should be room in the margins to title the graph.
4. Use the fullest scales possible. Do not confine a graph to one corner of the paper.
5. Use a semilog graph paper for exponential quantities. Label the x axis as 1×10^2 for 100s, 1×10^3 for 1,000s, etc.

Examples

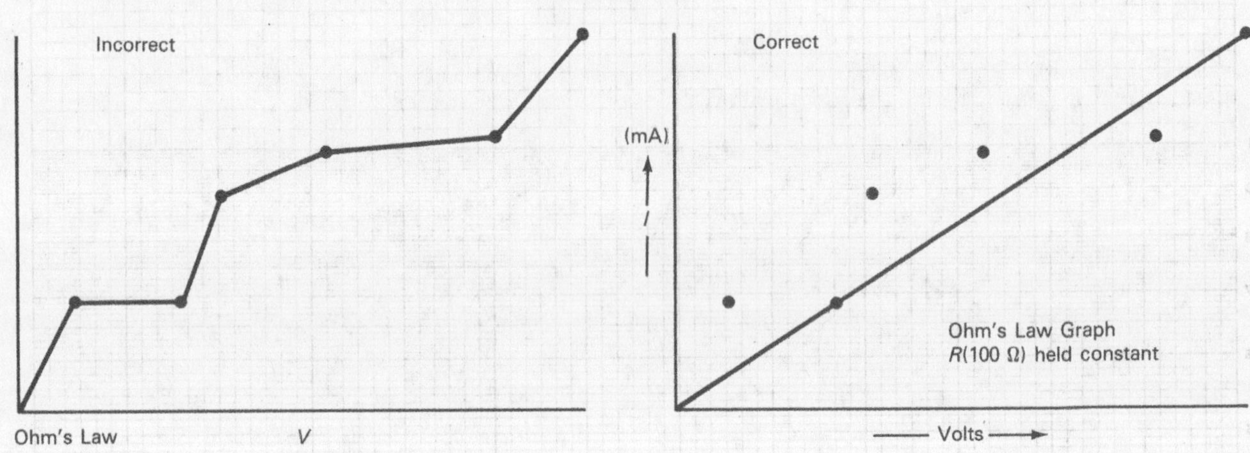

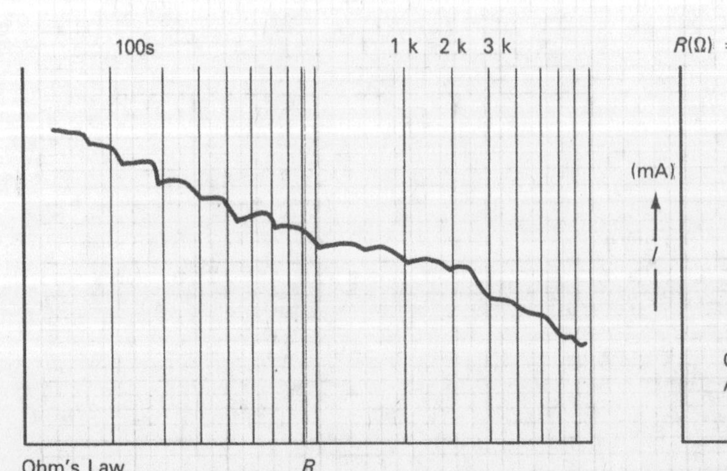

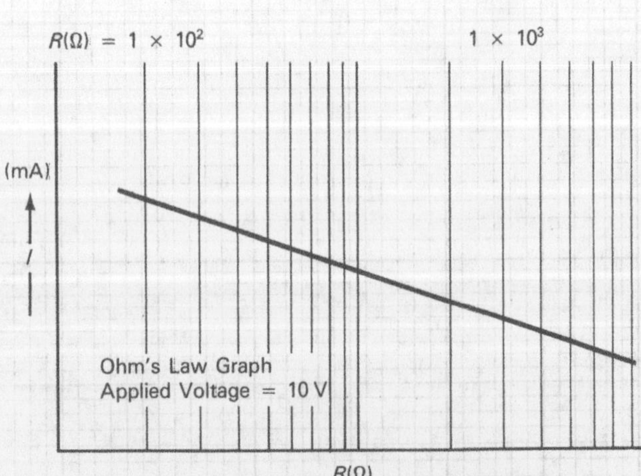

Appendix E Oscilloscope Graticules

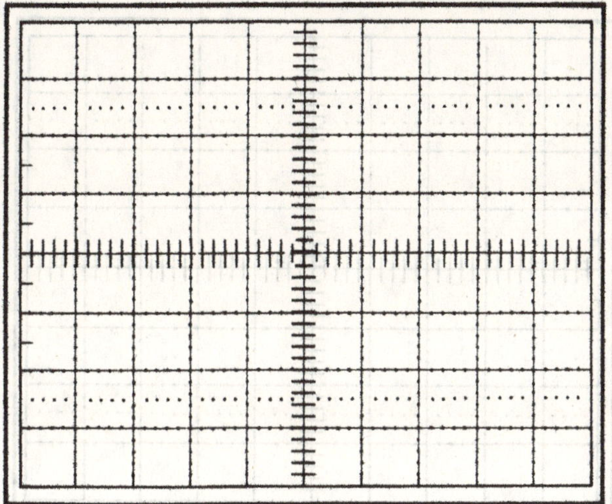
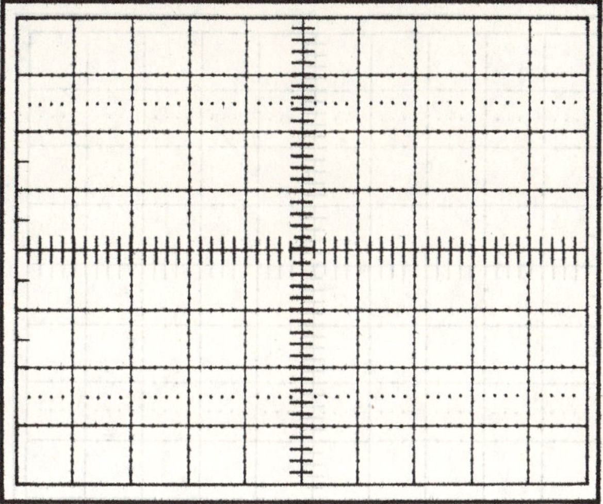
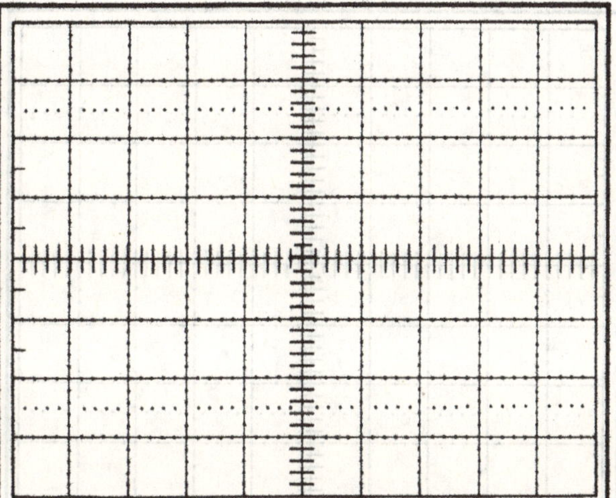
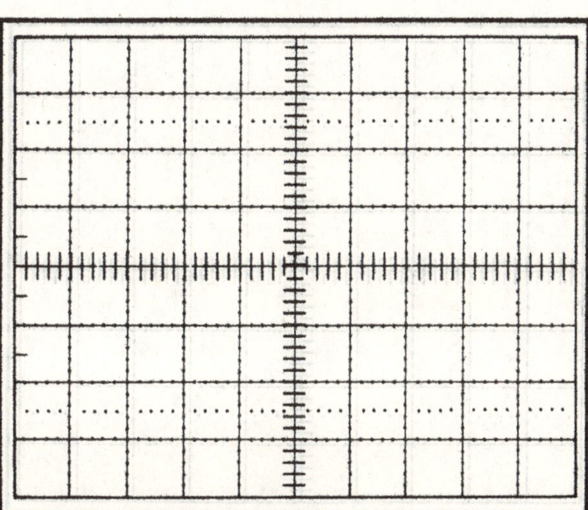

Appendix E Oscilloscope Graticules

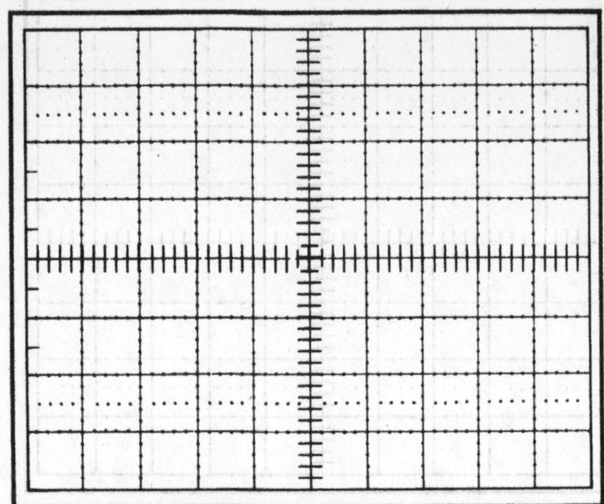

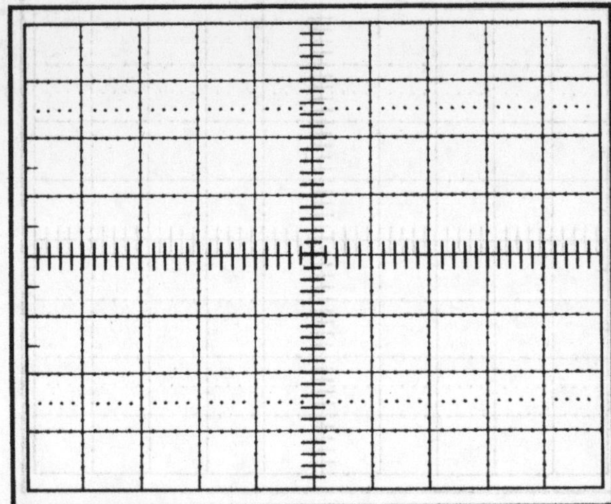

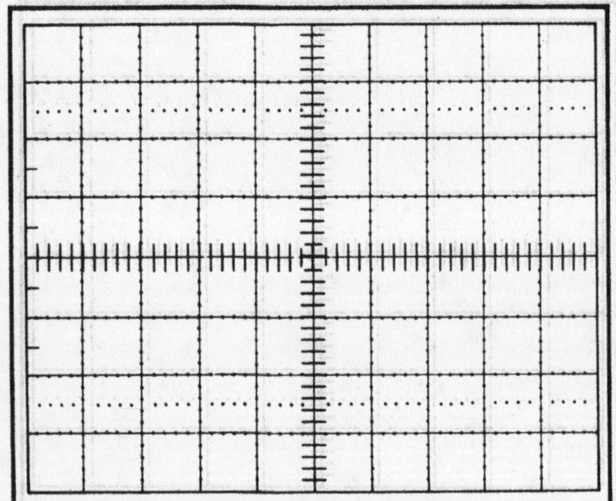

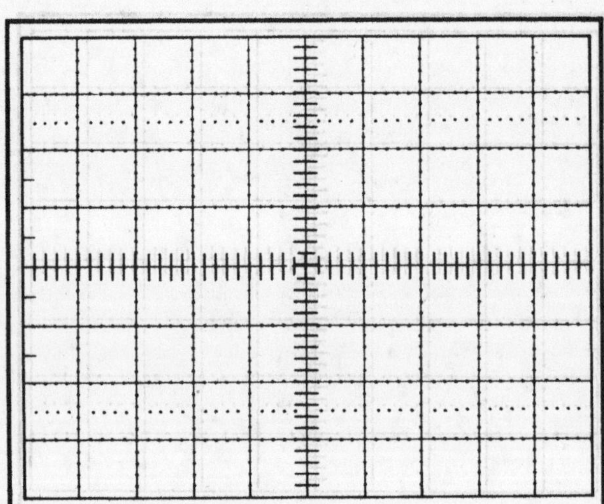

Appendix E Oscilloscope Graticules

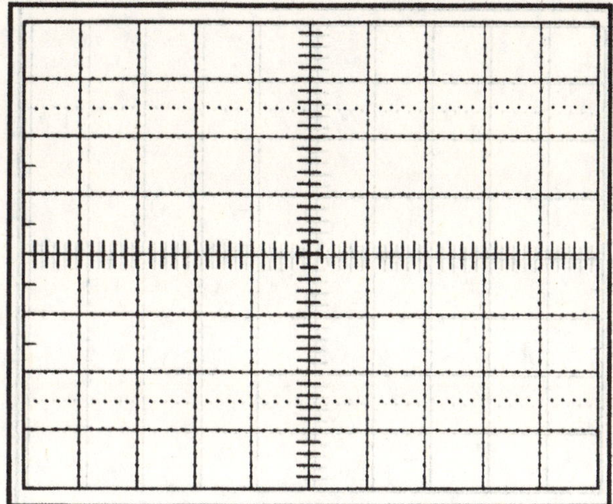

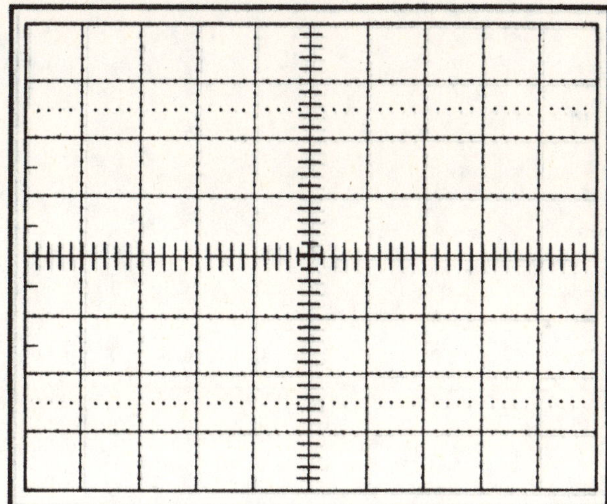

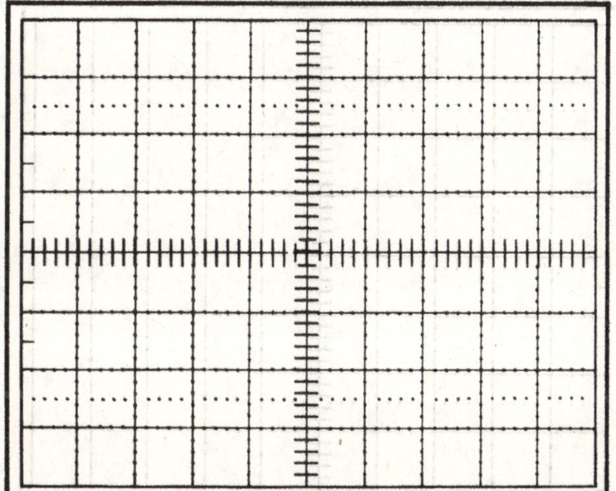

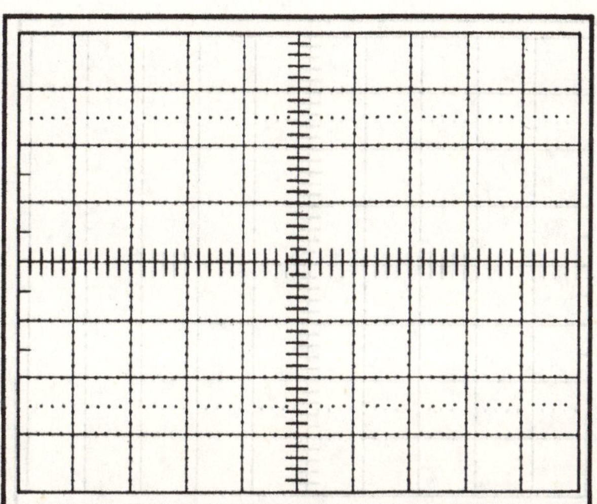

Appendix E Oscilloscope Graticules

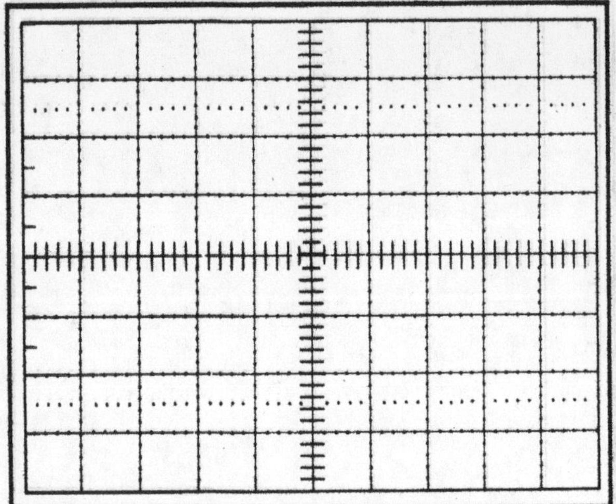

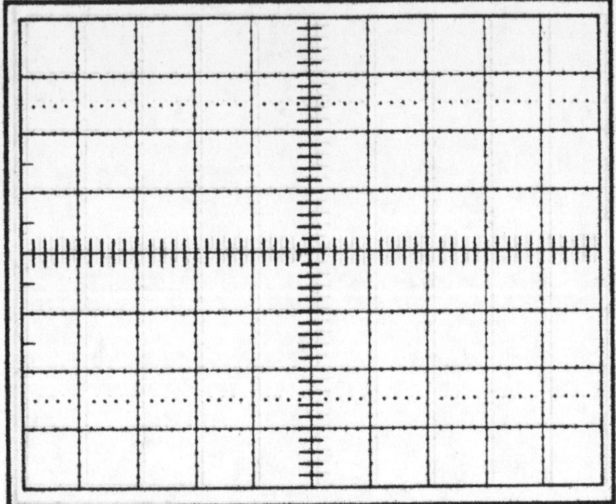

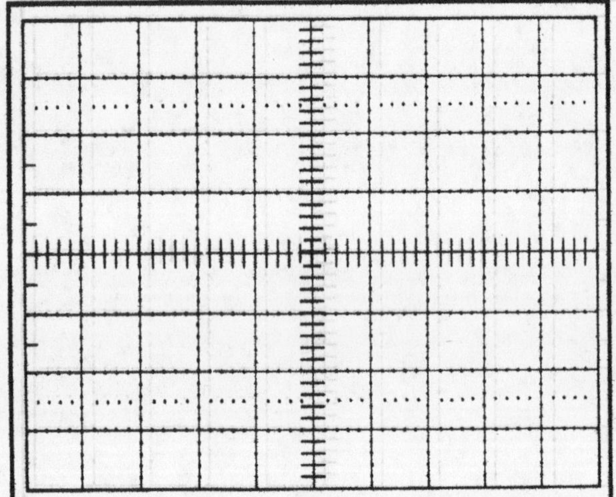

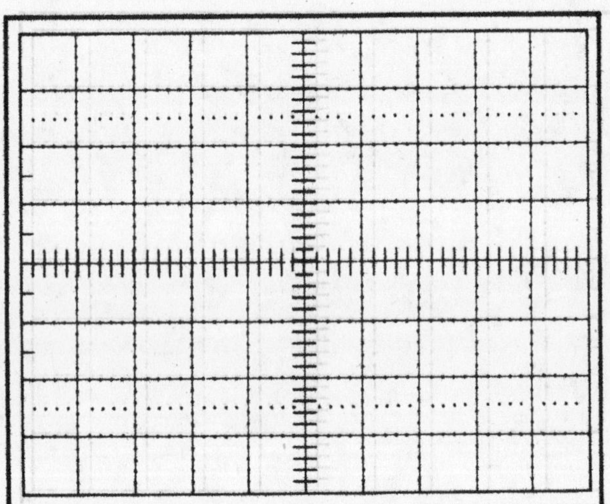

Appendix E Oscilloscope Graticules

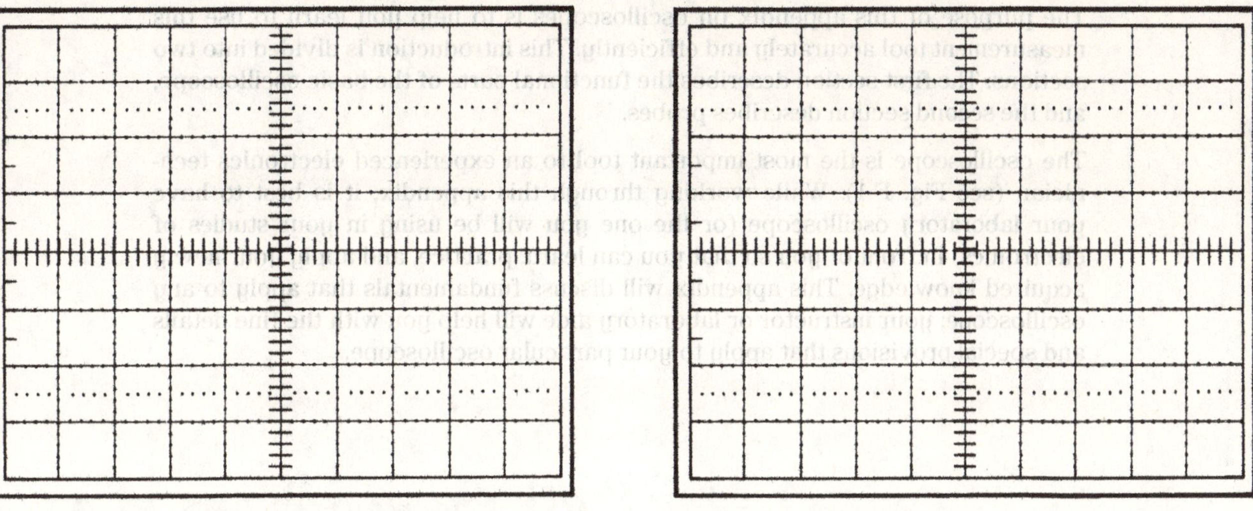

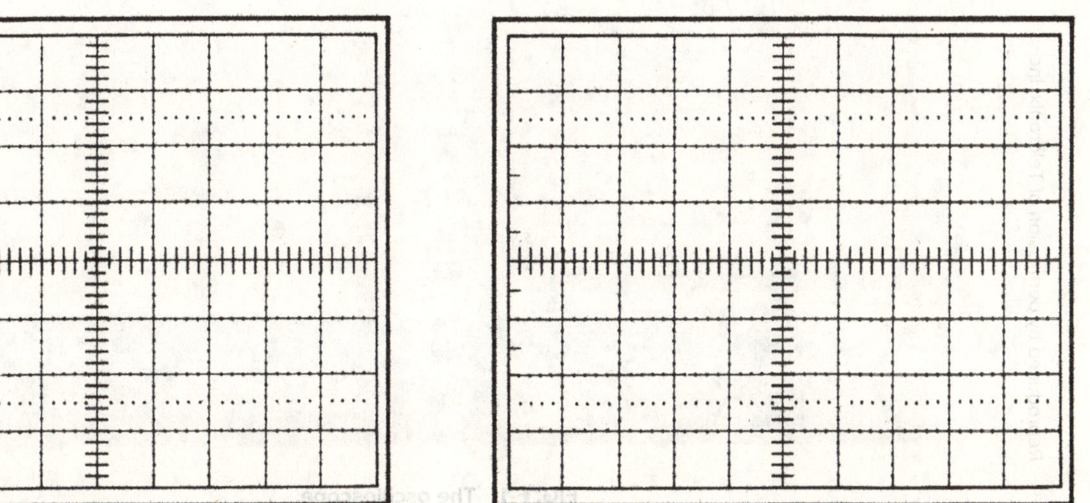

Appendix F The Oscilloscope

The purpose of this appendix on oscilloscopes is to help you learn to use this measurement tool accurately and efficiently. This introduction is divided into two sections: The first section describes the functional parts of the basic oscilloscope, and the second section describes probes.

The oscilloscope is the most important tool to an experienced electronics technician (see Fig. F-1). While working through this appendix, it is best to have your laboratory oscilloscope (or the one you will be using in your studies of electronics) in front of you so that you can learn, practice, and apply your newly acquired knowledge. This appendix will discuss fundamentals that apply to any oscilloscope; your instructor or laboratory aide will help you with the fine details and special provisions that apply to your particular oscilloscope.

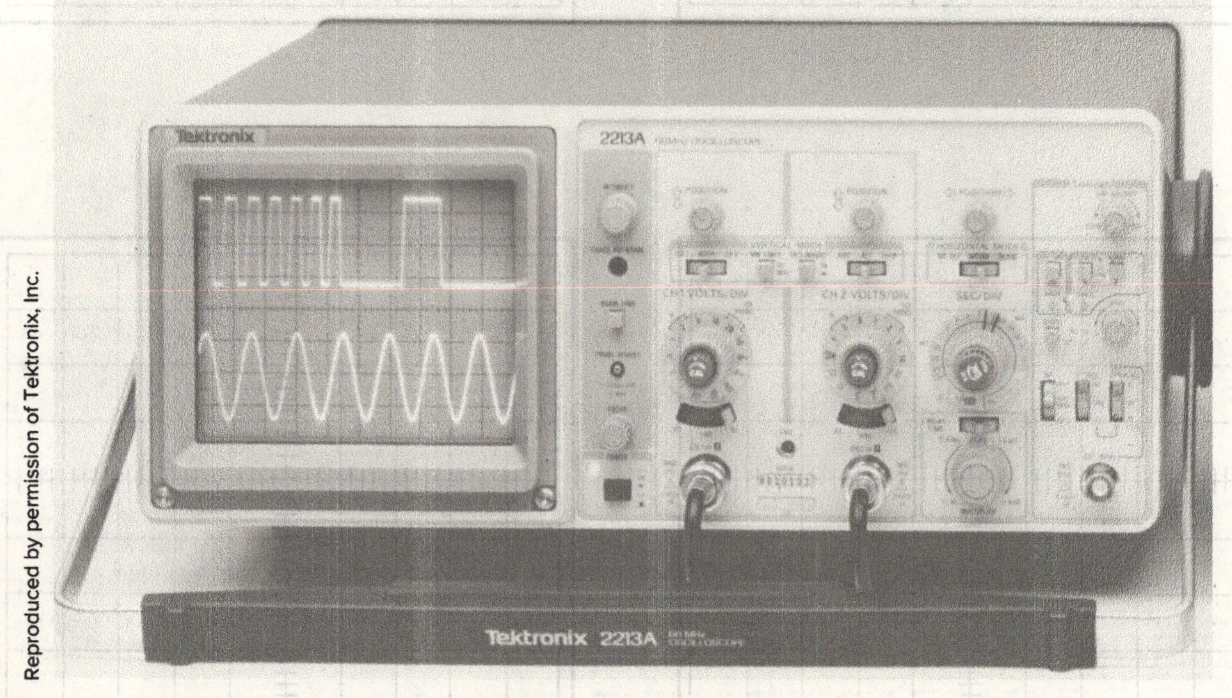

FIG. F-1 The oscilloscope.

Part 1: Functional Oscilloscope Sections

There are four functional parts to the basic oscilloscope: the display system, the vertical system, the horizontal system, and the trigger system.

Display System. Ths display system is a coordinated system of controls. It includes a cathode-ray tube (CRT), intensity control, and focus control. The CRT has a phosphor coating inside it. As an electron beam is moved across the phosphor coating, a glow is created, which follows the beam and persists for a short time. A grid, also known as a graticule, is etched or painted on the screen of the CRT. This grid serves as a reference for taking measurements. Figure F-2 shows a typical oscilloscope graticule. Note the major and minor divisions.

The intensity control adjusts the amount of glow emitted by the electron beam and phosphor coating. The beam trace should be adjusted so that it is easy to see and produces no halo. The focus control adjusts the beam for an optimum trace.

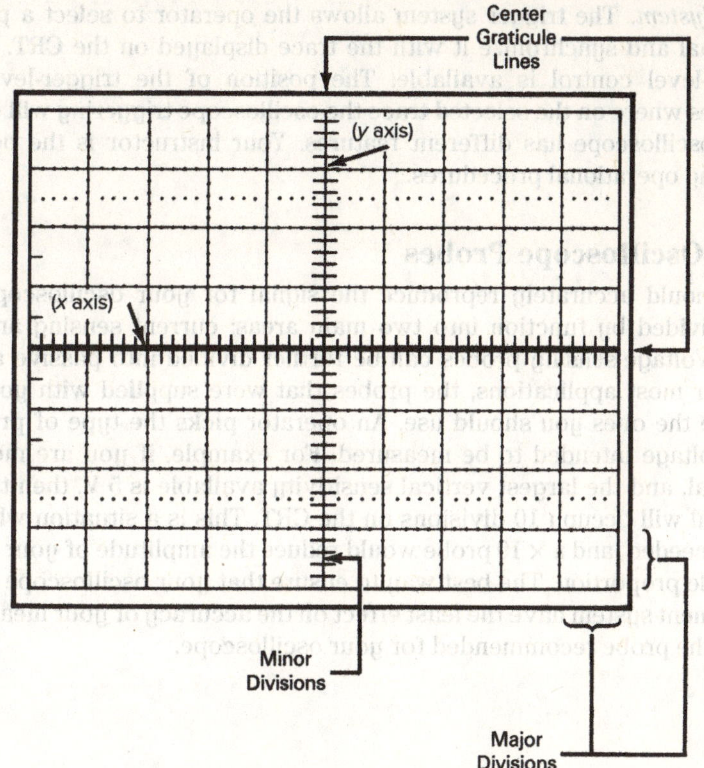

FIG. F-2 Oscilloscope CRT.

Vertical System. The vertical system controls and develops the deflection voltages that are displayed on the CRT. In this system, you typically find controls for vertical position, vertical sensitivity, and input coupling. The vertical position controls the placement of the trace. Adjusting this control will move the entire trace either up or down along the y axis of the CRT. This control is adjusted as needed to help the operator make accurate measurements. The vertical-sensitivity control, also known as the volts-per-division switch, controls the y axis sensitivity. The range is usually controlled from 1 mV to 50 V per division. For example, if you were to observe a trace occupying four divisions on the CRT, and the volts-per-division switch was turned to 1 mV/div., then the measured voltage would equal 1 mV × 4 divisions, or 4 mV. If the switch were turned to the 50 V/div. position and a 4-division trace was observed, then the oscilloscope would be measuring 200 V. The input-coupling switch lets the operator determine how the circuit under examination is connected to the oscilloscope. The three positions of this switch are ground, DC coupling, and AC coupling. When the switch is in the ground position, the operator can adjust the position of the trace with no input signal applied to the oscilloscope. This function is used primarily to align the oscilloscope to a reference point prior to taking a measurement. When in the DC-coupling position, the oscilloscope allows the operator to see the entire signal. However, when the input-coupling switch is in the AC-coupling position, only the AC signal components are displayed on the CRT. All DC components are blocked when the switch is in the AC position.

Horizontal System. As the oscilloscope trace is moved across the CRT (from left to right), it moves at a rate of speed that is related to frequency. The horizontal system is dominated by two main controls: the horizontal-position control and the time-base control. The horizontal-position control performs the same task as the vertical-position, but utilizes the x axis. The time-base, or seconds-per-division, control is used to select the appropriate sweep necessary to see the input signal. Ranges typically found on the time-base control extend from 0.1μs to 0.5 μs per division on the CRT.

Trigger System. The trigger system allows the operator to select a part of the input signal and synchronize it with the trace displayed on the CRT. Normally, a trigger-level control is available. The position of the trigger-level control determines where on the selected trace the oscilloscope triggering will occur.

Each oscilloscope has different features. Your instructor is the best source for varying operational procedures.

Part 2: Oscilloscope Probes

Probes should accurately reproduce the signal for your oscilloscope. Probes can be divided by function into two main areas: current sensing and voltage sensing. Voltage-sensing probes can be further divided into passive and active types. For most applications, the probes that were supplied with your oscilloscope are the ones you should use. An operator picks the type of probe based on the voltage intended to be measured. For example, if you are measuring a 50-V signal, and the largest vertical sensitivity available is 5 V, then that particular signal will occupy 10 divisions on the CRT. This is a situation where attenuation is needed, and a $\times$ 10 probe would reduce the amplitude of your signal to a reasonable proportion. The best way to ensure that your oscilloscope and probe measurement system have the least effect on the accuracy of your measurements is to use the probe recommended for your oscilloscope.

Appendix G Diode Data Sheet

1N4001 thru 1N4007

Designers▲ Data Sheet

LEAD MOUNTED SILICON RECTIFIERS

50-1000 VOLTS DIFFUSED JUNCTION

"SURMETIC"▲ RECTIFIERS

. . . subminiature size, axial lead mounted rectifiers for general-purpose low-power applications.

Designers Data for "Worst Case" Conditions

The Designers▲ Data Sheets permit the design of most circuits entirely from the information presented. Limit curves – representing boundaries on device characteristics – are given to facilitate "worst case" design.

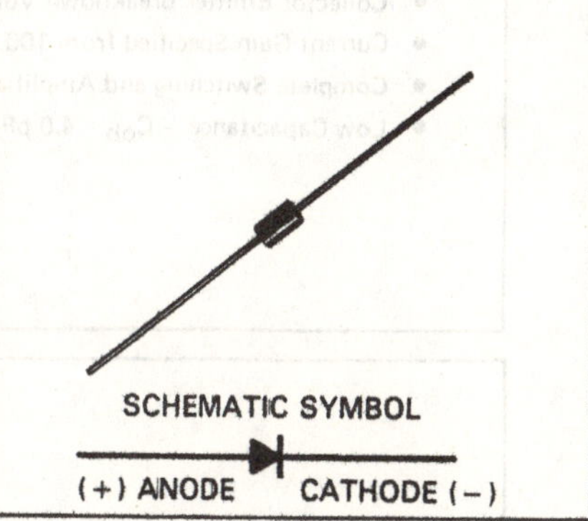

***MAXIMUM RATINGS**

Rating	Symbol	1N4001	1N4002	1N4003	1N4004	1N4005	1N4006	1N4007	Unit
Peak Repetitive Reverse Voltage Working Peak Reverse Voltage DC Blocking Voltage	V_{RRM} V_{RWM} V_R	50	100	200	400	600	800	1000	Volts
Non-Repetitive Peak Reverse Voltage (halfwave, single phase, 60 Hz)	V_{RSM}	60	120	240	480	720	1000	1200	Volts
RMS Reverse Voltage	$V_{R(RMS)}$	35	70	140	280	420	560	700	Volts
Average Rectified Forward Current (single phase, resistive load, 60 Hz, see Figure 8, $T_A = 75^oC$)	I_O	1.0							Amp
Non-Repetitive Peak Surge Current (surge applied at rated load conditions, see Figure 2)	I_{FSM}	30 (for 1 cycle)							Amp
Operating and Storage Junction Temperature Range	T_J, T_{stg}	-65 to +175							^oC

***ELECTRICAL CHARACTERISTICS**

Characteristic and Conditions	Symbol	Typ	Max	Unit
Maximum Instantaneous Forward Voltage Drop ($i_F = 1.0$ Amp, $T_J = 25^oC$) Figure 1	v_F	0.93	1.1	Volts
Maximum Full-Cycle Average Forward Voltage Drop ($I_O = 1.0$ Amp, $T_L = 75^oC$, 1 inch leads)	$V_{F(AV)}$	–	0.8	Volts
Maximum Reverse Current (rated dc voltage) $T_J = 25^oC$ / $T_J = 100^oC$	I_R	0.05 1.0	10 50	μA
Maximum Full-Cycle Average Reverse Current ($I_O = 1.0$ Amp, $T_L = 75^oC$, 1 inch leads	$I_{R(AV)}$	–	30	μA

*Indicates JEDEC Registered Data.

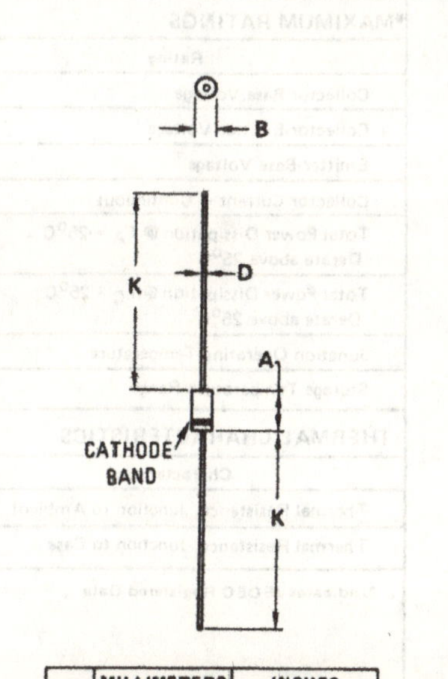

DIM	MILLIMETERS MIN	MILLIMETERS MAX	INCHES MIN	INCHES MAX
A	5.97	6.60	0.235	0.260
B	2.79	3.05	0.110	0.120
D	0.76	0.86	0.030	0.034
K	27.94	–	1.100	–

CASE 59-04

Does Not Conform to DO-41 Outline.

MECHANICAL CHARACTERISTICS

CASE: Void free, Transfer Molded

MAXIMUM LEAD TEMPERATURE FOR SOLDERING PURPOSES: 350°C, 3/8" from case for 10 seconds at 5 lbs. tension

FINISH: All external surfaces are corrosion-resistant, leads are readily solderable

POLARITY: Cathode indicated by color band

WEIGHT: 0.40 Grams (approximately)

▲Trademark of Motorola Inc.

DS 6015 R3

Appendix H Transistor Data Sheet

2N3903 (SILICON)
2N3904

NPN SILICON ANNULAR TRANSISTORS

. . . designed for general purpose switching and amplifier applications and for complementary circuitry with types 2N3905 and 2N3906.

- Collector-Emitter Breakdown Voltage – BV_{CEO} = 40 Vdc (Min)
- Current Gain Specified from 100 μA to 100 mA
- Complete Switching and Amplifier Specifications
- Low Capacitance – C_{ob} = 4.0 pF (Max)

NPN SILICON SWITCHING & AMPLIFIER TRANSISTORS

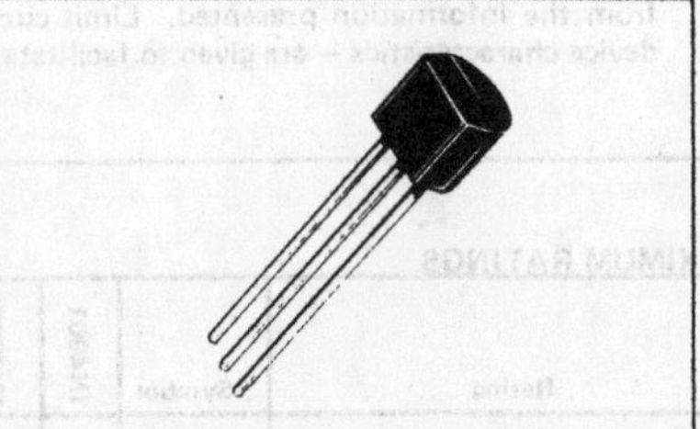

*MAXIMUM RATINGS

Rating	Symbol	Value	Unit
Collector-Base Voltage	V_{CB}	60	Vdc
Collector-Emitter Voltage	V_{CEO}	40	Vdc
Emitter-Base Voltage	V_{EB}	6.0	Vdc
Collector Current – Continuous	I_C	200	mAdc
Total Power Dissipation @ T_A = 25°C Derate above 25°C	P_D	350 2.8	mW mW/°C
Total Power Dissipation @ T_C = 25°C Derate above 25°C	P_D	1.0 8.0	Watts mW/°C
Junction Operating Temperature	T_J	150	°C
Storage Temperature Range	T_{stg}	-55 to +150	°C

THERMAL CHARACTERISTICS

Characteristic	Symbol	Max	Unit
Thermal Resistance, Junction to Ambient	$R_{\theta JA}$	357	°C/W
Thermal Resistance, Junction to Case	$R_{\theta JC}$	125.	°C/W

*Indicates JEDEC Registered Data

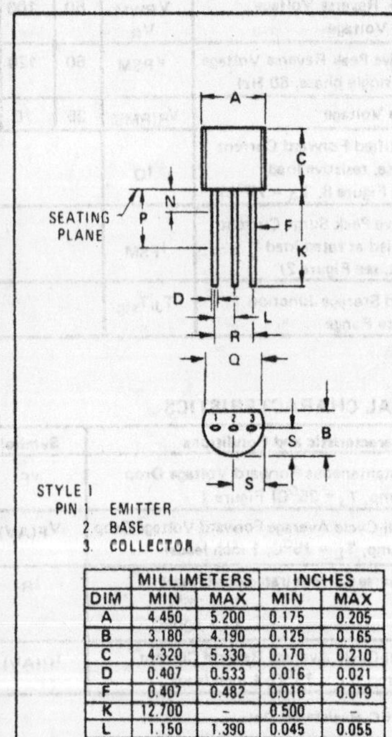

DIM	MILLIMETERS MIN	MILLIMETERS MAX	INCHES MIN	INCHES MAX
A	4.450	5.200	0.175	0.205
B	3.180	4.190	0.125	0.165
C	4.320	5.330	0.170	0.210
D	0.407	0.533	0.016	0.021
F	0.407	0.482	0.016	0.019
K	12.700	–	0.500	–
L	1.150	1.390	0.045	0.055
N	–	1.270	–	0.050
P	6.350	–	0.250	–
Q	3.430	–	0.135	–
R	2.410	2.670	0.095	0.105
S	2.030	2.670	0.080	0.105

CASE 29-02
TO-92

Appendix I Component List

Resistors

All resistors 0.25 W, 5% unless indicated otherwise.

(1) 10 Ω
(1) 15 Ω
(1) 47 Ω, 2 W
(1) 56 Ω
(1) 56 Ω, 2 W
(1) 68 Ω
(3) 100 Ω
(1) 100 Ω, 1 W
(1) 120 Ω
(2) 150 Ω
(2) 150 Ω, 1 W
(5) 220 Ω
(1) 270 Ω
(2) 330 Ω
(1) 330 Ω, 1 W
(7) 390 Ω
(3) 470 Ω
(5) 560 Ω
(3) 680 Ω
(3) 820 Ω
(4) 1 kΩ
(3) 1.2 kΩ
(2) 1.5 kΩ
(3) 2.2 kΩ
(1) 2.7 kΩ
(1) 3.3 kΩ
(1) 3.9 kΩ
(3) 4.7 kΩ
(1) 5.6 kΩ
(1) 8.2 kΩ
(4) 10 kΩ
(1) 12 kΩ, 1 W
(4) 22 kΩ
(1) 22 kΩ, 2 W
(1) 27 kΩ
(1) 33 kΩ
(1) 39 kΩ
(2) 47 kΩ
(1) 68 kΩ
(2) 82 kΩ
(1) 86 kΩ
(2) 100 kΩ
(1) 100 kΩ, 1 W
(1) 150 kΩ
(1) 220 kΩ
(1) 470 kΩ
(1) 1 MΩ
(1) 1.2 MΩ
(1) 3 MΩ
(1) 3.3 MΩ

Capacitors

All capacitors 25 V or greater.

(1) 0.0068 μF
(1) 0.068 μF
(4) 0.01 μF
(2) 0.1 μF
(4) 10 μF
(2) 25 μF
(2) 47 μF electrolytic
(2) 47 μF
(2) 4 μF or 1 μF
(1) 100 μF

Inductors

(2) 33 mH
(1) 100 mH
(1) 1 H (or optional value)
(4) Varying values from 10 mH to 0.5 H
(1) 0.5 H
(1) 0.01 H

Potentiometers

(1) 1-kΩ, 1-W linear taper
(1) 5-kΩ, 1-W linear taper
(1) 100-kΩ, 1-W linear taper
(1) 1-MΩ, 1-W linear taper
(1) 20-kΩ, 1-W linear taper

Batteries

(4) D cells
(4) D-cell holders

Diodes

(4) 1N4004 or equivalent
(4) LEDs
(1) Zener, 5 V (1 W)
(2) IR#S1M solar cells or equivalent photodiode

Transistors

(1) 2N3638
(2) 2N3904

FET

(1) 2N3823 or equivalent

Op Amps and ICs

(1) 741
(1) 7408
(1) 7432

Vacuum Tube (Optional)

6J5 or equivalent

Transformers

(1) 60 Hz with two or three taps

Bench Equipment

Ammeter with 30-mA capacity; VOM/DMM
Voltmeter: DVM, VTVM, VOM, DMM
DC power supply, 0–30 V (±15 V for op-amp experiment)
High-voltage power supply with 120:12.6 V center-tap @ 6.3-V filament transformer
Galvanometer or microammeter movement
Signal generator (sine wave, to 1 MHz preferred)
Oscilloscope (solid-state, auto-trigger, dual-trace, with operator's manual preferred)
Frequency counter
Breadboard
AC signal generator

Miscellaneous Parts

(1) Circuit board; proto springboard or breadboard
(2) SPST switch
(1) SPDT switch
(1) 0–1-mA meter movement
(1) 50-μA meter movement
(6) Test leads 3 red, 3 black
(1) Decade box
(1) Magnetic compass
(1) Heavy-duty horseshoe magnet, 20-lb plus pull
(1) Sheet-metal shield, 6 × 6 in.
(1) Grease pencil
2–3 ft, thin insulated wire
Iron filings
No. 18 steel nail
(4) lightbulbs: 25, 60, 100, and 150 W
(1) Clear glass functioning fuse (any current rating)
(1) Clear glass nonfunctioning fuse (any current rating)
(1) Neon bulb, approximately 60 V